ROADMAP TO SUCCESSFUL HEALTH ECOSYSTEMS

ROADMAP TO SUCCESSFUL DIGITAL HEALTH ECOSYSTEMS

A Global Perspective

Edited by

EVELYN HOVENGA, RN, PHD, FACS, FACN, FIAHSI

Professor Digital Health, Australian Catholic University, Melbourne, VIC, Australia

CEO & Director eHealth Education Pty Ltd, Melbourne, VIC, Australia

Global eHealth Collaborative (GeHCo), Melbourne, VIC, Australia

HEATHER GRAIN, ADIP HIM, DIP TDD, GDIP IS, MHI, FAIDH, FMU, FIAHSI

Director Course Development, eHealth Education Pty Ltd, Melbourne, VIC, Australia

Global eHealth Collaborative (GeHCo) , Melbourne, VIC, Australia

LG Informatics Pty Ltd, Melbourne, VIC, Australia

ELSEVIER

ACADEMIC PRESS

An imprint of Elsevier

Academic Press is an imprint of Elsevier
125 London Wall, London EC2Y 5AS, United Kingdom
525 B Street, Suite 1650, San Diego, CA 92101, United States
50 Hampshire Street, 5th Floor, Cambridge, MA 02139, United States
The Boulevard, Langford Lane, Kidlington, Oxford OX5 1GB, United Kingdom

Library of Congress Cataloging-in-Publication Data
A catalog record for this book is available from the Library of Congress

British Library Cataloguing-in-Publication Data
A catalogue record for this book is available from the British Library

ISBN 978-0-12-823413-6

For information on all Academic Press publications
visit our website at https://www.elsevier.com/books-and-journals

Publisher: Stacy Masucci
Acquisitions Editor: Rafael Teixeira
Editorial Project Manager: Susan Ikeda
Production Project Manager: Swapna Srinivasan
Cover Designer: Matthew Limbert

Typeset by STRAIVE, India

Contents

6. The knowledge-driven platform: Strategic technologies for a platform ecosystem approach

Thomas Beale and Evelyn Hovenga

7. Are professional guidelines and regulatory standards fit for purpose?

Evelyn Hovenga

8. Health data standards' limitations

Evelyn Hovenga and Heather Grain

9. Quality data, design, implementation, and governance

Evelyn Hovenga and Heather Grain

10. Guideline and knowledge management in a digital world

Evelyn Hovenga

11. Modelling clinical knowledge

Heather Leslie

12. Identity—What is in a name?

Heather Grain

13. Electronic health records and essential technology paradigms

Evelyn Hovenga, Thomas Beale, and Heather Grain

14. Health data security and privacy: Challenges and solutions for the future

Kassaye Yitbarek Yigzaw, Sílvia Delgado Olabarriaga, Antonis Michalas, Luis Marco-Ruiz, Christiaan Hillen, Yiannis Verginadis, Marcela Tuler de Oliveira, Dagmar Krefting, Thomas Penzel, James Bowden, Johan Gustav Bellika, and Taridzo Chomutare

15. The digital needs of genomics resulting from pandemics

Kathleen A. McCormick

16. Health informatics in the Middle East and North Africa

Najeeb Al-Shorbaji and Dari Alhuwail

17. A framework for regional health information systems interoperability: The Asia eHealth Information Network (AeHIN) experience

Alvin B. Marcelo, Kristin Chloe Pascual, Jai Ganesh Udayasankaran, Fazilah Shaik Allaudin, and Boonchai Kijsanayotin

18. The road to interoperability: openEHR modelling and implementation

Lu Xudong, Nan Shan, Cai Hailing, Cheng Yexuan, Li Mengyang, Wang Li, and Min Lingtong

19. Evidence-based biomedical information systems: The road ahead

Rune Pedersen and Luis Marco-Ruiz

20. Norway, Sweden, and Finland as forerunners in open ecosystems and openEHR

Hanna Pohjonen

21. The Brazilian digital health system: Building the digital transformation to engage country citizens

Heimar de Fatima Marin, Juliana Pereira de Souza Zinader, Frank James da Silva Pires, and Jacson Venancio de Barros

22. Benefits of using SNOMED CT in the UK National Health Service (NHS)

Monica Jones

23. Primary health reform for the digital era

Noah Crampton, Anson Dong, Payal Agarwal, Catherine Ji, and Onil Bhattacharyya

24. Caribbean/PAHO—Jamaican case study

Monica Jones

25. Addressing workforce knowledge gaps in digital health

Evelyn Hovenga and Helen Hovenga

26. Emerging digital health ecosystems

Evelyn Hovenga, Helen Hovenga, and Heather Leslie

About the editors and authors

Book contributors—in order of appearance—contact email addresses are provided for all corresponding authors.

(1) Evelyn Hovenga RN PhD FACS FACN FIAHSI e.hovenga@ehe.edu.au

Evelyn is a parttime Professor of Digital Health at the Australian Catholic University, manages eHealth Education, an RTO, the not-for-profit Global eHealth Collaborative (GeHCo), and continues to work as a Digital Health Consultant. She retired as Professor of Health Informatics in 2007 following a 25-year career in this discipline with a focus on standards development as these apply to EHRs, semantic interoperability, and terminology and is an Honorary Senior Research Associate at the Centre for Health Informatics & Multiprofessional Education, University College London http://www.chime.ucl.ac.uk/. Evelyn started her career as a registered nurse, has health executive, public service, educational and research experience, obtained a PhD in Health Administration (Nursing Informatics), initiated and hosted the first National Health Informatics Conference (HIC) in Melbourne in 1993, is one of founders of HISA and the Australasian College of Health Informatics, and a founding Fellow of the International Academy of Health Sciences Informatics (FIAHSI) Geneva, she is widely published. Evelyn is an honorary member of the International Medical Informatics Association's Nursing Informatics SIG as a result of representing Australian nurses from 1984 for many years, as a member and Past Chair of this group.

(2) Heather Grain ADip HIM, Dip TDD, GDip IS, MHI, FAIDH, FMU, FIAHSI H.grain@ehe.edu.au

Heather has international expertise in the development, implementation, management, and governance of digital health systems and data, including terminologies and classifications and in design and delivery of research and education in tertiary and vocational education environments. Heather is the only person who has held leadership roles in health informatics at the International Standards Organisation (ISO), Health Level 7 International (HL7), the organisation responsible for the development of health information exchange standards (computer to computer) and SNOMED INTERNATIONAL—the organisation responsible for clinical terminology development to represent meaning in information systems. She currently leads the ISO Health Information Governance Ad Hoc Group. Heather is the course and ISO's development manager for eHealth Education, and designed the Electronic Health Records Online Learning (eHRoL) clinical coding training tool and associated learning management system and is a Board member for the not-for-profit Global eHealth Collaborative (GeHCo). She understands the different requirements of governments, insurers, organisations, clinical domains, and researchers and the need for harmonisation to develop appropriate skills and education strategies. Her experience from the academic as well as the vocational perspective offers approaches to education, which are flexible and practical for any learning environment or methodology.

(3) Carol Hullin RN BN (Hons) ML (LLM) PhD carolhullin@hotmail.com

Carol is a leader of developing countries using Digital Innovation for Human Dignity. She completed a Master of Law LLM at Pontificia Universidad Catolica de Chile focusing on the legal implications for the privacy of health data using the Digital environment of healthcare delivery following which she was appointed as the Dean of Faculty of Health Sciences & Professor of Digital Health at the University of Temuco, Chile, where she was instrumental in establishing the Centre of Digital Health. At present, she is a senior lecturer of Digital Health, the University of Tasmania, Australia, and provides global strategic advice to several public and private entities around the world. Carol research was in mobile computing. Carol has several years of experience as the ehealth leader for the World Bank during which time she visited numerous developing countries, including the Caribbean; she introduced Evelyn Hovenga to meet with the Ministries of Health in Jamaica and Dominic, which lead to the inclusion of openEHR into the IS4H framework and for Jamaica to be the first countries to implement this. She loves being a mother of two gorgeous daughters and embraces her family in Chile and Australia.

(4) Thomas Beale thomas.beale@arssemantica.com

Thomas has an Electrical Engineering (communications) and Computer Science academic background and 25 years' experience in EHR technology and standards, information modelling, semantic interoperability, e-health strategy, and open health computing platforms. He is the principal of ARS Semantica, creator of the Archetype Definition Language (ADL), now an ISO standard, the Basic-Meta Model (BMM) alternative to XMI, used in openEHR, HL7 CIMI, and the ADL Workbench. Thomas is the principal architect of the openEHR Task Planning Specification, the openEHR Reference Model, and Asciidoctor-based documentation toolchain in openEHR specifications. He has widely shared his knowledge on computable guidelines, clinical workflow, decision support, FHIR comparison with openEHR, initial foundations for clinical workflow, open platforms, and more via his website, blogs, and other publications. Thomas has worked in e-health since 1994, has been involved in international e-health standards development (OMG, HL7, CEN, IHTSDO) since 1998, and has worked on the openEHR architecture since 2001 with a focus on knowledge-based standards development. He has consulted for the US Dept of veteran Affairs (VA), Intermountain Healthcare and various Ministries and Departments of Health, including in Australia, the UK, and Brazil, and is a Board member of the openEHR Foundation and he leads its specification programme.

(5) Heather Leslie MB BS (MD), FRACGP, FAIDH, GAICD, FIAHSI heather.leslie@atomicainformatics.com

Heather is the Principal of Atomica Informatics, and the Clinical Program co-lead and CKM Clinical Knowledge Administrator for openEHR International. Heather has pioneered an approach to collaborative clinical information modelling based on the openEHR specifications and led the development of the Clinical Knowledge Manager online tool since late 2007. This library of archetypes is capable of covering the breadth of clinical medicine, research, and related purposes. Initial work has prioritised core clinical concepts used in any clinical system, expanding to increasingly specialised areas over time. Archetypes have been translated into 30 languages so far. The work is being supported and contributed to by well over

2500 users from 102 countries, including a community of more than 900 reviewers covering all health domains. The international CKM currently supports 89 projects and incubators. She is an expert on clinical knowledge governance and has trained and mentored clinical modellers globally.

(6) Kassaye Yitbarek PhD Yigzaw@ehealthresearch.no

Kassaye is a Senior Researcher at the Norwegian Centre for *E*-health Research. He graduated from the University of Tromsø—The Arctic University of Norway with a PhD in Computer Science (2017). He holds an MSc in Telemedicine and E-health from the same university and BSc in Electrical Engineering from Hawassa University in Ethiopia. He has been actively involved in various e-health research projects with a research focus on the security and privacy aspects of health data reuse infrastructures. Dr. Yigzaw is currently actively involved in the research and development activities of the ASCLEPIOS Project funded by the European Commission's H2020 research and innovation programme.

(7) Sílvia Delgado Olabarriaga

Silvia is Assistant Professor of the Department of Epidemiology and Data Science of the Amsterdam University Medical Centers, Location AMC, University of Amsterdam, in the Netherlands. She leads the e-science research group, supervises PhD students, and teaches courses at the AMC Graduate School and Medical Informatics Bachelor and Master Programmes. Dr. Olabarriaga is mainly interested in the usage of advanced information technology for the enhancement of biomedical research. In the past year, she has been actively involved in various research projects with a focus on distributed computing and research infrastructures, as well as in applications of machine learning to biomedical research problems such as medical images and text. Silvia is currently involved in various cybersecurity activities in the scope of applications for healthcare and biomedical research, including projects funded by the European Commission's H2020 research and innovation programme.

(8) Antonis Michalas PhD

Antonis received his PhD from Aalborg University, Denmark. He currently works as an assistant professor at the Department of Computing Sciences, at Tampere University, Finland, where he also co-leads the Network and Information Security Group (NISEC). The group comprises PhD students, professors, and researchers. The group members mainly conduct research in the areas of applied cryptography, provable security, and privacy. His roles at the university include research, teaching, student supervision, and research projects coordination. Dr. Michalas has published a significant number of journal and conference papers in the field and has participated as a speaker in various conferences and workshops.

(9) Luis Marco-Ruiz PhD Luis.Marco.Ruiz@ehealthresearch.no or luis.marco-ruiz@plri.de or luismarco@gmail.com

Luis is a data engineer who focuses his work on the adoption of information standards and terminologies for clinical data reuse and clinical decision support. He holds an MSc in Applied Statistics and a PhD in Health Science. Since 2007, Dr. Marco-Ruiz has participated as a developer, advisor, and researcher in private- and public-funded projects in Norway, Germany, the UK, and Spain. In 2013, he joined the Norwegian Centre for E-health Research where he works in the development of national data reuse infrastructures and advises the Norwegian health authorities on the adoption of clinical information standards and biomedical ontologies. In addition, he works as a semantic interoperability specialist at Hannover Medical School developing a cross-institutional data reuse network.

(10) Christiaan Hillen

Christiaan is an ethical hacker with a strong academic background in information science (MSc from Radboud University) working in IT security as security engineer (Secura B.V.) to help organisations get in control of their data. In a prior life, he trained as an exercise therapist and thus has the combined knowledge of the IT security and medical domains. Branching out into the legal domain and getting to grip with the legal framework surrounding data protection, such as the GDPR, has made Christiaan a valued partner in large IT projects in healthcare such as the ASCLEPIOS project, as well as in the academic community. He continues to contribute in peer-reviewed scientific research where his expertise lies in novel ways of combining the needs to process large medical datasets for the benefit of healthcare, whilst keeping to the principles of privacy by design to protect the patients.

(11) Yiannis Verginadis PhD

Yiannis is an Assistant Professor in Management Information Systems in the Department of Business Administration at the Athens University of Economics and Business (AUEB). In addition, he has a long-term cooperation, as a senior researcher, with the Institute of Communication and Computer Systems (ICCS), Athens, Greece. He has more than 15 years of experience in ICT projects in several research areas, such as management of information systems, workflow management, complex event processing, cloud computing, and cloud security. He holds Diploma and Doctoral degrees in Electrical and Computer Engineering from the National Technical University of Athens, Greece. His PhD was on Inter-organizational Workflow Management Systems in e-Government. He has served as a co-chair or member of more than 25 review committees in journals and conferences and published more than 60 papers in international peer-reviewed journals and conferences.

(12) Marcela Tuler de Oliveira

Marcela is a PhD candidate at Amsterdam University Medical Centres, Location AMC. She has a Master degree in Electrical and Telecommunications Engineering from the Graduate Programme in Electrical and Telecommunications Engineering of the Universidade Federal Fluminense, Brazil. She graduated in Telecommunications Engineering from the same university in 2017 and participated in a student exchange programme at the University of Florida, Florida, USA. Her main areas of interest are New-Generation Networks, Sensor Networks, Network Security, Blockchain, and Computational Intelligence.

(13) Dagmar Krefting

Dagmar is professor for computer science at the University of Applied Sciences Berlin (HTW Berlin) and deputy head of the Department of Medical Informatic of the University Medical Center Göttingen (UMG), Germany. She has long-standing expertise in building collaboration platforms for clinical research with a focus on image and biosignal-based methods. Her research interests encompass security, interoperability, and usability aspects of scientific data sharing, as well as quality assurance, reliability, and reproducibility of biomedical analysis methods and scientific workflows. In the field of biomedical data sciences, she is in particular interested in interpretable machine learning methods for complex analysis of multidimensional biosignal recordings. She is founding head of the Center of Biomedical Image and Information Processing (CBMI) of the HTW Berlin and represents the UMG within the German Medical Informatics Initiative.

(14) Thomas Penzel

Thomas graduated from physics (1986), human biology (1991), and physiology (1995) at the University Marburg, Germany. In 2006, he moved to Berlin where he is the director of research of the Interdisciplinary

Sleep Medicine Center at the Charité—Universitätsmedizin Berlin (Germany). He received several international awards and chaired several international conferences. He is president of the German Sleep Society and Adcom member of IEEE EMBS. He is editor-in-chief of the journal Sleep and Breathing, and editorial board member of other journals. He has published more than 300 journal papers, 80 book chapters, and edited several books. His research interests are sleep medicine, biomedical signals, and wearables for sleep recording.

(15) James Bowden

James completed his Bachelors of Engineering at Queen Mary University of London in 2014. He worked as a commercial software developer and engineer in London until 2019, when he moved to Berlin. James is now a researcher at the Center for Biomedical Image and Information Processing at University of Applied Sciences Berlin where he is working on the ASCELPIOS Project funded by the European Commission's H2020 research and innovation programme.

(16) Johan Gustav Bellika

Johan has worked in the intersection between medicine, medical research, and informatics since 1992. He has a master and Phd in computer science from University of Tromsø—The Arctic University of Norway. He has worked with development, research and giving master-level courses in Telemedicine and ehealth topics. He is now professor in Medical informatics at the department of clinical medicine, faculty of health sciences at University of Tromsø—The Arctic University of Norway and the Norwegian Centre for *E*-health Research. His current research focus is on technology to enable privacy preserving reuse of health data, cybersecurity within the health sector and supporting a learning healthcare system. Prof. Bellika is the CRIO (chief research informatics officer) in the Norwegian primary care research network called PraksisNett and is heading the work developing, implementing, and operating the IT system part of the PraksisNett.

(17) Taridzo Chomutare PhD

Taridzo works as a Senior Researcher at the Norwegian Centre for *E*-health Research, where he previously held a System Developer position before moving onto research. He graduated from the University of Tromsø—The Arctic University of Norway with a PhD in Computer Science (2014). He also holds an MSc in Telemedicine and E-health from the same university. He has worked with telemedicine and e-health for over a decade, working with mobile devices, applications, motivational groups, and social networks for people with diabetes, and related diseases such as obesity. His recent interests extend to natural language processing (NLP), deep learning, and closely related security and privacy aspects, especially relevant to unstructured healthcare data in electronic health records (EHRs).

(18) Kathleen A. McCormick PhD, RN, FAAN, FACMI, FHIMSS scimind2@gmail.com

Kathleen is a pioneer in Nursing Informatics, Genomics, Evidence-Based Guidelines, and Healthcare Policy. After serving 30 years in the United States Public Health Service where she served at the National Institutes of Health and the Agency for Healthcare Research and Quality, she retired as a 06 Captain. During the next 17 years, she was the Vice President and directed Genomic Informatics programmes at SRA, SAIC, and Leidos supporting the DoD, NIH, and the CDC. She has received many awards in Vision, Leadership, and Nursing Informatics. She has over 125 publications and 10 books in Healthcare and Nursing Informatics that are on the Amazon best 100 books in Medical Informatics. The most recent book is by Saba, VK, and McCormick, KA. Essentials of Nursing Informatics, 7th Ed. She is currently Principal and Owner of SciMind, LLC, a consulting company in Healthcare and Nursing Informatics.

(19) Najeeb Al-Shorbaji PhD FIAHSI
shorbajin@gmail.com

Najeeb is from Jordan where he works as an independent consultant and advisor on knowledge management and eHealth. At the national level, he is the President of the Jordan eHealth Development Association (eHDA), the President of the Jordan Library and Information Association (JLIA); at the regional level, he is the President of the Middle east and North Africa Association of Health Informatics; and at a global level, he is the Vice-President of the International Medical Informatics Association (IMIA). He is a part-time Lecture and Researcher at the PLRI Medical Informatics Institute at Braunschweig Technical University, Germany, and advisor to e-Marefa Database and the Arab Citation and Impact Factor (Arcif). Dr. Al-Shorbaji retired from WHO/HQ in August 2015 as Director, Knowledge, Ethics, and Research Department where eHealth was a unit in that department. Since January 2000, he has over 210 publications and participated in over 150 conferences.

(20) Dari Alhuwail PhD FAMIA FIAHSI
dari.alhuwail@gmail.com.

Dari is a member of the WHO Digital Health Technical Advisory Group, and an Associate Professor at the Department of Information Science in the College of Life Sciences, Kuwait University. He is also the co-founder of the Kuwait Health Informatics Network and a member of the Kuwait Health Informatics Association. He also serves as a Health Informatics Consultant at Dasman Diabetes Institute where he co-founded the Health Informatics Unit and is an Adjunct Faculty member at the School of Medicine at the University of Dundee, supervising graduate students in health informatics. Dr. Alhuwail obtained his PhD in Information Systems with a specialising in Health Informatics from the University of Maryland, Baltimore County, and his Master's and Bachelor's from the University of Arizona. He has over 12 years working in industry and in various research laboratories around the world.

(21) Alvin B. Marcelo MD PhD admarcelo@up.edu.ph

Alvin is a general and trauma surgeon by training who is currently senior vice-president and chief information officer of the St. Luke's Medical Center. He is also the executive director of the Asia eHealth Information Network (www.aehin.org). Prior to this, he served as senior vice-president and chief information officer of the Philippine Health Insurance Corporation (PhilHealth). From 2003 to 2010, as the director of the University of the Philippines Manila National Telehealth Center and chief of the Medical Informatics Unit, Dr. Marcelo established the Master of Science in Health Informatics programme and conducted local and international research in the field of eHealth and health information systems development. He took his postdoctoral fellowship in medical informatics at the National Library of Medicine in Bethesda, Maryland, with research interests in telepathology, mobile computing, and bibliometric analysis of MEDLINE content. He is the Philippine representative to the Asia Pacific Association for Medical Informatics (APAMI) and the International Medical Informatics Association (IMIA). Dr. Marcelo is certified in the governance of enterprise IT (CGEIT—www.isaca.org), The Open Group Architecture Framework (TOGAF—www.opengroup.org), Archimate, and COBIT5 Foundation/Implementation.

(22) Kristin Chloe Pascual

Kristin has been providing capacity-building support to the Asia Health Information Network for multiple projects with PATH, GIZ, and WHO. Currently, Kristin is coordinating the activities of the openIMIS Regional Hub Asia. She is also working as a part-time communications consultant for the Asian Development Bank to support the establishment of the Standards and Interoperability-Lab Asia. Her academic background is in the field of Development Communication with specialisation in science communication (scientific and technical information processing, publication editing, and writing). She is currently taking her masters in Community Development at the College of Social Work and Community Development, University of the Philippines.

(23) Jai Ganesh Udayasankaran MSc, MBA

Jai is a senior manager who has implemented digital health solutions in tertiary, secondary, and mobile hospitals in India. He is a member of the Governing Committee of Asia eHealth Information Network, Expert Committee of Global South eHealth Observatory, World Health Organisation's Digital Health Roster of Experts, and Health Informatics Sectional Committee of Bureau of Indian Standards. He had been the vice-chair of the Working Group on Health and Medical Informatics Education of International Medical Informatics Association. He has been a resource person for several national eHealth convergence workshops and HIS capacity-building missions in South East Asia and Western Pacific regional countries. He is a visiting faculty at Department of Health Information Management, Manipal Academy of Higher Education in India, and conducted the key stakeholder consultation for Digital Health Applied Leadership Programme in South and South-East Asia. He has advised Health Information Systems Program-South Africa for HMIS capacity development services.

(24) Fazilah Shaik Allaudin MD

Fazilah holds an M.D. from the National University of Malaysia and an MBA in Healthcare Administration from the University of Toledo, Ohio. She is professionally certified in Enterprise Architecture (TOGAF 9.0) and in Governance & Management of Enterprise IT (COBIT 5.0).

(25) Boonchai Kijsanayotin MD, PhD

Boonchai earned his medical degree from the Chulalongkorn University and Board-Certified Internal Medicine and Preventive Medicine from Thai Medical Council. He spent 15 years' working at rural public hospital as an internal medicine clinician. While he was working as a clinician in the hospital, he was also responsible for the Hospital Information System (HIS) and oversaw the hospital telemedicine station, one of Thai Ministry of Public Health (MoPH) Telemedicine Project. In 2008, he received his MSc & PhD in Health Informatics from University of Minnesota, USA. Currently, he is the health informatician and research manager at Thai Health Information Standards Development Center (THIS), an affiliated agency of the Health Systems Research Institute (HSRI), Ministry of Public Health, Thailand. He is also the chair of the Asian eHealth Information Network (AeHIN), vice president of Thai Medical Informatics Association (TMI), International representative of TMI for International Medical Informatics Association (IMIA), member of Thai Medical Council and Royal College of Physicians of Thailand. His working areas and research interest include eHealth in developing country, especially Thailand, National health data standards and interoperability, Health Information Exchange (HIE), Electronic Health Records (EHR), Socio-technical aspect of Health Information

Technology and Health/biomedical informatics education in Thailand. His current work is the development of Thai Medicines Terminology (TMT), the semantic standard for Thai pharmaceutical products.

(26) Lu Xudong lvxd@zju.edu.cn

Xudong is the Professor of Medical Informatics in Biomedical Engineering Department of Zhejiang University in China since 2012, the Visiting Research Professor of Information Systems Group of Industrial Design Department of Technical University Eindhoven/the Netherlands since 2013. He is also the Secretary General of Chinese Medical Software Association since 2014 and openEHR Management Board Member since 2018. His research interests include clinical data modelling, clinical data integration, health big data analytics, clinical decision support, and clinical process intelligence. He has led several Chinese national projects, including 'Research on Key Technology of High-end Electronic Medical Record Systems' since 2012 and 'Chinese Coronary Disease Big Data Platform and its Development' since 2016. He has started openEHR-related research since 2011 and led the openEHR Localisation in China. Till now, he has got many achievements around openEHR, including CLEVER, an openEHR CDR product, which has already been used in many projects.

(27) Nan Shan

Shan Nan is an informaticist focusing on clinical decision support theories and clinical practices. He received dual PhD degrees from Zhejiang University in 2018 and Technical University of Eindhoven in 2020, respectively. Since 2009, he has been working on computerised clinical guideline acquisition, representation, and execution. He has conducted several decision support systems development and evaluations in both Chinese and Dutch hospitals. His recent research interest is combining industry standards (e.g. BPMN) and healthcare-oriented standards (e.g. openEHR) to facilitate the widely adoption of decision support applications.

(28) Cai Hailing

Hailing Cai is the Engineer of Medical Informatics in Biomedical Engineering Department of Zhejiang University in China since 2017. His research interests include clinical data modelling, clinical data integration, and e-Health system development and implementation. He has undertaken openEHR-related research since 2014 and participates in the openEHR Localisation in China.

(29) Cheng Yexuan

Yexuan Cheng is the postgraduate of Medical Informatics in Biomedical Engineering Department of Zhejiang University in China since 2020. She has been participating in openEHR-related projects and research.

(30) Li Mengyang

Mengyang Li has a PhD of Medical Informatics from the Biomedical Engineering Department of Zhejiang University in China since 2015. His research interests include clinical data modelling, clinical data integration, and clinical information retrieval. He has joined many openEHR-related projects and research, such as CLEVER, an openEHR CDR product, which has already been used in many projects.

(31) Wang Li

Li Wang is an informaticist focusing on the area of medical informatics. He received his PhD degree in biomedical engineering from Zhejiang University in 2017. He has a great understanding of openEHR and has achieved substantial results on its persistence solution. Now he is working at the college of computer science and software engineering of Shenzhen University and still contributing to the area.

(32) Min Lingtong

Lingtong Min is working at the Northwestern Polytechnical University and focusing on the research of clinical information modelling and high-resolution earth observation. He received his PhD degree in biomedical engineering from Zhejiang University in 2019. Since 2012, he has been working on clinical information models and has achieved rich results on its practices, such as electronic medical records, electronic health records, and specific disease models.

(33) Rune Pedersen PhD rune.pedersen@ehealthresearch.no

Rune's PhD in information systems is from the University of Tromsø, the arctic University of North Norway (UIT). He is currently working as a department director at the Norwegian Centre for ehealth research, and associate professor at UIT. Research interests are in the scope of standardisation and patient pathways. Semantic interoperability, with a focus on clinical information models and clinical terminology, has been a continuous research focus since 2007.

(34) Hanna Pohjonen PhD hanna.pohjonen@rosalieco.fi

Hanna is an eHealth management consultant, at her own consultancy company Rosaldo Oy, Finland, and openEHR Ambassador Finland, has worked in 31 different countries in Europe, North America, Middle East, and Asia. Her focus is on big regional and national healthcare IT/eHealth projects, where data sharing across various systems, organisations, and even country borders is essential. She has provided consultancy about healthcare information systems and IT architectures, vendor neutral archiving, document-based and structured data sharing (XDS, openEHR), shared workflow and open ecosystems. Besides consultancy Hanna is also a visiting lecturer in Tallinn University of Technology. Before consultancy, she did research work both in the UK and in Finland, worked in a university hospital in different positions and was responsible for eHealth-related financing under the Ministry of Trade and Industry in Finland. She also represented Finland in the eHealth area in the European Commission.

(35) Heimar de Fatima Marin RN, MS, PhD, FACMI heimarfm@gmail.com

Heimar has devoted her professional career to improving patient care using information and communication technologies. She is Alumni Professor at the Federal University of Sao Paulo (UNIFESP and a former fellow in Clinical Computing at the Center for Clinical Computing at Harvard Medical School. In 2004, she was elected at the American College of Medical Informatics. Former positions included President of Brazilian Society of Health Informatics (2002–2008); Chair (2009–2012) of the International Medical Informatics Association Nursing Informatics Special Interest Group. As a professor, she has mentored over 23 PhD students, 48 master students, and 180 specialists in health and nursing informatics. Dr. Marin is the Editor in Chief of the *International Journal of Medical Informatics* (IJMI) and Scientific Coordinator at ICT in Health Care (NIC.BR), a research conducted at the Regional Center for Studies on the Development of Information Society (CETIC.BR).

(36) Juliana Pereira de Souza Zinader

Dr. Juliana Zinader has 13 years of experience in Health Informatics. Bachelor, Master in Computer Science, and PhD in Science/Bioengineering. Dr. Zinader conducted a post-doctoral internship at the Faculty of Medicine of Ribeirão Preto/USP. She is also Specialist in Health Informatics by the Sírio-Libanês Hospital. Currently, she is Assistant Professor at the Institute of Informatics of the Federal University of Goiás. She also holds the position of General Coordinator of Innovation in Digital Systems—CGISD/DATASUS/SE. At DATASUS, Dr. Zinader coordinates ICT innovation actions, which include the Connect SUS Programme, the National Health Data Network (RNDS), and the Digital Health Strategy for Brazil 2020–2028 (ESD28).

(37) Frank James da Silva Pires

Mr. Frank Pires da Silva has seven years of experience in Health Informatics. Bachelor of Information Systems. Specialist in Health Informatics by the Sírio-Libanês Hospital, he serves at the Digital Government Secretariat of the Ministry of Economy, member of the Information Technology Analyst committee. Currently, at the SUS Informatics Department (DATASUS), Ministry of Health, Mr. Silva holds the position of General Coordinator of Information and Operation Systems and Deputy Director of DATASUS, being responsible for the architecture of the National Health Data Network (RNDS).

(38) Jacson Venancio de Barros

Dr. Jacson de Barros is the Director of the Department of Informatics of SUS—DATASUS, Ministry of Health and responsible for the implementation of Digital Health in Brazil. Dr. Barros has over 30 years of dedication to the health informatics projects in the country. He was CIO of Hospital das Clínicas at FMUSP—the biggest teaching hospital in the country. Graduated in Electrical Engineering and Master in Medical Sciences at FMUSP and currently attending the Graduate programme (doctorate) at the Faculty of Medicine of USP, having as a research project, the use of BigData in health, integrating care, and administrative data of public health establishments to adopt predictive models that will support clinical decision-making.

(39) Monica Jones BSc (Hons) PGCE PGDip MBCS CITP MInstLM mcmjones@btinternet.com

Monica is the Chief Data Officer for the HDRUK Data Research Hub for Cancer—DATA-CAN, Associate Director HDR North and also the Executive Lead for Population Health Management for the Yorkshire and Humber Care Record (YHCR). She is a highly motivated technical expert and leader with extensive Healthcare and Life Sciences knowledge and experience. She is ISEB-qualified Enterprise and Solutions Architect with a proven track record of designing and delivering complex, high value solutions to exceptional standards. She has many years' experience defining and directing current and future information systems. As the Managing Director of Animo Consultancy Ltd., she has provided consultancy services for CRUK, MRC, Genomics England, NHS England, and the MOD. An effective communicator used to working on major public and private sector initiatives, who is able to integrate teams to achieve their maximum potential. International clinical terminology and data standards (SNOMED CT, ICD-10, HL7 FHIR) expert promoting safe and effective health information for better patient outcomes. She is the Deputy Chair of the NHS Digital Data Design Authority (DDA) and Chair of the YHCR Data Architecture Design Authority (DADA).

(40) Noah Crampton

Department of Family and Community Medicine, University of Toronto.
Family Medicine Department, Toronto Western Hospital.

(41) Anson Dong

Department of Family and Community Medicine, University of Toronto.
Family Medicine Department, Sinai Health System.

(42) Payal Agarwal

Department of Family and Community Medicine, University of Toronto.
Institute for Systems Solutions and Virtual Care, Women's College Hospital.

(43) Catherine Ji
Department of Family and Community Medicine, University of Toronto.
Family Medicine Department, Toronto Western Hospital.

(44) Onil Bhattacharyya onil.bhattacharyya@wchospital.ca
Department of Family and Community Medicine, University of Toronto.
Institute for Systems Solutions and Virtual Care, Women's College Hospital.

(45) Helen Hovenga MBA, Grad. Dip. HRM & IR, BA (Psych)

Helen is a talented HR Executive who has worked with Executives, CEO's and Boards in developing and delivering transformation, the ideal Culture and the People Strategy with passion, courage, innovation, curiosity, kindness, drive, and effective performance outcomes. She is the Global Chief People Officer at Telix Pharmaceuticals Limited, Melbourne, Australia, her prior permanent role was as Executive Director, People & Culture with Peter MacCallum Cancer Centre for 3500 Health practitioners and Research employees. As well as having broad and deep HR expertise, she is a deep thinker, high achiever, quick learner, influencer, early adopter and futurist with diverse experience across large, complex organisations and sectors, including Public Health, Finance, Manufacturing, Mining, Automotive and service industries such as Retail. Swinburne. Helen's passions are in change, transformation, agility, developing and implementing Strategy, people leadership, stakeholder partnerships, high performance, digital solutions, and improved commercial outcomes.

Acknowledgements

This book is dedicated to all Health Informatics champions, the health clinicians, administrators, terminology, technology, researchers, academic and standards developer professionals from around the world who have contributed to our collective knowledge and our life-long health informatics learning journey, through many debates and discussions. These health informaticians voluntarily share their knowledge and skills while attending and participating in meetings, workshops, conferences, dinners, and social encounters in a great variety of settings and locations. We are grateful for this scientific collaborative knowledge sharing we have experienced over more than 35 years, including the many contributors to this text, without which we could not have compiled this book. We are indebted to our local, external national, and international collegial networks and to all those who believed in us, who have supported our vision and rewarded us for our relenting persistence towards realizing our vision.

A special acknowledgement to Thomas Beale and Heather Leslie who were instrumental in initiating this project during discussions at Medinfo 2019 in Lyon. The four of us made a great multidisciplinary team, complementing each other's knowledge domain in order to reach a shared understanding of the many complexities, scientific principles, and world realities covered in this text. It has been a very challenging journey for all of us and serves as a great example of any health informatics project—full of respect, skill, and sharing.

We also thank those close to us for accommodating our many absences and late nights.

Evelyn J.S. Hovenga
Heather Grain

Preface

The healthcare service supply and demand principles do not meet traditional economic market principles. Health benefits include many intangible in-kind features. The delivery of healthcare services should primarily be driven for the good of those receiving care and the public interest. Targeting cost reductions is not to make a profit but to ensure the sustainability of the health system. This is achieved by balancing service demand with the available resources to achieve the best possible health outcomes. A well-designed and functioning national digital health ecosystem provides a most cost-effective foundation and a lucrative market for many stakeholders, supporting international trade. The health industry is enormous in economic terms. The share of national GDP for health spending ranges from 4.2% (Turkey) to 17.1% (USA) for individual nations with a world average of 8.8% [1].

These sentiments provide a solid philosophical underpinning to the content presented. Our driver for compiling this text is our collective and extensive real-world national and international experience working in this field. We have witnessed many failures, primarily due to poor high-level decision making by executives who chose not to make use of expert advice and are unaware of the long-term consequences of their decisions. These initial errors at procurement are often perpetuated and even amplified, with end-users having to do the best they can with what is provided.

The authors have been participants in many cutting-edge developments and innovations, learning and building knowledge along the way. However, we are exhausted from witnessing perpetual knowledge gaps resulting in the ongoing promotion of myths and incomplete information. Understanding digital health requires being at the intersections and overlaps of many knowledge domains—computer science, software engineering, technical networks, medical and health sciences, information science, linguistics, philosophy, operational and data management, governance, and research disciplines, as these apply to any health-related domain, and within an ethical context. It can only be successful with the adoption of an interdisciplinary team approach where the combined skills and knowledge best match project requirements within a national supportive infrastructure.

There is a lot to learn from the COVID-19 pandemic and the financial aid provided by Governments that amounted to $11.7 trillion or close to 12% of GDP according to the IMF as of September 2020 [2]. A study of 12 government economic disbursement programmes for both individuals and small and medium-size enterprises (SMEs) in seven countries found three stand-out critical structural features of financial infrastructure determinants of programme effectiveness. These were:

1. Digital payment channels
2. The presence of a basic digital identification system with broad population coverage
3. Simple data on individuals and businesses that are tethered to the ID

Their presence in country-level financial infrastructure enabled these programmes to be optimally designed and delivered quickly. When one or more of these infrastructure features was not present, countries had to make trade-offs between the design ambition (scope, scale, and specificity of beneficiaries targeted), of their programme and delivery success [2].

We argue that these same principles apply to all national health systems. Adoption provides a unique opportunity to improve the delivery and success of national digital health transformation programmes. Global and national health and clinical professional infrastructures need to be designed to cost effectively and sustainably facilitate the provision of digital transfers of timely, accurate, and complete patient information channels according to the desired data supply chain to:

(a) follow individuals throughout their health life journey in support of patient-centred, value-based health service delivery
(b) meet the needs of all decision-makers along the way
(c) contribute to population health, knowledge transfer, and discovery.

Digital patient information transfers require:

1. The presence and adoption of standard identifiers for people, providers, types of services, health status, social determinants, and all types of resources.
2. Links between service availability by location, delivered services, individual and population health, whilst providing support for work and communication flows.
3. Links enabling effective resource management and performance monitoring.
4. Global, national, and professional leadership overseeing a major health transformation supported by well-connected digital health ecosystems plus supporting foundational infrastructures.
5. Extensive high-level clinical engagement enabling domain knowledge governance facilitating maximum advanced data analytics and knowledge computation.
6. Development, adoption, and maintenance of a harmonized set of technical and data standards.
7. A well-connected digital ecosystem based on an open knowledge-based platform, where all relevant data, information, and knowledge are future proof, stored in computable formats in vendor/technology neutral repositories, accessible to authorized individuals/applications able to meet the needs of every stakeholder's use case.

Unfortunately, we continue to witness a significant knowledge divide between health-related disciplines, information and technology practitioners, and their supporting global and national entities. This is severely compromising any number of digital health solutions. Best practice is often ignored because the workers or industry representatives are unable to collaborate effectively or just don't want to listen to the experts. Many traditional technical leaders just want to 'crack a whip' and build something to solve today's problem, which may occur only from their own, rather than from an ecosystem perspective. Pushing for rigorous best practice by the few is an uphill battle. Some things will be made to work, but the sheer disjointedness of the current players, artifacts, visions, and politics will always result in expensive, slow, and or mediocre quality solutions, no matter how brilliant the people working on it are. We don't have a suitable infrastructure able to support or maintain rigorous processes for many digital health-related activities being undertaken.

Our work has been shaped by the many challenges, confrontations, debates, and negotiations we have engaged in whilst working with leaders and team players from numerous countries, with experience and expertise associated with a great variety of disciplines. Our shared vision for digital was formulated 20–30 years ago. Our collective learning and experience from the years that followed is the basis for this book.

The text aims to identify, expose, and explore the many shortcomings, hard truths and real-world issues that we all face as a foundation for a future roadmap. Digital transformation is a slow and complex process as we need to consider many legacies we need to build on to transform today's healthcare environment to a successful global digital health ecosystem. This roadmap will vary for every jurisdiction, although we would like to see adherence to key foundational principles by all.

A primary focus for transformation needs to be on our health data assets. Best practice management and governance of any digital health ecosystem needs to focus on links between the semantics (adopted meaning) of languages/terminologies in use, and the use of information and communication technologies. How do these links best ensure that data/information/knowledge collected, processed, and presented for use, best meets functional requirements?

For example, the validity and reliability of the results obtained from activities, such as performance monitoring, data analytics, artificial intelligence applications, are dependent on the quality (accuracy, completeness, and timeliness) of all data, information, and knowledge processed. We explore issues associated with health data activities within a digital health ecosystem, identify and explain some of the technical health data interchange methodologies used for this purpose. Many technical solutions for health data interchange between systems and devices are explored together with supportive processes and infrastructures. Their suitability for continuation is then assessed based on their ability to meet the 'fit-for-purpose' test. This landscape needs to be viewed from a global open market perspective.

We know that it is critical to have a suitably prepared healthcare and health IT workforce although the evidence detailing the benefits is scarce. A major component of our roadmap is the need to work collaboratively with all members of the health industry's workforce along the entire transformation journey. The same applies to the population at large as everyone needs to be able to make the best possible use of all digital and medical advances to optimize and best manage their own health. Collaboration, workforce, and education issues need to be addressed.

All of these precursors are then summarized from the perspective of the roadmap, the need for change, the interdependencies and processes to consider, to successfully and globally transform the way the health industry is organized and managed. A successful digital health ecosystem needs to be based on a very different way of local and global thinking, to meet public health needs. It needs to focus on collaborative relationships across disciplines, sectors, and jurisdictions.

We suggest that the detail of this roadmap should be based on Adner's timeline to success [3], in terms of:

- why things go wrong,
- the many identified co-innovation and chain adoption risks,
- ecosystem mapping that considers roles and relationships required to line up the many entities,
- facilitate decisions to be made at every level within the health industry at the right time and place based on the best available information and knowledge,

- adopt and implement game-changing strategies in a manner that sequences success and considers the many interdependencies identified throughout this text.

Adner [3] provides a wide-lens toolbox considered crucial for multiplying our odds of success. His first step is about gaining a clear view of our collective *value blueprint*. This assists with articulating our shared vision of our value propositions and the way we can achieve it. Once we have a clear agreed vision of what a global digital health ecosystem needs to look like, we can begin to ask key questions about the role(s) each one of us wants to play and when to make the change. There will need to be natural ecosystem leaders. We need to consider how best to bring the necessary elements together in a world of numerous interdependencies, and how to identify and evaluate benefits to be realized.

Evelyn Hovenga
Heather Grain

References

[1] OECD. n.d. OECD Health Statistics, 2019 [cited 2020 24 February]. Available from: https://www.oecd.org/health/health-data.htm.

[2] O. White, A. Madgavkar, T. Sibanda, Z. Townsend, M.J. Ramirez, COVID-19: Making the Case for Robust Digital Financial Infrastructure, McKinsey & Company, 2021. Available from: https://www.mckinsey.com/industries/financial-services/our-insights/covid-19-making-the-case-for-robust-digital-financial-infrastructure?cid=other-eml-alt-mip-mck&hdpid=b68c6516-89c9-4b1c-b9af-2e1ed5bf4f74&hctky=12004972&hlkid=7113361bd-49c494da9c00a8ad6823d91.

[3] R. Adner, The Wide Lens, Penguin Group, New York, 2013.

Foreword

This is a time of great challenge, but also a time of great innovation, for global health care.

It is timely that this book has been launched during the COVID-19 pandemic. While placing healthcare systems, healthcare planners, and healthcare providers and their patients under great pressure, the pandemic has also been a time of rapid digital health innovation. We have seen many reforms including widespread introduction of teleconsultations, improvements in digital health data exchange through the greater use of electronic health records and e-prescriptions, advances in point-of-care testing, development of digital vaccine certificates and passports, and the adoption of digital tools to support virtual care.

We have also experienced the use of big data in public health at a scale never seen before to understand national and global pandemic responses and the impact on populations. At the same time, advanced digital communication capability has become ubiquitous in many parts of the world through widespread adoption of videoconferencing for personal and professional use, and online access to essential healthcare information has allowed rapid dissemination of critical advice to populations while also improving digital literacy for many. The changes keep coming, and our lives will never be the same.

This book provides us with a set of roadmaps on ways that we can continue to develop successful digital health ecosystems, with a focus on meeting the needs of patients and their healthcare providers, while also allowing innovation to flourish.

Edited by two long-standing and internationally renowned Australian champions of digital health innovation, Evelyn Hovenga and Heather Grain, and drawing on the expertise of leading digital health researchers and clinicians from around the world, this book will be an invaluable resource for all aspiring health informaticians and a source of wisdom and insights for those already working in the field.

This book provides us with the templates we need to support ongoing developments in digital health and highlights areas where we all need to pay attention to ensure that digital health transformations lead to better healthcare provision and better health outcomes for everybody.

Professor Michael Kidd AM, FAHMS

Deputy Chief Medical Officer and Principal Medical Advisor, Australian Government Department of Health & Professor of Primary Care Reform, The Australian National University

CHAPTER

1

Transforming health care

Evelyn Hovenga[a,b,c]

[a]Digital Health, Australian Catholic University, Melbourne, VIC, Australia [b]eHealth Education Pty Ltd, Melbourne, VIC, Australia [c]Global eHealth Collaborative (GeHCo), Melbourne, VIC, Australia

Issues and challenges

Today's global health systems consist of:

- multiple data silos,
- chaotic and fragmented data,
- poor data integrity,
- an inability to consistently and reliably share or link data,
- poor knowledge management strategies leading to operational inefficiencies,
- resource wastage, and
- significant delays regarding the widespread use of new knowledge.

All of the above are the result of continuing competition between large technology providers, insufficient collaboration, and business models promoting 'vendor lock-in'. This makes healthcare service providers dependent on their technology providers to maintain their system's currency whilst creating a perceived inability to change technology providers. The recent global pandemic has demonstrated the value of global collaboration.

Important challenges that need to be addressed include the realisation that the time has come for a major transformation to better engage with this new digital era to enjoy the significant benefits this offers. This requires global leadership and a lot of multidisciplinary collaboration. Another challenge is uniform standards adoption within and between different health-related entities to overcome incompatibility between different platforms and devices in use. Many digital healthcare support solutions in use, lack relevant legal norms to mitigate risks regarding personal information and privacy breaches associated with data transfer and data linkages. In addition, some users have difficulty using currently available technologies [1].

A major impediment to progress is a poor understanding of the many scientific principles associated with the design, development, implementation, and use of health information systems, including electronic health records, and the importance of effectively managing our health data and knowledge assets. There is an abundance of 'myths' reflected in the scientific and a lot of

Roadmap to Successful Digital Health Ecosystems
https://doi.org/10.1016/B978-0-12-823413-6.00020-3

'grey' literature, including social media, due to a poor understanding of digital health complexities. The literature reveals ambiguity resulting from, for example, the non-differentiation of standards, specifications, criteria, and guidelines. We aim to bust popular myths.

Another major issue is the monitoring and evaluation processes adopted for digital health interventions. A poor understanding of computer science foundations and evolutions over time, and the scientific principles of software engineering, makes it difficult to assess the quality of the evidence presented, yet many high-level decision-makers are making use of questionable evidence with significant costly repercussions. There is an urgent need for clinical specialists to fully engage in digital health by governing their own knowledge domain in a manner that facilitates sharing and transparency, supporting evidence-based practice, timely adoption of new clinical advances, and patient-centred care.

Revolutionary changes are needed to reform healthcare delivery systems globally. There is a need to realise the vision of sustainable national health systems and healthcare access for everyone, no matter their location. This vision is best facilitated and enabled by effective national digital health ecosystems. Attaining effectiveness is likely to become more and more difficult over time without the facilitation of an effective, high-level foundational infrastructure supporting a global digital health ecosystem.

This text explores current limitations, barriers, interdependencies, and possible solutions for this to be achieved. We argue for the need to adopt an open knowledge-based 'platform', a public health enterprise system architecture that any health information technology developer can deploy. We would like to see the adoption of new healthcare solutions. This requires a new way of building healthcare IT systems. Healthcare systems' architectures need to ensure that data can exist independently of any particular application so that data from different sources become comparable. This also encourages fast plug-and-play application developments, thus stimulating local development and economic growth.

Some may argue that we're well along the way; others point to the many inconsistencies and the lack of obvious progress. As was demonstrated by the recent COVID-19 pandemic, there is a need to have the ability to make rapid changes. For example, data transfers and staff communication flow changes associated with new health service delivery methods, resource allocations, material and supplies logistics, and patient pathways. What does a digital healthcare ecosystem need to look like to be sustainable in meeting the healthcare delivery needs of any nation's population?

The WHO/ITU toolkit provides three key criteria that collectively form the recipe for success [2]. These are as follows:

1. **Sustained senior government leadership** and **committed financing for digital health**. These are considered to be prerequisites for a successful digital health strategy.
2. **Effective governance mechanisms** at all levels, that engage stakeholders, who have clearly defined roles, to ensure efficient decision-making for a national digital health strategy.
3. **A national ICT framework** that facilitates alignment between health and ICT sectors, to promote connectivity and interoperability, establish common standards, and enable appropriate policies and regulations in digital health.

These criteria were adopted by its members. Several, current, internationally influential organisations were reviewed to assess the current global status. Findings regarding global and national leadership and their overall performance towards their ability to meet these criteria are presented across several of the following chapters.

What is an ecosystem and how does this apply to digital health?

The word 'ecosystem' has traditionally been used to describe a community of living interdependent autonomous organisms interacting with each other, within any physical or natural environment. Such ecosystems are sustained by conserving both energy and matter for a diverse range of organisms, each operating within their own niche or role and extensive recycling. Two characteristics tend to lead to greater resistance and resilience in the face of change:

(1) a high diversity of organisms, and
(2) autonomy, known as 'low coupling' in the IT world.

This resilience sustains the ecosystem in the face of disturbances or disruptive events, thus creating greater long-term stability and a balanced adaptive system. Ecosystems reflect constantly changing and evolving dynamic processes. These fundamental concepts are applicable to any type of ecosystem, where success is determined by how well the autonomy and adaptiveness of its diverse entities function within that domain and its physical environment. Relationships between diverse entities within an ecosystem determine their collective interdependencies.

Health and well-being is a fundamental concern for everyone. Every living individual needs to function within their own ecosystem consisting of their genetic and biological make-up, physical location, family and social connections, local community, prevalent societal, political, commercial, and economic factors. The plethora of personal and environmental interdependent factors collectively determine any individual's ability to meet their fundamental human needs and contribute to their society's economic productivity.

Any national healthcare ecosystem is made up of a diverse range of stakeholders, each with a unique role to play. Many of these stakeholders are also part of a diverse range of global ecosystems, such as international professional organisations, the World Health Organization, or other global entities associated with the health industry. This ecosystem is also constantly changing and evolving with an information or knowledge sharing focus. There are numerous interdependencies between these stakeholders. A vast number of interactions (complex systems) collectively make up this ecosystem.

> Digital transformation is about making use of digital technologies to facilitate and optimise all supporting infrastructures and types of interactions within any healthcare ecosystem.

There is a need to, for example, enable existing healthcare delivery system silos to link with each other's healthcare delivery system silos within a region. Primary care delivery services need to be able to digitally link with systems managed by their secondary and tertiary care service providers. Its purpose is to facilitate integrated care via networks across professions and organisations where care responsibilities are distributed amongst members of multidisciplinary teams. Achieving such integration requires different management competences from those promoted as requirements for traditional management [3]. It requires strong non-hierarchical leadership supporting innovation.

Primary and community care services are best placed to provide patient-centred, preventative, holistic, well-coordinated care within a community, with a focus on achieving care continuity for individuals managing multiple co-morbidities requiring any number of support services [4]. Continuity of care requires relational (patient and one or more care providers), management, and informational continuity to be effective. Integrated care models require coordination between policy-makers and managers at regional and local levels; single entry to social and healthcare services; case management; an individual care plan; an autonomous functional assessment instrument; and a shared electronic health record [5]. Primary health and social care needs to be accessible to any citizen in order to optimise population health which in turn supports national productivity. As has been demonstrated by the management of the COVID-19 pandemic, telehealth delivery strategies have a significant role to play.

Digital health roadmap development process

Developing a roadmap to achieve thriving national digital health ecosystems requires both a top-down approach and a bottom-up approach, simultaneously but within an agreed national framework. There is a need to identify agreed high-level value propositions as these apply to all contributing stakeholders from a global or jurisdictional perspective, and to consider individual healthcare provider roadmaps. These need to be explored and documented to identify current foundational infrastructure gaps. We are well aware that a plethora of bureaucratic, supportive, or obstructing overarching infrastructures exist at all levels of health care—many of these need to be transformed. We need to consider the place of individual stakeholders within existing local, regional, and global communities of various types, as well as infrastructures that collectively make up a digital health ecosystem.

A roadmap details how best to fill identified gaps or transform current practices. We have the skills, the know-how, and the technical capability. Their collective use is dependent not only upon their adoption but also on the workforce capacity. All digital transformations need to build on actual, and be in tune with desired, global digital ecosystem infrastructures. The WHO and OECD have reported requirements, based on terminology standards; others have identified research collaboration needs and/or the need to enable greater use of standard clinical trial protocols. We need to examine if the many current arrangements continue to be fit for purpose. In addition, decisions need to be made regarding investments and their continuing support for requiring the use of any set of existing standards, guidelines, and catalogues, many of which are likely to be unfit for purpose within the current technology landscape in this digital era.

Such a global context provides valuable information and knowledge. We need to explore, consider, and recommend innovative and transformative solutions to stimulate new investments and overcome known barriers and limitations. Before we develop a roadmap to enable a digital health ecosystem to emerge, we need to consider what a comprehensive health services landscape would look like. An agreed vision of a successful outcome can be used as a guide to implement the many roadmap components. Widespread global adoption of a knowledge-driven platform requires software developers to think differently. Traditional database schema will no longer matter. In such an environment, application data will be committed and accessed via relevant service APIs rather than any direct access to a database.

This text explores the impact of international and national drivers on digital health adoption, population health, and patient safety, within the context of national healthcare systems and service delivery performance. In so doing, there is a need to consider how these outcomes are influenced by international or national political, cultural, financial, technical, structural, social, industrial, academic, global value propositions and economic factors. Another key factor is the individual egos and life philosophies of those with positions of power or in leadership positions. Any one of these factors can inhibit progress and may influence interdependencies between those who need to work collaboratively in any number of various kinds of partnerships.

We intend to expose innovation blind spots that have the potential to undermine successful digital health transformation. As noted by Adner [6] p. 3

> Even the most brilliant innovation cannot succeed when its value creation depends on other innovations.

Successful digital transformation is very much dependent upon numerous innovation ecosystems where 'success of a value proposition depends on creating an alignment of partners who must work together in order to transform a winning idea to a market success' [6]. Establishing such alignments within a global health context is extremely challenging.

Our proposed roadmap aims to describe how best to deliver a successful national digital health ecosystem able to realise a well-defined vision. This ecosystem needs to be structured as an enabler to meet desired value propositions and the global health industry needs as a whole. A successful digital health ecosystem needs to be of benefit to the world's population. Health issues we need to resolve include [7]:

- An increasing demand for services, including mental health and community care, that is overwhelming healthcare providers—most notably nurses and midwives;
- Patients who are desperate to take control of their own health decisions;
- Payers struggling to keep healthcare costs down;
- Ability to mine the information available with confidence as to quality and meaning transmission of that data.

The issues identified above can best be addressed by well-connected healthcare systems, enabled by a digital health ecosystem, capable of meeting the information needs of all decision-makers, with the intent to ensure better health outcomes for all. Resolving these issues requires all data users to know how to assess the quality and utility of available data and information.

Value-based health care

There is a growing interest in value-based health care where patient health outcomes determine financial rewards for providers. Value-based health care requires an ability to measure meaningful outcomes against the cost of delivering healthcare services. Digital health ecosystems enabling the collection, management, sharing, linking, and processing of clinical and health data, information, and knowledge are essential to successfully realise desired value propositions. These data-related activities require digital assistance that facilitates the coordination of all events associated with individual patient journeys and best practice progress support. This digital assistance is dependent upon access to an Internet-connected platform able to provide all required e-services. The delivery of healthcare-related e-services, along with supporting intercommunication amongst authorised healthcare professionals and the sharing of virtual healthcare records, is in turn dependent upon the technical connectivity and system interoperability at every level.

With the rapidly increasing amount of new clinical knowledge, it is impossible for individual health professionals to rely on memory and past experience when making decisions about recommending the use of specific treatment or care protocols [8]. Many practice guidelines are made available in PDF formats, some many pages long. Such guidelines need to be presented in computable formats enabling system integration, to support care pathway engineering and decision support systems. Algorithms used for such purposes need to be able to use up-to-date knowledge services, be governed, and be openly available. Clinical knowledge needs to be curated, stored, and disseminated in computable formats, made available where and when required, and be usefully presented to those who need to make clinical decisions.

The electronic collection of data and the use of digitally supported health data processing, able to support task and information flows determine how well our many and varied value propositions are able to be realised. We need to consider how the use of any combination of digital technologies can best support patient-centred healthcare service delivery and guide individual patient journeys, based on an individual's unique characteristics in their primary location. This requires personalised healthcare assistance, improved preventative care, timely interventions, the collection and evaluation of practice-based outcomes that inform future practice and provide new knowledge. All of these functions are dependent on the choices made regarding technology support.

Key research findings following a review of current evidence and ways forward [9] indicate that the creation of sustainable and effective health data ecosystems requires the adoption of the following principles:

- Value propositions to be represented by health data in the context of interdependencies and interactions between stakeholders (actors) and sectors based on their relationships and resource availability;
- Building on existing efforts and momentum to address data quality and technical requirements;
- Making the most of recent data protection advancements and the sharing of data models and frameworks;
- Workforce capacity building so that health service professionals are in the best possible position to get the most out of health data;
- Increasing public awareness, acceptability, and engagement with health data.

Success measurements

Success needs to be measured within the context of suitable high-level supportive infrastructures. It can be measured by:

(1) The degree to which common desired short- and long-term beneficial outcomes are being realised following the digital transformation, such as [10]:

  - **(a)** Citizen access to national and local data and technology services that enable them to see and manage their own records; undertake a wide range of transactions with care providers; and increasingly manage their own health, care, and well-being.
  - **(b)** Care professionals' timely access to the information, data, analysis, and decision-support systems they need to deliver safe and effective care.

(2) The availability and use of workforce capacity and enablement to meet ongoing changing requirements.

(3) Strong managerial leadership within individual organisations and at all levels within the health system hierarchy, building upon established foundations, enabling them to progressively move forward and make continuous improvements.

Existing foundations

There are numerous international and regional eHealth, digital health, health professional, health research and standards development networks. Collectively, the organisations making up these networks contribute greatly to a plethora of ever-increasing developments and innovations, forming part of our current global digital health ecosystem. These network initiatives have a 60- to 70-year history. The International Medical Informatics Association began as a new Technical Committee (TC) of the International Federation for Information processing on Medical Information Processing during the 1960s under the leadership of Dr Francois Gremy [11].

Dr Lindberg, who became the Director of the US National Library of Medicine in 1984, published the first text on the computer and medical care in 1968 [12]. Early hospital information systems focused on the provision of departmental administrative support. Their growth in the United States resulted in the promotion of patient-centred, bedside terminals and system integration approaches during the 1980s [13, 14]. This early work has greatly influenced current large proprietary US enterprise EMR systems, who acquired multiple systems from smaller companies over the years, and whose large systems are now extensively in use in many countries.

One of the earliest research topics of the European Commission's Research and Development, Advanced Informatics in Medicine (AIM) Framework, included electronic health record systems. One of its 1989 funded projects, the Good Electronic Health Record (GEHR) project, was a precursor for the development of, what is now known as, the openEHR specifications as governed and managed by the not-for-profit (NfP) openEHR Foundation and its large global community of interest, openEHR International [15].

This early initiative was followed by the European Union's adoption of an eHealth action plan in 2004 when all member states were called upon to develop tailored national and regional eHealth strategies to primarily meet their own needs with the intent for all to develop and use the same standard of health care and support patient and healthcare provider mobility [16]. Considerable progress has been made by many around the globe, with similar research and development initiatives and incentives, to establish the many prerequisite requirements for the establishment of a global digital health ecosystem [17, 18].

Many of these research outcomes have been implemented, have been built on, and are in constant use. We continue to learn from each other's real-world experiences, whilst developing various solutions towards improving our data sharing and linkage capacity. We now know that the adoption of semantic data exchange protocols is essential. A key finding from one systematic review is that semantic data exchange can only be successful using a common platform or a common system architecture of which the Reference Model (RM) is fundamental [19].

> Gartner believes that truly effective and sustainable open architectures will need a capability for vendor-neutral data persistence, such as utilizing a common schema or set of archetypes and rules for managing structured and unstructured data. Providing open messaging standards (for example, FHIR®, HL7) for data exchange in specific use cases will only go so far in meeting the architectural challenges of digital citizen-centric care delivery [20].

One component example in use in the health industry is the openEHR Foundation's reference architecture [21–26] that has adopted a two-level modelling approach making use of its archetypes (formalisms) freely accessible via its clinical knowledge manager [27]. Its open platform establishes the totality of technical conformance points used for the development of purchasing tender documents and internal development. This also enables local entities to develop niche applications to adhere to this platform's technical requirement, and in the process enable their content to be accessible by other applications and retain its meaning.

Most independent commercial research efforts have been directed towards a variety of devices and applications commonly referred to as the Internet of things (IoT), big data, data analytics, and artificial intelligence. For example, heart rate measures, sleep tracking, ECG, and blood oxygen level monitoring can be provided from patients to physicians through the latest Apple Watch.

The Australian Government funded the establishment of a Digital Health Cooperative Research Centre early in 2018 [28]. These many and varied research and development efforts focus on solving very specific problems with the aim of contributing to a digital health market rather than to a well-connected digital health ecosystem using an open platform. Again, the focus on health data, the need to adopt health data science principles and to consider data quality issues, came far too late during this process, compromising high-quality connectivity.

These examples of existing foundations are not exclusive. There are many other examples of local foundational achievements, as many are in the process of developing various solutions to resolve the well-known interoperability issues. Collectively, these many initiatives highlight numerous overlaps. Many digital health, small- and large-scale pilot, 'proof-of-concept,' or case studies never manage to reach a critical mass due to various constraints. Olu et al. [29] identified such constraints as consisting of poor coordination, weak health systems, lack of awareness and knowledge about digital health, poor infrastructure, including unstable power supply, poor Internet connectivity, and a lack of interoperability between numerous digital health systems. In addition, global coordination is happening in silos due to the absence of an agreed high-level structural framework for a well-functioning digital health ecosystem. Traditional research reporting mechanisms are too slow and take too long to be widely adopted.

Despite the value of health data, there are no agreed metadata frameworks or information models guiding consistent health and social determinants' national and global data collections. There is much to learn from these experiences. Most importantly, this reality demonstrates the need for a common conceptual framework and the adoption of a coordinated approach enabling many local initiatives to be undertaken in a manner that, in the end, contribute productively to benefit the global population's health and welfare.

We need to examine these developments from a global perspective based on extensive knowledge sharing. There is a need to determine how best to make use of these significant contributions and build on these continuing, collective research and development journeys. There is a need to take advantage of our collective knowledge and wisdom gained to date. We need to avoid further fragmentation, as this will delay the ultimate ability to realise maximum benefits to global population health.

Global and regional value proposition drivers

There is international recognition that the adoption of digital technologies has the potential to make a significant contribution towards advancing universal health coverage (UHC), increased and faster knowledge acquisition, and other desired health outcomes. Every individual needs to be able to receive any health service as and when required. The attainment of sustainable national healthcare delivery systems is about realising a desired balance between using available resources and having a healthy population. A roadmap for establishing a global digital health ecosystem, that enables nations and healthcare service organisations to choose and use the best available digital technologies, is about facilitating the realisation of this vision. Research to date emphasises that the realisation of benefits to be derived from optimum health data use at scale requires a simultaneous focus on the technological and structural conditions that ultimately determine what is possible [9].

The World Health Organization (WHO)'s draft action plan has a total of 19 proposed evaluation monitoring indicators applicable to any nation, the first of which is to identify the existence of an entity governing digital health [30]. This was followed by a global strategy on digital health to promote healthy lives and well-being for everyone, everywhere, at all ages [31]. A digital implementation investment guide (DIIG), collaboratively developed by HRP, UNICEF, UNFPA, and PATH, was released in 2020 [32]. This document states that 'greater investment is needed to elevate the role of digital health in health systems, so that the positive impact on the health of individuals and population can be fully realized'. In recognition of the large-scale digital transformation need, its members requested WHO to provide guidance for scaling up digital health implementations at the seventy-first World Health Assembly. The DIIG refers to the following principles for digital development throughout their publication:

- Design with the use
- Understand the existing ecosystem
- Design for scale
- Build for sustainability
- Be data-driven
- Use open standards, open data, open source, and open innovation
- Reuse and improve
- Address privacy and security
- Be collaborative

WHO has provided a normative guide to digital health, that includes the promotion of evidence-based digital health interventions [33]. The investment guide details how to practically invest in and implement its recommendations. These guidelines are based on a critical evaluation of the evidence about emerging digital health interventions in terms of contributions made to health system improvements, based on an assessment of the benefits, harms, acceptability, feasibility, resource use, and equity considerations.

Unfortunately, these guidelines are not based upon data quality principles relevant to data collection at the source, including the use of reasons for missing data (flavours of null) to represent situations such as 'asked but not answered' or for unidentified unconscious patients. Data need to accurately represent the world's reality. This is essential for the enablement of reliable and accurate data collection, local data processing, and big data analytics. Such activities provide the necessary foundational information on which to act towards achieving these high-level value propositions.

A digital health enabling environment

Digital health builds on the energy (power availability), ICT infrastructures, and the use of digital technologies, which collectively represent foundational enablers for innovation and a global digital health ecosystem. Effective and timely public health management is best enabled by the availability of and access to a real-time global health data hub. This requires governance, numerous collaborative agreements, and compliance with agreed risk management regulatory requirements. There need to be appropriate levels of protection in place to ensure privacy, security, and safety for all. Cross-border data flows are crucial to the digital health economy. Given that every nation has its own privacy and security legislation, this is difficult to achieve.

Political commitment is essential to making use of lessons learned on digital health and to drive the development of a sound national action plan aimed at transforming healthcare service delivery methods. The design of a national action plan should endorse:

- the WHO global digital health strategy;
- the establishment of a knowledge governance and management approach to identify valid evidence, communicate risks, and mitigate threats associated with the use of digital technologies; and
- the adoption of sustainable financing models.

The WHO 2019 draft action plan [30] notes that *digital technologies and improved data capacity with good ethical standards are important catalysts for accelerating achievement of the 2030 SG3 targets. Actionable and relevant data provide the foundation to monitor progress.* Five digital health enabling environmental characteristics are described as follows:

(1) National and user adoption of an action plan that incorporates people-centred digital transformation as a national priority with support for change management policies.
(2) Prioritisation of investment on data and digital health by aligning an investment plan for digital health technologies, including governance and coordination whilst aligning capacity building support across partners.

(3) Adoption of standards, regulations, and policies that ensure that the public is protected, the quality of health delivered by digital technologies is preserved, and that ethics, cybersecurity, and data protection are considered, and that support is provided during the early stages.
(4) Development of strategies to select and purchase fit-for-purpose solutions with clear impact on lowering the cost of primary healthcare delivery
(5) Build local talent on digital transformation through partnerships and collaboration and promote health innovation for the global good.

Contributing factors to the growth of digital health solutions are increasing broadband coverage and unprecedented growing levels of population access to ICT. This enables the adoption of patient-centred care provision, transforms record keeping, data sharing, and usage, and enhances communication between patients and their providers, and more efficient resource allocation. Connectivity between different technology systems and software applications enhances continuity of care. This can only be achieved with well-suited national digital health governance.

Good governance is about exercising political, economic, and administrative authority adopted to manage national digital health. Good governance needs to be respectful of the many relationships between governments, citizens, markets (supply chain and health ICT vendors), public and private sectors, professional organisations, national and international institutions. Governance effectiveness may be viewed as crucial to a nation's prosperity including its digital health ecosystem.

A national digital health roadmap

There are numerous pockets of successful digital health innovations, representing a huge global market. A national digital health roadmap needs to guide innovators through the complexities associated with any national health system. We are aware of many hurdles and barriers as well as many foundational values, such as patient safety, ethical professional practice, confidentiality, and security, that need to be met. Digital disruption and innovation have the greatest impact on health-related research and development strategies and all patient experiences associated with prevention, diagnosis, treatment and care and follow-ups [34]. Digital innovations that affect patients also have an impact on every healthcare provider's workplace, workflow, patient interaction, and communication patterns.

The purpose of a national digital roadmap is to provide guidance for the establishment of a development pathway towards effective national, regional, and organisational infrastructures. Such a pathway needs to clearly identify and consider the many interdependencies showing what can and needs to be done before specified milestones can be realised. These are needed to facilitate our collective ability to work in harmony towards a common goal. The adoption of digital health disruption and innovative initiatives must be suitably enabled. There is a need to facilitate the development of individual roadmaps, big or small, for any number of components detailed in an overarching roadmap, to collectively build a successful national digital health ecosystem.

The roadmap to success needs to consider a number of different influences that shape the required transformation. First, there needs to be agreement about what multiple lower-level ecosystems need to look like or be able to support. We need an agreed vision of where we need to be and by when. This determines the overall strategic directions to be adopted. It is also important for every stakeholder to consider:

- long-term objectives and the points along the way representing short-term objectives;
- potential risks in terms of investments versus adopting innovative technologies;
- existing and emerging technologies, skills, and know-how that can be leveraged for the required transformation;
- the possible speed of change each stakeholder is willing to be part of; and
- the adaptability and flexibility of stakeholders in charge of key interdependent components that need to be harmonised in a timely manner.

Once a vision with specified essential and desirable outcomes is agreed, then a roadmap representing an implementation strategy can be developed. Such a roadmap needs to provide details that specify how to build the necessary foundations relevant to enable the desired endpoints to be realised. These foundational infrastructures can then be built on locally with various key initiatives that embed the whole of a national government's foundational ecosystem components. Then, any number of extensions can be developed and implemented. Every healthcare delivery service and supporting organisation needs to develop their own digital transformation roadmap. This needs to be in harmony with national and global directions. Each and everyone within the healthcare industry needs to become a well-functioning part within any country's digital health ecosystem. National ecosystems also have a need to be able to connect with a global health ecosystem.

A future vision

Government investments support economic and social priorities. These need to complement investments made by industry and other stakeholders. A national digital health ecosystem architecture and governance infrastructure provides its health system credibility and stability, thus enabling the desired impacts and value propositions to be realised. There are a number of desired outcomes that collectively contribute to a digital health transformation agenda.

Foundational infrastructure components are interdependent on each other and reliant on a well-functioning high-level supportive leadership and governance structure. Each visionary statement describes desired success factors. These represent an overall vision that needs to be implemented in a manner that best facilitates the successful implementation of a national digital health roadmap.

A decomposition of a proposal for a successful national digital health ecosystem identifies six visionary components. They cannot be considered in isolation for implementation purposes as they are interdependent upon each other and all are required simultaneously for operational success. Individual components of each visionary statement represent any number of interdependencies that need to be considered for implementation in some sequence to be described in the roadmap as desired outcome measures. Each heading represents a summary of desired functionalities and actions, and they are briefly defined as follows:

- **Global and National Governance Infrastructures Supporting Digital Health**—Strong government leadership, with delegated expert support, at every level is required to ensure patient safety throughout any individual's lifelong healthcare journey. This is dependent upon global and national compliance with agreed, and, in some instances, mandated, international laws, a set of key standards, guidelines, and policies. Their development, adoption, maintenance, review, and updates need to be supported by suitable governance processes.
- **Harnessing Clinical Data Using Advanced Technologies**—Highly granular clinical data are huge assets supporting knowledge discovery, evidence-based practice, clinical trials, lifelong learning, for capturing the full benefits of decision support systems, robotics, and artificial intelligence (AI), and are essential for precision medicine and patient-centred care. These values are best facilitated by smart data collection at points of care, and the use of standard open digital platforms, supporting data use life cycles, data access, data linkages, sound knowledge management and statistical reporting.
- **Digitising Healthcare Delivery Operations and Automating Processes**—Automated global public health information sharing, supply chain, and reporting enable the provision of more timely and comprehensive information for decision-makers. This is dependent upon consistent use of agreed data standards and requires the streamlining, centralising, and, in some instances, globalising many routine and core organisational operational processes. This transformation is expected to cause major disruption in most organisations during the transformation period but is expected to greatly improve national health system sustainability by automating many routine reporting processes.
- **Transitioning to Open Market-Driven Technologies**—Current global reliance on a few competitive, commercially managed, high-level infrastructure technologies, such as clouds, standard platforms, or large EMR systems, is a continuing high-risk and costly strategy for many countries. This is best mitigated by the adoption of openly available foundational technologies, governed and managed for the public good to form a publicly funded well-functioning national infrastructure. This approach provides a strong focus on applications well-suited to agreed national platforms, including any type of health information system, telecare, telehealth, wearables, health monitoring devices, and health analytics. This drives local economies, supports local application developers and provides the necessary market-driven competitiveness.
- **Digitising Collaborative Relationships**—Strong national leadership can ensure that an effective digital health ecosystem is more supportive of timely, complete, and accurate information sharing, communication, and collaboration between researchers, healthcare providers, and their patients/clients, irrespective of location. Researchers benefit from timely knowledge sharing. Patients/clients need support, for them to effectively make use of new technologies and their health information. Enabling them to partner in care and treatment option decision-making requires them to acquire the necessary health and digital health literacy.

- **Building a Digitally Talented Health Workforce**—Living and working within a supportive digital health ecosystem requires changes to roles and boundaries of practice for many healthcare professionals. Many new jobs will be created for people to enable them to drive or manage technical innovation, workplace transformations and implement new digitally enabled processes. Over time, a shift in roles is expected, and roles will increasingly transition from processes operated by people, to processes designed and audited by people with profound implications for individual healthcare delivery organisations. This needs strong national leadership, alignment, and integration between the health and education sectors.

Any roadmap needs to clearly articulate a vision from which value propositions to be realised can be determined, and provide a clearly defined pathway for the development, commercialisation, and commissioning of digital health products associated with any of the pillars above. We are currently experiencing many misalignments between the 'supply'-side stakeholders (innovators, policy-makers, entrepreneurs, investors) and the 'demand'-side stakeholders (health policy-makers, regulators, health technology assessment, purchasers) who operate under different and conflicting logics [34]. Misalignments tend to be the result of vested interests of existing suppliers who do not want change or who seek to delay change to sustain current profitable business models of benefit to them. As a general rule, vendors do benefit from knowing planned national directions as this enables them to innovate accordingly.

Value propositions for supply-side stakeholders differ greatly between them. Each value proposition also differs regarding specific value propositions that guide demand-side stakeholders. As a consequence, there is a mismatch between technologies entering the market and those that are highly desirable and/or more cost-effective. A successful national digital health roadmap needs to deliver value for all stakeholders whilst ensuring that new products, applications, and technologies are fit for purpose by benefiting population health first and foremost.

Change agents and innovators need to have a sound understanding of the local workforce and service availability as well as the local health, housing, community and social care real-world and regulatory environments and their key value propositions to assure alignment. This means that the roadmap needs to be viewed from many different perspectives, yet enables these to be fully harmonised to optimise outcomes for everyone. We need to plan for a five-to-ten-year journey and start now.

Our proposed roadmap is developed based on our collective multidisciplinary knowledge, design, architectural and systems thinking, research results, practical experiences, case studies, business and technical analysis as detailed in the chapters to follow. Content is structured to address the model for global and national digital health reformation to meet the outcome needs as expressed in our future vision previously identified. Each chapter makes a contribution to assist the reader to gain detailed insights into health system complexity. Underlying known theoretical and practical principles associated with health information and digital technologies will be identified in support of the arguments presented. This will be followed by identifying what is currently in use and presentations of current world realities, complete with a critique of benefits, limitations, and risks.

Our aim is to enable the minimisation of risks and costs associated with this digital transformation journey we need to undertake, to transition to a fully functioning successful digital health ecosystem based on the best available evidence. Activities to be included in the roadmap need to be initiated nationally under strong leadership, echoed regionally and locally to gain maximal impact more quickly and cost-effectively. We have the technology and know-how to make it happen.

The chapters to follow will explore these many actions and issues identified thus far in more detail to:

- document the challenges,
- identify what actions need to be taken,
- what to avoid doing,
- to document current and potential risks, and
- the likelihood of their impacts on all concerned.

This knowledge forms the foundation for an appropriate and practical roadmap.

References

[1] S. Tian, W. Yang, J.M.L. Grange, P. Wang, W. Huang, Z. Ye, Smart healthcare: making medical care more intelligent, Global Health J. 3 (2019) 62–65.

[2] WHO/ITU, National eHealth Strategy Toolkit 2012, 2019, [cited 2019 17 April]. Available from: https://www.itu.int/pub/D-STR-E_HEALTH.05-2012.

[3] R. Miller, K.V. Stein, The Odyssey of integration: is management its Achilles' Heel? Int. J. Integr. Care 20 (2020) 7.

[4] B. Baird, H. Reeve, S. Ross, M. Honeyman, M. Nosa-Ehima, D. Omojomolo, Innovative Models of General Practice London: The Kings Fund, 2018, [cited 2020 30 March]. Available from: https://www.kingsfund.org.uk/publications/innovative-models-general-practice.

[5] A.I. González González, A.M. Miquel Gómez, D.R. Morales, V.B. Sierra, A.L. Romero, J.M. Rodilla, The development of a successful integrated care model/El desarrollo de un exitoso modelo de atención integrada, Int. J. Integr. Care 15 (2015) 158–160.

[6] R. Adner, The Wide Lens, Penguin Group, New York, 2013.

[7] U.K. Kotla, G. Jain, Digital Healthcare Ecosystem—White Paper, Infosys Limited, 2018. [cited 2020 24 February]. Available from: https://www.infosys.com/digital/white-papers/documents/digital-healthcare-ecosystem.pdf.

[8] J.L. Shellum, R.A. Nishimura, D.S. Milliner, C.M. Harper Jr., J.H. Noseworthy, Knowledge management in the era of digital medicine: a programmatic approach to optimize patient care in an academic medical center, Learning Health Syst. 1 (2017), e10022.

[9] S. Marjanovic, I. Ghiga, M. Yang, A. Knack, Understanding value in health data ecosystems, RAND Health Quart. 7 (2018).

[10] NHS-Digital, Information and Technology for Better Care: Health and Social Care Information Centre Strategy 2015–2020, 2015, [cited 2019 3 May]. Available from: https://digital.nhs.uk/article/249/Our-Strategy. www.hscic.gov.uk.

[11] C.A. Kulikowski, C. McGrew, An informatics approach to chronicling the history of IMIA, Stud. Health Technol. Inform. 192 (2013) 1130.

[12] D.A.B. Lindberg, The Computer and Medical Care, C. C. Thomas, 1968.

[13] R.A. Korpman, Patient care information systems: looking to the future, Software Healthc. 2 (1984).

[14] R.A. Korpman, Patient care automation: the future is now. Part 1. Introduction and historical perspective, Nurs. Econ. 8 (1990) 191–193.

[15] D. Ingram, Origins of openEHR, 2002, [cited 2020 24 February]. Available from: https://www.openehr.org/about/origins.

[16] L.D. Serbanati, F.L. Ricci, G. Mercurio, A. Vasilateanu, Steps towards a digital health ecosystem, J. Biomed. Inform. 44 (2011) 621–636.

[17] ONC, A Shared Nationwide Interoperability Roadmap v1.0, Office of the National Coordinator for Health Information, 2014.

[18] ONC, A 10-Year Vision to Achieve an Interoperable Health IT Infrastructure, Office of the National Coordinator for Health Information, 2014.

[19] L. Ahmadian, M. van Engen-Verheul, F. Bakhshi-Raiez, N. Peek, R. Cornet, N.F. de Keizer, The role of standardized data and terminological systems in computerized clinical decision support systems: literature review and survey, Int. J. Med. Inform. 80 (2011) 81–93.

[20] T. Gornik, Industry News—Application-Centric and Data-Centric Interoperability, openEHR Foundation, London, 2019. [cited 2019 3 February]. Available from: https://www.openehr.org/news_events/industry_news.php?id=264.

[21] K. Atalag, T. Beale, R. Chen, T. Gornik, S. Heard, I. McNicoll, openEHR-A Semantically-Enabled, Vendor-Independent Health Computing Platform—White Paper, 2017.

[22] S. Kropf, C. Chalopin, D. Lindner, K. Denecke, Domain modeling and application development of an archetype- and XML-based EHRS. Practical experiences and lessons learnt, Appl. Clin. Inform. 8 (2017) 660–679.

[23] L. Marco-Ruiz, J.A. Maldonado, V. Traver, R. Karlsen, J.G. Bellika, Meta-architecture for the interoperability and knowledge management of archetype-based clinical decision support systems, in: IEEE-EMBS International Conference on Biomedical and Health Informatics (BHI), 2014, pp. 517–521.

[24] L. Marco-Ruiz, D. Moner, J.A. Maldonado, N. Kolstrup, J.G. Bellika, Archetype-based data warehouse environment to enable the reuse of electronic health record data, Int. J. Med. Inform. 84 (2015) 702–714.

[25] openEHR-Foundation, openEHR Reference Model (RM)—Latest, openEHR Foundation, 2017.

[26] C. Pereira, S. Frade, P. Brandão, R. Correia, A. Aguiar, Integrating data and network standards into an interoperable e-Health solution, in: 2014 IEEE 16th International Conference on e-Health Networking, Applications and Services (Healthcom), 2014, pp. 99–104.

[27] openEHR/CKM, Clinical Knowledge Manager, 2019, [cited 2019 21 February]. Available from: https://ckm.openehr.org/ckm/#.

[28] DHCRC. Digital Health Cooperative Research Centre [cited 2019 23 May]. Available from: https://www.digitalhealthcrc.com/.

[29] O. Olu, D. Muneene, J.E. Bataringaya, M.-R. Nahimana, H. Ba, Y. Turgeon, H.C. Karamagi, D. Dovlo, How can digital health technologies contribute to sustainable attainment of universal health coverage in Africa? A perspective, Front. Public Health 7 (2019).

[30] WHO, Action Plan for the Global Strategy on Digital Health 2020-2024, 2019, Available from: https://www.who.int/docs/default-source/documents/annexi-action-plan-for-the-global-strategy-on-digital-health.pdf.

[31] WHO. Digital health: World health Organization; Available from: https://www.who.int/health-topics/digital-health#tab=tab_1.

[32] WHO, Digital Implementation Investment Guide (DIIG): Integrating Digital Interventions Into Health Programmes, World health Organization, 2020. Available from: https://www.who.int/health-topics/digital-health#tab=tab_1.

[33] WHO, Guideline: Recommendations on Digital Interventions for Health System Strengthening, World Health Organization, Geneva, 2019. [cited 2019 18 April]. Licence: CCBY-NC-SA3.0IGO. Available from: https://www.who.int/reproductivehealth/publications/digital-interventions-health-system-strengthening/en/.

[34] Oxford-Academic-Health-Science-Network, Digital Health Roadmap: A Guide for innovators developing digital health products, 2019, [cited 2020 5 March]. Available from: https://www.healthandwealthoxford.org/wp-content/uploads/2017/12/DigitalHealthRoadmap_Overview.pdf.

CHAPTER

2

Global and national infrastructures supporting digital health ecosystems

Evelyn Hovenga[a,b,c] *and Heather Grain*[b,c,d]

[a]Digital Health, Australian Catholic University, Melbourne, VIC, Australia [b]eHealth Education Pty Ltd, Melbourne, VIC, Australia [c]Global eHealth Collaborative (GeHCo), Melbourne, VIC, Australia [d]LG Informatics Pty Ltd, Melbourne, VIC, Australia

Digital health foundational needs

Many initiatives have been undertaken over many years to form existing health system and structural foundations upon which to build a digital health ecosystem. Many systems tended to be monolithic with little regard for the bigger picture in terms of any number of actual or potentially desirable relationships of future needs. Most national infrastructures were developed to address any number of specific challenges or policy initiatives. The result is a lot of fragmentation, many silos, redundancies, reinventions, and systems that are no longer fit for purpose. National digital infrastructures need to focus on optimising day-to-day service delivery operations to suit the population at large within an ecosystem consisting of many stakeholders. Many foundational infrastructure needs are being met in an ad hoc manner by commercial and philanthropic initiatives. These in turn may be the result of jurisdictional policy developments and legislatures.

The availability of technologies that underpin successful digital health ecosystems is shifting economic and market imperatives. Business models and the use of intellectual property are in need of change. Transformation to a data and knowledge economy requires focused and well-funded national implementation strategies requiring a transition over time. Digital health commercialisation needs to focus on driving local economies, rather than the continuing support for technology giants. Ideally, national governments and healthcare provider organisations support multiple small and medium enterprises (SMEs) and the local workforce, to meet local/national needs in a competitive environment based on openly available foundations towards achieving a well-connected and functioning ecosystem able to meet the needs of all stakeholders.

Prior to initiating such changes, one needs to identify the existing foundational infrastructure in order to develop a change management strategy based on necessary transitional arrangements in a manner that minimises healthcare service disruptions to progressively fill

Roadmap to Successful Digital Health Ecosystems
https://doi.org/10.1016/B978-0-12-823413-6.00008-2

identified gaps. The study of any national, regional, or organisational digital health infrastructure can take the form of:

- Investigating strategic directions, planning, development, and deployment efforts, to reveal moments of social shaping due to new technological innovations as these become taken for granted (black box phenomena).
- Examining activities of maintenance, upgrade, and breakdown. This focuses on activities that hold together the functioning of the infrastructure and the ecosystem as a whole.
- Tracing the adoption and use of any new technological artefact.

When evaluating the performance of any digital health infrastructure, there is a need to focus on the challenges identified previously, desired outcomes, or the ability to meet various stakeholder specific value propositions such as the ability to:

- integrate new technologies and innovations providing flexibility;
- incorporate heterogeneous technologies;
- serve a diverse set of uses by adopting the 'enter once, use many times' philosophy;
- reuse data for public health, machine learning applications, and advanced analytics;
- share data with patients, carers, and individual providers in a timely, accurate, and complete manner providing actionable intelligence;
- assimilate clinical with non-clinical data;
- track edits;
- decentralise data storage into data lakes or data hubs supporting cross-institutional data sharing capacity;
- convert clinical guidelines into computable formalisms;
- identify adverse events compromising patient safety;
- automate reporting to external organisations;
- support the provision of patient journey tasks, data, and communication flows.

All of the above functionalities require future proof foundations and the availability of a common platform, that can be used by any software developer, supporting local economic growth.

Policy initiatives and associated funding mechanisms adopted by any national health system tend to dictate healthcare delivery service possibilities. National health policies are, to a greater or lesser extent, influenced by international developments relative to public or population health needs or ideological and technological advances. In some jurisdictions, health industry digital vendors and associated professional organisations do have significant influence on national policy developments as well. The adoption of digital health solutions has the potential to transform the effectiveness of national health systems enabling them to better meet population healthcare needs in a sustainable manner. National health infrastructures need to be able to support their desired national and organisational digital health ecosystems.

In summary, global and/or jurisdictional foundational digital ecosystem infrastructure needs include:

Hard Technologies

- Reliable electricity supply;
- Telecommunication including fibreoptic cables and grid computing (distributed computer resources);

- Broadband network, cybersecurity;
- Householder and organisational Internet access, dependent upon a variety of hard technologies;
- Open online platforms and system architectures.

Soft Technologies

- Identifiers—person, individual provider, organisational provider, service type by location, equipment, catalogues, directories;
- Metadata standards and registries, reporting frameworks and data governance mechanisms;
- Mandated standards, including terminologies, and domain ontologies, legislative and regulatory requirements pertaining to any aspect of digital health;
- Catalogues' directories;
- Knowledge management.

Agreements

- Any number of agreements between entities to facilitate data, information and knowledge sharing and linking;
- Collaborative working relationships between communities of interest and stakeholders;
- International and complimentary national legislation, regulatory, and governance requirements.

The Global Digital Health Partnership, a collaboration of governments and territories, government agencies, and the World Health Organization, was established in early 2018, to support the effective implementation of digital health services [1]. This international forum provides an opportunity for transformational engagement, global collaboration, and sharing of evidence to guide its members towards more effective digital health service delivery. Its five work streams are:

(1) Cybersecurity.
(2) Interoperability—AI for health care.
(3) Evidence and evaluation—sharing insights.
(4) Policy environments.
(5) Clinical consumer engagement—citizen access to health data.

The GDHP has produced a white paper on each of these topics. The interoperability work stream 2020 white paper outlines four strategic categories and associated policy recommendations to support policymakers to implement artificial intelligence (AI)-driven technologies into their health system. This relates to the recognised potential of the adoption of data-driven technologies. This requires data to be accurate at its source and to maintain context and meaning throughout the data supply chain supporting any number of use cases.

The four categories for their policy recommendations are as follows:

- Leadership and oversight supporting research, funding, and workforce development.
- Focus on the entire ecosystem based on the requirement to aggregate and link data, adopt public and private initiatives to address skills and funding gaps and a robust research to deployment pipeline.

- Ensure national standards and regulatory processes address safety and efficacy of all AI-driven technologies.
- Engagement with all stakeholders focusing on building trust and to complement healthcare professional workflows.

The most prevalent barrier to interoperability and data usability was identified as the lack of EHR capability to take action on exchanged data. The creation of a Global Master Standards Guide is recommended to address the interoperability challenge.

The GDHP guiding principles are as follows:

- Equality—equal opportunity to participate, contribute, and share in the lessons learnt.
- Cooperation—thoughtful, constructive, and respectful debates.
- Transparency—engagement openness.
- Responsibility—active contribution guided by their annual work plan, ensure that meeting outcomes are effectively and efficiently undertaken and that decisions are made in a transparent and fair manner without discrimination or bias, acting in the public interest.

The following chapters document the current international landscape, standards, policy drivers, and supplier requirements to be considered for a supported legal framework for a digital health ecosystem and potential benefits to be realised, based on these authors' experiences and publicly available evidence. There is a tendency for politicians to present their current national digital health environment with rose-coloured glasses focusing on desired outcomes rather than on foundational requirements for change in order to maximise potential benefits at minimal additional costs.

Digital health ecosystems

Digital health ecosystems apply to individual healthcare organisations, networks, or regions as well as any nation (jurisdiction) as a whole. Ecosystems represent supportive enablers of the delivery of healthcare services. The potential of digital health requires national or regional initiatives and accompanying investments, guided by a robust strategy that integrates financial, organisational, human, and technological resources [2]. Digital health refers to 'the field of knowledge and practice associated with any aspect of adopting digital technologies to improve health, from inception to operation' [3]. Successful implementation requires collaboration towards achieving a common goal. There is a need to think globally and act locally by filling capacity gaps and building on the related current digital health journey, that began during the early 1960s alongside scientific computing developments.

An ecosystem is a complex adaptive system consisting of diverse components that interact locally and are subject to the process of natural evolution and selection within its containing environment, such as a national health system. Fig. 1 presents the many dimensions of complex systems when compared with simple systems. Patterns or properties at higher levels within this construct emerge within any ecosystem from localised interactions and selection processes at lower levels [4]. An ecosystem generally refers to groups of interacting entities (representing a 'community of interest') that depend on each other [5]. Interacting entities need to be able to transfer and share data, information, and knowledge to suit a variety of

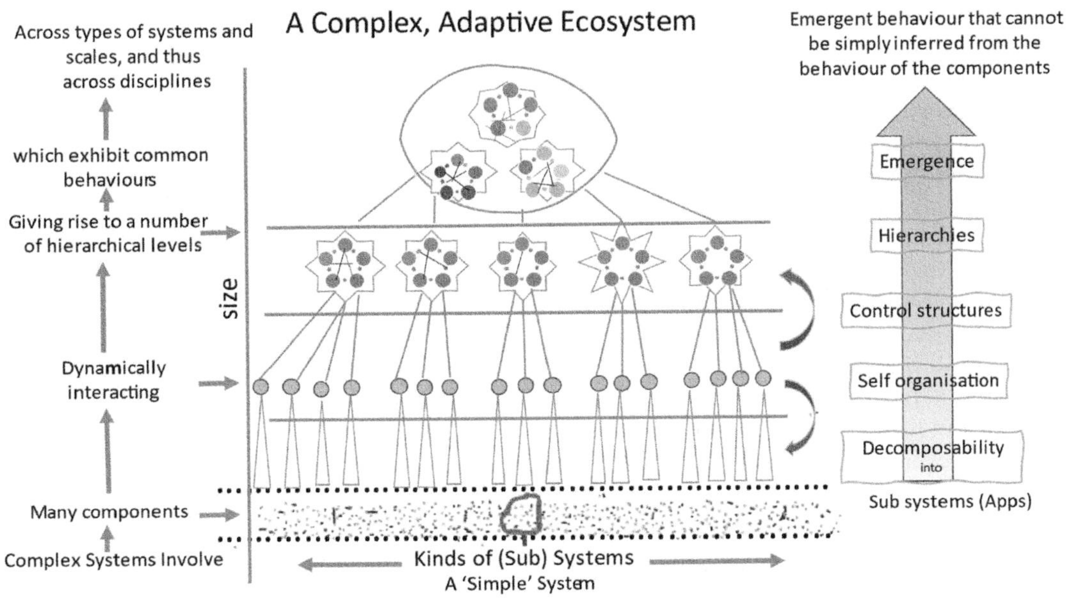

FIG. 1 Dimensions of simple and complex systems.

purposes. In a digital environment, this requires such communication to take place between information systems, where information sent in one digital format needs to be able to be read and processed by the receiving information system. Such transactions are referred to as being 'interoperable'. This functionality is a foundational requirement for a successful well-connected digital health ecosystem.

Connecting many and varied information systems with numerous other complex systems requires an agreed high-level structural framework or architecture. Such a framework needs to specify what characteristics each system needs to meet, in order to facilitate the necessary electronic data/information or knowledge exchanges and support how such resources need to be used in order to support the many actions and decisions required for the delivery of healthcare services. In the absence of the necessary interoperability capacity, digital health ecosystems will continue to compromise optimum healthcare service delivery. Interoperability is of fundamental importance and is being covered in considerable detail in the many chapters that follow.

Fig. 1 demonstrates at a high conceptual level why health IT messaging standards that focus exclusively on content transactions fail to achieve sustainable interoperability to the level required for a successful adaptive digital health ecosystem able to support even the most detailed knowledge processing requirements. Within any ecosystem, it is necessary to consider all possible people, equipment/supplies, communication, information, and knowledge flows between the constituent entities within and between multiple networks, as well as subsequent data/information processing requirements.

Every patient journey has its own unique associated communication and information flow needs that must be supported by digital technologies in a manner that facilitates care continuity. Patient journeys are also associated with logistics and resource management by healthcare service providers. Patient characteristics and their physical locations (their collective local, regional, and national disease burden) determine local and network demands for health service provision.

In the absence of an agreed ecosystem framework, any large healthcare delivery organisation or environment has no choice but to define its own information system architecture. This requires them to think about their own data, processes, and knowledge assets, in depth. The outline of how interoperability will be achieved at any interface point between any of these ecosystem entities must be part of that architecture. Only then can any published standard be considered for use, if it truly fits and provides for the use of a language of interchange that will be in wide use for the same purpose (use case).

The only way to engineer standards that will result in sustainable semantic interoperability is to define an open architecture (platform) for use as the agreed point of reference. 'Standards' will just be pieces of that specification that apply at interface points. To achieve semantic interoperability, common knowledge resources must be defined and used across the entire domain—i.e. ontologies, terminologies, definitions, and models of higher-level artefacts such as data sets and guidelines, as explained in some detail in the following chapters. These technical components need to be considered when developing a roadmap to a successful digital ecosystem. Any environment in which information system interoperability is a desired requirement is a complex system, operating on multiple hierarchical levels, with emergent properties such as 'interoperability' at each level as shown in Fig. 1.

The health industry represents an extensive and complex community of stakeholders who need to cooperate to create value. Cooperation requires communication (explicit knowledge transfer), enterprise memory (tacit knowledge), and timely coordination between workers and institutions (logistics). Collectively, these entities are influenced by their many interdependencies; in total, they represent all properties of a sustainable health ecosystem, without which their desired functional realisation becomes difficult. Only with a digital ecosystem can the human ecosystem of healthcare facilities efficiently achieve their intended outcomes. Digital health ecosystems are complex due to its many unique subsystems whose interactions and interdependencies need to be adaptive. These are difficult to describe and manage due to the large volume of 'use cases'.

A systematic exploration of the literature concerning the terms digital health, innovation, and digital ecosystems resulted in the development of a conceptual framework for a digital health innovation ecosystem [6]. This review identified a variety of interpretations in use for the term 'digital ecosystem', which reflected a number of different concepts and the term's conceptual complexity. These authors then applied the primary concepts identified to the health sector, and formulated a definition for the term 'digital health innovation ecosystem' as follows:

> a network of digital health communities consisting of interconnected, interrelated and interdependent digital health species, including healthcare stakeholders, healthcare institutions and digital healthcare devices situated in a digital health environment, who adopt the best-demonstrated practices that have been proven to be successful, and implementation of those practices through the use of information and communication technologies to monitor and improve the wellbeing and health of patients, to empower patients in the management of their health and that of their families [6].

Early research described digital ecosystems as being robust, self-organising, and scalable architectures that can automatically solve complex and dynamic problems [7, 8]. Digital health optimisation requires a holistic national approach based on good governance for successful implementation and sustainability [9]. We argue that these are not the only requirements.

A digital health ecosystem requires a digital health strategy that includes the adoption of health service-oriented system architectures providing a platform that enables stakeholders to connect with each other to communicate, share, link, and process data, information, and knowledge. Although necessary, the ability to share and link data is insufficient on its own to support modern distributed and team-based care. We propose that support for process (workflow, dataflow) and coordination (logistics) are also essential. Three strands supporting desired system functionality need to be considered:

1. Process/pathway support enabling evidence-based medicine/practice to become a reality;
2. Coordination enabling workers separated by time and place to cooperate;
3. Sharing and linking data to support patient-centred care and precision medicine.

Collectively, such adoption and use make up an ecosystem supporting knowledge development and leading to significant benefits for all [10]. Romanelli refers to business ecosystems and notes that small and medium enterprises (SMEs) tend to develop information technology infrastructures able to manage big data. They tend to have more flexible IT infrastructures and fewer legacy issues. They can adapt quickly by focusing on innovation to gain higher efficiencies. Other broad groups of ecosystems referred to in the literature are 'innovation'- and 'platform'-focused ecosystems [5].

An innovation ecosystem consists of an evolving set of actors, activities (relationships), and artefacts (resources) that collectively make up an adaptive complex system as shown in Fig. 1. These three properties represent the large number and diverse nature of health system participants and resources necessary for innovation to occur.

> An innovation ecosystem is the evolving set of actors, activities, and artifacts, and the institutions and relations, including complementary and substitute relations that are important for the innovative performance of an actor or a population of actors [11].

Fig. 2 provides a pictorial representation of a digital health ecosystem where patient data are at the centre. This shows the many different stakeholders who need to make use of patient data for multiple purposes designed to support patient-centred healthcare service delivery. This image also depicts the many possible data supply chains from collection (detailed source data) to high-level aggregated big data required to serve multiple purposes.

Successful digital health ecosystems need to maximise the value generated by systems in use. This requires interoperability, reliability, flexibility, changeability, reconfigurability, evolvability, adaptability, usability, maintainability, testability, modularity, resilience, robustness, and sustainability. Collectively, these 'ilities' represent system lifecycle properties [13]. The ability to meet these types of requirements is dependent upon the platform ecosystem adopted. This determines the degree to which a system can undergo configuration modifications to meet new functional requirements, in a simple cost-effective manner. Tiwana [14] notes that software platform ecosystems must be orchestrated rather than managed. Their strength is the diversity of external innovators whose work must also be integrated with the platform to enable the realisation of diversity advantages in practice.

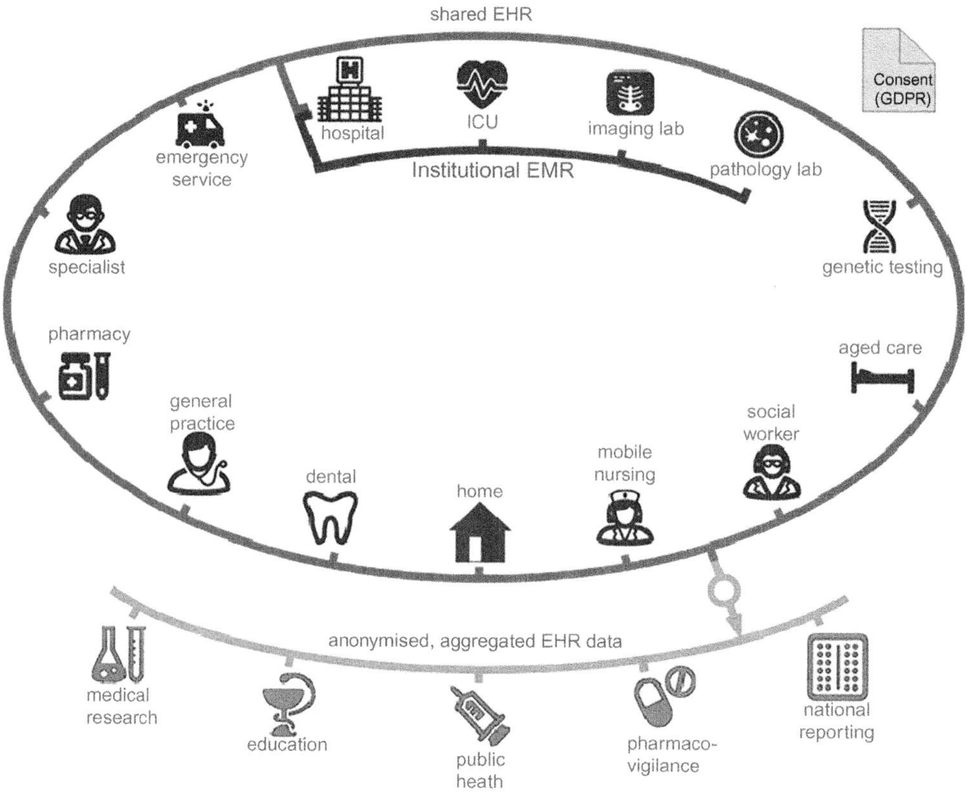

FIG. 2 A patient-centred, knowledge-based digital health ecosystem [12].

There is a need for a radical rethink regarding cybersecurity, privacy, data and knowledge ownership, information sharing, trust, taxation, need for transparency, and openness. These issues are particularly critical when foundational services are used to support digital health service delivery. Given the enormous vendor market and the technology giants who are influencing any digital health market, it is considered to be far preferable for every nation to establish and make use of its own health data cloud service, based on open market and open platform principles. Such a strategy maximises opportunities for local innovations whilst maintaining interoperability capacity throughout the digital health ecosystem.

What is the blueprint for policy and resourcing that would allow any region in any country to start to create a regional digital health system? What technology is needed for this? Ideally, we adopt a knowledge-enabled open health computing platform, that is incrementally buildable and deployable, and enables knowledge engineering to provide intelligence in a variety of formats.

What is a suitable concrete strategy for any region to implement this? What types of organisations are best placed to provide the necessary governance needed at every level within the health industry? Who should lead such initiatives?

Every nation has its own historical foundations representing entrenched legacies, knowledge, and skill gaps that need to be improved, transformed, and/or accommodated. Achieving digital health success is also subject to misuse or loss of past investments, egos, political grandstanding, investment priorities, and entrenched mythical beliefs that frequently conflict with scientific principles and proven theoretical concepts. Each of these impedes innovation due to the need to develop 'workarounds' as the means of overcoming such obstacles. Shaping an adaptive digital health ecosystem requires strong leadership at every level. Such leaders need to be agile, have the authority to be innovative, committed and make good use of collaborative hierarchical (non-hierarchical) networks.

Current national examples

Australia

Australian governments have at least 50 years of digital health history. During the 1970s, the Victorian Government was using advanced technologies that enabled the monthly electronic collection of financial (budget vs actual of all staff categories by department) and activity data (HOSPOWER) from more than 160 hospitals throughout that state.

National standards development began during the early 1990s, that was financially supported by the Australian Government, unlike today's situation as was well-documented in 2019 [15]. Telehealth around the country was booming especially in Queensland, but its continuation was not funded until there was an urgent need as a result of the recent pandemic [16]. Major initiatives were undertaken to build workforce capacity, a health online strategy was developed [17, 18], hundreds of primary healthcare applications were piloted, and requirements for electronic health records and decision support systems were explored and published, only to be dismantled by subsequent governments. These activities led to nowhere as public service officers changed, and a lot of tacit corporate public service knowledge was lost. Many external expert committees provided advice to multiple governments only to never see their recommendations successfully implemented. This author has witnessed and participated in many of these 'stop-start' digital health activities and learned a lot along the way.

Today's Australian Government has published a digital service standard to 'ensure digital teams build government services that are simple, clear and fast'. Its seventh criterion requires such teams to 'use open standards and common platforms', criterion 8 states 'make source code open (by default)', and criterion 9 states 'make it accessible' [19]. These standards have not been adopted for Australia's health system. The Australian Digital Council was formed in September 2018 to drive better cross-government collaboration on data and digital transformation [20]. Its task is to align services and strategies across all states and territories to achieve smarter service delivery and improved data and digital policy outcomes.

A similar activity is required to harmonise Australia's digital health system. As is the case in many countries, Australia's health system has a fragmented infrastructure, a plethora of separate data collections in need of amalgamation, harmonisation, and streamlining.

Australia's national digital health strategy was launched in 2018 [21]. Its tag line is 'safe, seamless and secure'. Its vision's value proposition (desired outcome) is *better health for all Australians enabled by seamless, safe, secure digital health service and technologies that provide a range of innovative, easy to use tools for both patients and providers.* It has a strong focus on interoperability and data quality, although these concepts are poorly described reflecting limited understanding of the complexities associated with that 'interoperability' concept. The development of this strategy included the use of an extensive consultation process that resulted in a focus on three projects, MyHealthRecord, Telehealth, and Electronic prescribing. These reflect user perspectives which are listed as the primary guiding principles. There is no mention of any national supportive infrastructure or of the need for adopting any set of standards.

This Australian strategy focuses primarily on automating what we do today, rather than exploring how we can more efficiently and cost-effectively undertake these activities within a nationally supported digital health ecosystem. Such an analysis would enable us to consider possibilities of transforming current practices by developing innovative solutions.

Healthcare providers are looking for instant access to a patient's information, especially in an emergency, to support earlier diagnosis, better disease management, and reduce their administrative burden. Digital services also need to support the development of new medicines and treatments based on identifiable best practice and new medical advances. These strategic priorities relate to any nation's health system, as do the following desired priority outcomes intended to be achieved by 2022 despite the changing implementation strategies.

(1) Health information that is available whenever and wherever it is needed.
(2) Health information that can be exchanged securely and accurately without loss of meaning.
(3) High-quality data with a commonly understood meaning that can be used with confidence.
(4) Better availability and access to prescriptions and medicines information.
(5) Digitally enabled models of care that drive improved accessibility, quality, safety, and efficiency.
(6) A workforce confidently using digital health technologies to deliver health and care.
(7) A thriving digital health industry delivering world-class innovation.

Australia's primary national focus has been the implementation of the MyHealthRecord which is essentially a repository of many documents, uploaded from numerous clinical software products in use by healthcare service providers [22]. The Australian Digital Health Agency (ADHA) publishes a register of conformity listing software products and versions that have been assessed for conformance with national digital health requirements.

One step now being implemented is to establish a national health information framework by attempting to address the many identified fragmented health data repositories, data sharing, and linkage limitations, by building on the Australian Institute of Health and Welfare's past achievements [23]. This framework needs to address current territoriality and competitive short-term thinking about data use and sharing. Current data access limitations are a major barrier to collaboration within clinical and research communities and the private sector.

The Australian Government has now established an Office of the National Data Commissioner to provide oversight and regulation of Australia's new national data sharing and release framework [24] in response to the 2017 Productivity Report on its inquiry into the benefits and costs of options for increasing availability of and improving the use of public and private sector data by individuals and organisations, which included a number of references to health data [25]. A review and update of Australia's privacy legislation have recently been undertaken to overcome some of the barriers to data access.

ADHA has published a recent report that was commissioned to provide advice on how interoperability standards development, maintenance, and management can be improved to meet health sector needs in Australia. This report presented the case for the adoption of standards as it argued that:

> Interoperability is impossible without high quality standards that are maintained across their lifecycles and are continuously (re-) engineered for co-existence. [15]

And noted that 'failure to embrace standards will endanger, slow the timeliness of and/or increase the cost of big data analytics in health care'. A National Health Interoperability Roadmap is now being developed.

Australia has a long way to go to achieve this aim due to the many weaknesses identified. Public funding for standards development and maintenance is urgently required. This needs to be followed by the development of a comprehensive and realistic national implementation strategy incorporating significant transitional arrangements to be orchestrated over the next ten years or so.

National Health Service Digital (UK)

The UK's NHS Digital's 2015-2020 strategy's overarching objective is to have revolutionised the way technology, data, and information are used to transform the delivery of England's health and social care service [26]. This consisted of five objectives:

(1) Ensuring that every citizen's data are protected.
(2) Establishing shared architecture and standards so everyone benefits.
(3) Implementing services that meet national and local needs.
(4) Supporting health and care organisations to get the best from technology, data, and information.
(5) Making better use of health and care information.

It adopted the following principles to guide the determination of the right standards to be used across their system for their digital, data, and technology standards framework [27]:

- That these standards should be based on international standards and only specialised where it is necessary for these standards to be adopted, such as using the NHS number as the primary identifier.
- That these standards are open standards.
- That these standards address the user needs of patients and care professionals.
- That we have a clear evidence base of these standards being useful, usable, and used.
- That we have considered how these standards apply across health and care.

The UK's INTEROPen group was established to implement their vision for a health and care community through which digital information seamlessly flows [28]. This group represents a community of individuals, industry, standards organisations, and providers who collaborate on identifying, developing, and adopting technical standards for interoperability.

This community includes all professional colleges in recognition that clinical leadership and engagement are one of the most critical success factors.

United States

The US Department of Health and Human Services and Food and Drug Administration (FDA) have created a precision FDA community who drive this initiative. A 2014 presidential directive to the National Institute of Standards and Technology (NIST) has resulted in the development of a Cybersecurity Framework (CsF) based on a collection of cybersecurity standards and industry best practices, to broadly apply across all organisations. The CsF provides guidance for risk management and identifies effective standards, guidelines, and practices [29]. The US National Institutes of Health created the Big Data to Knowledge (BD2K) initiative in 2013. This aimed to assist organisations planning to share their research data and for them to take better advantage of the exponential growth of biomedical research data sets [30, 31].

Financial incentives to adopt certified EHR technology systems were the result of US legislation introduced in 2009. As a result of this programme, 97% of reported hospitals had certified EHR technology in 2014, up from 72% in 2011 [32]. That same year, the Office of the National Coordinator (ONC) for Health Information was established by the Health Information Technology for Economic and Clinical Health (HITECH) Act. The ONC provides regulatory resources and an annual report to Congress [33] along with other information and a number of publications. It has been the US Government's nationwide priority to increase the interoperable exchange of health information across the care continuum and individuals. Unfortunately, the requirement for the use of standardised information representation or information modelling was not included. The ONC laid out a draft vision for a future health IT ecosystem in 2014 for public comment [34]. President Obama launched the Precision Medicine initiative in 2015 noting that:

> Until now, most medical treatments have been designed for the "average patient." As a result of this "one-size-fits-all" approach, treatments can be very successful for some patients but not for others. Precision Medicine, on the other hand, is an innovative approach that takes into account individual differences in people's genes, environments, and lifestyles. [35]

ONC's second draft of the Trusted Exchange Framework (TEF) and Common Agreement (CA) was released in April 2019 [36]. This document provides distinct components, including privacy and security protections for all participating entities, and for the creation of technical and legal requirements for sharing Electronic Health Information (EHI) at a nationwide scale across disparate Health Information networks (HINs). Their proposed exchange modalities for exchanging EHI include Qualified Health Information networks (QHIN) Targeted Query, QHIN Broadcast Query, and QHIN Message Delivery. These technical frameworks aim to facilitate core use cases for interoperability, including electronic access to and use of EHI by individuals.

Despite the clear focus on data collection, sharing, and use, there is no mention of any shared information model or the need for governance of source data. This omission reflects the general lack of knowledge or appreciation of health data science principles as well as clinical data quality needs. There are instructions regarding the imposition of fees associated with data access and use, apparently regarded as being of greater importance than data quality. The ONC published a standardised set of health data classes and constituent data elements in early 2020, known as the United States Core Data for Interoperability (USCDI), for nationwide, interoperable health information exchange [37]. Data class descriptions and applicable standards supporting data elements are provided as USCDI v1 on their website or in a PDF publication. The digital health ecosystem needs to be able to support these and many other continuing developments and consider how best to overcome such critical foundational shortcomings.

New Zealand's digital health strategic framework

New Zealand's health strategic directions are based on five themes, people powered, closer to home, value and high performance, one team, and smart system. This has formed the foundation for its digital health strategic framework [38]. This is comprehensive and includes numerous technical details. Unlike other national strategies, it includes a long list of health sector architectural principles as well as many technical details. The New Zealand Ministry of Health has an internal cloud computing policy that may be used as a guide for other organisations to follow. Only accredited Connected Health access suppliers are authorised to provide a secure network link between their cloud environment and Connected Health, an umbrella term for IT systems and environments securely sharing information. Connected Health is a standards-based, commercial model for the delivery of universal connectivity across the New Zealand (NZ) health sector. Its common framework includes network connectivity standards approved by the NZ Health Information Standards Organization (HISO). All approved open, non-proprietary standards are freely available for health providers and their industry partners.

Its digital health sector architecture, standards, and governance section has mandated minimum requirements that individual health organisations are expected to meet when procuring new digital services. In general terms, these include:

- Conformance and in some instances compliance with the Ministry of Health published HISO standards, roadmaps, and architecture guidelines and integrate with its mandated national digital services.
- Integrated to support consistent, and where possible seamless, user experience and avoid unnecessary data duplications and functions.
- Stay within agreed supplier support thresholds as a minimum by system maintenance and regular upgrades.
- Cloud delivery to be considered as a preference for all digital services.
- Regular assessments of Health Information Security and Cloud Risk Assessment frameworks
- Data governance and conformance with their Health Information Governance Guidelines.
- Provision for service retirement to include maintaining data access.

The NZ Government has appointed a Chief Data Steward whose role is, to support the use of data as a resource, and who is driving their data content standardisation programme. HISO is one of its participants.

HISO has recently published their interoperability roadmap for its health and disability system [39]. This is intended to accelerate their shift to a fully interoperable digital health ecosystem. It provides a great case study for a critical evaluation,and serves as a possible exemplar for the digital health roadmap presented in this text.

No doubt other countries have developed similar digital health strategies. These strategic criteria are fundamental to our ability to fully connect every system within a domain's ecosystem. Somehow such strategic criteria do not appear to have been recognised as required by implementations at local and regional levels in most countries of the world.

National digital health governance

Australia is not alone in having a fragmented digital health governance infrastructure. Governance is distributed between the state and territory governments. There are also differences in how these governance responsibilities are distributed within each state and territory, to health networks or individual healthcare delivery organisations. Digital health governance activities also form a part in the management of private healthcare delivery organisations or corporations as well as health IT vendors. The result is a severely fragmented system and many data/information silos even within healthcare delivery organisations.

A prerequisite to the development of a national digital health governance strategy is the need to identify existing limitations, new opportunities, and priority areas for investment. The strategy needs to include implementation steps that collectively enable all governance principles to be met. Successful implementation of such a strategy is highly likely to require major infrastructure transformation. Digital transformation requires well-targeted investments. All implementation costs need to be considered relative to costs associated with delays, incomplete implementations, and usage limitations. Every nation needs to adopt a shared vision for its digital health ecosystem to guide both top-down and bottom-up approaches adopted by its stakeholders; these approaches need to be well balanced towards meeting the vision's objectives.

Leading national digital health

A key component making up any national digital health environment is effective leadership and governance responsible for the establishment of an enabling environment through multi-sector engagement. Leadership and governance entities need to take responsibility for directing and coordinating all activities associated with the establishment of the necessary foundational infrastructure and the development of a digital health strategy at the national level. Clinical leadership and engagement are critical as both buyers and users collectively possess the necessary market power. Desired outcomes of a successful national digital health ecosystem include:

- the need for political support to ensure alignment with health goals;
- agile stakeholders who are aware and engaged;
- policy and legislative compliance; and
- a capable workforce.

Leadership and governance are required to develop a shared vision, secure the necessary resources, develop a strategic plan, set priorities, and detail the various digital health component implementation phases including the setting of mandates and provision of incentives regarding their adoption. This requires access to expertise with the right mix of technical, health domain, and interpersonal knowledge and skills. The latter is essential for effective communication amongst all stakeholders to address the many diverse behavioural drivers, support numerous decision points, and gain effective collaboration and teamwork.

Digital health implementation components include the development of investment strategies, overarching infrastructures, choosing any number of standards for adoption and incorporation into standard procurement strategies, all types of healthcare delivery services, and the acquisition of software applications. Leadership is about supporting and empowering transformational changes to be realised via the implementation of recommendations and the monitoring of results to ensure that expected benefits can be achieved.

Effective leadership needs to start globally amongst many stakeholders, especially clinicians. It is about providing a strong approach to the identification of all factors that contribute to successful outcomes and their implementation, whilst fully engaging with the most relevant stakeholders and their short- and long-term outcome experiences. This is applicable at every level within a national healthcare delivery system.

Conclusion

Some of the work activities associated with setting and maintaining data standards, facilitating data sharing and linkages need to be undertaken globally to enable more cost-effective digital support for clinical trials, the timely collection of evidence of best practice and to improve the global management of public health, including pandemics.

There are many individual country initiatives that form a foundation upon which to build any number of digital health ecosystems which collectively make up a global digital health ecosystem. Global success needs to be based on numerous collaborative cross-country efforts. Many existing international networks play important leadership and investment roles. It is of some concern that many stakeholders within these organisations do not appear to have acquired new digital skills required for their own organisational digital transformation. These issues are explored in greater detail in their relevant chapters.

The chapters to follow explore the most relevant and influential global entities, and their current contributions to the evolution of public/population health, standards, guidelines, and regulatory developments. These entities represent significant global stakeholders whose contributions need to be reviewed in terms of their potential to reform and optimise their contributions as a key component of the roadmap to a successful global digital health ecosystem. These organisations have the authority to provide the necessary drivers for change.

References

[1] GDHP. Global Digital Health Partnership [Available from: https://gdhp.org/.

[2] WHO/ITU, National eHealth Strategy Toolkit, 2012, [cited 2019 17 April]. Available from: https://www.itu.int/pub/D-STR-E_HEALTH.05-2012.

[3] WHO, Global Strategy on Digital Health 2020-2040, 2019, [cited 2019 17 April]. Available from: https://extranet.who.int/dataform/183439.

[4] S.A. Levin, Ecosystems and the biosphere as complex adaptive systems, Ecosystems 1 (1998) 431–436.

[5] M.G. Jacobides, C. Cennamo, A. Gawer, Towards a theory of ecosystems, Strateg. Manag. J. 39 (2018) 2255–2276.

[6] G.E. Iyawa, M. Herselman, A. Botha, Digital health innovation ecosystems: from systematic literature review to conceptual framework, Procedia Computer Sci. 100 (2016) 244–252.

[7] G. Briscoe, P. De Wilde, Digital Ecosystems: Evolving Service-Oriented Architectures, arXiv e-prints, 2007.

[8] G. Briscoe, S. Sadedin, P. Wilde, Digital ecosystems: ecosystem-oriented architectures, Nat. Comput. 10 (2011) 1143–1194.

[9] A. Marcelo, D. Medeiros, K. Ramesh, S. Roth, P. Wyatt, Transforming Health Systems Through Good Digital Health Governance, Asia Development Bank, 2018. [cited 2020 2 March]. Available from: https://www.adb.org/sites/default/files/publication/401976/sdwp-051-transforming-health-systems.pdf.

[10] M. Romanelli, Towards sustainable ecosystems, Syst. Res. Behav. Sci. 35 (2018) 417–426.

[11] O. Granstrand, M. Holgersson, Innovation ecosystems: a conceptual review and a new definition, Technovation 90-91 (2020) 102098.

[12] openEHR, An Open Domain-Driven Platform for Developing Flexible e-Health Systems, openEHR Foundation, 2019. [cited 2019 6 February]. Available from: https://www.openehr.org/.

[13] MIT-SERG. System Lifecycle Properties ('ilities'): MIT Strategic Engineering Research Group (SERG); Available from: http://strategic.mit.edu/changeability.php.

[14] A. Tiwana, Chapter 2. Core concepts and principles, in: A. Tiwana (Ed.), Platform Ecosystems, Morgan Kaufmann, Boston, 2014, pp. 23–48.

[15] D. Rowlands, J. Parker, A Health Interoperability Standards Development, Maintenance and Management Model for Australia, ADHA, 2020. Available from: https://www.digitalhealth.gov.au/sites/default/files/2020-12/Standards_Development_Model_v1.1_2020.pdf.

[16] S. Hall Dykgraaf, J. Desborough, L. de Toca, S. Davis, L. Roberts, A. Munindradasa, A. McMillan, P. Kelly, M. Kidd, "A decade's worth of work in a matter of days": the journey to telehealth for the whole population in Australia, Int. J. Med. Inform. 151 (2021) 104483.

[17] F. Australia. Parliament. House of Representatives. Standing Committee on, A. Community, Health on line: report into health information management and telemedicine/House of Representatives Standing Committee on Family and Community Affairs, The Committee, Canberra, 1997.

[18] NHIM-Advisory-Council, Health online: a health information action plan for Australia, Health Inform. Manag. 29 (2000) 179–181.

[19] Australian-Government, Digital Service Standard Canberra: Commonwealth of Australia, 2019, [cited 2019 9 May]. Available from: https://guides.service.gov.au/digital-service-standard/.

[20] Australian-Digital-Council, State of the Data and Digital Nation, Commonwealth of Australia, Canberra, 2019. [cited 2019 8 April]. Available from: https://www.pmc.gov.au/resource-centre/public-data/state-data-and-digital-nation.

[21] Australian-Government, Australia's National Digital Health Strategy: Safe, Seamless and Secure, Australian Government, Canberra, 2018. Available from: https://conversation.digitalhealth.gov.au/.

[22] ADHA. MyHealthRecord: Australian Digital Health Agency; [Available from: https://www.myhealthrecord.gov.au/.

[23] AIHW, National Health Information Strategy—Draft Framework for Discussion, 2020, [cited 2020 28 February]. Available from: https://consultation.aihw.gov.au/executive-secretariat/national-health-information-strategy-online-consul/.

[24] ONDC. Office of the National Data Commissioner: Australian Government; Available from: https://www.datacommissioner.gov.au/.

[25] Productivity-Commission, Data Availability and Use, Overview and Recommendations. Report No. 82, Canberra, 2017, [cited 2019 5 June]. Available from: https://www.pc.gov.au/inquiries/completed/data-access/report.

[26] NHS-HSCIC, Information and Technology for Better Care: Health and Social Care Information Centre Strategy 2015-2020, Health and Social Care Information Centre, 2015. Available from: https://digital.nhs.uk/about-nhs-digital/corporate-information-and-documents/our-strategy.

[27] NHS, BETA—NHS Digital, Data and Technology Standards Framework, National Health Service, United Kingdom (UK), 2020. Available from: https://digital.nhs.uk/about-nhs-digital/our-work/nhs-digital-data-and-technology-standards/framework.

[28] INTEROPen, IHE, HL7 FHIR & openEHR—A Smorgasboard of Standards: An Action Group to Accelerate the Development of Open Standards for Interoperability in the Health and Social Care Sector, 2019, [cited 2019 13 March]. Available from: http://www.interopen.org/2018/07/19/ihe-hl7-fhir-openehr-a-smorgasbord-of-standards/.

[29] K. Saranto, C. RonQuillo, O. Velez, Nursing competencies for multiple modalities of connected health technologies, in: J. Murphy, P. Weber, W. Goosen (Eds.), NI2016 Post Conference Proceedings: Informatics Competencies for Nurses in the Next 10 Years, IOS Press, 2016.

[30] L. Ohno-Machado, NIH's Big Data to Knowledge initiative and the advancement of biomedical informatics, J. Am. Med. Inform. Assoc. 193 (2014).

[31] R. Margolis, L. Derr, M. Dunn, M. Huerta, J. Larkin, J. Sheehan, M. Guyer, E.D. Green, The National Institutes of Health's Big Data to Knowledge (BD2K) initiative: capitalizing on biomedical big data, J. Am. Med. Inform. Assoc. 21 (2014) 957–958.

[32] R.S. Huckman, M. Uppaluru, The Untapped Potential of Health Care APIs, Harvard Business Review, Mulesoft, 2015.

[33] ONC, Report to Congress on Health IT Progress, Office of the National Coordinator for Health Information Technology (ONC), USA, 2016.

[34] ONC, A 10-Year Vision to Achieve an Interoperable Health IT Infrastructure, Office of the National Coordinator for Health Information, 2014.

[35] P. Obama, Precision Medicine Initiative, 2015, [cited 2020 4 March]. Available from: https://obamawhitehouse.archives.gov/precision-medicine.

[36] ONC, Trusted Exchange Framework and Common Agreement (TEFCA), 2019, [cited 2020 9 March]. Available from: https://www.healthit.gov/topic/interoperability/trusted-exchange-framework-and-common-agreement.

[37] ONC, United States Core Data for Interoperability (USCDI), 2020, Available from: https://www.healthit.gov/isa/united-states-core-data-interoperability-uscdi.

[38] Ministry-of-Health-New-Zealand. Digital Health: New Zealnd Government; Available from: https://www.health.govt.nz/our-work/digital-health.

[39] H. Ministry-of-Health-New-Zealand. HISO 10083:2020 Interoperability Roadmap: NZ Health Information Standards Organization; 2020 Available from: https://www.health.govt.nz/publication/hiso-100832020-interoperability-roadmap.

CHAPTER

3

Global collaborative leadership challenges and economic drivers

Evelyn Hovenga[a,b,c] *and Carol Hullin*[d,e,f]

[a]Digital Health, Australian Catholic University, Melbourne, VIC, Australia [b]eHealth Education Pty Ltd, Melbourne, VIC, Australia [c]Global eHealth Collaborative (GeHCo), Melbourne, VIC, Australia [d]Centre of Digital Innovation of Latin America, Temuco, Chile [e]United Nations, Washington, DC, United States [f]College of Business and Economy, University of Tasmania, Sydney, Australia

WHO is leading global digital health reformation

The World Health Organisation (WHO) has adopted a long-term vision of universal health access, sustainable health systems, and optimum global public health management. The realisation of these desired outcomes represents high-level value propositions that can best be achieved with the support of digital health. Strong leadership from multiple contributing entities is required to develop and reach agreement about what constitutes their global vision for a digital health ecosystem. Key foundational and structural components need to be identified and established to guide and support the digital transformation agenda. At its 66th World Health Assembly held in 2013, a resolution on eHealth standardisation and interoperability was adopted by its Member States [1]. A 2020 Global Digital Health virtual summit hosted from Riyadh, Saudi Arabia, with collaboration from the G20, focused on a Global Strategic Partnership in Digital Health to fight pandemics.

During the health crisis, the World Economic Forum declared that the success of digital health is improving human health by using the technological power of saving people's lives during COVID-19 [2]. Some encouragement to digital health innovation is desperately needed from the public sector. The World Bank has invested more than $12 billion for COVID-19 vaccine, tests and treatments for citizens of developing countries [3], adding to more than $4 billion allocated by the private sector of the International Finance Corporation (IFC).

The success of digital health implementation is highly dependent upon how well the high-level authorities understand:

- user requirements at every level,
- health service needs to be met at any point of care,
- how critical health data standards are governed and processed at every level,

Roadmap to Successful Digital Health Ecosystems
https://doi.org/10.1016/B978-0-12-823413-6.00001-X

- how data are used,
- how knowledge is made available and applied, by the health system and industry as a whole, and
- the value of Internet access for everyone.

Every stakeholder, from every level, needs to be concerned with acquiring, screening, analysing, interpreting, and using data at the right time within the right context. Engaging in these activities, plus the application of prior knowledge attained by stakeholders, influences the performance of all health-related activities, including automation and advanced analytics for population and public health. All high-level authorities need to act accordingly.

An ultimate aim is not only to facilitate improvements in the delivery of health services and population health. There are two primary needs to be considered when reviewing or establishing global supportive infrastructures:

1. to ensure patient safety can be maintained throughout any health service delivery process and
2. health system pandemic and disaster preparedness.

The latter requires a real-time system for timely disease surveillance, along with a sound global prevention strategy, the ability to increase healthcare service capacity at short notice and ongoing research and development activities. This approach can achieve sustainable models of care at any level of any patient's digital health journey.

At present, the profound fragmentation within and between national health systems does not allow people to actively participate in digital health implementations at the point of care. This is preventing the empowerment of people, patients, families, and communities, to self-manage at home and in any other location. These digital health gaps are putting healthcare services around the world under pressure. Delivering efficient and effective services supported by technological innovations that integrate services provided by all stakeholders has become a global priority for high-level authorities' decision-making and evidence-based public policymakers. Appropriate use of digital health technologies can enhance patient safety and information security.

For example, the use of a semantic interoperable electronic medical/health record is a minimal requirement expressed by any given country, as this allows decision-makers at the national level to better understand their own health systems. This enables access to information reflecting real workflows among their primary, secondary, and tertiary providers, adding to the ease of access to people-related data in real time that facilitates precise decision-making from any location and the forming of an informed community of practice. Timely availability of this type of information enhances policy analysis, able to lead to useful well-targeted policy initiatives.

Today, a mandatory adoption of key global standards and guidelines designed to optimise ecosystem connectivity and patient safety is *a critical requirement* that largely remains nonexistent within the health industry, despite all the scientific evidence showing the importance of these. All these requirements are in principle no different from those developed by the banking system, who adopted the global use of Bank Identification Codes (BICs) and the SWIFT code standard developed by the Society for Worldwide Interbank Financial Telecommunication (SWIFT) for international financial transfers or the airline or telecommunication industries. Each of these industries collaborated globally to solve their connectivity issues and to streamline their operational processes. Similar global efforts are required for the health industry in order to integrate processes and systems and improve business collaboration.

Such initiatives ensure that the digital journey of every patient is secured and trustworthy throughout any individual's lifelong receipt of healthcare services, while supporting all healthcare providers at every point of care. This requires global and national compliance with agreed, and in some instances mandated, international laws, a set of key standards, guidelines, and policies. Their design, development, adoption, maintenance, review, and updates need to be supported by suitable governance processes. This needs strong government leadership with delegated expert support to transform from a capital-intensive IT infrastructure to a highly variable infrastructure and cost structure, improving reliability and flexibility that enables faster responses to changes in supply and demand throughout any digital health ecosystem.

At present, the unique opportunity exists to bring together all the fragmented parts of the health system to respond effectively to the global pandemic and frequent increasing demands for innovations such as using Telehealth, which has seen an exponential growth from 11% in 2019 to more than 76% now using this method on an ongoing basis for patient care [2]. Government standards for hardware and software platforms can guarantee that mHealth applications can connect with each other and other mobile tools [4]. The establishment of the Global Digital Health Partnership (GDHP) [5], mentioned in the previous chapter, is a welcome development.

Prior to COVID-19, some recommendations resulting from the August 2020 Global Digital Health Virtual Summit could serve as the necessary stimulus to seriously activate global digital health transformation towards a well-functioning global digital health ecosystem. These recommendations are reproduced in Box 1, and they should serve as a guide to the reader of this text. Its focus is very much on high-quality governed health data and system interoperability. The many concepts referred to are explained in some detail in the chapters that follow.

BOX 1

Panel Recommendations from the Riyadh Global Digital Health Summit [6]

1. Implement data-driven and evidence-based protocols for clear and effective communication with common messaging to build citizens' trust.
2. Work with global stakeholders to confront propagation of misinformation or disinformation through social media platforms and mass media.
3. Implement a standard global minimum data set for public health data reporting and a data governance structure tailored to communicable diseases.
4. Ensure countries prioritise digital health, particularly improving digital health infrastructure and reaching digital maturity.
5. Enable health and care organisations by providing the necessary technology to collect high-quality data in a timely way and promote sharing to create health intelligence.
6. Cultivate a health and care workforce with the knowledge, skills, and training in data and digital technologies required to address current and future public health challenges.
7. Ensure surveillance systems combine an effective public health response with respect for ethical and privacy principles.
8. Develop digital personal tools and services to support comprehensive health programmes (in disease prevention, testing, management, and vaccination) globally.
9. Maintain, continue to fund, and innovate surveillance systems as a core component of the connected global health system for rapid preparedness and optimal global responses.

This chapter aims to identify the most significant influencing and enabling global and national governing and funding entities, their specific digital health relationships and governance structures or the lack thereof. Other chapters focus on the development and use of health data standards, one of the primary precursors for any effective digital health ecosystem, and the data processing relationships with digital health technologies. First, we need to explore global entities, their driving forces and inhibitors, impacting on digital health transformational activities.

An analysis of these global entities' historical underpinnings that formulated their role, memberships, funding, and associated organisational structures explains their current global realities and scopes of practice. Such information forms the basis from which we can explore how best to collaborate to resolve current in-depth fragmentation, role overlaps, and governance deficiencies that need to be of primary consideration for our road map to successful digital health ecosystems. These realities represent precursors to the identification of each entity's potential for enabling or driving the digital health transformation agenda by ensuring that the necessary high-level digital health infrastructures are globally available. Many of these entities may need to consider the need for their own internal transformation to better enable them to provide the necessary leadership. There is a need to break down existing barriers to successful digital health transformation and to minimise adverse consequences experienced by people, families, and communities as well as health service and associated business entities.

Digital health markets and trade

Evidence-based digital health products face a significantly different commercialisation pathway, an evolving regulatory landscape, and limited reimbursement potential. In addition, these digital health products require novel commercial models to penetrate risk-averse, budget-constrained, and variable procurement systems [7]. Digital health is a productive sector that needs to be supported to generate new paradigms for innovation in healthcare to improve people's lives. It has become evident during the global COVID-19 pandemic that markets with the necessary infrastructure, ability, and mechanisms to introduce new products and services of digital health can provide the best secure solutions for their customers, people, community, and society.

The health system's capacity has been challenged over the last few years due to several social, political, and economic factors impacting the market and trade patterns. The economic systems prior to the health crisis were experiencing difficulties moving from the analogue world towards highly technological environments such as Industry 4.0, the term used to represent the fourth manufacturing revolution with a focus on smart and autonomous systems fuelled by data and machine learning. Consequently, the situation was more fragile in terms of new investment for digital health. Global health innovation funding has increased exponentially over the last 10 years. StartUp Health reported the rapid reaction to COVID-19 has become an enduring wave of interest, adoption, and investment in health innovation companies [8]. Wealth is dependent upon population health and wellness; this people focus needs to be the driving force for digital health investments [4].

Market research [9] indicates that the value of the global digital health market in 2018 was USD 95.8 billion. The digital economy represents 15.5% of global GDP [10]. The digital health market is expected to grow at a compound annual growth rate of 27.7% from 2019 to 2025. The recent pandemic has accelerated this growth rate. The revenue forecast for 2025 is USD 509.2 billion. There is an expected rising demand for remote monitoring services, driven by the desire to track and access medical information. Digital health systems generated the second largest revenue in 2018. This is attributed to an increase in government initiatives to promote digital health and health information system adoption by hospitals. Associate installation, training, and integration services delivered accounted for a 47.7% market share. Prominent digital health market players are undertaking strategic initiatives to gain maximum market share globally [11].

New market entrants are focusing their technology developments on home-based remote patient monitoring, to support primary care service delivery including diabetes and lifestyle conditions, mental health, and on self-health management and changes to patient behaviours, mHealth, and data analytics and systems. Tech-enabled homecare leveraging AI (artificial intelligence) is expected to increase as this is far more cost-effective than any type of facility-based care. Mobile health applications can also support the coordination and quality of care, especially in rural and underserved communities, as well as support supply chain management to reduce delays in medical supplies [4].

Global trade has a very long history. The global desire for peace and security that emerged during the early 1940s has driven the creation of our current global economic system that intertwines foreign policy with trade. As such, the World Trade Organisation (WTO), established in 1995, has an important role to play towards resolving the extensive existing fragmentation that is having a negative impact on our ability to successfully manage global health with optimum digital support. The WTO needs to deal with a continuing redistribution of economic and political power in the world that directly influences world trade agreements.

World trade agreements have the capacity to influence the adoption of standards and/or the ability of suppliers to ensure that products meet local regulatory requirements, as required by purchasing countries. This is only one of many issues of relevance to trade policy. The digital health market needs to feature in trade agreement negotiations, including the use of intellectual property and patents, trade in knowledge, global value chains (GVCs), supply chain trade [12], and mapping regulatory requirements. The future of the trading system is mostly determined by how it deals with competition policy, investment, government procurement, state-owned enterprises, labour rights, the environment, the relationship between exchange rates and trade, and other topics that are either outside the current scope of the WTO law or for which the current rules are arguably incomplete [13] p.557. Most, if not all of these, factors are also of relevance to the digital health market.

Another associated issue is the need for global governance in some areas and more generally eHealth governance with a focus on cybercrime. The latter relates directly to the pharmaceutical industry that increasingly needs to combat illicit online pharmacies that pose a significant patient safety risk [14]. There is the potential for similar risks associated with medical devices. Any online ecosystem making use of the Internet needs to consider relationships with e-commerce services, domain name server, web hosting service, and other ISPs and search engines. Self-governed open information exchange has led to a lack of adequate

Internet governance. Subsequent risks tend not to be well managed by national legislative regulations [14] which collectively are of concern.

There is a need to capture technological and economic forces, within developed and developing countries, and the shift in the share of global purchasing power, to drive the transformation of global digital health interconnectedness. Spence [12] noted that

> The mobile-internet- and platform-centred open ecosystems, along with mobile payment systems and enabled financial services, have the potential to support inclusive growth patterns and expand the channels, opportunities and accessible markets for SMEs (Small to Medium Enterprises).

These same principles apply to digital health ecosystems. We need to be able to exploit the international potential of open platforms to improve global population health. This requires SMEs to have access to the necessary investments and global markets building on digital platforms and new trade regimes that increase the openness of digital health ecosystems.

The availability of technologies such as telecommunication infrastructures and cloud computing that underpin successful digital health ecosystems is shifting economic and market imperatives. Transformation to a data economy requires focused and well-funded national implementation strategies requiring a transition over time. Digital health commercialisation needs to focus on driving local economies, supporting multiple SMEs to meet local/national needs in a competitive environment based on openly available foundations. There is the need for a radical rethink regarding privacy, data and knowledge ownership, trust, taxation, need for transparency and openness, which are particularly critical when foundational services are used to support digital health purposes. It may be far preferable for every nation to establish and make use of its own health data cloud service supporting its health data governance requirements, based on open market principles.

Global health security

The Global Health Security Index (GHSI) claims to be 'the first comprehensive assessment and benchmarking of health security and related capabilities to prevent and mitigate epidemic and pandemics across the 195 countries that make up the States Parties to the International Health Regulations (IHR)' [15,16]. The GHS Index makes use of 34 indicators and 85 subindicators to assess individual countries. It is intended to be a key resource in the face of increasing risk of high consequence and globally catastrophic biological events. The GHS Index is a project of the Nuclear Threat Initiative (NTI) and the Johns Hopkins Center for Health Security (JHU) who developed this index with the Economist Intelligence Unit (EIU). This index (GHSI) consists of six categories:

1. prevention,
2. detection and reporting,
3. rapid response,
4. health system capability,
5. compliance with international norms, and
6. risk environment.

Countries need to be able to measure its readiness to deal with any biological risk. This initiative came about following the 2014 West African Ebola epidemic and the realisation that the world needs to be well prepared to manage any event that impacts on population health and health service delivery capability within any nation. These are events that could wipe out gains in sustainable development and global health because of their potential to cause national and regional instability, global economic consequences, and widespread morbidity and mortality.

GHS Index data are drawn from publicly available data sources. This was put to the test during the 2020 COVID-19 pandemic when the John Hopkins Institute's published data became the global source of information regarding the number of identified cases per country globally. An improved global digital health infrastructure is expected to provide more timely, accurate, and complete data used to inform not only this index but also global public health management, especially during a pandemic.

Digital health infrastructures: Research and policies

Sociotechnical themes adopted for cyberinfrastructure (computational or digital health infrastructures) research [17] to represent desired outcomes are as follows:

- Relationality—and interdependencies
- Integration of heterogeneity:
  - Human factors: interdisciplinary relations between computer and information scientists, and domain scientists who need to collaborate. These groups splinter into many others with diverse skills and responsibilities, for example data, funders, developers, etc.
  - Heterogeneous data factors.
  - Heterogeneous technology factors and their relationship to data processing capability.
- Sustainability—need for long-term resources—for example, maintenance needs, breakdown probability, need to future proof. This can best be achieved via the adoption of an open platform and open-source technologies designed to integrate standardised data formats, agreed upon technical development methods. A reliance of open standards and common formats supported by committed communities of practice provides for greater flexibility and a nonreliance on private vendors—such as Open Innovation 2.0—a quadruple helix ecosystem.
- Standardisation—a solution to managing all types of heterogeneity while enabling flexibility to best suit local circumstances or needs.
- Growth enablement—scaleability and extensibility of infrastructure from local to global.
  - Plan for change, facilitate innovation.
- Automation—a redistribution of human work and technologies.

These desired outcomes, as they relate to digital health, are explored in some detail throughout this text within the context of their relationship with digital health. The formulation of these desirable outcomes constitutes the foundation for policy developments and associated financial investments by any one of many global or national entities. Policies are formalised high-level statements, developed and adopted by a governance body. Policy initiatives detail a deliberate system of principles to fund and guide decisions aimed at achieving desired outcomes and long-term goals. Policies may outline roles, responsibilities, and expectations for specific programs and tasks, based on values and objectives that apply to a specific area.

Organisational policies regulate operations and are enforceable where compliance is mandatory. Government policies reflect a set of ideas or plans associated with funding initiatives intended to influence and determine its management of public affairs or inform legislature as the means by which order is maintained and citizen needs are met. The World Bank's Digital Development Global Practice works hand in hand with governments to help create strong foundations to enable the digital economy to thrive [10]. Every country needs to invest in people by prioritising education and building digital skills.

Entities influencing global digital health

We'll start by exploring current world realities. Many historical foundational organisational structures, terminologies, and technologies, designed prior to the digital era, continue to be in use today. These need to be examined to assess their continuing 'fit-for-purpose' status. We need to ask should or could any of these historical foundations now be eliminated, be transformed, built upon, or replaced in the interest of sustainable development to best suit the new digital age? There is a need to reconstruct basic paradigms by respecting human dignity supported by digital innovation and building entirely new foundations for our economic and social systems [18].

The COVID-19 pandemic experiences have highlighted many shortcomings. The economic and social recovery from this pandemic has required everyone to revisit their business models. Major changes may be required to better enable the required transformations to occur as contributions to the fourth industrial (digital) revolution. Global Infrastructure needs include support for managing any global public health crises including:

- New innovations and the integration of artificial intelligence into the public health response.
- Adoption of clear scientific methods for big data analysis of citizens' movement, disease transmission patterns, and health monitoring, to aid prevention measures and support decisions made.
- Transparent surveillance tools—ubiquitous instruments and powerful algorithms.
- Provision of real big data transmission management and capacity issues.
  - Connectivity between diverse types of industries—transport and population movement as well as community-based testing and healthcare providers.
  - Logistics—equipment, supplies, and workforce supply chain.
- Ethical and bioethical quarantine management for all, including people with special needs

Which global entities have the responsibility for making the high-level infrastructures available to benefit the population at large globally? This is required not only to manage any future pandemic but also to support the management of noncommunicable diseases, which constitute a large component of the overall health burden within any nation, along with other health determinants. A prevention strategy is most cost-effective [19].

Infrastructures enabling digital health

Numerous global entities, supported by national governments, influence opportunities and possibilities for enhancing the adoption and use of digital health. Health system operational processes are about managing the distribution of human work relative to health service demand anywhere anytime. These functions are supported by evolving specialised knowledge and a great variety of medical and digital technologies. Digital technologies making use of international and national infrastructures refer to any electrical devices that make use of data, information, and knowledge in a digital form (binary code) to perform useful functions designed to support anyone's health [20].

Successful digital health requires strong investment and smart leadership from a multitude of international organisations and at every level within the health industry. It also takes strong leadership and a fair amount of collaboration between entities, nations, and industry sectors, to establish and maintain the availability of enabling infrastructures required for users to have access to all foundational high-level infrastructures such as:

- A reliable energy source (electricity);
- Telecommunication and Internet access which are dependent upon a variety of hard technologies such as fibreoptic cables and grid computing (distributed computer resources);
- Soft technologies such as metadata standards and ontologies;
- Any number of agreements between entities to facilitate data, information, and knowledge sharing;
- Collaborative working relationships between communities of interest; and
- Legislative and regulatory governance.

These need to be established, governed, and maintained by global, national, and local entities, all of whom need to work collaboratively.

The many current fragmentations are being highlighted. This fragmentation is due to each entity implementing its own digital health strategy in the absence of a global or national high-level mandated supportive infrastructure. There tends to be a focus on individual organisational needs with little regard for the need to be able to connect with the external systems in place and in use by other health industry support services and care provider entities. We have witnessed a poor understanding of the importance of the need to be compliant with the use of relevant terminology or technical standards as a precursor to facilitate optimum interoperability. This has led to a great many distortions and inequities in the way health services are being delivered and public health is practiced, adding a layer of complexity that needs to be unravelled in order to enable successful digital health to be realised. Recovery today and resiliency tomorrow require global multistakeholder cooperation [20].

A major weakness of current approaches adopted to embrace exponential growth is the lack of formal recognition of the economic value of contributions made by those working in 'soft' industries with a composition of over 80% women and the subsequent lack of substantial investments in nonmasculine-dominated industries such as health and education. There is an urgent need for global gender equality [21] to address this. Contributions made to building workforce capacity, population health and welfare have enormous benefits for the global economy but are not included in GDP measures.

Global health is a 21st century imperative. As a desired outcome, global health is either an unfamiliar concept of minimal influence within any sovereign nation or actively resisted by some, diminishing opportunities for useful collaboration [22]. Fortunately, there are a number of different types of global entities or stakeholders who fulfil many different roles. Collectively they support global health in some way. The primary differences between them are their funding arrangements, value propositions, workforce knowledge and skill sets, and governance mechanisms and scope. These foundational requirements enabling global digital health, including the provision of person centred care, are summarised in Fig. 1.

The many and varied communities of interest (stakeholders) listed in Fig. 1 need to work collaboratively by freely sharing knowledge and skills to support activities undertaken for the public good at the lowest possible cost. This requires adherence to any number of international standards, guidelines, professional, ethical, legislative, and governance principles relative to both hard and soft technologies, to ensure patient/client safety and their health interests remain the highest priority at all times. The development of these foundational requirements as well as their adoption is very much dependent upon many and varied global and national authorities with such responsibilities. An overview of key global influential entities follows.

The United Nations and associated government-funded entities

The United Nations (UN) was established in 1945 following World War II, by 51 countries committed to maintaining international peace and security [23]. Their visionary aim was to develop friendly relations among nations, promote social progress, better living standards,

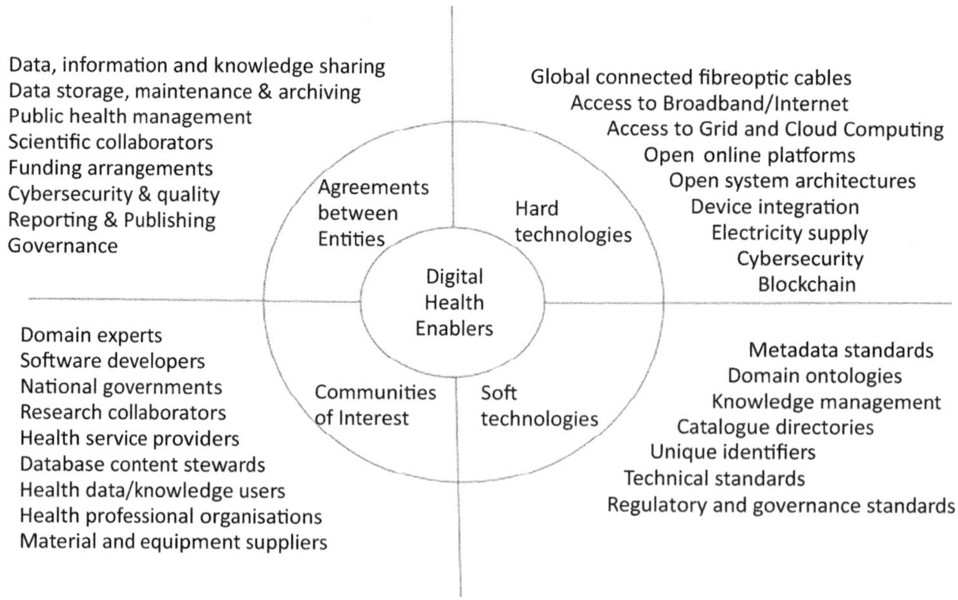

FIG. 1 Digital health enablers.

and human rights. In 1946, the UN's International Children's Emergency Fund (UNICEF) was established to help children and young people whose lives and futures were at risk [23]. The Universal Declaration of Human Rights, a milestone historical document, was proclaimed in 1948 [24].

Their work is continuing during this COVID crisis. By 2011, the UN's membership had grown to 193. The values enshrined in the UN Charter are peace, justice, respect, human rights, tolerance, and solidarity. The UN plays a crucial role in maintaining global order. This relates directly to the need to minimise conflict and associated trauma that directly impacts upon our collective ability to deliver healthcare.

The UN provides a global view of the world's diversity and is actively involved in promoting and protecting good health for everyone. It's chief architect and sponsor was the United States. This support is declining, thus other countries now need to fill these shortfalls. It is highly bureaucratic and appears to be stuck in a time warp. By 2020, the UN Secretariat was supporting a large governance infrastructure consisting of numerous regional commissions, other entities, a number of funded programs, and overseeing many field operations. It has failed to keep pace with various global changes, as reflected in its outdated sovereign security council membership that has failed to recognise the many economic and political changes that have occurred over the last 75 years in countries such as Japan, India, Brazil, Germany, China, or any African country [25]. Late 2016, when the current Secretary General, Antonio Guterres, past Prime Minister of Portugal, took the oath of office, he noted that the international community's most serious shortcoming was our inability to prevent crises. Despite its global leadership, the UN has not managed to respond effectively in accordance with its universal values.

The World Bank Group was established in 1944, in response to major human development disasters. It currently has 189 member countries or shareholders who are represented by a Board of Governors [26]. It works closely with the International Monetary Fund (IMF) also established in 1944 with the same membership [27]. Its primary purpose is the promotion of financial stability and monetary cooperation in addition to facilitating international trade and promoting employment and sustainable economic growth. The World Bank Group has two goals for the world to achieve by 2030, which directly influence the world's population health:

1. End extreme poverty.
2. Promote shared prosperity by fostering the income growth of the bottom 40% for every country.

In 2018, the United Nations and the World Bank Group signed a Strategic Partnership Framework that consolidates their joint commitment to cooperate in helping countries implement the 2030 Agenda for Sustainable Development. The World Bank provides financial and technical assistance to developing countries by bringing together knowledge and expertise across all major sectors of development, and this includes digital development. The World Bank's ICT work program focuses on expanding access to fast, affordable Internet, and developing reliable online platforms that promote improved service delivery, good governance, and social accountability. The partnership between these organisations is fundamental to their ability to achieve results at the desired scale. The Strategic Partnership Framework and associated work program is expanded upon in another chapter.

The World Health Organisation (WHO): Role

As from 1948, the World Health Organisation (WHO) has been actively involved in promoting health for everyone. Its top priorities at that time were to eradicate or at best minimise the incidence of malaria, tuberculosis, venereal disease, environmental pollution, and improve women's and children's health and nutrition. Many of these priorities remain, and others such as polio and small pox were eradicated in the interim. A number of new viral diseases, such as HIV/AIDs, Ebola, Zika, severe acute respiratory syndrome (SARS), and most recently the COVID-19 pandemic, are current WHO priorities along with a number of other frequently occurring noncommunicable health conditions such as cancer and diabetes. WHO has a central role in disease surveillance and response coordination.

In September 1978, in Alma Ata (now Almaty), then in Soviet Kazakhstan, WHO's World Health Assembly adopted the Alma Ata Declaration with a 'health for all' agenda [9]. This described health as a human right and a socioeconomic issue and it drew attention to the deep health inequalities that existed between developed and developing countries. This was followed by the Ottawa Charter for Health Promotion in 1986 which stressed the importance of action on public policy, environmental influences, and supportive action at community and individual levels. Since then, WHO has adopted a 'Health in All Policies (HiAP)' approach, a method for facilitating a more integrated and networked approach to policymaking.

A 2017 review of global case studies by WHO and the Government of South Australia [28] detailing progress made, provides ideas and guidance to policy professionals with responsibility to address complex social, economic, and environmental challenges reflecting the global commitment to sustainable development. This requires the ability to understand and address the underlying determinants of health. Enabling digital solutions needs to be considered for inclusion in new policy initiatives designed to meet these requirements.

WHO's responsibilities in need of digital transformation

WHO established and maintains the international health regulations which countries need to follow, to identify disease outbreaks, and to stop them from spreading. In addition, WHO assists in the prevention of chronic diseases. The Pan American Health Organisation (PAHO) and WHO have adopted digital health as a strategy to maintain healthcare for people living with noncommunicable diseases during COVID-19 as these continue to be the main cause of death and disability worldwide [29].

WHO has the responsibility for the development and maintenance of the international standard in use for defining and reporting diseases and health conditions, the International Classification of Diseases (ICD) and others within their family of classifications. This work is undertaken in conjunction with the International Federation of Health Information Management Association (IFHIMA) whose vision is 'a healthy world enabled by quality health information' [30].

WHO's ability to fulfil its roles and responsibilities has changed over the years, due to global changes such as decolonisation, which increased the number of developing countries requiring assistance. In addition, there have been changes in member views of WHO's value propositions that have resulted in associated changing funding arrangements impacting on WHO's ability to fulfil its original charter.

WHO has recognised that sustainable development requires poverty elimination. It has adopted strategies that build economic growth and address a range of social needs, including education, healthcare, social protection, provision of job opportunities, along with tackling climate change and protect natural environments, all detailed in its 17 sustainable development goals [31]. Investments in infrastructure are sustainable development goals; this includes transport, irrigation, energy, and information and communication technologies (ICTs) and is considered to be crucial not only to empowering communities but also to sustainable development.

WHO and digital health

WHO has established a separate department, agency, or national working group for digital health with links to not only its own members but also to other high-level agencies and communities of practice such as the IFHIMA [30]. The WHO's global strategy has identified this requirement as the first step towards enabling any country to achieve a successful digital transformation of revolutionary proportions. This is considered to be the first enabling action towards the attainment of a sustainable digital health ecosystem. It is important for members of such a group to have the right knowledge, skills, and personal characteristics for this to be successful.

WHO has developed a global strategy on digital health 2020–2040 [32] along with a draft action plan [33]. Its vision is to *improve health for everyone, everywhere by accelerating the adoption of appropriate digital health*. Its three guiding principles are as follows:

1. Advocate that successful digital health initiatives require a unified strategy.
2. Acknowledge that the digital health adoption process is a country's decision.
3. Promote the appropriate use of digital technologies.

Its first strategic objective is to promote global collaboration and advance the transfer of knowledge on digital health. The second is to build and consolidate digital health capacity that reflects national needs. This is followed by the need to commit and engage stakeholders to advance digital health in every country. The final strategic objective is to improve measurement, monitoring, research, and practice on digital health. Priority needs listed at the 1st meeting of the WHO Digital Health Advisory Group were as follows [34]:

- Data governance standards (data ownership, consent, ethics, monetisation, etc.);
- Interoperability standards/architecture (starter kit);
- Position WHO as a leader of digital transformation (disrupt culture, foresight, capabilities, research agenda, human-centred approaches);
- Role of WHO vis-à-vis digital public goods (best buys, investment guidance, performance qualification (PQ), regulatory alignment, etc.);
- New multisector engagement/partnership models;
- Mobilise and optimise funding for digital health.

A summary indicating functions, responsibilities, value propositions, and the beneficiaries of the work undertaken by these global entities who are potentially able to provide the necessary smart leadership required to drive the necessary cooperation for the establishment of a successful global digital health ecosystem is provided in Table 1. This provides an overview of current high-level leadership.

TABLE 1 Global entities able to lead global cooperation required for the establishment of a digital health ecosystem, their functions, values, and beneficiaries.

Global entity	Functions/responsibilities	Value propositions	Beneficiaries
United Nations	Maintains international peace and security Supports the World Bank Group Works closely with the International Monetary Fund (IMF)	Peace, justice, respect, human rights, tolerance, and solidarity	The world's population
World Bank Group	• Supports developing countries • ICT work program	• Sustainable development • Access to fast affordable Internet, reliable online platforms, improved service delivery, good governance, social accountability	Developing countries
World Health Organisation (WHO)	• Eradicates or minimises the incidence of infectious diseases, environmental pollution, improves women's and children's health and nutrition, and improves management of frequently occurring noncommunicable diseases • Establishes and maintains International Health Regulations (IHR) and standards for defining and reporting national health burdens • Disease surveillance and response coordination • Manages and maintains family of classifications • Global strategy on digital health 2020–2040	• Health in All Policies (HiAP)-integrated and networked approach to policymaking • Sustainable development • Build economic growth • Address social needs, incl. Education, healthcare, social protection, provision of job opportunities • Natural environment protection, tackle climate change • Invest in infrastructure, incl. Irrigation, energy, ICT, transport • Sustainable digital health ecosystem based on a unified implementation strategy • Multisector engagement/ partnership models	• The World's population • Industries contributing to the health sector, including those associated with digital health • Health science researchers • Health service providers
International Federation of Health Information Management Associations (IFHIMA)	Works closely with WHO in respect of the management of the family of classification terminologies	Quality health information	Governments/health funders Health Insurance Health service delivery organisations

The Pan American Health Organisation (PAHO/WHO)

PAHO is the specialised health agency of the Inter-American System with more than 110 years of history. It also serves as the Regional Office for the Americas of WHO, the specialised health agency of the United Nations, led by 52 member countries and territories. It promotes evidence-based decision-making to improve and promote health as a driver of sustainable development by setting regional health priorities mobilising action to address health problems [35]. One of their main areas of work is health systems and services focusing on health services and access, human resources for health, and medicines and health technologies.

The PAHO/WHO have developed an Information Systems for Health (IS4H) framework and toolkit [36]. Its conceptual framework is based on a set of constructs that has adopted a holistic view of information systems for health. It was based in parts on the adaptation from Mario Bunge's ontology from the information system field [37]. This toolkit provides the most comprehensive strategic plan and set of implementation guidelines. It has a public health focus with the following four strategic goals:

1. Data management and information technologies
2. Management and governance
3. Knowledge management and sharing
4. Innovation

This reference framework details expected results relative to each of its strategic goals, incorporated into its sustainable health agenda for the Americas 2018–2030, and represents:

> a mechanism for managing interoperable systems with open data that comes from different sources and that is ethically used, through effective ICT tools, to generate strategic information for the benefit of public health.

The IS4H Maturity Model v.1.0 is a tool for assessing the maturity levels of the information system for health and the organisation's capabilities to operate, interact, and benefit from it [37]. It details five progressive levels providing an awareness for planning. These levels are described in some detail for use as criteria against which to assess implementation progress.

- Level 1—Public health Information is organised.
- Level2—Public health information is managed.
- Level 3—Public health decisions are based on information.
- Level 4—Information governance and knowledge management is implemented.
- Level 5—Information systems for health implemented.

The IS4H toolkit is one of PAHO's 11 goals aimed at ensuring that every woman, man, and child can live a healthy and productive life. Key elements identified for these systems include:

- analysis of the level of maturity of the countries' IS4H, as a first step to identify gaps and needs;
- data management and governance;
- information and communication technologies;
- knowledge management for health;

- national capacity building for human resources and strengthening of infrastructure for data management and analysis;
- information systems for health, including electronic medical records and civil registries, that facilitate interconnectivity and interoperability through data integration processes in a systematic and routine manner.

PAHO/WHO supports the Caribbean Community (CARICOM), who share in economic, social, and cultural prosperity by promoting and supporting a unified, inclusive, resilient, and competitive community representing a single market and economy (CSME). A 2011 intergovernment agreement enabled the establishment of the Caribbean Public Health Agency (CARPHA) as its single regional public health agency. The Jamaican case study presented later made use of the IS4H toolkit.

Global public health governance and WHO funding

Global governance mechanisms for addressing epidemics only emerged following the SARS outbreak in 2002 2003. WHO was then authorised to issue travel warnings without first receiving approval from the relevant government, and it permitted to seek information from nonofficial sources to augment its disease surveillance capacities [38]. WHO's independent disease surveillance capacities were formalised in 2005, following a review and update of its International Health Regulations (IHR). This authorised WHO to issue Public Health Emergency of International Concern alerts based on both official and nonofficial information [39]. Another major IHR focus is to transform and network domestic health systems across borders for the purpose of fighting against infectious diseases.

WHO has the responsibility for providing 'best practice' guidelines, for monitoring national health systems' pandemic preparedness, disease surveillance, and response coordination. The global system for addressing infectious disease outbreaks that emerged before COVID-19 was clearly not fit for purpose. This became apparent during WHO's recent performance which resulted in WHO's inability to meet pandemic needs during the COVID-19 pandemic that led to many unnecessary deaths and global economic chaos.

Following the Alma Ata Declaration, developing countries, WHO's main funders, were deeply concerned about potential redistributive demands. They were responsible for significantly defunding WHO at that time and the majority of health aid for developing countries was rerouted via the World Bank, where it became part of the Bank's growing structural adjustment agenda. The WHO's operational budget was 'cut to skimpy levels' [40]. Contributing states shifted to voluntary and project-based funding instead to ensure the WHO was able to pursue their own preferred objectives. By the 2000s, donors directly controlled around 80% of the WHO's budget. Very limited financial support is available to help developing countries build up their domestic health systems, yet they are subjected to ongoing surveillance by the WHO and developed countries.

It is evident that developing countries have the vote as they are more numerous, yet the developed countries are required to provide most of the funding. Developed countries have politically moved the global culture from an equity agenda, to the adoption of a more capitalist ('Shareholder Capitalism') market-oriented focus. In 2020, China, who has adopted a 'State Capitalism' approach to the management of its economy, was the third biggest provider of compulsory, 'assessed', contributor to the WHO, but not a major contributor of voluntary funding [38]. There is no escaping the need to develop more durable, equitable, and better funded forms of cross-border cooperation to address the ever-present threat of pandemics.

Addressing this global governance issue is of critical importance for the adoption of digital technologies required to enable timely and effective management of global health embracing a person-centred approach. Successful digital health depends on how this will be resolved.

Inter-American Development Bank (IDB)

The IDB was founded in 1959 and is owned by its 48 member countries. The IDB is the leading provider of development financing to Latin America and the Caribbean [41]. Its operations are funded by the bank's issuing of bonds to investors. Its mission is to improve lives in Latin American and Caribbean countries by supporting efforts to reduce poverty and inequality in a sustainable, climate-friendly way. Funded projects are the result of a comprehensive proposal development and approval process. The IDB adheres to detailed procurement policies applicable to borrowing member countries. Navigating these is complex and time-consuming and may be influenced by external advice. Funded projects include numerous digital health initiatives undertaken since 1997.

Asian Development Bank (ADB)

The United Nations Economic Commission for Asia and the Far East was responsible for the establishment of the Asian Development Bank (ADB) in 1963 to foster economic growth and cooperation in this region [42]. ADB is owned 68 members of whom 49 are from the region. Its primary aim is to eradicate extreme poverty in the region by providing loans, technical assistance, grants, and equity investments to promote social and economic development.

The ADB facilitates policy dialogues, provides advisory services, and mobilises financial resources through cofinancing operations that tap official commercial and export credit resources. It assists its members and partners by providing loans, technical assistance, grants (including educational), and equity investments to promote social and economic development in its region. Procurement is managed via its supplier portal.

The Asia eHealth Information Network (AeHIN) resulted from the 2007 formation of the PAN Asian Collaboration for Evidence-based eHealth Adoption and Application which brought together 16 researchers from 10 Asian countries to implement 8 multinational research projects. This network now enables its members (government agencies, private and civil society organisations, development agencies) from 25 countries to share knowledge and experiences and maximise a regional approach for greater country-level impacts. AeHIN works closely with the Asian Development Bank (ADB) [43]. Their experiences are presented in a chapter to follow.

African Development Bank (AfDB) group

Twenty-three independent African States established the African Development Bank at their inaugural meeting in Lagos, Nigeria, in 1964 [44]. Their prime objective is to mobilise financial resources from within and external to the African continent, to support economic and social development in Africa as a means of contributing to poverty reduction. The AfDB now has 54 regional country members plus 28 nonregional country members with a strong focus on sustainable development. Its overarching objective is to spur sustainable economic development and social progress in its regular member countries.

Africa has embarked on a process of economic transformation along with regional economic integration. Africa continues to have massive infrastructure need. The AfDB's strategic directions indicate that infrastructure development, including transport and creating ecosystem services such as the improvement of water, energy, and food security, has a high priority. There is no mention of the need for a digital ecosystem which is not surprising given other fundamental priorities such as electricity, although it is highly desirable to consider digital transformation as a cost-effective support component and its ability to contribute to the sustainable development of other infrastructure needs. The AfDB's investments are guided by the 2015 agreed set of 17 sustainable development goals as part of their adopted economic transformation process [45].

G20: Group of twenty

The G20 represents an international forum for governments and central bank governors from 19 countries plus the European Union. This was established in 1999 to discuss policy pertaining to international financial stability [46]. This was expanded in 2008 to include heads of government or state as well as finance and foreign ministers to address global macrofinancial, socioeconomic, and development issues. Collectively this group manages around 80% of the world's economy, 75% of world trade, two-thirds of the world population, including more than half of the world's poor and around half the world's land area.

Organisation for economic co-operation and development (OECD)

The OECD is an international organisation established in 1961 to promote sustainable economic growth, jobs and prosperity, support multilateral trade, and devise policies to assist developing countries. By 2020, it had 37 country members whose governments, policymakers, and citizens aim to establish international norms and find evidence-based solutions to a range of social, economic, and environmental challenges [47]. OECD member countries and key partner countries represent about 80% of world trade and investment. The OECD fosters strong education, fights international tax evasion, and provides a unique forum and knowledge hub for data and analysis, the exchange of experiences, sharing best practice and to provide advice on public policies and global standard setting. Its work is undertaken by more than 300 thematic committees, expert, and working groups.

The OECD is a large and trusted source of comparative socioeconomic data and analysis used by many to support decision-making. The development of international agreements, guidelines, and standards is viewed as necessary so that everyone plays by the same rules and cooperates to reach shared objectives. The OECD maintains close relationships with parliamentarians through its Global Parliamentary Network used to provide access to OECD analysis and to serve as an exchange forum. The Business and Industry Advisory Committee (BIAC) and the Trade Union Advisory Committee (TUAC) represent their members' positions and interests at the OECD, along with major NGOs, community organisations, think-tanks, research organisations, and academia serve as consultative groups to the OECD to inform and craft policies.

The OECD's core values are as follows:

- OBJECTIVE: Our analyses and recommendations are independent and evidence-based.
- OPEN: We encourage debate and a shared understanding of critical global issues.
- BOLD: We dare to challenge conventional wisdom starting with our own.
- PIONEERING: We identify and address emerging and long-term challenges.
- ETHICAL: Our credibility is built on trust, integrity, and transparency.

The OECD has the potential to play a major role in enabling the successful growth of a global digital health ecosystem. Anyone is able to engage with the OECD via its Forum Network and drive the digital health agenda. Numerous publications are freely available via the OECD website which details its work according to six distinct areas of interest: Science, technology, and innovation policy; Industry and globalisation; Emerging technologies; Digital economy; Broadband and telecom; and Consumer policy. For example, its privacy framework relative to the digital economy [48] focuses on the practical implementation of privacy protection through an approach grounded in risk management and the need to address the global dimension of privacy through improved interoperability. This publication includes a list of eight fair information practices.

A summary of these global and regional entities, with the smart leadership potential to contribute to the establishment of successful digital health ecosystems is presented in Table 2. This provides an overview of current funding opportunities and the potential for collaboration to optimise the distribution of roles and responsibilities needed to transform the digital health high-level foundational infrastructure in a manner that best enables digital health ecosystems in any location to be realised and add value to healthcare service delivery.

World economic forum

The World Economic Forum (WEF) was established in 1971 as a not-for-profit international organisation for independent and impartial public-private cooperation. It strives to demonstrate entrepreneurship in the global public interests based on underpinning moral and intellectual integrity. Its new Davos Manifesto, a set of ethical principles to be used as a guide for companies in the age of the fourth industrial revolution, was launched in 2020 [49].

This is referred to as 'stakeholder capitalism' as opposed to the traditional 'shareholder capitalism' or 'state capitalism' as adopted for the Republic of China, for example. Shareholder capitalism with a single-minded focus on profits is becoming increasingly disconnected from the real economy and as such is considered to no longer be sustainable. The Davos Manifesto reflects stakeholder capitalism by requiring companies to pay their fair share of taxes, show zero tolerance for corruption, uphold human rights throughout their global supply chain, and advocate for a competitive level playing field, particularly in the 'platform economy' as is a requirement for successful digital health.

COVID-19 is expected to have long-term consequences for economic which, according to WEF, is likely to leave the world even less sustainable, less equal, and more fragile. This means that we must build entirely new foundations for our economic and social systems [18].

TABLE 2 Global and Regional digital health ecosystem enabling and contributing entities, their functions, values, and beneficiaries.

Global entity	Functions/responsibilities	Value propositions	Beneficiaries
Pan American Health Organisation (PAHO/WHO)	• Regional Office for the Americas of WHO • Promotion of evidence based decision-making • Setting regional health priorities driving sustainable development • Implements its Information Systems for Health (IS4H) framework • Supports the Caribbean Community (CARICOM) and the Caribbean Public Health Agency (CARPHA)	• Interoperable systems with open data • Effective ICT tools generating strategic public health information	• Latin America and the Caribbean governments and populations
Inter-American Development Bank (IDB)	• Provides project funding	Reduces poverty and inequality in a sustainable, climate-friendly way	Latin America and the Caribbean governments and populations
Asian Development Bank (ADB)	• Fosters regional economic growth and cooperation by providing advisory and investment services	Eradicates extreme poverty	Asian governments and populations
African Development Bank (AfDB) Group	• Sustainable development goals • Infrastructure development • Regional economic integration • Skill and technology • Governance and accountability • Private sector development	• Ends poverty and hunger • Sustainable water management, sanitation • Affordable energy access • Health and well-being • Quality education, life-long learning • Gender equality, empowers women and girls • Infrastructure and innovation • Manage climate change • Global partnerships	African nations and their industries and populations
G20	• A forum for governments and central bank governors to discuss policies with global impact	• Macroeconomic stability • Sustainable development • Women empowerment • Enhanced human capital • Increased flow of trade and investment	The world's population
Organisation for Economic Co-operation and Development (OECD)	• Promotes sustainable economic growth and supports multilateral trade • Devises policies to assist developing countries • Establishes evidence-based international norms to solve a range of social, economic, and environmental challenges • Provides a unique forum and knowledge hub for data, analytics, exchange of experiences • Fosters strong education • Provides advice of public policies and global standard setting	• OBJECTIVE: Our analyses and recommendations are independent and evidence-based • OPEN: We encourage debate and a shared understanding of critical global issues • BOLD: We dare to challenge conventional wisdom starting with our own • PIONEERING: We identify and address emerging and long-term challenges. • ETHICAL: Our credibility is built on trust, integrity, and transparency	Governments, researchers, and the world's population

The WEF has established Centres for the Fourth Industrial Revolution based in the USA, India, China, and Japan (Asia Pacific), in recognition of the many digital transformations and new governance requirements. These centres are codesigning and piloting policy frameworks and governance protocols to shape the future of technology governance, across six areas of focus, each of which is referred to as a 'platform'. Each of their focus areas has relevance to digital health, and they are as follows:

1. Artificial intelligence and machine learning
2. Internet of things, robotics and smart cities
3. Blockchain and distributed ledger technology
4. Data policy
5. Autonomous and urban mobility
6. Drones and tomorrow's airspace

Each of these developments requires multistakeholder collaboration, accountability, transparency, privacy, and impartiality as the means to build digital trust and confidence, fundamental conditional requirements for spurring innovation, adoption, and progress. Each of these technologies and associated issues will be referred to in other subsequent chapters. The results of this work have the potential to become key enablers for the realisation of successful digital health.

Philanthropic, not-for-profit and nongovernment organisations (NGOs)

There are a few very large and many other philanthropic foundations and not-for-profit organisations that in essence deliver services that governments are unable to provide, around the globe. These organisations work closely with WHO and others representing important contributors and global enablers for the growth of digital health ecosystems. Each program focuses on collecting or using its own set of data based on program delivery, monitoring, and evaluation needs. Effective public-private partnerships are important. There are two very large well-known philanthropic organisations, the Bill & Melinda Gates Foundation and the Rockefeller Foundation who support WHO activities financially.

The Bill & Melinda Gates Foundation began in 1994. It teams up with partners from around the globe to take on extreme poverty and poor health in developing countries and the failures of America's education system [50]. The Foundation is governed by its three trustees and a strong leadership team, and it funds innovative ideas that could help remove these barriers. Its Chief Executive Officer leads these efforts in the foundation's quest to promote equity for all people around the world, across five grant-making areas, four of which are global including a health program, supported by well over a thousand employees located in its head office and five regional offices. The foundation partners with all kinds of organisations. Many of these funded activities are having positive impacts on global health and have the potential to influence or contribute to digital health solutions. In 2020, $5 million was made available for applications for funding digital health projects.

This foundation partners with the Global Fund [51] along with governments, civil society, technical agencies, the private sector, and people affected by three specific diseases. The fund pools resources and was designed to accelerate the end of AIDS, tuberculosis, and malaria epidemics. Its technical and development partners offer technical expertise and participate by,

- supporting resource mobilisation and advocacy efforts,
- providing or supporting country coordination,
- assisting with stakeholder engagement, and
- monitoring and evaluating Global Fund-supported programs.

The Rockefeller Foundation has a long historical legacy that began in 1910. It aims to improve lives and the planet by supporting human potential to be unleashed through innovation with better use of science and data [52]. The foundation works to improve the well-being of people everywhere through collaboration with partners and grantees. This foundation's President leads this philanthropy focused on ending energy poverty, achieving health for all, nourishing the world, and expanding economic opportunity. In 2020, it established data.org as a platform for its partnerships to continue to build the field of data science for social impact. In 2019, it initiated Precision Public Health, to empower community health systems and frontline workers with the latest data science innovations with the aim to prevent 6 million deaths in 10 countries by 2030. These initiatives and many other funded activities may be viewed as significant contributions towards the growth of digital health.

Contract research organisation (CRO) services market

Companies offering research-based services on a contractual basis to many pharmaceutical and biotechnology companies, medical device industries, and any number of government research organisations make up a very fragmented CRO market. Their work includes drug and device research and development including commercialisation of outcomes, biologic assay development, preclinical research, clinical research, clinical trial management, and pharmacovigilance. CROs may also provide support for foundations, research institutions, and universities. Their size ranges from large, international full-service organisations to small, niche specialty groups.

One example is Novo Nordisk, a pharmaceutical company based in Denmark, with the passion, skills, and commitment to drive change to defeat diabetes and other serious chronic diseases, including haemophilia [53]. Its business philosophy is one of balancing financial, social, and environmental considerations, their 'triple bottom line' to benefit its shareholders, employees, and their communities. This is part of its company culture, known as 'the Novo Nordisk Way' that every employee must abide by. This company partners with a number of philanthropic organisations.

The conduct of clinical trials represents opportunities to conduct biomedical or health-related research studies on human beings. Such trials attract large investments around the globe, generate thousands of jobs, and constitute significant growth sectors. They are vital to the future of global health [54]. Clinical trial outcomes inform new treatments, interventions, or tests as well as the need for refinement of existing standards of care and clinical practices with many beneficial flow-on affects. Such benefits may be measured by identifying the number of quality-adjusted life years (QALY) and/or health sector wide benefits. They may be industry sponsored, mostly focused on the development and testing of new medicines or devices, or investigator initiated focusing on treatment protocols and clinical processes to identify 'best practice'.

Clinical trials connect researchers from around the globe based on partnerships and global collaboration. These connections are core to successful outcomes. Each clinical trial team needs to make use of shared IT platforms. A less fragmented, well-connected digital health infrastructure would be extremely beneficial to this sector by providing an increase in research opportunities, more timely accurate data, greater sample sizes and a significant reduction in the overall costs. A summary of contributions made by these four global entities indicating their potential for smart leadership towards digital health ecosystems is presented in Table 3.

TABLE 3 Global digital health ecosystem enabling and contributing entities, their functions, values, and beneficiaries.

Global entity	Functions/responsibilities	Value propositions	Beneficiaries
World Economic Forum (WEF)	• Demonstrates entrepreneurship in the global public interests • Promotes 'stakeholder capitalism'	• Moral and intellectual integrity • Ethical principles to guide companies transforming to the digital age • Human rights throughout global supply chain • Competitive level playing field for a 'platform economy'	Governments and the world's corporate sectors and the world's population
Bill & Melinda Gates Foundation	• Funds individual projects that address extreme poverty and poor health in developing countries and failures of America's education system • Works closely with WHO	• Equity for all people around the world	Each project's target population
Rockefeller Foundation	• Builds the field of data science for social impact • Precision public health to empower community health systems and frontline workers.	• The unleashing of human potential through innovation with better use of science and data	Each project's target population
Contract Research Organisation (CRO) Services	• Drug and device research and development • Clinical trials	• A variety of business philosophies	Pharmaceutical and biotechnology companies, medical device industries, government and private research organisations, universities, and foundations

Global digital health technology foundations

Technologies underpinning digital health and the ability to share data and knowledge include telecommunication networks, Internet access, and cloud computing. These represent critical foundations for successful digital health ecosystems. These technologies are being controlled by a few giant companies with access to global data. These global companies originated in the United States with business models that best fit with the shareholder capitalism philosophy. They are the five most valuable listed firms in the world, Alphabet (Google's parent company), Amazon, Apple, Facebook, and Microsoft with surging profits [55]. Each of these makes use of their own proprietary platform. They provide individuals and organisations with access to a number of different services made available free of charge or for a nominal fee. This control of data gives these Internet companies a lot of power influencing competition conventions and generating a new data economy.

Big data availability enables machine learning and the use of artificial intelligence (AI) techniques. Their collective surveillance systems span the entire global economy with profits flowing to a select few. Cloud computing services that have become very popular are made use of by many within the health industry for digital health data processing purposes at commercial rates, subject to relevant local regulations. This can become very costly for individual users in contrast with public access to the Internet and World Wide Web technologies which were made available to benefit the public at large.

The current focus on AI has raised a number of ethical, moral, quality, privacy, transparency, and other issues as detailed by, for example, the European Parliament [56]. This review highlights current gaps around the mechanisms of fair benefit sharing; assigning of responsibility; exploitation of workers; energy demands in the context of environmental and climate changes; and more complex and less certain implications of AI, such as those regarding human relationships. A number of organisations are working to address some of these issues.

The Alan Turing Institute is of the view that data science and artificial intelligence will change the world [57]. AI Global leverages technology to build helpful tools for responsible AI implementation. This nonprofit organisation is committed to a world making use of trustworthy, safe, and fair technology. It is supported by governments, industry, academia, and civil society [58]. The Future of Life Institute (FLI) has as its mission to catalyse and support research and initiatives for safeguarding life and developing optimistic visions of the future. They note that 'technology is likely to keep improving at an accelerating pace' and that we need to 'ensure that tomorrow's most powerful technologies are beneficial for humanity' [59]. The adoption of AI in health can be extremely beneficial, provided the associated issues are addressed in a manner that does not compromise patient safety or ethical principles.

Discussion

The previous chapter identified a number of key foundational infrastructure requirements. This chapter has identified the many global entities with the potential to influence digital health activities. Many governance, membership commitments, and funding changes over time were revealed. These influence the capacity of such organisations to meet their original roles and responsibilities or how well they have adapted to our constantly changing environment.

One pertinent example of an inability to retain currency is the apparent unchanging bureaucratic structures of the United Nations [25]. We have witnessed many national sovereign changes, along with a lack of recognition of positive changes that have occurred in a number of large sovereign nations over time. This is particularly evident within the UN's bureaucracy by its inability to make changes to its security council such as the continuing permanent membership denial of Japan, Germany, Brazil, India, or any African country.

Similar circumstances have influenced funding changes that have occurred over time for the WHO resulting in its reduced capacity to meet its continuing obligations. It is also of concern that a number of these influential organisations do not appear to have acquired new digital skills required for their own organisational digital transformation desirable for more streamlined operational data and information processing. For example, the WHO has yet to address data model requirements for reportable public health data.

These deficiencies have resulted in a number of large-scale changes such as the [22]:

- weakening of the WHO;
- rise of health interests at the World Bank (and IMF);
- emergence of powerful new private actors such as the Bill & Melinda Gates Foundation and the quasiprivate Global Fund;
- internationalisation of pharmaceutical research and Contract Research Organisation (CRO) markets;
- truncated success of International Health Development educational programmes;
- shift in the morbidity profiles made visible by such instruments as the DALY, the QALY, and the International Burden of Disease index; and
- the rise of a 'commerce/security/disease' nexus.

Collectively these potentially highly influencing global entities have established a new world disorder which is more difficult to navigate [25] for the purpose of overcoming existing limitations. The establishment of key global foundations for a well-functioning digital ecosystem have been compromised as a result. This has led to numerous costly inefficiencies at the coalface for many.

The global pandemic further demonstrated existing fragmentations along with a lack of strong leadership or a global collaborative response primarily due to narcissistic or ego-driven national leaders and excessive partisanships. Yet the global scientific research community and health professionals generally continue to demonstrate the power of collaboration, to overcome bureaucratic and political international failures or shortcomings.

The digital health market and trade is dominated by a small number of technology giants and electronic medical record (EMR) system vendors. Most are based in the United States, working closely with large hospital corporations and with HIMSS, the professional organisation that supports their marketing strategies as a benefit to its members. These strong relationships have influenced the US Government's national digital health strategy in line with strong support for its unsustainable shareholder capitalism. All have established financially lucrative business models they wish to retain. Their collective marketing and lobbying activities have included the gradual expansion of delivering HIMSS conferences and expos to Europe, Southeast Asia, and Australia. This in turn has resulted in the adoption and expansion of legacy systems at great expense to other nations' health systems and to the detriment of achieving optimum digital health success. This result is primarily due to widespread knowledge gaps among many decision-makers regarding scientific truths, as explained in the following chapters, and a desire to minimise risk by choosing what are considered to be tried and tested solutions.

Our road map to successful digital health needs to consider desirable reformations at a global level, in order to optimise scientific, professional, and commercial collaborations needed for the global adoption of the necessary foundational digital health infrastructure. The G20 has the greatest potential to identify which entity needs to take responsibility for what and drive the associated reformations needed for managing the establishment and continuing maintenance of high-level infrastructure requirements for the successful global functioning of a digital health ecosystem. This needs to have a strong focus on a digital infrastructure able to support timely data collection and processing as required to manage any pandemic, including both infectious and noncommunicable disease prevention and management, as one could argue that the increasing incidence of chronic diseases equals the same health and economic burdens imposed by pandemics.

The road map to be driven collaboratively needs to identify infrastructure requirements, including public-private partnerships (PPPs) delivering Technology-enabled Remote Health Care (TeRH) [60] that enable the following outcomes to be achieved:

- globalisation and localisation of digital health;
- a growing power of end users through improved technology accessibility;
- a growing focus on data ownership, ethics, rights, and equity;
- a transformation of health systems based on optimum use of digital data;
- a broadening of community-wide resiliency in any location;
- merging digital health sector outcomes;
- greater support for patient/citizen-centred care; and
- support for the implementation of cost-saving solutions.

Conclusion

The many and varied global entities listed have roles and responsibilities that directly or indirectly influence many options for global digital health transformations. Ideally they work cooperatively and collaboratively so as to avoid a duplication of efforts. There is an urgent need to harmonise activities and eliminate continuing fragmentation. Global success needs to be based on numerous collaborative cross-country efforts. Many existing international networks, including the many entities described in this chapter, play important leadership and investment roles. Successful digital health is dependent on multilevel international collaboration to overcome current translational interoperability friction.

The leadership provided by PAHO/WHO and others has resulted in the availability of and access to valuable guidelines, expert advice, education, and professional development for many. Funding organisations are able to prioritise funding for digital transformation projects, but they are not in positions of authority to actually adopt and implement a digital health strategy. Successful implementation requires strong leadership from those occupying authoritative positions locally, together with access to the necessary resources and investments to make it happen. Positive globalism requires leaders to delegate their responsibilities to leaders with the necessary expertise to oversee and manage the many activities associated with the successful implementation of digital health that is best suited to their environment.

Digital health transformation must first and foremost consider and aim to meet the needs associated with achieving and maintaining global population health and patient safety at every point of care. All of us need to make the best possible use of available technologies to reduce the national health burden and meet the health needs of individuals in a timely, appropriate, and cost-effective manner. These desired outcomes reflect measures of success.

References

[1] WHO, Sixty-Sixth World Health Assembly (WHA66.24) eHealth Standardization and Interoperability Resolution Geneva, World health Organization, 2013. Available from: https://www.who.int/ehealth/events/wha66_r24-en.pdf?ua=1.

[2] World-Economic-Forum, Why This is Digital Healthcare's Moment, 2021, Available from: https://www.weforum.org/agenda/2021/01/why-this-digital-healthcare-moment/.

[3] World-Bank, World Bank Approves $12 Billion for COVID-19 Vaccines, 2020, Available from: https://www.worldbank.org/en/news/press-release/2020/10/13/world-bank-approves-12-billion-for-covid-19-vaccines.

[4] C.Z. Qiang, M. Yamamichi, V. Hausman, D. Altman, Mobile Applications for the Health Sector, ICT Sector Unit World Bank, 2011. Available from: https://core.ac.uk/download/pdf/207034074.pdf.

[5] GDHP. Global Digital Health Partnership Available from: https://gdhp.org/.

[6] B. Al-Knawy, M. Adil, G. Crooks, K. Rhee, D.W. Bates, H.I. Jokhdar, et al., The Riyadh declaration: the role of digital health in fighting pandemics, Lancet (2020). Available from: https://www.thelancet.com/journals/lancet/article/PIIS0140-6736(20)31978-4/fulltext.

[7] ANDHealth, Digital Health; Creating a new growth industry for Australia, 2018, Available from: https://andhealth.com.au/wp-content/uploads/2019/04/Digital-Health_Creating-a-New-Growth-Industry-for-Australia.pdf.

[8] Startup+Health, With $8.5B Raised Globally in Q1, Health Innovation Funding Hits New Milestone—Report, 2021, Available from: https://healthtransformer.co/with-8-5b-raised-globally-in-q1-health-innovation-funding-hits-new-milestone-40008618740a.

[9] WHO, Alma Ata Declaration, World Health Organisation, 1978. Available from: https://www.who.int/publications/almaata_declaration_en.pdf.

[10] World-Bank, Digital Development-An Overview, 2020, Available from: https://www.worldbank.org/en/topic/digitaldevelopment/overview.

[11] Grand-view-research, Digital Health Market Size, Share & Trends Analysis Report by Technology, 2019, Available from: https://www.grandviewresearch.com/industry-analysis/digital-health-market.

[12] WTO, Technological Innovation, Supply Chain Trade, and Workers in a Globalized World, World Trade Organisation, 2019. Available from: https://www.wto.org/english/res_e/publications_e/gvcd_report_19_e.htm.

[13] C. VanGrasstek, 15 The future of the WTO in: The History and Future of the World Trade Organisation, World Trade Organisation, 2013. Available from: https://www.wto.org/english/res_e/publications_e/historyandfuturewto_e.htm.

[14] T.K. Mackey, B.A. Liang, Pharmaceutical digital marketing and governance: illicit actors and challenges to global patient safety and public health, Glob. Health 9 (2013) 45.

[15] A. Razavi, N. Erondu, E. Okereke, The Global Health Security Index: what value does it add? BMJ Glob. Health 5 (2020), e002477.

[16] E.E. Cameron, J.B. Nuzzo, J.A. Bell, Global Health Security (GHS) Index: Building Collective Action and Accountability, John Hopkins Centre for Health Security, 2019. Available from: https://www.ghsindex.org/wp-content/uploads/2019/10/2019-Global-Health-Security-Index.pdf.

[17] D. Ribes, C.P. Lee, Sociotechnical studies of cyberinfrastructure and e-research: current themes and future trajectories, Comput. Supported Coop. Work 19 (2010) 231–244.

[18] World-Economic-Forum, Now Is the Time for a 'Great Reset', 2020, Available from: https://www.weforum.org/agenda/2020/06/now-is-the-time-for-a-great-reset/.
[19] CDC, Health and Economic Costs of Chronic Diseases, Centres for Disease Control and Prevention, 2021. Available from: https://www.cdc.gov/chronicdisease/about/costs/index.htm.
[20] World-Economic-Forum, Insight Reports, 2021, Available from: https://www.weforum.org/reports.
[21] Victorian-Government, Gender Equality: What Is it and Why Do We Need It? 2021, Available from: https://www.vic.gov.au/gender-equality-what-it-and-why-do-we-need-it.
[22] V. Adams, D. Behague, C. Caduff, I. Löwy, F. Ortega, Re-imagining global health through social medicine, Glob. Public Health 14 (2019) 1383–1400.
[23] UNICEF. United Nations International Children's Emergency Fund, Available from: https://www.unicef.org/.
[24] United-Nations, Universal Declaration of Human Rights, 1948, Available from: https://www.un.org/en/about-us/universal-declaration-of-human-rights.
[25] The-Economist, Geopolitics: The New World Disorder: The Economist, 2020, June 18th Edition: Available from: https://www.economist.com/leaders/2020/06/18/the-new-world-disorder?utm_campaign=the-economist-today&utm_medium=newsletter&utm_source=salesforce-marketing-cloud&utm_term=2020-06-18&utm_content=article-link-2.
[26] World-Bank-Group. The World Bank, Available from: https://www.worldbank.org/.
[27] IMF. International Monetary Fund, Available from: https://www.imf.org.
[28] V. Lin, I. Kickbusch, B. Baer, Progressing the Sustainable Development Goals Through Health in All Policies: Case Studies From Around the World, World Health Organisation, 2017. Available from: https://www.who.int/social_determinants/publications/Hiap-case-studies-2017/en/.
[29] A. PAHO/WHO, Digital Health: A Strategy to Maintain Health Care for People Living With Non-Communicable Diseases During COVID-19, 2020, Available from: https://www.paho.org/en/documents/digital-health-strategy-maintain-health-care-people-living-noncommunicable-diseases.
[30] IFHIMA. International Federation of Health Information Management Associations, Available from: https://ifhima.org/.
[31] UN/WHO, United Nations Sustainable Development Goals, UN/WHO, 2021. Available from: https://www.un.org/sustainabledevelopment/.
[32] WHO, Global Strategy on Digital Health 2020-2040, 2019, [cited 2019 17 April]. Available from: https://extranet.who.int/dataform/183439.
[33] WHO, Action Plan for the Global Strategy on Digital Health 2020-2024—Draft, 2019, Available from: https://www.who.int/docs/default-source/documents/annexi-action-plan-for-the-global-strategy-on-digital-health.pdf.
[34] WHO, Digital Health Technical Advisory Group 1st Meeting Report, 2019, Available from: https://www.who.int/publications/m/item/dh-technical-advisory-group-1st-meeting-report.
[35] PAHO, Pan-American Health Organisation (PAHO) and Regional Office for the Americas of WHO, 2021, Available from: https://www.paho.org/en/who-we-are.
[36] A. PAHO/WHO, Information Systems for Health (IS4H) Framework & Toolkit, 2020, [cited 2020 4 March]. Available from: https://www.paho.org/ish/index.php/en/toolkit.
[37] A. PAHO/WHO, Information Systems for Health Standard Assessment Method, 2018, [cited 2020 4 March]. Available from: https://www.paho.org/ish/index.php/en/maturity-model.
[38] S. Hameiri, COVID-19: Why Did Global Health Governance Fail?, The Lowy Institute, 2020. Available from: https://www.lowyinstitute.org/the-interpreter/covid-19-why-did-global-health-governance-fail.
[39] WHO, International Health Regulations (2005) 3rd edition, World Health Organisation, 2016. Available from: https://www.who.int/ihr/publications/9789241580496/en/.
[40] N. Chorev, The World Health Organisation between North and South, Cornell University Press, 2012.
[41] IDB. Inter-American Development Bank, Available from: https://www.iadb.org/en.
[42] ADB. Asian Development Bank, Available from: https://www.adb.org.
[43] AeHIN. Asia eHealth Information Network [cited 2020 3 March]. Available from: http://www.aehin.org/.
[44] AfDB. African Development Bank Group, Available from: https://www.afdb.org.
[45] AfDB, Innovative e-Health Solutions in Africa Award: Investing in Smart Human Capital Innovations: Spreading Inclusive Growth Capacities in Africa, 2014, Available from: https://www.afdb.org/fileadmin/uploads/afdb/Documents/Publications/Innovative_e%E2%80%91Healthb_Solutions_in_Africa_Award.pdf.
[46] G20. Group of Twenty, Available from: https://g20.org.

[47] OECD, Discover the OECD, Organisation for Economic Co-operation and Development (OECD), 2019. Available from: http://www.oecd.org/general/Key-information-about-the-OECD.pdf.

[48] OECD, The OECD Privacy Framework, Organization for Economic Co-operation and Development, 2013.

[49] K. Schwab, Davis Manifesto 2020: The Universal Purpose of a Company in the Fourth Industrial Revolution, World Economic Forum, 2019. Available from: https://www.weforum.org/agenda/2019/12/davos-manifesto-2020-the-universal-purpose-of-a-company-in-the-fourth-industrial-revolution.

[50] Bill-&-Melinda-Gates-Foundation, Every Life Has Equal value, 2021, Available from: https://www.gatesfoundation.org/Who-We-Are/General-Information/Foundation-Factsheet.

[51] Global-Fund, The Global Fund Partnership, 2021, Available from: https://www.theglobalfund.org.

[52] The-Rockefeller-Foundation, The Rockefeller's Foundation History, 2021, Available from: https://www.rockefellerfoundation.org/about-us/our-history/.

[53] Novo-Nordisk, A Global Healthcare Company, 2021, Available from: https://www.novonordisk.com.au.

[54] MTPConnect, Clinical Trials in Australia: The Economic Profile and Competitive Advantage of the Sector, MedTech and Pharma Growth Centre, 2017.

[55] The-Economist, Regulating the Internet Giants: The World's Most Valuable Resource is No Longer Oil, But Data: The Economist, LEADERS May 6th Ed, 2017, Available from: https://www.economist.com/leaders/2017/05/06/the-worlds-most-valuable-resource-is-no-longer-oil-but-data.

[56] EPRS, The Ethics of Artificial Intelligence: Issues and Initiatives, European Parliamentary Research Service, 2020. Available from: https://www.europarl.europa.eu/RegData/etudes/STUD/2020/634452/EPRS_STU(2020)634452_EN.pdf.

[57] The-Alan-Turing-Institute, National Institute for Data Science and Artificial Intelligence, 2021, Available from: https://www.turing.ac.uk/.

[58] AI-Global, Artificial Intelligence (AI) Global, 2021, Available from: https://ai-global.org/.

[59] S. Pinker, S. Russell, Foundations, Benefits and Possible Existential Threat of AI, Future of Life Institute (FLI), 2020. Available from: https://futureoflife.org/2020/06/15/steven-pinker-and-stuart-russell-on-the-foundations-benefits-and-possible-existential-risk-of-ai/.

[60] K. Ganapathy, S. Das, S. Reddy, V. Thaploo, A. Nazneen, A. Kosuru, U.S. Nag, Digital Health Care in Public Private Partnership Mode, Telemedicine and e-Health, 2021.

CHAPTER

4

Fragmented global standards development organisations

Evelyn Hovenga[a,b,c], *Heather Grain*[b,c,d], *and Thomas Beale*[e]

[a]Digital Health, Australian Catholic University, Melbourne, VIC, Australia [b]eHealth Education Pty Ltd, Melbourne, VIC, Australia [c]Global eHealth Collaborative (GeHCo), Melbourne, VIC, Australia [d]LG Informatics Pty Ltd, Melbourne, VIC, Australia [e]ARS Semantica, London, United Kingdom

Why do we need standards?

Standards development organisations emerged in concert with the industrial revolution during the 1800s. The adoption of standards, such as standard sized screws, enabled continuous and streamlined manufacturing that resulted in significant associated economic benefits for manufacturers. These same principles apply to the current digital revolution. Standards serve as a basis of weight, measure, value, comparison, or judgment of recognised excellence from an established authority. Their adoption supports international trade.

Meeting the need for system connectivity and interoperability, to benefit both software developers and users, is dependent upon the adoption of standards. The use of standards within any health ecosystem needs to enable us to enter data once only and to enable its use many times along the data supply chain. This starts at the most appropriate point of data collection (source) and then progresses to these data being archived or discarded. Effective dataflows streamline information processing capacity and facilitate maximum automation while meeting the needs of multiple use cases within any nation's health system.

Some generic ICT standards are of relevance to digital health, and others need to be able to meet unique health industry requirements. Such standards need to be developed along with the emergence of digital health transformational activities. Standards are something considered by an authority or by general consent as a basis of comparison, as an approved model. They define rules and processes to aid in tasks for products and processes that people use, including rules about how products (such as software) are made and how quality control tests should be performed. Standards are intended to identify structures, methods, or things. They represent the level of quality or attainment used to ensure safety or efficiency. In many cases, these standards are accepted as normal or average such as a standard rate of income tax, but in other cases, they can be used to indicate best practice or directions to improve outcomes.

https://doi.org/10.1016/B978-0-12-823413-6.00025-2

This chapter identifies the many different standards development organisations providing this leadership. Their functions and focus along with known limitations are then explored. The intent is to make use of these findings towards the development of a roadmap for collaborative digital health standards developments to enable any national authority to select a key set of standards to be mandated and included in digital health product procurements, to best meet digital health ecosystem needs as described previously.

Standards applicable to digital health

Standards associated in some way with a digital health ecosystem may be clinical, regulatory, or technical. Their development processes vary as do their adoption, implementation, and benefits realised. The pace of new technology developments, relative to the development of standards, differs greatly. This has the potential to render newly developed standards obsolete or in need of review and an update fairly frequently. In addition, the timing of system acquisition, as well as the level of system maturity when acquired, makes the choice of necessary standards compliance needed for system connectivity and interoperability challenging and complex. The level of system maturity differs greatly between systems adopted within one or multiple connected organisations.

Another complicating factor is the need for systems to meet national regulatory requirements which may not have been considered during the standards development process. Every jurisdiction adopts its own standardisation approach impinging on the efficacy of development processes and standards adoption policies [1]. Individual stakeholders adopt variable, yet rapid digital transition processes, adding to connectivity complexities in terms of technical interoperability, economic efficiency, and public accountability regarding decisions made in the public interest.

There is no shared opinion on the role and need for standards in healthcare. There is an acknowledged requirement for the adoption of data standards associated with routine statistical reporting and to collect patient data that are sharable. Despite these requirements, the only solution practically considered has been about how to take data represented in one system, then using that systems data elements and structures, and then enable the conversion of that same data to the form required for information sharing and processing by another system. Such processes make use of a health data exchange protocol. Alternative data sharing approaches exist; their use is now gaining momentum.

Standards and interoperability: A common challenge

One common illusion is that we only need one standard to solve the interoperability issue. This appears to be the result of a prevalent knowledge gap regarding the relationship between data and the technologies in use for data processing. The following analogy regarding the use of email demonstrates this.

Have you ever tried to find information in an email you received some time ago? Difficult isn't it! This is because, though the message works well (the email itself is sent and received quickly and accurately), there is a limited structure to the message and everyone files their email and manages tasks generated by their email in different ways. Messages represent data exchange mechanisms, they contain data.

What needs to be appreciated is that:

- Source data (data collection at any point along the data supply chain) are the beginning of digital health—not data sharing
- If source data were consistently structured (though potentially at different levels of specificity) and consistently represented, then:
  - Maps matching terms and concepts from one system, with those from another system aren't needed, thus reducing costs and improving the quality of data throughout all systems.
  - System costs can then focus on functionality rather than design of structure and concept representation as:
    - this reduces duplication of effort for all vendors,
    - this enables functional innovation, and
    - this frees vendors and users to plug and play.

Some groups within the health informatics standards environment continue to focus on messaging as THE solution. This occurs largely through seeking a simple solution from the perspective of the way the organisation has solved solutions in the past. This is no longer fit for purpose to meet the needs of the current digital health 'industry 4.0' revolution promoting smart and autonomous systems fuelled by data and machine learning.

Semantic interoperability is usually understood as the ability of systems to share meaning or similar. Syntactic interoperability just means that the systems agree on how to share data or API calls concretely. For healthcare IT, and indeed most industries, the only interoperability of interest is the semantic kind; if you have not achieved that, then there is more work to do. The following definition of 'interoperability' means 'semantic interoperability'.

> Interoperability refers to the level of interoperability between distinct components of an information processing environment (applications, services, systems, etc.) that is proportional to their ability to correctly communicate their internal semantics to each other, without special measures, other than syntax or technology adaptation [2].

Digital healthcare requires a package of standards and innovation across the whole data continuum. This starts with the data, its specification, representation, and structure within the context of numerous legacy systems. Standards have evolved along with technical advances made since the beginning of the computing era when there were no mandates for standards compliance. Standards adoption was driven by market demands.

Foundational ICT standards

For those not familiar with these technical complexities, it is useful to reflect on the foundational computer connectivity protocols that enabled distributed systems to be connected when the only way for health data exchange was via the application layer (syntactic interoperability) as shown in Fig. 1.

Further historical computing innovative milestones influencing additional usage possibilities over time are shown in Fig. 2. This makes it clear that the configurations used for system development were influenced by the availability of these technological advances at the time systems were being designed.

ISO seven-layer communications model for Open Systems Interconnection (OSI) - 1979		Protocols & Services	Transmission Control Protocol (TCP) and Internet protocol (IP) Network Standards – 1974 Stanford University
Layer 7	Application	HTTP, FTTP, Telnet, NTP, DHCP, PING	Application
Layer 6	Presentation		
Layer 5	Session		
Layer 4	Transport	TCP, UDP	Transport
Layer 3	Network	IP, ARP, ICMP, IGMP	Internet
Layer 2	Datalink	Ethernet	Network Interface
Layer 1	Physical		

FIG. 1 Foundational computing connectivity protocols.

Computing Technology Innovations	Internet layers	Computing Technology Innovations
Z3-1941 Konrad Zuse	COMPUTERS	1976 Apple - Jobs & Wozniak
SAGE Air Defense - George Valley -1956	NETWORKS	1973 Ethernet - Metcalf & Boggs
ARPANET-Kleinrock, Baran, Davies & Roberts - 1969	INTERNET	1974 –TCP/IP Cerf and Kahn
www - Tim Berners-Lee 1990	WORLD WIDE WEB	1990 http://Tim Berners-Lee
Hypertext -Neil Larson -1986	BROWSERS	1993 Mosaic Marc Andreessen (NCSA)
Internet Archive- Brewster Kahle - 1996	SEARCH ENGINE	1998 Google - Brin & Page
Email- Ray Tomlinson- 1971	CONTENT	1987 Hypercard - Bill Atkinson

FIG. 2 Historical computing innovations influencing digital systems' maturity.

System design or architectures define the foundations upon which systems are further developed and the potential for data integration and system connectivity. This principle is no different from the foundations upon which buildings are constructed. Standards for such physical structures have also evolved to ensure the foundations are well suited to avoid rising damp, for example, and withstand environmental changes for the life of the building. In the health industry, the continuing use of legacy systems has resulted in unconnected and fragmented data silos and system incompatibility through the use of different standards, interfaces, and semantics [3].

Fig. 2 demonstrates that there were periods when new significant innovations became available within short periods of time. This made it impossible for systems being developed during this period to be up to date. The new innovations shown in Fig. 2 became internationally adopted industry standards. Any chosen combination of standards and guidelines makes up the local digital platform in use. Such a digital platform needs to ensure it is able to provide evidence that it meets all (gold standard) technical, clinical, usability, and cost requirements associated with:

- patient, electrical, and physical safety;
- information exchange and cyber security;
- the minimisation of implementation, maintenance, and replacement costs in a manner that supports sustainability;
- the optimisation of data sharing, data linkage, knowledge management;
- streamlining information sharing and automation;
- automated reporting and performance monitoring; and
- the accessibility of universally supported health service provision.

The timing of foundational system design and acquisitions crucially determined investments. We continue to live with some of these legacies. This understanding relates to a number of major legacy EMR systems whose foundational architectures were designed many years ago. Large investments were made; consequently, many such systems continue to be in use. Many adjustments to these systems have been made since, in an effort to maintain commercial advantage. There continue to be many limitations associated with data sharing due to fundamental design issues. That is, the logic layer that makes use of the systems' data is hardwired within proprietary legacy systems, which means optimising system functionality is limited, as customisation is expensive and can't be done in a modular fashion [3].

Literature regarding evidence of benefits following compliance with health informatics standards is scant. This is not surprising, given the associated complexity and numerous confounding variables that need to be considered when undertaken such research. Benefits to be achieved following standards compliance are dependent upon the set of standards adopted by the system to be evaluated. Deciding which set of standards are critical for any proposed information system is not a simple task as there are many standards to choose from. Critical areas for which standards have been developed that need to be considered for selection are as follows:

- Terminology and semantics,
- Communication protocols standards,
- Standards for output information (reporting),
- Standards for source systems information,
- Architecture (system/network design) standards,
- Security/safety standards (processes), and
- Information technology platform(s).

Standards development organisations (SDOs)

There are many diverse standards development organisations (SDOs), and they differ in terms of contributor motivation and their ability to balance opinions with decision-making. Many lack the sustainability in available expertise and engagement. There is some competition, and there are limited instances of coordination. A Joint Initiative Council (JIC) has been established to address this issue, but not all organisations are actively collaborating. Input from all its relevant stakeholders has not progressed to require coordination with other related standards bodies. Such coordination occurs at an informal level through individuals operating as formal or informal liaisons between organisations, but this process is very haphazard. At the simplest level, if an SDO wants to use an ISO (International Organisation for Standardisation) standard, they must purchase that standard. This is true even when members are part of the JIC.

Many SDOs are of the view that only their standards are needed or that they are the best. Standards development organisations produce a range of standards which suffer from problems including multiple overlapping or conflicting standards. In general, SDOs arrive at standards through stakeholder engagement to seek agreement on the best solution to an issue. Most SDOs, such as ISO [4] and HL7 (Health Level Seven) [5], require ballots with input from a balanced stakeholder group. The ballots must identify agreement within the community. Not all SDOs function in this way.

Who are these SDOs, what was their historical and is their current purpose, and how are they contributing to digital health? The following SDOs were selected based on their relevance to digital health and their international reach. Fig. 3 broadly indicates the scope of activity of some of the key digital health standards bodies. This diagram does not show all healthcare activities nor does it include all organisations, but it does show key organisations in specific areas.

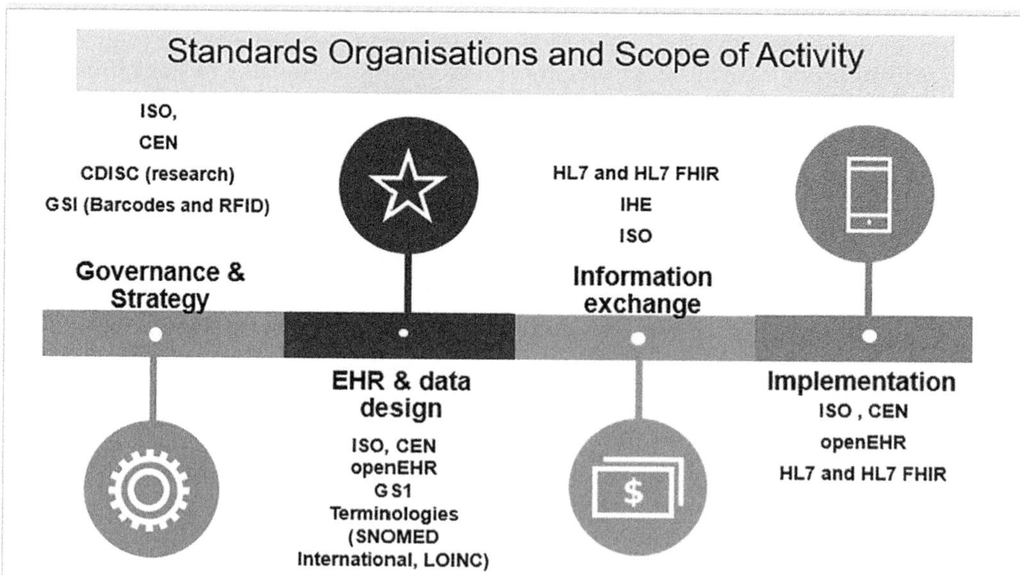

FIG. 3 Digital health standards organisations and the scope of activity.

International Electrotechnical Commission (IEC)

The IEC, founded in 1906, is the world's leading international standards development organisation for all electrical, electronic, and related technologies [6]. It works closely with both ISO and the ITU to ensure that international standards fit together seamlessly and complement each other. Joint committees ensure that all relevant expert knowledge is shared. Millions of electronic devices that use or produce electricity rely on IEC International Standards and Conformity Assessment Systems to perform, fit, and work safely together. All stakeholders are brought together through each country's member national committee to serve on anyone of many technical committees. The IEC's standards serve world markets and society generally. The IEC promotes world trade and economic growth by encouraging the development of products, systems, and services that are safe, efficient, and environmentally friendly.

International Telecommunication Union (ITU)

The ITU is the United Nations specialised agency for information and communication technologies (ICTs) founded in 1865 to promote cooperation among international telegraphy networks of the day [7]. Now the ITU facilitates international connectivity in communication networks. The ITU allocates global radio spectrum and satellite orbits and develops technical standards to enable networks and technologies to seamlessly interconnect and improve access to ICTs to underserved communities worldwide. Its membership includes 193 member states plus 900 companies, universities, and international and national organisations.

The ITU began looking into e-Health standardisation in 2003. This was given greater impetus following the 2012 World Telecommunication Standardisation Assembly when Resolution 78—'Information and communication technology applications and standards for improved access to e-health' was adopted, following the result of the preceding Global Standards Symposium where participants expressed the need for the ITU to take a leading role in e-health by collaborating with the WHO. The aim was to create a global, open environment for the development and promotion of e-health standards. In particular, for the adoption and success of e-health and telemedicine applications, where the conformance with standards demand requires an increase in system interoperability, privacy and security to be safeguarded, and the reduction in device costs through economies of scale.

The ITU-T committee's work includes the development of multimedia systems to support widespread deployment of e-health applications, artificial intelligence for health, and safe listening devices. It was noted that in addition to the need to provide solutions regarding technical matters, discussions around e-health also enfold a variety of other fields such as legal, ethical, cultural, economic, and regional. The ITU-T is of the view that its consensus-driven processes provide the right environment to harmonise and coordinate the development of a set of open global standards for e-health applications. The ITU/WHO developed a toolkit for assisting governments, industry, and civil society to foster faster standards adoption in 2012 [8]. This technical committee also works with other SDOs on an as-required basis.

GSMA (GSM Association) Mobile for Development Foundation Inc.

Groupe Speciale Mobile (GSM) began in 1982 to design a pan-European mobile technology [9]. By 1987, basic parameters of the GSM standard were agreed. This became the internationally accepted digital cellular telephony standards in 1989. The first ever SMS message was sent in 1992. By 1997, 100 countries were 'on air' globally, and by 2003, the GSMA membership came from more than 200 countries. GSMA Intelligence was launched in 2013 along with mobile money services across 84 countries. GSMA represents the interests of mobile operators worldwide in our mobile ecosystem, including handset and device makers, software companies, equipment providers and internet companies, and many other organisations. GSMA works closely with the European Telecommunications Standards Institute (ETSI).

In 2020, GSMA and O-RAN (Open Radio Access Network) announced their alliance to collaborate on opening up 5G networks. O-RAN's worldwide community operates in the radio access network industry. Their media release notes that 'Collaboration aims to accelerate industry-wide consensus on the adoption of open, interoperable interfaces and radio access network virtualisation'. GSMA has also signed an agreement with the Global UTM (Unmanned Traffic Management) Association (GUTMA) enabling the mobile and aviation industries to explore how best to work together and navigate the SDO industry.

GSMA's history demonstrates the value of global collaboration, as this is enabling the acceleration of the adoption RAN products and solutions taking advantage of new open virtualised architectures, software, and hardware, benefiting everyone around the globe, including anyone making use of or contributing to the adoption of digital health.

European Telecommunications Standards Institute (ETSI)

ETSI was created in 1988 following a signed MoU with GSM. It was officially recognised as a European SDO in 1993. ETSI has members and partners from 65 countries. It collaborates with many other global SDOs to create technical standards intended for global use [10].

ETSI is the founding partners of:

1. the Global Standards Collaboration (GSC) established in 1990, to enable many SDOs from around the globe to collaborate,
2. the Third Generation Partnership Project (3GPP), launched in 1998, developing specifications for advanced mobile communications including 5G, and
3. the oneM2M (Machine to Machine) project, launched in 2012. The M2M project is developing specifications for a common M2M service layer, that is the most efficient deployment of machine-to-machine communications systems that can be embedded within hardware and software to connect the wide range of devices worldwide with M2M application servers. These specifications impact diverse business domains including healthcare.

By 2012, ETSI had published more than 30,000 standards, when it launched the Cloud Standards Coordination initiative. Its final report for the European Commission was published in 2013 [11]. A special report on Cloud Standards Coordination Phase 2 was published in 2016 [12]. By 2018, more than 40,000 standards had been published, and in 2019, the first release of 5G was ready for implementation.

As major economic, political, and social changes are shaping new global markets, ETSI plus several other European SDOs have partnered to establish offices in China and India with the intent to increase cooperation regarding standardisation matters. ETSI has a Centre for Testing and Interoperability (CTI) through which a support service for its technical committees is provided to validate standards, test specifications, and document complex interaction using modelling techniques, tools, and specialised specification and testing languages. The CTI promotes best practices to achieve interoperability, as explained in their publication [13].

Institute of Electrical and Electronics Engineers (IEEE) Standards Association (SA)

IEEE began soon after electricity was invented around the 1880s. It is the world's largest technical professional organisation for the advancement of electrical, electronics, and computing technologies. Its members include computer scientists, software developers, information technology professionals, physicists, medical doctors, and many others in addition to IEEE's electrical and electronics engineering core, from more than 160 countries. IEEE SA is an international SDO, a member of both ANSI and ISO actively promoting innovation to enable the creation and expansion of international markets and to help protect health and public safety. Their work drives the functionality, capabilities, and interoperability of a wide range of products and services that transform the way people live, work, and communicate [14].

United Nations Economic Commission for Europe (UNECE)

UNECE was set up as one of five regional commissions of the United Nations. One of its sustainable development roles is about simplifying and standardising business processes and fostering the environmental and social sustainability of international trade [15]. This includes its development and maintenance of over 480 standards and recommendations on trade facilitation and electronic business, including the United Nations Rules for Electronic Data Interchange for Administration, Commerce and Transport (UN/EDIFACT). The adoption and use of these standards cuts trading costs, reduces bureaucracy, and implements the WTO Trade Facilitation Agreement. The UN/EDIFACT standard created in 1987 predates the HL7 messaging standards and is widely used in international supply chains, transport, and logistics. It includes an interactive Electronic Data Interchange (i-EDI) protocol.

International Organisation of Standardisation (ISO)

The British Standards Institution (BSI), as the world's first standards body (founded in 1901), is a founding member of ISO, an independent, nongovernmental international organisation that began in 1946. It has a current membership of 164 national standards bodies. Its primary aim of international standardisation is to facilitate the exchange of products, goods, and services through the elimination of barriers to trade. ISO has 323 technical committees (TCs) each with a number of working groups. Many of these TCs are relevant to the health and ICT industries, for example the Healthcare Organisation Management TC304 or Robotics

TC299. The World Standards Cooperation (WSC) was formed in 2001 to strengthen standards systems of its three SDOs: ISO, IEC, and ITU. ISO works with United Nations partners and specialised agencies, such as the UN Economic and Social Council (ECOSOC).

There is a Joint Technical Committee (JTC) between ISO and IEC for information technology standards developments. This JIC works not only with the IEC but also with ITU-T and IEEE in response to the necessary convergence of information technology, consumer electronics, and telecommunications required for this digital age. These three areas are now collectively referred to as ICT. From a standards perspective, this includes the specification, design and development, integration and interoperability of systems, services, tools, and applications. These concepts deal with the capture, representation, processing, security, transfer, interchange, presentation, management, organisation, storage and retrieval of information, and their related accessibility, cultural, linguistic adaptability, and societal aspects [16]. Costs associated with standards development projects are born by its members and business organisations who pay for their contributing experts' travel and time associated with each project.

From a digital health roadmap perspective, there is a need to identify and consider the foundational principles associated with all ICT standards development areas relative to their application for the health industry creating further complexities. The involvement of any number of technologies for this vertical convergence demands a closer cooperation between technology-oriented, service-oriented, and application-oriented experts, both in product and standards development. The business environment of the health industry needs to consider political, economic, technical, regulatory, legal, and social dynamics as these relate to its services, products, materials, disciplines, and practice. These were the drivers for the establishment of a Health Informatics Technical Committee (ISO TC215) in 1998. Each of these areas significantly influences the conduct of standards development processes and the content of resulting standards. Standards development processes are covered in greater detail in another chapter.

Organisation for the Advancement of Structured Information Standards (OASIS)

OASIS is a not-for-profit consortium of vendors and users that began in 1993 under the SGML –open name (Standard Generalised Markup Language) [17]. Its technical work scope has expanded. The adoption of its current name in 1998 reflected this development. It now has a very broad standards development agenda that encompasses cybersecurity, blockchain, privacy, cryptography, cloud computing, IoT, urban mobility, emergency management, and content technologies. Many of its open standards have subsequently been published by ISO, IEC, or ITU SDOs.

Its work is undertaken by any one of its numerous committees covering a long list of work categories, including three technical committees with a healthcare focus. One has a focus on enabling the interoperable exchange of healthcare privacy policies, consent directives, and authorisations, two others on emergency management adoption, and the other on enabling information exchange to advance incident preparedness and response to emergency situations. OASIS has a formal liaison relationship with ANSI (US), IEC, INCITS, ISO, the US Intelligence Agencies InfoSharing project (PM-ISE) plus numerous informal and other liaisons including an MoU with HL7, W3C, SNIA, ETSI, and the China Electronics Standardisation Institute.

A summary of these SDOs' areas of focus, relationship with other SDOs, and their values is presented in Table 1.

TABLE 1 Standards development organisations.

Global entity	Area of focus	Relationships	Values/beneficiaries
IEC	Electrical, electronics, and related technologies	ISO, ITU	Promote world trade and economic growth
IEEE	Electrical, electronics, and computing technologies l	ANSI, ISO	• Promote innovation supporting international markets, protect health and public safety • Drive functionality, capability, and interoperability
ITU	• Information and communication technologies • Global radio spectrum • Satellite orbits • Open global e-Health application standards • Multimedia systems	IEC, WHO Other SDOs as required	• International communication networks • Improve ICT access to underserved global communities • Support widespread deployment of e-health applications, AI for health, safe listening devices
GSMA	• Mobile technologies and ecosystem • Digital cellular telephony • Mobile money service standards	ETSI, O-RAN, Global UTMA	• Service mobile operators, handset and device manufacturers, software companies, equipment providers, Internet companies, etc. • Mobile & aviation industry
ETSI	• Advanced mobile communication • Machine-to-machine communication for device connectivity • Cloud standards coordination • Centre for testing and interoperability	Many global SDOs—GSC	Promote best practice to achieve interoperability supporting a global ecosystem
UNECE	• Promote pan-European economic integration	ISO/TC215	Maintains the UN/EDIFACT and many other standards promoting countries to achieve their sustainable development goals
ISO	• Facilitate the exchange of products, goods, and services through the elimination of barriers to trade	• World Standards Cooperation (WSC) (IEC, ITU, and ISO) • Un Partners, for example ECOSOC • 700 international and national organizations	• International community • Reducing technical barriers to trade

TABLE 1 Standards development organisations—cont'd

Global entity	Area of focus	Relationships	Values/beneficiaries
OASIS	• Supports open source and open standards, transparency • Member-driven technical committees • Provision of corporate and legal infrastructure and fiscal agency services enabling groups to run their own foundations • ISO Technical Advisory Group working with ANSI	• Many formal and informal liaisons. • MoU with HL7, W3C, SNIA, ETSI, and the China Electronics Standardisation Institute • Major multinational companies, SMEs, government agencies, NGOs • Universities, research institutions	• Global influence and OASIS standards adoption • Member-driven, new work encouraged from members to suit any of the numerous categories of focus. • Recognition in international policy and procurement • Standards to suit any industry

Health Informatics Standards Development Organisations (SDOs)

Health informatics standards development organisations began in Europe, the USA, Japan, and Australia. This was due to the then recognised growing need for multisystem connectivity and electronic data transfer within the healthcare industry and the proliferation of incompatible solutions [18]. Working groups were established during the early 1990s to work in the following areas:

- Healthcare information modelling and medical records.
- Healthcare vocabulary, terminology, semantics, and knowledge bases/representation.
- Classification and code systems/coding schemes.
- Data development for medical effectiveness research.
- Object-oriented healthcare information model.
- Pharmacopeia and toxic substances.
- Computerised electrocardiography and neurophysiology.
- Healthcare communications, messages, and syntax.
- Telematics relative to nursing, the medical record, case mix, resource management, primary care, education, and training.
- Medical imaging, multimedia, and broadband communication services.
- Communication with medical devices.
- Healthcare security, privacy, quality, and safety.
- Intermittently connected devices.
- Open systems.

SDOs share an evolving history alongside numerous computing and telecommunications networks research and development efforts. Many electronic, computer science, telecommunications developments, and new insights that occurred during the late 1980s and early 1990s greatly influenced this history and associated emerging standards. There was strong collegiality between individual members of SDOs in the USA and in Europe at that time. Now there are many national and international health informatics-related standards development organisations and a number of communities of practice who develop their own industry standards. Some of these are accepted as THE international standard and may in time be endorsed by a formal SDO. As part of the transformation to a new digital age, we are witnessing increasing collaborative efforts to ultimately benefit everyone around the globe, including the establishment of a standards profiling organisation. Despite these many historical efforts, we still have a long way to go to enable the realisation of a well-connected digital health ecosystem.

ISO's Health Informatics Technical Committee (TC215)

The Health Informatics Technical Committee 215 (ISO-TC215)'s scope is to 'facilitate the capture, interchange and use of health related data, information, and knowledge to support and enable all aspects of the health system' [19]. A health system relates to the organisation of people, institutions, and resources that deliver healthcare services to meet health needs of target populations.

This committee has 29 participating country members and 33 observing members. There are a number of working groups and task forces, each with their own specific scope, as well as a Genomics Informatics subcommittee. It is only in the last 6 years that ISO TC 215 Working Group 3 developed a framework for health informatics standards for semantic content. This framework is used to periodically review existing standards and identify gaps and prioritise the standards needed.

The ISO TC215 Health Informatics Committee has established an ad hoc group to undertake a similar process on information governance. The result of which is expected to be the identification of existing standards which should be used in specific situations and of gaps where standards are needed. This should be based on a shared vision and objectives to meet the needs of a complex adaptive ecosystem.

In addition to these ISO TC215 working groups, there are a number of joint working groups with members from other ISO technical committees. Its work is guided by the e-health reference architecture as defined in ISO TR 14639—Part 2 [20]. It aims to address the needs of all its diverse stakeholders. This TC's work is influenced by numerous developments of relevance to the health industry and digital health. It has significant liaisons with:

- WHO, the World Health Organisation;
- IEEE, in relation to the joint ISO/IEEE health device interface standards;
- ICN, International Council of Nurses;
- IEC/TC 62, electrical equipment in medical practice;
- IEC/SC 62A, common aspects of electrical equipment used in medical practice;
- IMIA, International Medical Informatics Association;
- ITU-T/SG 17; Security
- UN-ECE, UN—Economic Commission for Europe—EDIFACT, etc.;
- Regenstrief Institute;
- PCHA, Personal Connected Healthcare Alliance;
- ICH, International Council for Harmonisation of Technical Requirements for Pharmaceuticals for Human Use;
- EFMPA, European Federation of Pharmaceutical Manufacturers & Associations;
- HON, Health On the Net Foundation.

An informal Joint Initiative Council in 2007 was formed as a formal liaison group under ISO TC215 in 2015 to enable common, timely health informatics standards by addressing and resolving issues of gaps, overlaps, and counterproductive standardisation efforts [21].

Joint Initiative Council for Global Health Informatics Standardisation (JIC)

This joint strategy is being undertaken collaboratively between ISO/TC215, CEN/TC251, HL7, CDISC, SNOMED International, GS1, IHE, and DICOM, and their descriptions follow. This has resulted in an integrated work program with a focus on resolving overlapping or counteracting standards within the participating SDOs existing work programs. There is now a greater liaison between SDOs [22] although each SDO is working within its own vision concerning required standards, rather than a common view of global needs or a well-considered distribution of the standards development activities required to collectively meet global needs.

European Committee for Standardisation (CEN) Technical Committee (TC) 251

The CEN Technical Committee 251 officially began in 1990. Despite its European SDO status, this committee now prefers to cooperate with other SDOs at a global level to produce specifications and standards within the context of ensuring these meet European needs as identified when working closely with EU projects [23]. ISO/TC215 has adopted many CEN/TC251 standards. The 1991 Vienna agreement process is used to ensure that, for each subject area, only one standard is approved by ISO and CEN, and that this standard is adopted worldwide [24]. This committee's business plan 2019–2022 notes that:

> There is a bewildering number of initiatives, many overlapping, some diverging and others unaware of the other, leading to a confused, unstable and unproductive market-place for health informatics applications.

Its scope is the development of standards 'in the field of Health Information and Communications Technology (ICT) to achieve compatibility and interoperability between independent systems and to enable modularity. This includes requirements on health information structure to support clinical and administrative procedures, technical methods to support interoperable systems, as well as requirements regarding safety, security, and quality. Europe's new goal is a data-driven healthcare system'. It has two working groups, one with a focus on safe and secure use of information within the e-Health domain, as well as its governance and management, and the other on achieving compatibility and interoperability between independent systems.

HL7 International

HL7 International [5], a not-for-profit organisation, was founded in 1987 to address the need for standard nonproprietary structured data interchange protocols, to enable systems from multiple software vendors to connect. HL7 is a transaction-based standard that only addresses the application layer of the open systems interconnection (OSI) seven-layer model for data networks of the ISO standard. It is a major supplier of a 'family' of health data exchange standards, consisting of a suite of specifications based on HL7's Reference Information Model (RIM), its Clinical Document Architecture (CDA) series of standards, HL7 FHIR profile standard, and the Arden Syntax.

Its HL7 v2 standards were the most successful and are used widely. CDA is a document markup standard that specifies the structure and semantics of clinical documents for exchange purposes; HL7 FHIR is an interoperability standard intended to facilitate the exchange of healthcare information between organisations discussed in more detail in another chapter with a focus on health data standards. The Arden Syntax is a formalism for representing procedural clinical knowledge in order to share it. Required characteristics of such formalisms to maximise semantic interoperability are described in the chapter on health data standards to follow.

HL7 does not address database architecture, data standards, or physical architectures. HL7 is now an ANSI (American National Standards Institute)-accredited standards development organisation with the mission to provide standards that empower global health data interoperability. HL7 now has more than 1600 members from 50 countries. Its members represent healthcare providers, government stakeholders, payers, pharmaceutical companies, vendors/ suppliers, and consulting firms. Its standards development process is rigorous and needs to meet ANSI's accreditation requirements. Its standards are created by committees consisting of these members represented by IT and health people with various degrees of technical competence and health domain knowledge. This often results in its committee members talking past each other. There around 54 active working groups (committees), whose work is constant.

Prior to HL7's establishment, the Institute of Electrical and Electronic Engineers (IEEE) was undertaking the P1157 Medical Data Interchange (MEDIX) Standards Project as their attempt to bridge the islands of healthcare information through the development of international standards for healthcare data interchange. This work formed the basis for the HL7 v 3 methodology [25]. Its other early health-related project was IEEE P1073, a standard for medical device communications. Current IEEE initiatives of relevance to a digital health ecosystem include 'wearables and medical IoT (Internet of Things) interoperability and intelligence', 'big data governance and metadata management', e-Health, and a community-powered platform for open source projects. A number of white papers relevant to these initiatives are available from the IEEE website [14].

Object Management Group (OMG) health domain task force

This international not-for-profit technology software standards consortium was founded in 1989 to set standards and specifications in the area of distributed object computing supporting numerous vertical markets including healthcare [26]. Some of these have been ratified as ISO standards. OMG manages an open vendor-neutral, industry-wide standards and specification development process. Its standards are developed using tried and tested engineering or computer science methodologies by highly technically competent developers who may be weak on health domain knowledge.

A number of their standards are well known and widely used, such as CORBA (Common Object Request Broker Architecture) an object-based language—and platform-neutral interoperability standard, and a number of modelling specifications such as UML (Unified Modelling Language), BPMN (Business Process Model and Notation), and OCSMP (systems modelling). OMG members share their expertise to rapidly produce high-quality free and openly published technical standards, to provide a suitable foundation for vigorous, competitive markets to benefit all stakeholders.

In health, there have been three OMG efforts. The first, starting in the late 1990s, was referred to as the CORBAmed Roadmap, which later became the Health Domain Taskforce, or HDTF, and resulted in a collection of standards including PIDS (person identification service), RAD (resource access decision service), HILS (health information location service), COAS (clinical observations access service), and TQS (terminology query service). Although technically coherent offerings, these standards have not survived, partly due to not adequately taking into account the semantic complexity of health and partly because they were simply ahead of their time: the thinking of the health domain was not ready to understand services. A further reason (possibly the main one) was weak visibility in the healthcare domain, which was dominated by HL7, ISO, and CEN.

The second OMG health standardisation effort is recent and was a partnership with HL7. A group of standards under the umbrella of 'HSSP' (health services specification project) was begun in 2006. It resulted in RLUS (record location and update service), EIS (entity information service—a PIDS equivalent), and CTS2 (clinical terminology service), a modern version of TQS. These standards benefit from access to more representative (in health domain terms) input from HL7 members as well as other industry specialists. Only CTS2 experienced uptake; however, this now appears to be superseded by simpler standards from IHTSDO and HL7 FHIR.

The third OMG health standards effort is known under the umbrella term BPM + which seeks to integrate BPMN, case management model and notation (CMMN), and decision model and notation (DMN) and adds further standards relating to shared data representation and artefact bundling. This appears likely to gain real traction in healthcare, a sector that still seeks workable solutions for the problem of computable guidelines and care pathways. BPMN and CMMN offer a means of plan representation; DMN provides decision logic; and the new standards provide a connection between the former elements and data representation real-world state of subjects (e.g. patient heart rate, etc.).

In addition to managing these programs, the OMG publishes white papers. The recently approved information exchange framework (IEF) reference architecture is part of a family of specifications developed for policy-driven, data-centric information sharing and safeguarding (ISS) services. This targets the automation of key policy decision and enforcement points to enable responsible information sharing, a highly desirable function for the health industry.

Digital Imaging and Communications in Medicine (DICOM)

The American College of Radiology (ACR) and the National Electrical Manufacturers Association (NEMA) joined force in 1983 to form a standards committee to meet the combined needs of radiologists, physicist, and equipment vendors. Their first standard was released in 1985 and has been updated and expanded upon ever since in line with other relevant technical developments. They became known as DICOM in 1993 and have continued to grow. This is now *the* international standard for medical images and related information. DICOM is one of the most widely deployed healthcare messaging standards in the world and has revolutionised the practice of radiology enabling the replacement of X-ray film with a fully digital workflow. It has been published as the ISO 12052 standard. This result is a great example of strong clinical leadership.

GS1-Australia

The GS1 SDO began as the not-for-profit Australian Product Numbering Association (APNA) in 1978, changed its name to EAN Australia in 1993 to become a global system with one global standard GS1 in 2005. GS1 standards are designed to assist with the identification, capture, and trading partners information sharing. These unique identifier standards, including barcodes, QR codes, DataMatrix, and radio frequency identification (RFID), are now the global supply chain's language of business. GS1 standards are at the forefront of e-commerce and supply chain management initiatives. Its 21,000 + members span diverse industries including healthcare (7.7%).

Clinical Data Interchange Standards Consortium (CDISC)

The CDISC organisation was founded in 1997 by an experienced medical researcher, Dr. Kush, as an all-volunteer organisation to develop two draft standards and a glossary. It became an incorporated not-for-profit organisation supported by 32 charter member organisations in 2000. The majority of its 350 + organisational members of the CDISC SDO are contract research organisations (CROs) (34%), followed by technology service providers (20%) and pharmaceutical organisations (17%) [27]. CDISC's vision is to inform patient care and safety through higher quality medical research. Its mission is to develop industry-wide data standards to enable the harmonisation of clinical data and to streamline research processes from protocol through analysis and reporting, including the use of electronic health records, to facilitate the collection of high-quality research data. Its aim is to significantly decrease the time and cost of medical research and improve quality.

CDISC has developed a suite of foundational clinical research standards that focus on the core principles for defining data standards, including models, domains, and specifications for data representation. Its Biomedical Research Integrated Domain Group (BRIDG) Model version 3.2, a nonstatic model that bridges clinical research with healthcare, was approved as an ISO standard in 2015. Regulators, including the US FDA, the PMDA in Japan, and the European Medicines Agency (EMA), now require clinical trial applications to be compliant with CDISC standards. These foundational standards have been extended to represent data that pertain to specific disease areas, referred to as therapeutic area standards. These include disease-specific metadata, examples, and guidance on implementing CDISC standards for a variety of uses, including global regulatory submissions. Its data exchange standards are optimised to represent CDISC content and are sufficiently flexible to be used by information systems that are noncompliant with its foundational standards.

A standards profiling organisation

Integrating the Healthcare Enterprise (IHE) international

In 1998, the Radiological Society of North America (RSNA) and the Health Information Management Systems Society (HIMSS) convened a working group of their key members, industry representatives, standards experts, and others to drive the systems integration process. A year later IHE planning and technical committees were established. IHE has continually expanded and developed strong international support resulting in its incorporation as IHE International in 2007. National/Regional initiatives are undertaken by IHE North America (IHE-NA), IHE Europe (IHE-EUR) and IHE Japan (IHE-J), each with their own group of sponsors. Individual or small groups of health professional organisations sponsor problem-solving initiatives associated with their speciality domain.

IHE's aim is to be able to seamlessly pass information from system to system within and across departments and made readily available at the point of care. IHE's work is driven by their recognition that healthcare enterprises have not yet begun to realise the full potential of computer systems to reduce medical errors, improve the efficiency of care providers, and enhance the overall quality of clinical care [28].

IHE promotes the coordinated use of established standards such as DICOM, HL7, and W3C to address specific clinical needs. Its work is sponsored by its members including any number of professional organisations. IHE's initiatives are based on the identification of a common integration problem, from which an integration profile specification is developed by an intensive ongoing process of collaboration and communication among key parties. Such specifications are then implemented and tested to ensure that the desired state of system integration is achieved, known integration barriers are removed, and that the problem is solved. Vendors then publish IHE integration statements to document the integration profiles supported by their products. Users are able to reference integration profiles in requests for proposals to simplify the systems acquisition process. IHE's technical framework delineates standards-based transactions among systems (generically defined as IHE actors) required to support specific workflow and integration capabilities.

A summary of these health informatics SDOs' areas of focus, relationships, values, beneficiaries, and categories, plus this standards' profiling organisation, is presented in Table 2.

TABLE 2 Health informatics standards development organisations.

Global SDO	Area of focus	Relationships	Values/beneficiaries	Category
ISO TC215	e-Health architecture roadmap as described in ISO TR14639	WHO, IEEE, ICN, IEC. TC62, IEC.SC62A, IMIA, ITU-T/SG17, UN-ECE, Regenstrief Institute, PCHA, ICH, EFMPA, HON	• Informed health policy • Improved access to care • Evidence-based practice • Informed health service planning • Efficiency, productivity, cost-effectiveness • Improved quality of care to individuals and populations	Various
JIC	Joint projects making all [joint] standards available through ISO	ISO/TC215, CEN/TC251, HL7, CDISC, SNOMED International, GS1, IHE, DICOM	Collaboration to avoid overlaps and improve harmonisation to benefit the vendor industry	Various

TABLE 2 Health informatics standards development organisations—cont'd

Global SDO	Area of focus	Relationships	Values/beneficiaries	Category
CEN TC251	• All types of data use and reuse • Various types of artefacts • Health ICT applications, mobile health • Data communication • Risk management for IT infrastructure	ISO/TC215 CEN has numerous European partners, affiliates, and companion SDOs	Safe and secure use of information within the e-Health domain Governance and management System compatibility and interoperability	Interop
HL7 International	Exchange, integration, sharing, and retrieval of electronic health information that supports clinical practice and the management, delivery, and evaluation of health services	• Its global membership including vendors, government, and nonprofit organisations • HL7 International Affiliates (Countries)	Enabling interoperability in healthcare Software developers Healthcare organisations	Interop
OMG	Within healthcare: architectural standards for workflow and decision logic • Distributed object computing • Support multiple vertical markets Free and openly published technical standards and modelling specifications	• Global membership across all sectors • Growing interest from large provider orgs (e.g. US VA) and professional orgs (e.g. American College of Emergency Physicians (ACEP)) • Membership cohorts, including vendors, research/academic institutions • ISO, HL7, ITU-T • Healthcare Services Platform Consortium • Cloud Security Alliance (CSA) • CEN/TC287, CCSDS, WinnF • DMTF (Distributed Management Task Force)	• Tool developers • Software developers • Healthcare knowledge engineering organisations (via tool use) • Innovate customer experiences • Optimise resources • Adjust mission/ business needs • Break down silos • Mitigate risk to project budgets • Reduce cost per service	Architectural

TABLE 2 Health informatics standards development organisations—cont'd

Global SDO	Area of focus	Relationships	Values/beneficiaries	Category
DICOM	Facilitates interoperability of medical imaging equipment	SNOMED International HL7, IHE, ISO, IEC, JIRA, JIC, ISO TC215, LOINC BI-RADS terminology	Diagnostic information databases that can be interrogated by a wide variety of devices distributed geographically Medical imaging organisation vendors	Info models
GS1—Australia	Unique identifiers, including barcodes, QR codes, DataMatrix, and radio frequency identification (RFID)	ISO TC215	Global language of business Trade and industry associations	
CDISC	Clinical research data, protocols, data collection, tabulation, datasets, metadata	ISO TC215	Global meaningful clinical research Research experts regulators, FDA, NMPA, PMDA	Info models
IHE	Enable seamless and secure access to health information that is usable whenever and wherever needed	DICOM HL7 International	Improve the way computer systems in healthcare share information Clinicians, health authorities, industry, and health information users	Profiled interop

Communities of practice for digital health

Communities of practice are groups of people who participate and share a concern or passion for something they do or know about. Such activities do result in the development of standards by some of these communities, for example the 'open source' or clinical specialist communities; many are international. They most commonly consist of groups of volunteers, who may contribute financially for group sustainability, including any health professional organisation or research groups. Their regular interactions are about sharing ideas, strategies, determining solutions, or building innovations. This facilitates knowledge sharing among group members.

Successful digital health transformation is dependent upon many domain-specific multidisciplinary communities of practice where its members are practitioners within that domain. For example, the successful DICOM standard, referred to previously, was the result of a multidisciplinary community of practice established from members of two professional organisations, the American College of Radiology (ACR) and the National Electrical Manufacturers Association (NEMA), who agreed to work together to achieve a common goal.

Information Systems Audit and Control Association (ISACA)

In recognition of the criticality of computer system to any organisation's operations, a small group of individuals saw the need for a centralised source of computing information and guidance in 1967, formalised and incorporated as the EDP Auditors Association in 1969. An education foundation was established in 1976 to undertake large-scale research to expand the knowledge and value of IT governance and control. ISACA's community of practice comes from more than 180 countries and cover a variety of professional IT-related positions in the disciplines of IS/IT audit, risk, security, and governance plus educators, consultants, and regulators working in any industry. ISACA is a globally recognised organisation for information governance, control, security, and audit professionals. Its' COBIT governance framework is world-renowned.

Internet Corporation for Assigned Names and Numbers (ICANN)

The first registry of socket numbers for the Arpanet, the precursor to the Internet, became the Internet Assigned Numbers Authority (IANA) responsible for managing the coordination of domain names (DNS) [29]. The US Government transferred this responsibility to a nonprofit corporation based in the United States with global participation in 1998. The first expansion of the DNS system occurred in 2000. IANA's stewardship transition to ICANN was completed in 2016. Its mission is to coordinate policy development related to the Internet's system of unique identifiers while making use of an open and transparent multistakeholder policy development process. ICANN is supported by a number of related organisations and advisory committees.

World Wide Web Consortium (W3C)

The World Wide Web was invented in 1989 by Tim Berners-Lee, a British scientist, to meet the demand for automated information sharing between scientists in universities and institutes around the world, while he was working at CERN. The WWW was about merging evolving technologies of computers, data networks, and hypertext into a powerful and easy to use global information system. The W3 Consortium was established in 1994 by its inventor, at the Massachusetts Institute of Technology, Laboratory for Computer Science in collaboration with CERN, a European atomic physics laboratory [30,31]. An essential requirement was that the web should remain an open standard for all to use. No one should lock it up into a proprietary system. Further developments were, and continue to be, funded by its growing number of member organisations under a Community Contributor License Agreement with royalty-free patent licensing terms and permissive copyright for W3C community and business groups. By 2020, W3C had published numerous standards associated with the use of the WWW to suit multiple purposes.

SNOMED International

SNOMED-CT is the most comprehensive clinical terminology standard in use globally. Its development began in 1965 when the College of American Pathologists (CAP) published the Systematized Nomenclature of Pathology (SNOP) to describe morphology and anatomy. This became the Systematized Nomenclature of Medicine (SNOMED) in 1975, SNOMED Clinical Terms (CT) in 2002, and is in continuing further development to accommodate new content and meet user needs.

In 2007, nine charter national governments established the International Health Terminology Standards Development Organisation (IHTSDO), a not-for-profit organisation governed by its country members and advisory bodies. ITHSDO then purchased all intellectual property rights to all versions of SNOMED from CAP. The purpose of this acquisition and change of ownership was to promote its international adoption and use. IHTSDO took over the responsibility of adopting an unbiased and transparent mechanism to cope with new content, updates, and distribution of this terminology. In 2017, IHTSDO adopted SNOMED International as its trading name.

SNOMED International's decision-making sits within the organisation's structure and employees, with input from their owners (the governments who have purchased membership in the product). Though this is sensible in many ways as a terminology cannot be effectively managed by committee, the priorities of this organisation are governed by the interests of the owners. Though users and others can engage in discussions and opinions are sought—decisions are made centrally.

This can be a positive if well coordinated, but unfortunately this is lacking. The JIC, of which SNOMED International is a partner, was created to assist in scope management, reduce overlaps, and fill gaps. Analysis of the minutes of JIC meetings indicates that work programs are shared but not actively discussed [32]. Where coordination occurs, it is usually through individuals who are members of multiple standards bodies. SNOMED International manages this internationally recognised standard but is yet to be a formally designated SDO. This terminology is discussed in some detail in chapter covering data standards.

Logical Observation Identifiers Names and Codes (LOINC)

This common terminology was developed by the LOINC Committee established by Clem McDonald, an investigator at the Regenstrief Institute, a nonprofit medical research organisation associated with Indiana University [33]. This initiative was a response to address an identified need following an increasing desire to be able to send clinical data electronically from laboratories and other data producers to hospitals and other healthcare providers. Its aim was to develop a common terminology to serve this purpose as pathology systems in use presented their test results according to their own internal, idiosyncratic code values resulting in the receiving system's inability to understand these test results. At that time, and in the majority of instances this is still the case, systems were making use of HL7 message standards.

The Regenstrief LOINC Mapping Assistant (RELMA) is made available for searching the LOINC database and helping map local codes to LOINC codes. LOINC was built as an open standard. Regenstrief collaborates with several SDOs including HL7 International, IEEE, SNOMED International, and the Radiological Society of North America (RDNA) who own and maintain the RadLex, a radiology lexicon. LOINC is a widely used internationally recognised data standard owned and managed by a private organisation yet to be a designated SDO.

The Regenstrief Institute and Partners in Health, a philanthropic organisation, has also developed the Open Medical Record System (OpenMRS) as an open source medical system platform for developing countries [34]. This began in 2004 and continues to be supported by numerous organisations, including international and government aid groups, NGOs, and others. OpenMRS is a software platform and a reference application enabling the design of customised medical record systems. Although valuable for local users, these customised systems are not designed to interoperate. OpenMRS is based on a conceptual database structure based on a concept dictionary, such as diagnosis, tests, procedures, drugs, which are not dependent on specified information requirements to facilitate customisation for different uses. Its data model borrows heavily form the Regenstrief proprietary model. This system is widely used in African countries and elsewhere around the globe. It is supported by a global community of developers, implementers and users working towards a shared and open foundation for managing health information in developing countries.

The openEHR Foundation

The European Union established the Advanced Informatics in Medicine (AIM) initiative in 1988. Between 1990 and 1994, 38 projects were funded via AIM; the Good Electronic Health Record (GEHR) project was one of those [35]. The research consortium that undertook this project delivered the first GEHR object model for the health record. This represented an original, formal approach to electronic health record architecture, based on object modelling methods and founded on a comprehensive and systematic review of patient and clinical professional roles and requirements, across Europe, in relation to records. Another key attribute of the approach taken was empirical and iterative prototyping of an evolving architecture that emphasised implementation and practical testing of concepts at each stage. A project team established under the CEN TC251 made use of these deliverables to propose a prestandard health record architecture. This project's results were openly published within the public domain with IP remaining with the consortium based at the Centre for Health Informatics and Multi-Professional Education (CHIME), University College London (UCL).

Deliverables from a follow-up project resulted in the implementation of several successful pilot record servers across Europe. Two Australian consortium members, Dr. Sam Heard and Thomas Beale, also adopted a two-level modelling approach consisting of the record architecture itself and the clinical models or standards required for automatic processing of information which became known as the GEHR archetype system. These two similar research results were progressively harmonised within a common openEHR architecture. A 6-year trial by implementation was then undertaken by numerous consortium members in Europe and Australia.

This project's research team came together again some 10 years later and decided to establish an open source foundation, named openEHR, to continue to build on learning gained from their collective research, development, and practical experiences, adopting an empirically based evolution of rigorously defined software and systems, organised around the two-level strategy of the UCL object dictionary and the Australian archetype methodology. This developed into an international, online growing community of practice whose members share their work with this community. Its aim is to promote and facilitate progress towards electronic healthcare records of high quality, to support the needs of patients and clinicians everywhere. All publications are in the public domain.

The openEHR Foundation was formally established at UCL in 2003 as an asset-locked and not-for-profit company to engage with industry, national health IT programmes, and international standards bodies. This facilitated the program's second developmental phase during which time its international committee and further developments continued to grow. It has applied for JIC membership as its standards are increasingly being implemented internationally. The openEHR foundation is yet to become a designated SDO.

In 2013, a self-governance framework was adopted to guide adopters of the openEHR methodology. Its intellectual property is owned and protected by the openEHR Foundation. A separate trading entity was established in 2015 and openEHR transitioned to a new governance structure of openEHR as a Community Interest Company (CIC) in 2018. Access to the openEHR specifications, clinical models, mailing lists, wiki, and other IP is free to all. Paid individual, professional membership plus organisational and industry partnerships provide financial support for its ongoing activities. Financial member/partners participate in the openEHR's formal processes.

Its Clinical Knowledge Manager (CKM), first established by Dr. Sebastian Garde in 2004 at the Health Informatics Research Centre of Central Queensland University, is now the most advanced library of open source, atomic clinical content models in the world. These formalisms are explained in some detail in the chapter on health data standards.

European Connected Health Alliance (ECHAlliance)

In 2017, following the birth of the Digital Health Society in Estonia, a vision paper titled 'Free Movement of Data', making reference to this Fifth Freedom of the European Union, was published [36]. Under the Estonian EU Presidency, a call for action was launched to bring together all stakeholders of the digital ecosystem to sit around the table to drive this agenda. The Digital Health Observatory (DHO) and the Digital Health Society (DHS) movements provided a thought leadership platform to generate innovative ideas, exchange knowledge, experiences, best practices, success stories, and design a route for large-scale deployment. The European Commission has adopted a digital single market strategy. European task forces are addressing key challenges including the development of a convergence roadmap based on interoperability standards [37]. These actions have led to the establishment of the European Connected Health Alliance (ECHAlliance), a not-for-profit organisation [38].

This ECHAlliance was established as a global connector for digital health to facilitate multistakeholder connections around ecosystems. It aims to drive sustainable change and disruption in the delivery of health and social care. This global community works closely with the European Commission, has many working groups, conducts hundreds of ecosystem meetings every year, manages international events, and makes use of their online platform. Anyone can become a member of the ECHAlliance, via one of a number of different types of membership categories, enabling access to tools, resources, and facilities needed to boost your network, to drive improvements in health and social care. This work has generated more than 40 regionally focused, permanent, multistakeholder partnerships. They aim to break down silos, transform healthcare delivery, and create economic growth. These ecosystem partnerships do this by:

- Working together to implement innovative solutions that improve the quality of health and well-being of citizens within their specific region, and
- Transforming the effectiveness of the healthcare system, the scope for wealth creation and business opportunities.

The European Union (EU) has developed an innovative technological platform to enable the reuse of EHR data for clinical research via a 4-year project EHR4CR (Electronic Health Records for Clinical Research) [39–41]. This project highlights the need to adopt one common platform and to specify legislative and standards requirements that comply with (match) the common platform requirements to be successful. There are many EU examples of extensive collaborative unification between European experts in different fields (clinicians, biologists, IT specialists, ethical, and legal), a large number of organisations and countries, with impressive results.

Another major European Union initiative was the Horizon 2020 Research and Innovation Programme funding a number of different projects between 2014 and 2020 [42]. The Horizon 2020 programme was initiated to drive economic growth and create jobs while ensuring the production of world-class science and to remove barriers to innovation. The results of and experiences gained from some of their projects are of great relevance to the establishment of a European digital health ecosystem. For example:

- The EPIC EU-Pacific Partnerships for ICT [43],
- CrowdHEALTH which is managed by an industry consortium,
- MIDAS—Meaningful Integration of Data, Analytics and Services,
- C3—Cloud A Federated Collaborative Care Cure Cloud Architecture for addressing the needs of multimorbidity and managing poly-pharmacy,
- CPaaS.io City Platform as a service—integrated and open,
- POWER2DM—Predictive model-based decision support for diabetes patient empowerment,
- mHealth4Afrika—Community-based ICT for maternal healthcare in Africa,
- Q4HEALTH—Quality of service and prioritisation for emergency services in the LTE RAN stack,
- TO-REACH—Transfer of Organisational innovations for Resilient, Effective, equitable, Accessible, sustainable and Comprehensive Health Services and Systems,
- DIA-VIT, a noninvasive glucose self-monitoring device for diabetes patients,
- GUIDES—GUideline Implementation with DEcision Support systems,
- EU*US—e-Health works to improve global workforce development

The CrowdHEALTH international research listed above intends to integrate high volumes of health-related heterogeneous data from multiple sources to support policymaking decisions by delivering a secure ICT platform. They have produced numerous deliverables which are listed at their website and a number of publications [44–48].

INTEROPen UK

A group of leading organisations and individuals created INTEROPen in the United Kingdom in 2016 as an action group to accelerate the development of open standards for interoperability in the health and social care sector [49]. It has a strong focus on the development of HL7 FHIR profiles to meet local requirements.

A summary of the standards developed by these communities of practice, their area of focus, relationships, and values is presented in Table 3. Some will argue that those organisations owning and managing the SNOMED-CT, LOINC, and openEHR standards need to be formally recognised as SDOs.

TABLE 3 Communities of practice standards.

Global CoP	Area of focus	Relationships	Values/beneficiaries	Category
ISACA	IS/IT audit, risk, security, and governance	EDP Auditors, Global Chapters & Members Training Partners	Inspire confidence enabling innovation through technology IS/IT professionals and enterprises	
ICANN	Internet domain name and addressing system	Global Internet community	One world, one Internet	Infrastructure
W3C	Open standards ensuring long-term web growth	Numerous liaisons listed at https://www.w3.org/2001/11/StdLiaison.html Community of practice	Open web platform Web for all Web on everything Implementation on a Royalty free basis	Infrastructure
SNOMED International	Global standards for health terms	ISO TC215 National Governments	Improving health of humankind	Terminology
LOINC	Identification of health measurements, observations, and documents Open data standards	JIC, ISO TC215, CDISC, CEN TC251, IHE, GS1, HL7, SNOMED International	Optimal health catalysed by a seamless network of health information systems	Terminology
openEHR Foundation	• Life-long interoperable electronic health records (EHRs) • Computable health data • Trusted privacy, accountable information sharing balance	Community of practice Industry partners Organisation partners Clinical Modellers (CKM) Professional members	Open, vendor neutral platform for EHRs and interoperable clinical and research data Technical rigour, clinical engagement and trust	Architectural, content
ECHAlliance	• Facilitating multistakeholder digital health connections around ecosystems	Global network of digital health alliances Multistakeholders partnerships	Innovative solutions improving quality of citizen health and well-being, healthcare system effectiveness, wealth creation, and business opportunities	

Discussion

This exploration of standards development activity by SDOs and communities of practice relevant for the transformation for this digital era has revealed that despite a number of positive efforts to collaborate, many of these activities continue to be undertaken in isolation and in the absence of an agreed global vision to best meet digital health ecosystem standards requirements globally. A number of overlaps and continuing fragmentation have been exposed.

The common national desires, to realise the benefits of person-centred healthcare supported by a fully integrated system, have increased individual and country attendance at SDO meetings and teleconferences. Despite this fairly new development, health informatics standards developers are very aware of an overall lack of stakeholder support, from governments, industry, and the many health disciplines who need to be at the table to contribute to resolving the issues identified.

Standards on good ideas are of great interest but less likely to provide a usable, implementable tool and belong in the realm of academia rather than be part of the standards development process. Some standards are useful and academically valuable, by investigating problems and potential solutions, but that doesn't make them of demonstrable value to those who need to implement systems. These documents do not give conformance requirements that help you choose your software and ensure your systems are efficient or safe. They require too much effort to work out what they are for, or how to use them, etc.

Health informatics standards need to be developed in a coordinated fashion, using tried and tested engineering, computer science, ontology, and linguistic methodologies where appropriate by those with the necessary expertise.

Grand View Research valued the global digital health market at $US95.8 billion with an expected compound growth rate of 27.7% from 2019 to 2025 [50]. The size of this market warrants appropriate investment. Yet there continues to be a lack of clarity regarding financial support for these standards' development activities. This leads to a lack of the availability of the right expertise for individual standards development projects. This in turn can lead to a lack of credibility of resulting standards by the business community. Until significant funding is provided to inform standards priorities and best practice, this current situation is unlikely to change. Current investments are grossly insufficient for the scale of the task.

Use of standards

There is a lack of knowledge about how standards are used, where they are used, their impact and value despite the size of the digital health market. This gap in evidence regarding the impact of standards compliance makes it difficult for standards developers to have a complete view of industry requirements.

Health informatics standards are not used as standards are in other industries. For example, governments require standards conformance for helmets worn when riding motorcycles. Some health informatics standards can impact on the safety of using healthcare systems, yet such standards are rarely incorporated into legislation. Health quality and safety organisations tend to focus on performance data standards but have ignored the need for technical standards to ensure patient safety. This may be because decision-makers are unaware of them, or because software vendors lobby against them or a lack of technical expertise.

These realities are impeding the possibility of arriving at an agreed set of standards required for a well-functioning digital health ecosystem. The ISO TC215 technical committee has developed an e-health reference architecture to guide its standards development activities. It would be useful for other SDOs to make use of this reference as a guide for their own standards development activities.

Some people feel that a standard without research evidence is not of value. Standards that express untested solutions can be developed and tested based upon the experience of experts. This approach was proven to be a bad idea due to performance failures resulting in costly legacies, as demonstrated following the use of this approach for the development of HL7 v3, HL7 Clinical Document Architectures (CDA).

There has been a dearth of quality research regarding the use of standards in health informatics. Despite this, there are experts with significant experience and knowledge. Leveraging the expertise of those who have done it is a viable, comparatively low cost alternative.

Who pays to support the digital health economy?

Most health informatics standards at ISO, HL7, IHE and others are written by volunteers who are not paid to produce the standards; in some cases, they have to pay to belong to the community. This can lead to standards' writers being:

- those desperate to try and make the tasks they are responsible for work better (project focus).
- national experts prepared to be involved as they enjoy the community and see it as a way to give back to the health informatics world.
- those seeking to learn from the experts within these standards communities and wishing to contribute their disciplinary knowledge.
- those paid by their employers, software developers, government entities, to push a specific line or to meet an immediate local need.

Financial support for the development and testing of standards able to offer real benefits, for not only the health informatics community but also society at large, is needed. Only then is it possible to ensure an equal distribution of independent expertise, who are enabled to engage in this process, and for the results to be valued through the successful adoption and global implementation of key foundational global standards. This requires funding not only for the standards development process but also for implementation guidelines in recognition of a need to transform transitionally over time.

At the ISO TC215 SDO, it has been pleasing to note that contributions to comment on ballots for draft standards have increased by over 300% of late. This indicates higher relevance and the considered importance of the activities being undertaken and the types of standards being produced by all participating ISO members. The realisation of a common global vision for healthcare making the best possible use of available technologies requires widespread adoption of agreed sets of standards that facilitate data sharing, aggregation, and linkages. These 'big data' advantages are dependent upon effective data governance and semantic interoperability between health systems.

Conclusion

It is clear that the adoption of international standards enabling semantic interoperability within digital health ecosystems has many potential benefits yet to be fully identified and costed. Large-scale automation of routine reporting is expected to transform health workforce roles and boundaries of practice while significantly improving both the speed and quality of data collection, processing, and reporting. This impact alone is expected to significantly improve productivity, reduce adverse events, and improve health outcomes. This equally applies to the management of clinical trials which represents a billion dollar industry and improvements in the adoption of evidence-based practice. The ability to better manage continuity of care will reduce the incidence of hospitalisation for those dealing with chronic disease and complex comorbidity treatments, including cancer care. There will be greater opportunities for the use of more accurate big data. Collectively these benefits are expected to outstrip the cost of standards development and maintenance.

A digital health ecosystem roadmap needs to focus on global and national foundational infrastructure requirements that enable local innovators and developers to meet their specific needs within that context. There are a number of principles that need to be considered by all concerned to ensure a well functioning, global, complex, adaptive digital health ecosystem may be realised sometime in the future. A primary requirement is the international adoption of suitable standards in support of a global health IT market.

The previous chapter explored a number of global entities with an ability to drive this agenda and generate the necessary investments, to establish the global foundational infrastructure and governance required to facilitate cross-sector and cross-jurisdictional collaboration among those with the necessary expertise. This chapter explored the global SDOs and communities of practice who collectively drive international trade and market growth opportunities associated with the need to digitally support patient-centred healthcare service delivery and population health and well-being, by developing standards.

There is a need for standards development workforce capacity building among users especially clinicians with the collective expertise to cover all clinical specialties and associated work roles. Foundational standards must be mandated and included in procurement requirements. High-level decision-makers need to be fully aware of the value of standards adoption and invest accordingly. The health ICT market needs to be driven by its users, rather than a few mega technology vendors and stimulate local software developments that best meet local needs within a well-connected digital health ecosystem.

Standards development and the widespread adoption of the 'right' standards have the potential to significantly reduce health ICT and health service delivery costs and improve population health. Choosing the wrong set of standards is likely to impede progress. These activities should be based upon scientific principles and best practice evidence collected from those who have undertaken the tasks before, documenting and sharing that practice with standard developers. This global evidence-based approach is considered to be a more cost-effective method than for each country, new initiative, or a professional group starting from scratch and redeveloping the wheel, based on their own local version of reality. These potential savings and contributions to world trade are not being recognised. In part this is due to the production of too many standards which are:

- Developed with insufficient input from the best available experts,
- too academic (brilliant but not useful),
- represent a single version of the world—pushing a specific group's reality,
- set a bar that is too hard to meet—without an implementation plan to progressively improve, and
- untested for efficacy.

Global adoption of some standards, such as a universal electricity plug or standard screw sizes, is known to aid trade and economic growth. The same principles apply to the global adoption of digital health standards. These not only aid trade and economic growth but they also potentially have a significant impact on patient safety and population health. Healthy populations in turn further contribute to global economic growth and the realisation of sustainable national health systems.

The next chapter describes known limitations of global standards required to best support future digital health ecosystems enabling the delivery of patient-centred health and welfare services.

References

[1] ITU, Standards and eHealth, 2011, Available from: https://www.itu.int/dms_pub/itu-t/oth/23/01/T23010000120003PDFE.pdf.

[2] T. Beale, What is Interoperability?: Woland's Cat's Blog on Computing, Semantics and Health IT, 2021, Available from: https://wolandscat.net/2021/04/17/what-is-interoperability/?fbclid=IwAR31x-K_gd0ZhYVYLcx-K9JJHsAuE7wslTi1czZz56PEsxOFjJTvp2exgX1E.

[3] A. McBride, D. Roberts, R. Dunscombe, How Will You Design Information Architecture to Unlock the Power of Data?, EY, Ernst & Young ABC Pty Ltd, 2020. Available from: https://www.ey.com/en_gl/health/how-will-you-design-information-architecture-to-unlock-the-power.

[4] ISO/TC215, Health Informatics, 2021, Available from: http://www.iso.org/iso/iso_technical_committee?commid=54960.

[5] HL7, Health Level Seven International, 2019, [cited 2019 6 February]. Available from: http://www.hl7.org/.

[6] IEC. International Electrotechnical Commission, Available from: https://www.iec.ch/.

[7] ITU, Digital Health and Standards, 2021, Available from: https://www.itu.int/en/ITU-T/e-health/Pages/default.aspx.

[8] WHO/ITU, National eHealth Strategy Toolkit, 2012, [cited 2019 17 April]. Available from: https://www.itu.int/pub/D-STR-E_HEALTH.05-2012.

[9] GSMA. GSM Association, Available from: https://www.gsma.com.

[10] ETSI. European Telecommunications Standards Institute Available from: https://www.etsi.org.

[11] ETSI, Cloud Standards Co-ordination—Final Report, 2013, Available from: https://ec.europa.eu/digital-single-market/en/news/cloud-standards-coordination-final-report.

[12] ETSI, Cloud Standards Coordination Phase 2: Cloud Computing Standards Maturity Assessment: A new snapshot of Cloud Computing Standards, 2016, Available from: https://www.etsi.org/deliver/etsi_sr/003300_003399/003392/02.01.01_60/sr_003392v020101p.pdf.

[13] ETSI (Ed.), Solve the Challenge of Interoperability! Interoperability Best Practices, second ed., ETSI, 2020. Available from: https://portal.etsi.org/Portals/0/TBpages/CTI/Docs/IOT_Best_Practices.pdf.

[14] IEEE-SA. IEEE Standards Association, Available from: https://standards.ieee.org/.

[15] UNECE, United Nations/Electronic Data Interchange for Administration, Commerce and Transport (UN/EDIFACT), United Nations Economic Commission of Europe (UNECE), 2017. Available from: https://www.unece.org/info/media/news/trade/2017/celebrating-30-years-of-successful-data-exchange-for-trade-and-logistics-using-the-unedifact-standard/doc.html.

[16] ISO, JTC 1 Strategic Business Plan, 2015, Available from: https://isotc.iso.org/livelink/livelink/fetch/2000/2122/687806/ISO_TC_215__Health_informatics_.pdf?nodeid=1001750&vernum=-2.
[17] OASIS. Organization for the Advancement of Structured Information Standards, Available from: https://www.oasis-open.org/.
[18] G.J.E. De Moor, C.J. McDonald, J. Noothoven van Goor, Progress in Standardization in Health Care Informatics, Studies in Health Technology and Informatics, IOS Press, Amsterdam, 1993.
[19] ISO, ISO/TC215 Health Informatics, 2021, Available from: http://www.iso.org/iso/iso_technical_committee?commid=54960.
[20] ISO-TC215, ISO/TR14639:2014 Briefing Note: Strengthening National Health Systems Through a Capacity-Based eHealth Architecture, International Organization of Standards (ISO), 2014. Available from: https://www.iso.org/files/live/sites/isoorg/files/archive/pdf/en/14639-brochureversionv7.pdf.
[21] JIC, Joint Initiative Council for Global Health Informatics Standardisation, 2021, Available from: http://www.jointinitiativecouncil.org/.
[22] H. Grain, Terminology Standards ISO TC215 WG3 Chair, 2017.
[23] CEN, European Committee for Standardization -Technical Committee 251 (CEN/TC251), 2021, Available from: https://standards.cen.eu. https://standards.cen.eu/BP/6232.pdf.
[24] ISO-CEN, The Vienna Agreement: Guidelines for the implementation of the Agreement on Technical Cooperation between ISO and CEN, seventh ed., 2016. Available from: https://boss.cen.eu/ref/VA_Guidelines_implementation.pdf.
[25] Ringholm, History of the HL7 RIM- Whitepaper: Ringholm BV, 2011, Available from: http://hl7book.net/docs/04500_en_History_of_the_HL7_RIM.htm.
[26] OMG. Object Management Group, Available from: https://www.omg.org/.
[27] CDISC, Clinical Data Interchange Standards Consortium, 2019, [cited 2019 23 May]. Available from https://www.cdisc.org/.
[28] IHE-International, Integrating the Healthcare Enterprise (IHE), 2019, [cited 2019 22 May]. Available from: https://www.ihe.net/.
[29] ICANN. Internet Corporation for Assigned Names and Numbers, Available from: https://www.icann.org.
[30] CERN-Accelerating-Science, A Short History of the Web, 2021, Available from: https://home.cern/science/computing/birth-web/short-history-web.
[31] W3C. World Wide Web Consortium, Available from: https://www.w3.org/Consortium/facts.
[32] JIC, Joint Initiative Council for Global Health Informatics Standardization, Minutes, 2020.
[33] LOINC, Logical Observation Identifiers Names and Codes from Regenstrief: Regenstrief, 2019, [cited 2019 25 February]. Available from: https://loinc.org/.
[34] OpenMRS. Open Medical Record System (OpenMRS): Regenstrief Institute, Available from: https://openmrs.org.
[35] D. Ingram, Origins of openEHR, openEHR Foundation, 2002. Available from: https://www.openehr.org/about/origins.
[36] ECHAlliance, The Digital Health Society Declaration, 2017, [cited 2019 18 February]. Available from: https://echalliance.com/page/DHSdeclaration.
[37] ISO/TS22789, Health Informatics—Conceptual Framework for Patient Findings and Problems in Terminologies, 2010, [cited 2019 2 December]. Available from: https://www.iso.org/standard/41141.html.
[38] ECHAlliance, The Global Connector for Digital Health, 2020, [cited 2020 24 February 2020]. Available from: https://echalliance.com.
[39] A. Beresniak, A. Schmidt, J. Proeve, E. Bolanos, N. Patel, N. Ammour, M. Sundgren, M. Ericson, T. Karakoyun, P. Coorevits, D. Kalra, G. De Moor, D. Dupont, Cost-benefit assessment of using electronic health records data for clinical research versus current practices: Contribution of the Electronic Health Records for Clinical Research (EHR4CR) European Project, Contemp. Clin. Trials 46 (2016) 85–91.
[40] P. Coorevits, M. Sundgren, G.O. Klein, A. Bahr, B. Claerhout, C. Daniel, M. Dugas, D. Dupont, A. Schmidt, P. Singleton, G. De Moor, D. Kalra, Electronic health records: new opportunities for clinical research, J. Intern. Med. 274 (2013) 547–560.
[41] G. De Moor, M. Sundgren, D. Kalra, A. Schmidt, M. Dugas, B. Claerhout, T. Karakoyun, C. Ohmann, P.-Y. Lastic, N. Ammour, R. Kush, D. Dupont, M. Cuggia, C. Daniel, G. Thienpont, P. Coorevits, Using electronic health records for clinical research: the case of the EHR4CR project, J. Biomed. Inform. 53 (2015) 162–173.

[42] Horizon-2020, EU Research and Innovation Programme 2014-2020, European Commission, 2020. Available from: https://ec.europa.eu/programmes/horizon2020.

[43] Horizon-2020, EU-Pacific Partnerships for ICT RDI: European Commission CORDIS EU Research Results, 2019, Available from: https://cordis.europa.eu/project/id/687794.

[44] CrowdHEALTH, European Union's Horizon 2020 Research and Innovation Programme Under Grant Agreement No. 727560, European Consortium; 2017–2020, 2020. Available from: https://www.crowdhealth.eu/.

[45] P. Gallos, S. Aso, S. Autexier, A. Brotons, A. De Nigro, G. Jurak, A. Kiourtis, P. Kranas, D. Kyriazis, M. Lustrek, A. Magdalinou, I. Maglogiannis, J. Mantas, A. Martinez, A. Menychtas, L. Montandon, F. Picioroaga, M. Perez, D. Stanimirovic, G. Starc, T. Tomson, R. Vilar-Mateo, A.M. Vizitiu, CrowdHEALTH: big data analytics and holistic health records, Stud. Health Technol. Inform. 258 (2019) 255–256.

[46] S.A. Lete, C. Cavero, M. Lustrek, D. Kyriazis, A. Kiourtis, J. Mantas, L. Montandon, Interoperability techniques in CrowdHEALTH project: the terminology service, Acta Inform. Med. 27 (2019) 355–361.

[47] S.A. Lete, C. Cavero, A. Magdalinou, J. Mantas, L. Montandon, Advanced interoperability techniques: structure mapping service in CrowdHEALTH project, Acta Inform. Med. 28 (2020) 52–57.

[48] U. Wajid, C. Orton, A. Magdalinou, L. Montandon, Generating and knowledge framework: design and open specification, Acta Inform. Med. 27 (2019) 362–368.

[49] INTEROPen, IHE, HL7 FHIR & openEHR—a Smörgåsbord of Standards: An Action Group to Accelerate the Development of Open Standards for Interoperability in the Health and Social Care Sector, 2019, [cited 2019 13 March]. Available from: http://www.interopen.org/2018/07/19/ihe-hl7-fhir-openehr-a-smorgasbord-of-standards/.

[50] Grand-view-research, Digital Health Market Size, Share & Trends Analysis Report by Technology, 2019, Available from: https://www.grandviewresearch.com/industry-analysis/digital-health-market.

CHAPTER

5

Standards for digital health, known limitations, and procurement

Thomas Beale[a], *Heather Grain*[b,c,e], *and Evelyn Hovenga*[b,d,e]

[a]ARS Semantica, London, United Kingdom [b]eHealth Education Pty Ltd, Melbourne, VIC, Australia [c]LG Informatics Pty Ltd, Melbourne, VIC, Australia [d]Digital Health, Australian Catholic University, Melbourne, VIC, Australia [e]Global eHealth Collaborative (GeHCo), Melbourne, VIC, Australia

Introduction

The previous chapter introduced significant foundational computing standards that have influenced the design and development of many legacy systems that continue to be in use today. This demonstrated the importance and impacts of historical innovative computing and associated technical milestones. It was noted that these historical developments have in turn contributed to today's continuing existence of numerous unconnected data silos and system incompatibility. Today's issues are commonly referred to as a lack of system connectivity (technical) or semantic interoperability (technical relative to data and knowledge processing capability) within large healthcare organisations and beyond to suit a larger ecosystem. These limitations are due to many factors, including the following:

- Fragmented standards development organisations resulting in overlapping and competing standards,
- There are many and varied standards,
- Standards may not be compatible with each other,
- Many developers only access mandated standards,
- Many standards are not widely used, or implementations are noncompliant,
- Most industry standards remain proprietary,
- There are many categories of standards, based on purpose and desired functionality,
- Many standards are no longer suitable to meet new digital era needs despite updates,
- Backwards compatibility may now compromise system usability,
- Software engineering methodologies have seen many advancements,
- Data and content representational methods have seen many changes,
- Most terminologies were developed to suit a number of historical statistical purposes,

https://doi.org/10.1016/B978-0-12-823413-6.00019-7

- Many terminologies in use have limited clinical data and knowledge management suitability,
- Many lessons have been learned based on experience,
- Today's technical know-how requires us to change our thinking to maximise benefits now achievable to better suit the needs of patient-centred care, smart automation, machine learning, and artificial intelligence.

This situation has led to a strong global desire to address the need for semantic interoperability as the means of overcoming this extensive data fragmentation. There is a continuing perception that the solution is the development of many new standards. We argue that a foundational digital health ecosystem platform architecture is the primary factor required for long-term success. In this approach, the platform is the standard, and today's health informatics technical standards would either be subsumed into the ecosystem or else discarded. A common platform acts as an integrating point for existing standards and other architectural features, ensuring their mutual coherence. The latter is sorely lacking in the current disjointed standards landscape.

A number of established technical and data standards development organisations (SDOs) have adopted the interoperability challenge by establishing new technical committees. In addition, a number of other organisations and communities of practice have also engaged in the development of some highly specialised health informatics standards. Significant contributions to this landscape have resulted from the wide use of the Internet, the adoption of open source principles, the World Wide Consortium (W3C) [1] and their adoption of the Community Contributor License Agreement with royalty-free patent licensing terms, and permissive copyright for community and business groups. We are increasingly appreciating the benefits of knowledge sharing and collaboration as a result of these developments. Standards on quality, governance, and workforce capacity requirements are not getting the necessary traction to make a difference. Issues associated with these types of standards are discussed in other chapters.

This chapter explores the value of technical standards and the part they play in the roadmap to successful digital health ecosystems. We need to be able to distinguish between 'standards', what these represent, and the concept of 'standardisation' as a process or policy directives, before we explore available standards, their characteristics, and how to make use of them. Selected standards need to have the potential to support adaptive digital health ecosystems. There is also a need to combine differing disciplinary scientific truths and perspectives that need to be reflected in digital health standards to ensure that the needs of both software engineering and content representational and terminological imperatives can be met, as these are inextricably linked.

Health informatics standards: Focus, purpose, and good practice principles

Standards development organisations have foci, relationships with other SDOs, and values/beneficiaries as listed in Tables 1–3 in the previous chapter. Standards, as referred to in other chapters, relating to the topics mentioned are produced for different purposes including the following:

- Data representation (such as formalisms, audio, video, signals, and images);
- Concept representations (including SNOMED-CT [2] and ICD [3]);
- Data specification (such as ISO 11179 [4] and HL7 Value Set Definition [5], and health record structures [6]).

- Conformance and quality (such as Map Quality—the ability to analyse whether a map is safe for use [7]; Systems and Software Quality Requirements and Evaluation (SQuaRE)—system and software quality models [8]);
- Information exchange—message structures and strategies (HL7 [9]);
- Security [10] and privacy (includes legislation and regulatory requirements);
- Workflow and rules representation (e.g. OMG BPMN and CMMN [11]);
- Programming interfaces—API standards, e.g. IHE [12] XDS and PIX;
- Maturity measurement—to assist understanding of the pathway to improvement and for decision-making regarding purchases and developments to protect investment in systems and data.

Standards development good practice principles are summarised by Rowlands [13] as:

- **Openness**—Standards development processes are open to any interested party on a nondiscriminatory basis.
- **Transparency**—Information on current work programs, proposals, and how to participate is available to all interested parties.
- **Representation**—There is balanced participation in the standards development process by interests that will be significantly affected by a resulting standard.
- **Impartiality**—Standards development processes do not give privilege to or favour the interests of any particular party.
- **Consensus**—There should be general agreement to the publication of a standard characterised by [14]:
  (i) the absence of sustained opposition to substantial issues by any important part of the concerned interests and.
  (ii) a process that involves seeking to take into account the views of all parties concerned and to reconcile any conflicting arguments. Note: consensus need not imply unanimity.
- **Market need and net benefit**—A new or significantly revised standard should only be generated when a market need has been clearly defined and net benefit to the international community can be reasonably demonstrated.
- **Timeliness**—The degree of urgency with which a standard is required should be considered but should not override other principles.
- **Internationality**—International standards pertaining to health interoperability should be used in preference to national and subnational standards.
- **Compliance**—Standards are developed for compliance with relevant regulation, including competition law.
- **Coherence**—Interoperability standards are developed for coherence with other health sector and digital health standards and industry developments as appropriate.
- **Availability and support**—Standards for health sector interoperability should be freely available.
- **Support**—Approaches to disseminating supplementary knowledge associated with a standard should be considered during its development.

This list is missing the most important criterion for quality, namely a methodology appropriate to the kind of artefact being developed, and competent professionals working under the methodology to undertake the development. For example, information model standards need to be developed using recognised modelling principles, and by individuals who understand these principles. In general, consensus processes have to be limited to requirements prioritisation; contrary to popular opinion, democracy has no place in the development of technical models or architectures. The following sections refer to technical standards which differ from, for example, concept representation, quality, maturity measurements, data, procedural, best practice, and governance standards which are covered in other chapters.

Standards versus standardisation

One of the essential properties of a sustainable regional or national IT infrastructure is that its formal artefacts—APIs, information models, model formalisms, query language, etc.—are *consistent* across the infrastructure and *mutually coherent*. The former property means that there are no competing models of common architectural elements, for example, 'clinical document', or even simple types such as 'Coded term'; instead, there is an internal 'gold standard' for each component, and any alternate representations are dealt with at the periphery as translation questions.

The latter property of coherence means that all the components follow a common underlying design. This will ultimately devolve to the use of common types/models, e.g. for clinical data types (quantities and terminology codes, etc.), but also to common assumptions related to wider semantics, for example where terminology subset definitions are stored and managed (either in the terminology service or as part of the EHR environment, but not in both).

Coherence is usually achieved by the various components being co-evolved within the same architecture, but need not be: in some cases, distinct architectural elements (e.g. clinical notes, clinical guidelines, extract, transform, and load (ETL)) from different design lineages (generally, from different organisations) can be successfully integrated with only light adjustments, for example to leaf data types of one or other (in computer science, a tree is a widely used abstract data type that simulates a hierarchical tree structure, with a root value and subtrees of children with a parent node, represented as a set of linked nodes—see https://en.wikipedia.org/wiki/Tree_(data_structure)). *Coherence fails to be achieved when architectural components and languages from significantly different design paradigms are forced together, typically in an unplanned integration exercise.* The differing type systems, design concepts, and philosophical assumptions will always rear their heads in the engineering and deployment phases, and at great expense.

A common error is to assume that consistency and coherence among data or technical standards can be achieved by the use of published standards, the more international the better. In fact, *de jure* standards have *almost the opposite raison d'être*, which is to define the technical specification of an agreed way for two foreign components or products, which *assume no mutual consistency or coherence* of internal design, to connect. Such standards are intended to allow separately produced items to be used together, for example an Android phone in a BMW car (USB and other data standards), a high-end video card in a server computer (PCI bus standard), and a Tamron lens in a Nikon SLR camera.

Quite often, standards provide an acceptably workable, but inefficient means of connection, because they are the result of negotiated compromises. The more the international and/or cross-industry, the greater the compromise.

For example, most computer monitors will connect to a laptop or docking station via a VGA connector. However, VGA limits the speed and resolution of the connection. Some monitors contain a Dvi-X (see https://en.wikipedia.org/wiki/Digital_Visual_Interface) connector for high-fidelity video communication to the monitor; however, the bulky physical cable end does not suit modern laptops, and in recent years, the HDMI standard is more commonly supported by laptops and monitors. It is notable, however, that Apple computers commonly ignore industry standards and use their own video connectors, such as FireWire (aka IEEE 1394, developed by Apple, Sony, and Panasonic in the 1980s–1990s) and Thunderbolt (see https://en.wikipedia.org/wiki/Thunderbolt_(interface)) (developed by Apple and Intel).

Standards are compromise models of shared interfaces

These latter specifications allowed for essentially ideal communications and have been used on Apple Mac laptops for a decade. The cost of use for Apple Mac users is the well-known collection of expensive adapter cables acquired to plug into generic monitors, projectors, and other devices at conferences and board meetings, at which those with less beautiful computers quietly chuckle to themselves as they connect directly via an HDMI cable, while the Mac user discovers with horror that a precious white adapter cable had been left at home in another travel bag. The benefit for Mac users, outside of the conference situation, is bullet-proof, maximum-quality video communication between a top-end computer and a top-end monitor at home or work—also manufactured by Apple.

The general lesson is that externally developed standards are likely to be suboptimal compared to a proprietary interface that is engineered with knowledge of the internal design of the connecting components (even if published openly, as was the case for Thunderbolt). A vendor is often faced with a choice of making the product universally 'plug-n-play' or maximally efficient, but not both; and in fact, many simply support 'both', e.g. DVD+Bluray players and laptops having multiple video standard connectors.

Apart from methodological approaches, technical standards for digital health (and indeed most industries) are an attempt to find common ground for interfaces between nonstandard (i.e. proprietary) products or sometimes to define the safety properties of a particular kind of component. Examples in other industries are SCSI (small computer system interface) bus, PCI (peripheral component interconnect) bus, and car window glass properties. In health IT, many standards are to do with communicating data; some relate to device safety and data safety (for example, digital signing and attestation of health records).

Inevitably, the final content of a health informatics standard (right down to attribute names within technical models) is the result of negotiation among industry vendors (those for whom standards represent a cost) and also between vendors and domain users (those for whose working lives standards would improve, at least theoretically). Box 1 details the lesson learned from these limitations.

BOX 1

Lesson learned from standards development process limitations

Lesson: any published technical standard is a result of negotiations, and as such, is never going to be a model of the data or APIs needed by some particular health IT project. Models and specifications for an infrastructure need to be developed based on the needs of the users served by the infrastructure. Ideally, the architecture will be based on a model-driven paradigm such that definitions of new semantics can be routinely consumed.

In many industries, including digital health, most 'standards' are data or API standards rather than physical connections such as cables and buses, and they address *data interoperability* rather than signal-level communication. Since it is always assumed that it will be easy to transform data to something else using a layer of software, *far less care is taken to create data standards that are both high quality* and widely accepted. The approach taken with most data standards is to try to get agreement on message or document data items and assume that adopters will engineer the inevitable layers of data conversion in order to use such standards, which have only limited correlation to the design of any particular product. This is primarily the case in digital health, and indeed much of the economic activity in healthcare IT is simply converting data streams from one standard to another and/or to proprietary specifications for eventual consumption inside a product.

Health data standards need to be taken far more seriously to meet the needs of smart automation and machine learning, as explained in the chapter on health data standards.

Why use health informatics standards?

Standards are one of those things automatically taken to be universally 'good' to the extent that many professionals in procuring organisations cannot articulate why. There are certainly some reasons to use standards within a regional procurement environment, as follows:

- **Potential to purchase wider range of products** which also use these standards (e.g. one could procure an HL7v2.5 based lab data messaging product for an environment also using HL7v2.5) [9];
  - API-based—products using the same API specification will talk to each other;
  - Data/message-based—products using (exactly) the same message specification will talk to each other.
- **Potential to share data** and/or other artefacts across the region border, assuming other regions (e.g. EU countries) use the same standards *in the same way*.

Experience in the last 20 years of health informatics standards has shown that neither of these is even remotely guaranteed to be the case and might not even be all that important. The long-term experience with HL7v2—a standard that works well enough, when tamed—is that

BOX 2

Lessons learned from past 20 years' experience of using HL7 messaging standards [1]

Firstly, an 'HL7 standard' often identifies only a 'standard family', and that without profiling (i.e. constrained versions), the standard is not:

(a) directly usable, and
(b) confers no interoperability between products.

In other words, to say that a product is 'HL7v2-compliant', 'HL7v3-compliant', or 'FHIR-compliant' means very little. It would indeed be better if the industry used the terminology 'HL7v2-capable', etc., to indicate just that the basic architecture of a product might be capable of handling specific versions of the named standard (family).

only with long-term, careful national profiling, and/or use of highly specified IHE profiles [12] can it be said that a given component obeys a certain version of the standard. For example, in Australia, there are numerous detailed HL7v2.x profiles for specific uses, developed over 20 years by HL7 Australia and, up to 2015, were published by Standards Australia [15]. These continue to be maintained by HL7 Australia. Without these, HL7v2 is not usable in its raw form. Every country that uses HL7v2 has its own [16] similar national set of painfully derived standards.

The HL7v3 experience was far worse. Major national efforts, including the United Kingdom (Connecting for Health) [17], Netherlands (NICTIZ) [18], and Canada (Canada Health Infoway) [19], had to create their own highly customised national versions of HL7v3 (i.e. quite different from the US version), to the extent that there was no hope of being able to purchase any product from any of the other countries using HL7v3, built using that country's variant form. Lessons learned from both these experiences are described in Box 2.

Profiling is not necessarily a bad thing. Indeed, the ISO 13606-1 standard [20] and the openEHR specifications [21] work on a kind of profiling basis, via the use of archetypes [22]. However, both of the latter have formally and clearly defined information models, in contrast to the HL7vX standards.

There are reasons to be judicious about or even avoid international standards, including the following:

- **Coverage**: they might not cover the semantic or technical area of interest; e.g. HL7v2 has no solution for API;
- **Fitness for purpose**: they might have been designed for another purpose; e.g. HL7 CDA was designed as a way of sharing clinical notes; some authorities have tried to use it as an EHR persistence model, with poor results;
- **Lack of integration**: a major problem with standards such as the various HL7 standards families, X12, ISO standards, and the numerous terminology standards (many nation-specific),

is that they were not designed as a coherent whole, and interpretation and integration of the standards are left up to the procuring body, at great expense.

Organisations that achieved standardisation without *de jure* standards include the following:

- The US Dept of Veterans Affairs (VA), which used (and still uses) the VistA product in its 160 + hospitals; similarly, it created its own enterprise form of HL7v3, called the Federal Health Information Model (FHIM) (browsable at https://fhims.org/content/420A62FD03B6_root.html).
- Large health management organisations such as Intermountain Healthcare, Mayo Clinic, Kaiser Permanente, and more recently US VA and US Dept of Defense (DoD). Procurements at these locations in the last 5 years primarily achieve a kind of standardisation within the infrastructure by virtue of a mega-suite product purchase—in the above examples, Cerner or Epic. *De jure* standards in such environments remain at the periphery, as an API added above the existing products.
- Since the failure of UK NPfIT in 2011 [17], the UK NHS has used various nationally developed protocols and formats in order to share GP data outside of its own system (e.g. EMIS, TPP); today, this kind of interoperability is finally being addressed by UK specific FHIR profiles created by INTEROPen [23], a UK FHIR profiling group.
- Outside of healthcare, communications inside tech companies such as Facebook, Amazon, and LinkedIn are achieved by the use of *openly available technologies* such as Google Protocol Buffers, Kafka, and various message queue systems, but they create their own internal semantic standards.

It could probably be argued that the current IT systems of some healthcare organisations constitute a form of 'standardisation without standards', including those large health management organisations mentioned above.

There are of course many reasons to use good standards, and indeed, standards such as DICOM [24], HL7v2.x (in nationally profiled forms) [9], CEN/ISO 13606 [6], ICDx [3], SNOMED CT [2], and numerous generic representational standards (UML/XMI, XML Schema, OWL, RDF, JSON, ISO 8601, etc.) are part of today's technological fabric. It is assumed that with sufficient time and experience, national FHIR profiles may eventually approach the level of stability of national HL7v2-based standards.

Standards are not architectures

Jurisdictional bodies often prescribe standards for a country or region, assuming that they will function as architectures for systems, for example, to be built by corporate consortia. This approach was adopted by the failed Victorian Government's HealthSMART programme launched in 2004 [25] and the UK's National Programme for IT (NpfIT [REF]) in 2003–13.

A combination of marketing by standards organisations and industry hype (particularly in the case of HL7v3 and FHIR) has resulted in some vendor companies (not the major vendors) treating these standards as if they were architectures. This assumption is generally a mistake.

Overlapping or conflicting standards

There are many examples where standards relevant to digital health are overlapping or worse, conflicting.

Example 1:
HL7 messaging standards have different versions within v2. Clinical Document Architecture (CDA) and Fast Healthcare Interoperability Resources (FHIR) are all created by one standards body but do not yet share a single agreed set of terms or data elements, though they are now working on resolving this issue [26].
Example 2:
ISO TS 21564:2019 Health Informatics—Terminology Resource Map Quality Measures (MapQual) [7] was developed by international experts in healthcare data and terminologies, including input from SNOMED International, and defines the requirements for quality maps to be used in healthcare. Yet the SNOMED CT Starter Guide, section 12 Mapping has not been updated to reference this standard [27].

Standards development processes

HL7 standards [9] are mostly produced in committees emerging as the result of ad hoc discussions on domain semantics that are captured in the current message formalism of the day, such as HL7v2, HL7v3, HL7 CDA, and most recently HL7 FHIR. HL7 has over 50 such committees.

HL7 standards (famously HL7v3, CDA, and also FHIR to a lesser extent) do not generally obey recognised modelling methodologies, and indeed, in HL7 FHIR, 'modelling' itself (including 'inheritance') is seen as something to be avoided: each separate resource definition treated as its own thing, with the minimal assumption of commonality (e.g. among 'XXXRequest resources, Admin resources). As a result, HL7 standards do not look like technical architectures, but rather like a message (HL7v2, v3), a document (CDA) (information exchange objects) or predefined query result definitions (FHIR). HL7 is largely technical where its suite of standards are the result of work undertaken by various sized groups of participants, who have variable levels of expertise and motivations. HL7 FHIR [5] is now becoming part of the digital health environment, and where used correctly and judiciously, may confer benefits, although this won't be known until this standard has been widely in use for 10–15 years or so. Committee-produced standards also exhibit significant inconsistency [28].

This approach differs significantly from the openEHR specification and clinical model standards development processes adopted. These standards are produced by a community of users, testers, and application developers who have to adhere to governance processes managed by highly qualified experts. It makes use of transparent input from experts as a competency-based engineering process that includes peer review.

Standards produced in OMG [11] are developed differently, following a more engineering approach in which the specification is created by an engineering group, either from one company (i.e. from an existing product) or as a development effort using engineers from multiple companies. At least two interoperating implementations are required before an OMG standard

BOX 3

Lesson learned from standards development experiences

Lesson: the architecture of a health IT infrastructure for a region or large healthcare provider (such as US VA, UK NHS, and Scottish NHS) is not the same as de jure standards, and generally speaking is not likely to be that close. One reason is that a modern infrastructure architecture is based on services, knowledge artefacts, and information models, but the extant digital health standards in use are not built on this basis, but rather on use-case specific 'content' definitions—messages, documents, etc., as well as being committee-produced to satisfy specific use cases.

can be accepted. Major standards such as BPMN, UML, and many others have followed this route and are part of numerous products today.

However, most OMG health IT standards, as described in the previous chapter, have not achieved wide uptake (BPM + may prove to be an exception). They are in almost all cases better architected than HL7 standards. Nevertheless, the latter's committee approach, for all its shortcomings, usually reflects a better knowledge of the health domain. Box 3 details lessons learned from these experiences.

Maintaining and updating content-related standards, including terminologies and data standards, require different approaches. There are also different objectives for those who write the standards. It is important to remember the potential motivations of these writers. For example:

- **Academic**—publication recognition often results in clever and informative work that is rarely implemented.
- **Industry**—implementable with minimal change and minimal cost to the vendor, the outcome is often barely considered. These works are often pragmatic and uncoordinated but they are used. The lowest common denominator, a common HL7 problem.
- **Experts**—those who have seen it all before and seek a solution with the biggest impact for the minimal effort. They seek to make their task easier next time, but possibly also seeking self-aggrandisement. Their ideas are likely to be incredibly useful but resisted by industry.
- **Government**—seek solutions within their period of power, minimal cost with maximal outcomes, but often without a clear long-term objective or understanding of the long-term cost implications. Result in usable but short-term solutions or solutions which wander along an indistinct pathway.
- **Healthcare workers**—want the systems to work. They often cannot see the potential solutions and are fearful of the effort of change. This results in small progressive steps based upon the past rather than on innovation.
- Some **standards bodies** have a history of having to solve problems. In some cases, that has led to an environment where the priority focus is not on the best solution but on a solution, any solution.

Standards development workforce

Workforce skills needed to undertake standards development tasks effectively and safely rely on limited education and training provided by SDOs. In addition, this may be included in some professional development activities provided by professional bodies, but such activities rarely cover health informatics. Those engaged in developing and writing health informatics standards need to have the appropriate technical competency and appropriate health domain knowledge. Knowledge and skill requirements change along with technical advances and the adoption of new strategic infrastructures such as the adoption of platform ecosystem system architectures. Such a foundational change would require existing SDOs to change their foci and purpose. Many existing standards will no longer be fit for purpose.

Few standards bodies have a standards development process based upon a plan of need and gap analysis. In addition, there are insufficient active contributors to standards development for three main reasons:

1. writing standards is very difficult and time-consuming and requires highly specialised technical competencies not often available among the pool of standards meeting participants;
2. writing standards is not valued by the healthcare industry, as a result of poor design and/or they were found to be hard to use, such as HL7 CDA and HL7 v3. Not valuing standards, their development and maintenance processes, is evidenced by the lack of support, which may in part be due to a lack of knowledge about the existence, purpose, or potential benefits of making use of standards;
3. writing standards is voluntary, i.e. not funded.

These issues can be overcome in part by the use of a national educational approach aimed at attracting the right expertise and numbers to participate in standards development, testing, and review processes, by mandating compliance with key standards and including such compliance in routine procurement processes.

Doing standards differently

An example of an organisation that takes a different approach to standardisation, one that starts from a platform architecture rather than individual interoperability use cases, is openEHR [16]. Its goal is to enable the development and deployment of open, interoperable, and computable patient-centric health information systems. Its development, maintenance, and governance are supported by an international community of members and contributors, including individuals from multiple language groups, and cultures as well as with a diversity of technical, clinical, and informatics backgrounds. These specifications are openly available, and all processes, including quality criteria and terms of reference, are transparent. Its goals include the following:

- **quality in health information**: to enable data quality, validity, reliability, consistency, and currency of clinical data across the data lifecycle from creation to archival, and across enterprises and sectors;
- **support industry technologies**: to actively support widely used ICT technologies, e.g. major programming languages and frameworks;

- **de jure standards integration**: to provide means for the specifications to be useful to users of related de jure standards, e.g. by providing additional transformation or mapping specifications and/or implementation guides;
- **manage impact of change**: to ensure the preservation of validity of clinical data created according to previous releases of the openEHR specifications.

Its governance structure consists of a Specifications Editorial Committee (SEC) consisting of no less than 5 and no more than 40 members representing major stakeholders, particularly implementers, who are suitably qualified and share an interest in maintaining the specifications into the future on behalf of the community. Change is managed and released as applicable to its separate components such as the reference model, the archetype model, querying, conformance, and others. The whole community is able to make use of a global problem tracker, which is used to report issues. Decisions are normally consensus-based; wider consultation and voting are used when no agreement is available. The SEC Chair and Co-Chairs perform facilitation roles. Their responsibilities include maintaining the roadmap for future releases, problem report/change request processing, communication to the wider openEHR community, and publishing the specifications and related materials.

Health informatics standards: A fragmented, political landscape

An unfortunate truth about standards for digital health is that they are produced by organisations, as described in the previous chapter, whose members often compete for visibility and dominance. As a consequence, far less cooperation has occurred than might have been expected. The primary organisations implicated are HL7 [9], IHE [12], SNOMED International [2], and to a lesser extent, OMG [11] (HDTF, BPM +) and ISO (TC 215) [29]. We have omitted the long list of other terminology publishers, such as WHO, WONCA [30], and Regenstrief [31], here, as their terminologies are explored in the chapter on health data standards.

All of these organisations produce some standards that are widely used. These standards were not only not co-designed, but usually not even based on the same type of system or modelling paradigm. Many health informatics standards, no matter how individually useful (e.g. HL7v2.3, v2,5, CDA, DICOM, IHE PIX, PDQ, XDS, SNOMED-CT, and LOINC), are difficult to integrate, and integration has sometimes proven prohibitively expensive. One single example characterises the problem: expression languages for writing decision logic, for clinical decision support (CDS), guidelines, for example. The available possibilities include the following:

- HL7 CQL,
- HL7 FHIRpath syntax,
- OMG FEEL (a subsyntax of DMN, the OMG Decision Modelling Notation standard),
- OMG SysML2 (in development),
- Arden,
- HL7 Gello,
- GLIF3,
- ProForma, and
- many others.

BOX 4

Lesson learned from the existence of this fragmented landscape

Lesson: if standards from different standards organisations or from different product families from an SDO are to be used in a jurisdiction, e.g. to influence industry offerings, significant costs and time should be planned for.

Developers of guidelines-driven decision support systems generally do not have an obvious choice from among the above. Some of the above are effectively abandoned. The question is: how would an implementing authority know if the latest available offering will not also be abandoned, rather than becoming a widely used *lingua franca*? Box 4 provides the lesson learned from this reality.

Foundational standards harmonisation

Much has been written about the holy grail of achieving 'semantic interoperability' between distributed systems [32–39], which is a foundational requirement for a successful digital health ecosystem. This can be an emergent outcome over time once a workable set of foundational standards are agreed upon. In principle, this would require SDOs to collaborate on a set of foundational standards—a 'core architecture'—whose components complement each other for the following categories:

- **Unique identifiers**—as explored in the identity management chapter.
- **Data types**—as described in the health data standards chapter.
- **Domain information models**—as explained in the art of modelling chapter.
- **Terminology, concept representation, and definitions**—refer health data standards and quality data chapters.
- **Open platform architecture**—enabling specific messages and API calls to be understood automatically—as described in the chapter on the knowledge driven platform.
- **Governance**—covered in a number of other chapters.
- **Workforce capacity**—refer the chapter on educating the workforce.

Such harmonisation is unlikely to take place within the Health IT space, due to a combination of SDO politics and competition, and weak technical competency. Alternatively, every jurisdiction needs to stipulate which standards associated with the above categories need to be mandated within its jurisdiction. These need to be well maintained and governed.

Domain models and languages need to be defined and shared. In health IT, these include things like:

- **common information models**, i.e. clinical data, basic structures to represent observations, orders, and actions;

- **domain terminology**, i.e. formal representation of ontological aspects of the domain, i.e. invariant truths, particularly the taxonomy of natural kinds;
- **domain content models**, i.e. domain-created definitions of information that is created due to particular business activities, such as taking a blood pressure, reviewing a diabetic patient, and ordering a medication;
- **care pathways and guidelines**, consisting of plans and decision logic for best practice of larger-scale activities;
- **languages of representation** of terminology, content models, plans, decision logic.

These concepts, and how they link to semantic interoperability, are explored in greater detail in the chapter on Health data standards limitations and are referred to in other chapters.

The first and last in the above list represent technical concerns and require the input of both domain and IT professionals. They are also the basis of software—tools, platform back-ends, and applications, which if built correctly will consume domain models, that is the middle three concepts in the list above. These artefacts may be built by domain experts using tools based on the two technical items as described in the chapter on the art of modelling knowledge concepts. Ideally, these are consumed at runtime by deployed software, enabling such systems to continually track the domain rather than ossify into obsolescence, as is the usual case today. Collectively these standards can provide a cost-effective central point of agreement to reduce system development and information exchange costs.

How should a procuring organisation use standards today?

The strategic approach to using standards is, firstly, to identify any standards which may be used to guide purchasing or the determination of system or workforce knowledge and skill requirements enabling informed decision-making.

Secondly, **not to base the internal architecture of the infrastructure on standards in a direct way**. The reasons for this are as follows:

- To avoid *dependency* of the architecture on the standards, particular the future directions—which are out of control of the using organisation;
- To *avoid bringing the standards integration problem* into the heart of the platform, which is unavoidable if standards are used as architecture because no single standard covers all the needed components of the platform;
- To *enable uninhibited development* of the semantics of the intended infrastructure, free from the constraints of particular standards, which tend to have a 'straitjacket' effect on architecture by limiting possible data types, imposing certain rules, etc.

Having accepted this premise, it can be assumed that various e-health standards will in fact be used somewhere in the system, in order to achieve interoperability with external components, data partners, or other entities, or to conform with government requirements. The implementations of these standards must be limited to dedicated layers/interfaces/components, such that they do not influence the functioning of the core system semantics or other critical attributes such as scalability or performance.

Despite the caveats above, there is nothing wrong with adopting and/or adapting ideas, type definitions, or other elements of standards as part of an architecture, *if they solve problems or reduce work, and can be integrated into the architecture in a coherent way*. Doing so may help the infrastructure deal more easily with formal standards within the integration layers. The key to this kind of adoption is to realise that usage of elements of standards inside the platform architecture should be understood as a *copy* or *fork* (a fork is a copy of a repository that allows you to freely experiment with changes without affecting the original project. A forked repository differs from a clone (copy) in that a connection exists between your fork and the original repository itself. In this way, your fork acts as a bridge between the original repository and your personal copy where you can contribute back to the original project using Pull requests to tell others about and discuss changes) of those elements, since it is likely that the direction of technical progress of the platform over time will not follow exactly the path of change of the standard(s).

Conclusion

This exploration of health informatics standards, their development, and their use has led to a number of lessons we need to take seriously. Any digital health ecosystem needs to first and foremost be based on a shared vision; it needs to have an agreed overall purpose as this determines the foundational infrastructure that needs to be made available. Such a foundation determines the key standards to be mandated and governed for the list of key categories of standards, identified as foundational pillars that facilitate growth, using local bottom-up approaches over time, with all participants 'singing from the same song-sheet' but meeting the needs of its own stakeholder use cases. This means priorities can be established locally based on need and available resources, including workforce capacity to enable the ultimate collective realisation of a well-functioning digital health ecosystem. Choosing the right platform is key. This is described in more detail in the following chapter. Other chapters to follow focus on content (data, information, knowledge, and wisdom) standards that information systems need to be able to represent and process concepts.

References

[1] W3C, World Wide Web Consortium. Available from: https://www.w3.org/Consortium/facts.

[2] SNOMED-International, Leading Healthcare Terminology, Worldwide. Available from: http://www.snomed.org/.

[3] WHO, International Classification of Diseases (ICD), World Health Organisation, 2018. Available from: http://www.who.int/classifications/icd/en/. (Cited 18 August 2018).

[4] ISO/IEC-11179. Information Technology—Metadata Registries (MDR) Part1–6 Edition 2-2004, ISO. Available from: http://metadata-standards.org/11179/.

[5] HL7-International, Fast Healthcare Interoperability Resources Specification (FHIR). Available from: https://www.hl7.org/implement/standards/product_brief.cfm?product_id=343.

[6] ISO13606: The ISO 13606 Standard Explained, 2019. Available from: http://www.en13606.org/information.html.

[7] ISO/TS_21564: Health Informatics—Terminology Resource Map Quality Measures (MapQual), International Organisation for Standardisation, 2019. Available from: https://www.iso.org/standard/71088.html. (Cited 6 December 2019).

[8] ISO/IEC25010. Systems and Software Engineering—Systems and Software Quality Requirements and Evaluation (SQuaRE)—System and Software Quality Models, International Organization for Standards, 2011. Available from: https://pdfs.semanticscholar.org/57a5/b99eceff9da205e244337c9f4678b5b23d25.pdf. https://www.iso.org/standard/35733.html.
[9] HL7, Health Level Seven International. Available from: http://www.hl7.org/. (Cited 6 February 2019).
[10] ISO-13606-4. Health Informatics—Electronic Health Record Communication—Part 4: Security, 2019. Available from: https://www.iso.org/standard/62306.html.
[11] OMG, Object Management Group. Available from: https://www.omg.org/.
[12] IHE-International, Integrating the Healthcare Enterprise (IHE). Available from: https://www.ihe.net/. (Cited 22 May 2019).
[13] D. Rowlands, J. Parker, A Health Interoperability Standards Development, Maintenance and Management Model for Australia, ADHA, 2020. Available from: https://www.digitalhealth.gov.au/sites/default/files/2020-12/Standards_Development_Model_v1.1_2020.pdf.
[14] ISO/IEC Guide 2:2004, Standardization and Related Activities – General Vocabulary, https://www.iso.org/standard/39976.html (Accessed 24 November 2021).
[15] Standards-Australia, eHealth Publications from IT/14, Standards Australia. Available from: http://www.e-health.standards.org.au/Home/Publications.aspx.
[16] openEHR, Specifications. Available from: https://specifications.openehr.org/.
[17] T. Justinia, The UK's national programme for IT: why was it dismantled? Health Serv. Manage. Res. 30 (2017) 2–9.
[18] NICTIZ, Dutch Competence Centre for Electronic Exchange of Health and Care Information. Available from: https://www.nictiz.nl/.
[19] Canada-Health-Infoway. Available from: https://www.infoway-inforoute.ca/en.
[20] ISO-13606-1, Health Informatics—Electronic Health Record Communication—Part 1: Reference Model, ISO, 2019.
[21] openEHR-Specifications, REST API Specifictions (ITS-REST) Component—Latest Release 1.02 24 April 2021. Available from: https://specifications.openehr.org/releases/ITS-REST/latest.
[22] openEHR, Archetype Technology Overview, OpenEHR Foundation. Available from: https://specifications.openehr.org/releases/AM/latest/Overview.html.
[23] INTEROPen, IHE, HL7 FHIR & openEHR—A Smorgasboard of Standards: An Action Group to Accelerate the Development of Open Standards for Interoperability in the Health and Social Care Sector, 2019, Available from: http://www.interopen.org/2018/07/19/ihe-hl7-fhir-openehr-a-smorgasbord-of-standards/. (Cited 13 March 2019).
[24] DICOM, Digital Imaging and Communications in Medicine. Available from: https://www.dicomstandard.org. (Cited 23 May 2019).
[25] K. McDonald, Why the Axe Fell on HealthSMART: PULSE+IT, 2012, Available from https://www.pulseit-magazine.com.au/index.php?option=com_content&view=article&id=1002:why-the-axe-fell-on-healthsmart&catid=16:australian-ehealth&Itemid=328.
[26] HL7-Vocabulary-Workgroup, 2020. Available from: https://confluence.hl7.org/display/VOC/Vocabulary+Work+Group+Policy+Statements.
[27] SNOMED-International, SNOMED-CT Starter Guide, SNOMED International, 2017. Available from: https://confluence.ihtsdotools.org/display/DOCSTART.
[28] T. Beale, A FHIR Experience: Consistently Inconsistent, 2019, Available from: https://wolandscat.net/2019/05/05/a-fhir-experience-consistently-inconsistent/. (Cited 8 May 2019).
[29] ISO/TC215. Health Informatics. Available from: http://www.iso.org/iso/iso_technical_committee?commid=54960.
[30] WONCA, World Organization of National Colleges, Academies and Academic Associations of General Practitioners/Family Physicians. Available from: https://www.globalfamilydoctor.com/.
[31] LOINC, Logical Observation Identifiers Names and Codes from Regenstrief, Regenstrief. Available from: https://loinc.org/. (Cited 25 February 2019).
[32] E. Hovenga, S. Garde, Electronic health records, semantic interoperability and politics, Electron. J. Health Inform. 5 (2010) 2.
[33] A. Moreno Conde, D. Moner, W. Cruz, M. Santos, J. Maldonado, M. Robles, D. Kalra, Clinical information modeling processes for semantic interoperability of electronic health records: systematic review and inductive analysis, J. Am. Med. Inform. Assoc. 22 (2015).

[34] C. Martínez-Costa, M. Menárguez-Tortosa, J.T. Fernández-Breis, An approach for the semantic interoperability of ISO EN 13606 and OpenEHR archetypes, J. Biomed. Inform. 43 (2010) 736–746.

[35] S. Janaswamy, R.D. Kent, Semantic interoperability and data mapping in EHR systems, in: 2016 IEEE 6th International Conference on Advanced Computing (IACC), 2016, pp. 117–122.

[36] A. Goossen-Baremans, S. Collins, H.A. Park, Semanticification in connected health, Stud. Health Technol. Inform. 232 (2017) 133–151.

[37] S. Garde, R. Chen, H. Leslie, T. Beale, I. McNicoll, S. Heard, Archetype-based knowledge management for semantic interoperability of electronic health records, Stud. Health Technol. Inform. 150 (2009) 1007–1011.

[38] T. Beale, What is Interoperability?: Woland's Cat's Blog on Computing, Semantics and Health IT, 2021, Available from: https://wolandscat.net/2021/04/17/what-is-interoperability/?fbclid=IwAR31x-K_gd0ZhYVYLcx-K9JJHsAuE7wslTi1czZz56PEsxOFjJTvp2exgX1E.

[39] K. Atalag, T. Beale, R. Chen, T. Gornik, S. Heard, I. McNicoll, openEHR-A Semantically-Enabled, Vendor-Independent Health Computing Platform—White Paper, 2017.

CHAPTER

6

The knowledge-driven platform: Strategic technologies for a platform ecosystem approach

Thomas Beale[a] *and Evelyn Hovenga*[b,c,d]

[a]ARS Semantica, London, United Kingdom [b]Digital Health, Australian Catholic University, Melbourne, VIC, Australia [c]eHealth Education Pty Ltd, Melbourne, VIC, Australia [d]Global eHealth Collaborative (GeHCo), Melbourne, VIC, Australia

Introduction

Previous chapters have focused on standards. This chapter focuses on how these relate to the system architecture or 'platform'. Many challenges that need to be addressed were previously highlighted. This chapter first presents the rationale underpinning the need to transform our traditional computing infrastructures to better meet the needs of this new digital era. We then examine key foundational technical requirements and explore a number of options to resolve these. This is followed by a description of what a knowledge-driven platform looks like, and how this relates to data and knowledge management, procurement processes, and desirable outcomes from a business perspective.

Why a knowledge-driven platform?

Those in charge of managing any healthcare delivery organisation's IT environment are well aware of the symptoms that indicate when the time has come for a major renovation. These symptoms indicate that the current IT environment is no longer sustainable, and may be described as a combination of any of the following world realities:

- Growing size and complexity of system maintenance backlog;
- Changes are taking longer to effect;
- Growing costs of maintaining and adding data integrations;
- Difficulties in scaling systems, e.g. to more users, transactions, etc.;

https://doi.org/10.1016/B978-0-12-823413-6.00006-9

- Rising costs and data errors due to unavoidable frequent data migrations;
- The costs of data transformation to satisfy a new tool or application often approach or outweigh the cost of the application itself;
- Establishing semantic consistency across different applications and databases is difficult and worsening—i.e. knowing if a given data point means the same thing in each system;
- Representation of real-world 'singletons' such as patient record and medication list is via multiple fragmented pieces spread across the environment, e.g. hospital and GP locations both having partial, unreconciled medication lists;
- Inconsistency in use of terminology across applications and systems;
- The lack of apparent incremental options for achieving 'obvious' updates to UI and UX, e.g. making applications available on tablets for the ward or smartphone for patients;
- The introduction of major new functionality looks complex and expensive and as if it will recreate many of the problems of the existing environment in a new area; and
- A growing number of users report a negative user experience.

Many of these world realities are referred to and/or have their impact explained in some detail in a number of other chapters. This list represents issues that need to be analysed from a technical perspective in order to find solutions.

What's the real problem?

There is an underlying sickness that permeates large diverse IT environments, even when they include numerous individually good applications or technologies, and excellent quality developers. It is the unavoidable result of long-term organic growth and purchasing, resulting in a *heterogeneous environment*, in which the one common characteristic is *inconsistency*. This exists in more than one place:

- Application user interface and user experience;
- Data representation across different systems; and
- Concept representation across different systems.

The first is obvious to any user who moves across different applications and/or institutions, and should be uncontroversial. Secondly, the difficulties with using data across systems are also well-understood and are the main generator of cost in system integration. However, the third point above is at least as important if not more so. It is caused by the (often radically) differing information models and terminologies within different software components and products, of which the second problem—inconsistent and irreconcilable data—is just the most visible symptom. Further, the semantics of most applications and databases are generally not explicitly defined or documented, so the problem is not just incoherence, but ambiguity, i.e. *lack of explicit definition*.

We can term this overall problem as something we may call *semantic incoherence*. An important question is: *how much does semantic incoherence really matter?* This depends on the number of semantic entities and inter-relationships needed to represent healthcare data, rules, and so on. We can make a relatively simple characterisation of this via the following figures:

- The SNOMED CT terminology contains approx. 350,000 concept nodes and 1.36 million relationships (2020);

- An estimate based on the openEHR international archetypes (around 10,000 data points in 650 archetypes) and Intermountain Healthcare's largely non-overlapping Clinical Element Model library [1] (7500 atomic models) points to around 50,000 clinical information data points required for recording of clinical notes in general medicine, assuming we have about 1/3 coverage today.

This does not take into account clinical practice guidelines (CPGs), business rules, or clinical knowledge hidden inside the hundreds of thousands of application screen forms, which collectively probably dwarf the above-mentioned 50,000 content data points. We have to assume that the number of discrete and complex knowledge artefacts that encode clinical semantics for computational use—excluding terminology—is in the 100,000 range and might easily be 500,000–1 million.

We can reasonably assume that a rate of change of 5–10% will affect the clinical knowledge base required to properly express routine clinical semantics in IT solutions. These numbers are significant, and it is therefore clear that semantic inconsistency, and indeed absence (most likely numerous concepts are not represented at all), is likely to be a central problem in any large, organically evolved health IT environment. It is not hard to draw causal connections between the underlying fact of inconsistent semantics and technology and the list of symptoms above: almost everyone is a result of one or both problems. In summary, we can say that the real problems of an organically assembled health IT environment are likely to be both *technological and semantic incoherence.*

What are the options?

The commonly identified options for addressing an unsustainable health IT environment which no longer continues to be an economically or technically viable solution for ongoing needs are as follows:

- A monolithic *mega-suite* purchase;
- Separate purchases of so-called *best-of-breed commercial systems*, followed by internal or separately contracted integration works, based on published standards.

We propose here another option:

- Moving core services, data, and capabilities to a new paradigm based on a *knowledge-driven platform (KDP)*, within a coherent services infrastructure and modern application development technology environment.

It is recognised that even the target state (say in 5–10 years) of a large-scale health IT environment may well contain mega-suite and/or best-of-breed solutions, for numerous reasons, including needed functionality, contractual obligations, and limited funding for the future. The platform strategy should therefore be understood as applying to the essential core services, business functions, and data of the enterprise in such a way as to alleviate the majority of the symptoms being encountered with the current mode.

The knowledge-driven platform, as a central paradigm present in the desired future state of the environment, should not be understood as advocacy for a revolution in order to get there. Gradual means are indeed possible, and it is just that the steps taken will be quite different than those under one of the other strategies.

A knowledge- and platform-based strategy

Following the arguments above, the central problems that give rise to most of the symptoms described above are incoherence in semantics and technology across the environment.

If we assume that the technological incoherence problem is solvable, we may concentrate on the semantic problem as described in some detail in previous chapters. The first challenge is to explicitly define the domain semantics, typically either undefined or hidden as assumptions inside software and database schemas. Doing so involves identifying suitable formalisms, e.g. model languages, rule languages, and ontology meta-models, and then creating maintainable models in these formalisms that may be governed and maintained into the future.

Considering the numbers provided above (and assuming that terminology is represented independently), it is not feasible to express such large numbers of domain definitions directly into databases and application software, as described by Beale [2] p.16–32. It would imply UML models of 10–50 k classes, which is an order of magnitude at least above what can be sustainably managed by that means. This would imply software with 10 k Java (or whatever other language) classes. Not only that, but the rate of change will cause endless software and database change, which represents a significant risk of data errors and loss of disruption to users.

The obvious solution is to *remove the vast majority of semantic definitions from software and databases* and move them to an independent representation and existence, thus providing a *single authoritative definition*. This solution is in fact the general principle underlying the third strategy above, that is the knowledge-driven platform (KDP) approach. The other aspects of the KDP, such as componentisation and service orientation, are essential to the *platform* notion, but could conceivably be provided in a purely technological fashion, while failing to address the semantic problem. The result of that will not do much to address many of the symptoms listed earlier, since they relate to inconsistent data.

Further, the healthcare-specific paradigms of patient-centric health record and process-based health service delivery are likely to be crucially important to the success of a paradigm shift in health IT provision in a complex environment, such as a digital health ecosystem. The EHR could conceivably be provided in a weak semantic environment, but it is difficult to see how process-based health IT could, since it is completely reliant on knowledge artefacts in the form of care pathways and guidelines, which are already considered by the industry to be separate entities, like terminology.

In summary, we state that:

> any approach to solving the unsustainability of an enterprise or regional healthcare IT environment, and the underlying conditions of technological and semantic incoherence, that is not based on separating out knowledge from software and databases and explicitly representing it, is most likely to fail.

This is true, even if it uses, for example, best-of-breed products to solve individual challenges acceptably, because the inconsistency problem will never be solved with current health IT strategies and it will only be shifted to new locations.

There are of course always advocates of the comprehensive mega-suite, arguing that it brings semantic coherence by virtue of being one product. Even if this were true (in fact knowledge representation within such products is generally poor), this direction introduces major new problems: the sheer cost, up to two orders of magnitude higher than other approaches, and the near-total loss of freedom on the procuring side, not to mention the difficulties to be overcome to link to a larger digital health ecosystem.

What is a platform?

The notion of *platform* is common across many industries. The variety we are concerned with here is the information systems platform, that is software that may be deployed and used by other software components such as applications.

Platform: Conceptual definition

The word 'platform' indicates the notion of a common base on which higher-level components can be built, obviating the need for those components (typically applications) to privately solve needs that are provided for by the platform. These are typically functions, such as reliable persistence, versioning, identification, communication, various administration functions, certain types of reference data, and so on. In order for applications and other components to be able to use a platform, they need to know:

(a) what functions it provides and
(b) how to connect to it.

To be usable, it must therefore have a *published description*, which includes the *formal specification of its interface*. To actually be used, it must also of course have one or more implementations, a question to which we will come later. In order for an information systems platform to be a more attractive option than a private implementation of a system, we need to assume that a platform for a major domain, such as healthcare delivery and research, is likely to provide non-trivial functionality. It then follows that its architecture needs to follow the basic principle of *separation of concerns*, consisting of multiple *components*, each providing a well-defined function within the overall functionality. Based on this minimalist analysis, we propose the following definition for a platform:

> a published architecture whose component interfaces form the basis for independent application development.

Here, a *component* is any coarse-grained element of an architecture that corresponds to an identifiable *business function or kind of information*, for example patient record, terminology, and medications list. This component-centrism is what distinguishes a platform from a non-platform technology solution. The latter, including in mega-suite and best-of-breed applications, are by and large black boxes, both from a technical and a purchasing point of view.

A number of obvious questions pose themselves, including the following:

- What is the scope of a particular platform? What is included and what is not?
- How do we cut the functionality up into components, i.e. where are the component boundaries?
- How does the platform grow over time?
- How are new components added, and existing components improved?

We will deal with these important questions later, but first we must delve a bit further into the question of what a 'component' is technically, as well as how to specify an 'interface' properly. To do so, we will assume that the question of how many components and their functional scope is solvable for a specific domain. We can then consider the general properties of components and their interfaces.

If we think of a component as a black box that performs certain functions, say persistent patient health records, then the *interface* of the component can be expressed in terms of a formal interface definition, often known as an Application Programmer's Interface (API). This is the set of logical function and procedure calls that can be made by a calling application or other component. Such calls include typed arguments and, in the case of functions, return types. Pre- and post-conditions for each call are commonly specified, as well as exceptions that may be raised under certain calling conditions. We can visualise a component with its interface like a machine such as an audio mixer, with gauges and visual indicators, and controls, as shown below in Fig. 1.

Formally speaking, the 'interface' of this simple machine consists of something like the following:

- Procedures (i.e. commands and controls):
  - Set_volume (level: Decimal)
  - Set_treble (level: Decimal)
  - Set bass (level: Decimal)
- Functions (i.e. queries, or outputs):
  - Output: Decimal
  - Eq: Map <frequency: Decimal, level: Decimal>

This particular arrangement follows what is known as command/query separation, which separated side-effect-generating calls (pure procedures), from non-side-effect-generating calls (functions).

FIG. 1 An example of a visualised component.

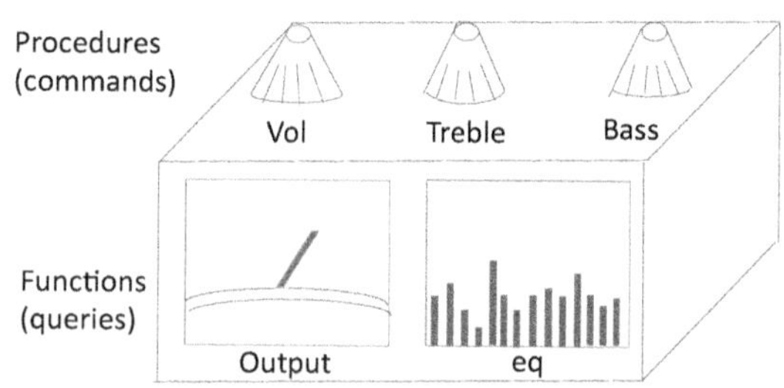

Returning to the health IT domain, the following is an example of an API function call that creates a new EHR, i.e. patient health record, that might be defined within an EHR service.

```
create_ehr_with_id (
      an_ehr_id: UUID[1],
      an_ehr_status: EHR_STATUS[0..1]
): UUID
Pre_valid_id: not has_ehr (an_ehr_id)
Post_ehr_created: has_ehr (Result)
```

We can dissect the above call into its constituents as described in Table 1.

This formal representation is stating the same information as found in a modern REST API, a common form of API used on the web, except this is presented in abstract terms.

For non-technical readers, the details of the above will not matter much, but one general feature is important: an API call cannot be specified without an *information model* to supply the parameter and return types. In the above, the types EHR_STATUS and UUID are data-types—but what are their definitions? In order to supply the type definitions of all calls defined in all of the component APIs of a platform, there must be a *platform information model*. This may be logically compartmentalised following the service components, or defined as a common model.

Following these considerations, we can now say that the published description of a platform includes at least the following elements:

- **Component architecture**: arrangement of components defining functional elements of the platform;
- **Interface descriptions**: interfaces of components enabling various modes of interaction by applications, defined by APIs (including usage protocols);
- **Information model(s)**: i.e. models from which data are instantiated in real systems; these models encode the most basic universal concepts relevant to the domain of the platform, e.g. 'clinical statement' and 'order'.

TABLE 1 API call constituents.

Call	`create_ehr_with_id`	This call creates a new EHR in the EHR persistence service, with .....
Arguments	`an_id:UUID [1]` `an_ehr_status: EHR_STATUS[0.0.1]`	A UUID that will be used as the EHR's *ehr_id*
Pre-conditions	Valid_id: `not has_ehr (an_id)`	No EHR with *ehr_id = an_id* currently exists
Post-conditions	Ehr_created: `has_ehr (an_id)`	An EHR with *ehr_id = an_id* has been created
Exceptions	`Ehr_already_exists`	EHR with *an_id* already exists
	`Auth_error`	Caller authorisation failure

These elements provide a structural definition sufficient to enable technical access by a client application or other component. We can visualise this *technical platform* as shown in Fig. 2 as follows:

A digital health service architecture

We do not propose to describe a definitive platform for digital health; indeed, there is no one platform that will work equally for say the US Dept of Veterans Affairs, the Stockholm region in Sweden, and a major research project. However, to make some of the above more concrete, we can state some of the core services needed in a digital health platform based around a patient-centric health record. Here, we assume the openEHR architecture [3] as the core, along with a number of other common services:

- openEHR Clinical Data Repository (CDR);
- openEHR knowledge service: archetypes, templates, and terminology subsets;
- openEHR Query service;
- openEHR Task Planning and Guidelines;
- Enterprise Demographics/MPI (Master Patient Index) service or registry/ies;
- Terminology service;
- IHE services, either wrapped for openEHR or native
  - ATNA—system log
  - PIX and/or PDQ as needed
  - XDS, if PDF or HL7 CDA or other documents need to be served
- Integration engine processing messages (e.g. lab results and radiology reports):
  - HL7 v2.x, FHIR, CDA, DICOM
- Integration services/APIs for standards including:
  - HL7 v2.x, FHIR, CDA
- Access control;
- Consent management;

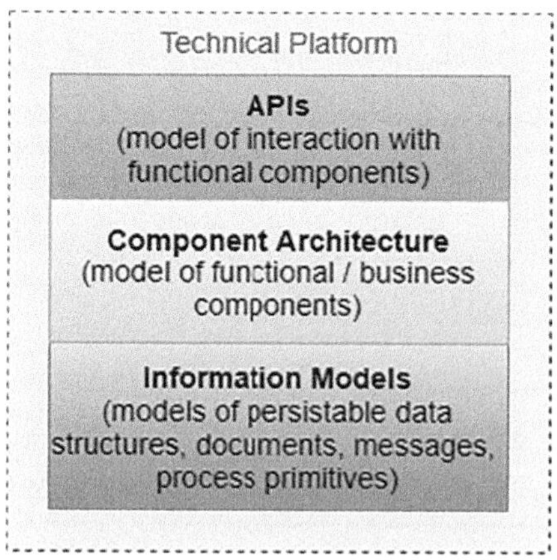

FIG. 2 Technical platform.

- Administrative functions, including
  - EHR archiving;
  - EHR deletion & move.

There are many more services that could be included, such as booking and scheduling, referrals, and so on. Different organisations may define such components with different scopes. One key issue in the context of establishing a platform architecture for a major enterprise is that of health data standards. A common assumption is that everything must conform to the latest standards. This is an error. Thirty years' experience in health IT has shown that a rigid obsession with data standards has often been counter-productive, for various reasons. Data standards have been found:

- not to have full coverage;
- not to generally work easily together, and significant data standards-integration efforts are required;
- to invariably contain competing and inconsistent type systems and semantics; and
- not to provide a good basis for architecture, being the result of largely committee-produced common message definitions.

We therefore make the strong recommendation to treat the use of data standards as a secondary consideration when developing the platform architecture, and instead to limit them to the system periphery. The essential priority is platform internal coherence, not standards adoption. Standards matter at the touch-points between the platform as a whole and for non-platform data and applications. This reasoning does not necessarily apply to knowledge representation standards, described below, since these will be central to many components, such as guidelines/clinical decision support (CDS).

Since the patient-centric health record (EHR) and any clinical process are two central architectural elements of any digital health platform for healthcare provision, as distinct from research or analytics, we now describe these two components in more detail.

The patient-centric health record

The first and most central design concept for a health computing platform is the patient-centric health record. This can be distinguished from the institution-centric record, as found in typical EMR products in hospitals and general practice. The patient-centric record acts as a longitudinal record of all major observations, clinical assessments, conditions, and procedures, as well as providing a picture of current patient state, active orders, and interventions. A patient-centric record is independent of an institution and provides a logical place for a single source of truth medication list, allergies, vaccination history, and so on, as well as being a place to control long-term access via consent mechanisms.

Institution-level records represent care episodes, and they remain useful since a large amount of detailed information may be recorded in them that is not needed past the final phase of a care episode. Some of the contents of the patient-centric health record may be partly created by data transferred from one or other episodic institutional record. In turn, it may be used as a source of data for secondary analytic and research uses. Patient records represent key components of any digital health ecosystem as shown in the diagram [4] presented in Chapter 2. The relationship between electronic health records and essential technology paradigms is covered in more detail in another chapter.

Process-driven health IT

Healthcare is primarily understood in terms of a process by healthcare professionals in the sense that any given action, such as a physical exam, lab test order, or diagnosis, is normally situated within a process of best practice for the kind of condition or complaint presenting in the patient.

A care process for a problem may be formalised in terms of *plans*, ideally conforming to published *guidelines*. Since plans contain decision points based on patient state and other factors, process-based care also involves *rules* and *decision logic*. There is a third major element, which is the connection between referenced *subject variables*, such as diabetic status, systolic blood pressure, and data sources, sometimes known as *bindings*. We term this here as the *subject proxy*. We can visualise these major elements as shown in Fig. 3 as documented in the openEHR specifications [5].

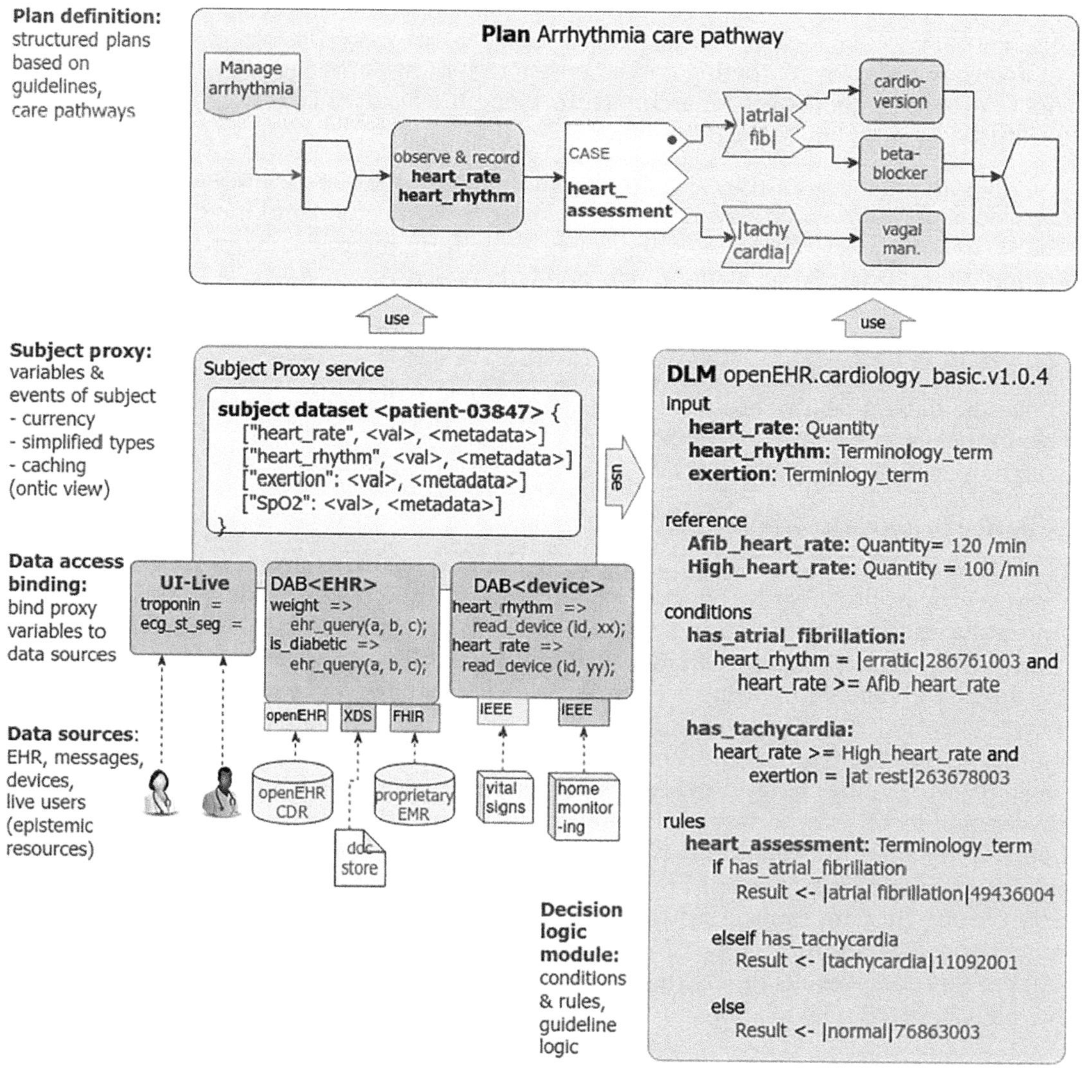

FIG. 3 Conceptual model of computable guideline representation.

For many conditions, including complex and/or high-risk conditions such as ARDS (acute respiratory distress syndrome), sepsis, and complicated childbirth, the plans and rules are defined by a published care pathway aimed at addressing the overall condition for the patient. Knowledge formalism(s) capable of expressing plan and decision logic are therefore essential to enabling such pathways and their constituent guidelines to be represented in a computer-processable way, and therefore integrated directly into clinicians' cognitive workflow at the point of care. Pure logic guidelines are also in common use, for determining risk based on prior patient observations and recommending medication or other treatments.

An advanced digital health platform therefore requires support for formally expressing care pathways, guidelines, and plans, as well as executing them to enable their direct use at point of care.

Representing domain semantics

A central challenge for a platform in most domains is *how the semantics of the domain will be represented in the platform*. For example, in health IT, where are entities such as 'blood pressure measurement', 'Apgar result', and 'breast cancer treatment plan' represented? One approach is to encode them directly into the interface and information models (or database schemas). As described earlier, for complex and information-rich domains such as healthcare, this entails numbers of information model entities that are not only very large (O(10*k*)–O(100*k*)[a]), but also constantly growing.

The question then is whether such large numbers of entities should be directly modelled in the main information model(s) of the platform. To try to do so has been the classical approach of IT for decades, but invariably becomes unsustainable, meaning that deployed products gradually stop reflecting new requirements and domain specifics, as their vendors can no longer keep up with the numbers. Experience shows that classical information models, whether expressed in UML (Unified Modelling Language), RDBMS (Relational Database Management System) schemas, or another form, are manageable for up to around 150 classes or tables.

Size of information model is not the only problem and might not even be the main one. The fact of constant change in domain content and process models means that a single information model or schema that tries to directly incorporate everything from the domain *will never become stable*. This is the last thing anyone from software engineering or product deployment wants to hear.

The alternative is to express domain semantics, i.e. the *domain knowledge* relevant to the purpose of the platform, as *independent formal entities*, which may be used to configure technical information entities, or may be directly consumed by the platform. For example, if a platform information model contains the formal type of document, definitions of the contents of the thousands of particular kinds of documents used in healthcare could be stated as formal models that defined particular headings and possible contents of each kind of document. Similarly, a workflow definition or guideline rule-set might be directly consumable by the platform. This approach enables the definition of a *highly stable technical platform*, on which

[a] O(N) describes how many steps an algorithm takes based on the number of elements that it is acted upon.

back-end software and databases can rely, while providing separate means of expressing the ever-growing and changing domain semantics.

Not every platform is defined in this way. A *knowledge-driven platform* is one whose domain semantics are *explicitly* defined by knowledge artefacts rather than being implicitly expressed in software and database schemas. Such a platform for healthcare includes the following additional elements:

- *Domain content models*, describing data-sets and their elements used in the domain;
- *Computable guidelines and Care pathways*, expressing diagnostic and therapeutic plans, and associated decision logic;
- *Terminology value-sets*, expressing allowable values of coded elements within other models;
- *Portable queries*, used to interrogate data, based on knowledge and information models, and terminology; and
- *Terminology sub-sets*, the choice of available terminologies and/or limited subsets for use in the platform.

Underlying this are pure domain descriptions in the form of *Terminologies and Ontologies*, for formally describing domain entities (universals) and their natural relationships, i.e. domain 'truths' as described in the chapters covering health data standards and clinical modelling.

The knowledge artefacts mentioned above are created as *an ongoing activity* when the platform architecture is put to use. These models [6] provide the semantics to implementations of the platform services and other components. A major part of defining the platform for an organisation or geography is consequently the establishment of the knowledge modelling activity, and engaging domain specialists to work within it. We can think of this as establishing a *knowledge factory* run by domain experts, rather than IT staff.

We can visualise this form of the platform as shown in Fig. 4 as follows:

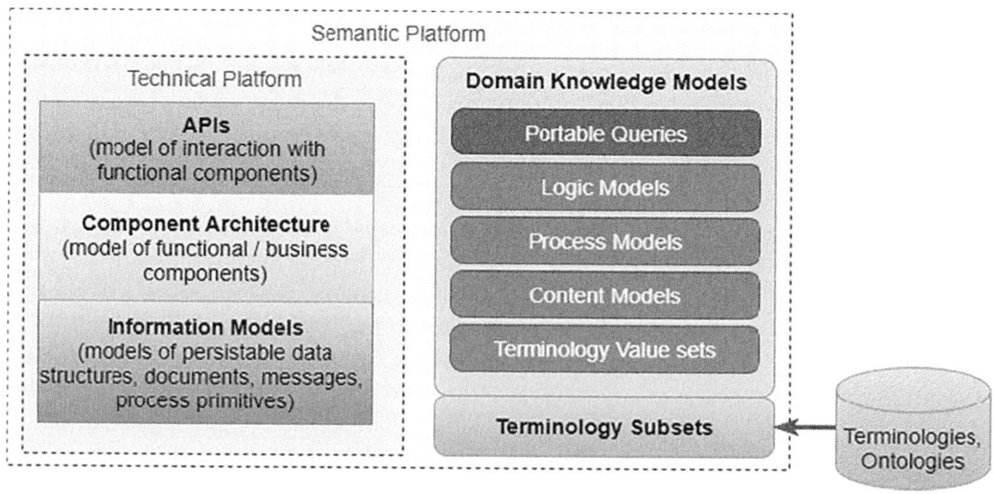

FIG. 4 Semantic platform separating IT from domain knowledge.

The use of formal knowledge artefacts separate from the software and database allows for the possibility for domain knowledge to be *consumed at runtime by implementations*. Back-end services and databases can then be constructed with no direct dependency on domain-level models, enabling them to be deployed and maintained with no modification required as new knowledge models are added. This is the same principle on which today's rule and workflow engines operate, generalised to the totality of an information systems environment.

A major advantage of a knowledge-based platform is the possibility of *portable queries*. These are queries based on:

(a) the logical information model and
(b) relevant knowledge models,

rather than the concrete back-end database schemas, the latter of which may be optimised in structures quite distant from those of the domain.

Portable queries will run on any implementation of the same platform specification, even if the database of one is designed as a 3NF (third normal form) classical schema, and the other is a highly optimised path-based hierarchical, object or other no-SQL persistence engine.

Portable model-based queries open the door to a viable clinical decision support (CDS) productisation, since a CDS product may now have its data retrieval queries written once for all implementation environments, rather than having to solve this problem for each one. The cost of this latter problem has all but killed CDS as a viable industry to date.

Knowledge-representation formalisms

A foundational aspect of knowledge representation is the set of *formalisms* used to express the domain knowledge artefacts. Well-designed formalisms are crucial to any platform's longevity and success since they determine the sophistication and coherence of what may be expressed in the models. In particular, the formalism for each kind of artefact must be as *close as possible to the cognitive models* of the designers and users of that artefact. Cognitive dissonance forces model-builders to, first mentally translate into unfamiliar primitives their familiar concepts in the manner of old programming languages such as Assembler and C, weakening the ability to clearly express domain semantics or later understand them. Similarly, a lack of type-coherence across all knowledge artefacts will act as sand in the wheels of any instantiation of the platform.

Typical formalisms that may be used for domain knowledge representation in a digital health platform include the following as described in Table 2.

Formalisms of course need to be implemented in tools to be of any use, and the tool outputs need to be usable either to generate software or else directly in execution environments.

Thus, in order to create a knowledge-driven platform, the following infrastructure is needed:

- *Knowledge engineering environment:* the large-scale use of knowledge models requires the use of domain-oriented authoring tools, intelligent repositories, and governance procedures;
- *Knowledge artefact translation:* tools used to translate knowledge models into developer-usable artefacts such as data schemas, JSON, APIs, and source code used for application development.

TABLE 2 Formalisms used to represent knowledge in an digital health platform.

Artefact	Formalisms
Clinical data items: a reusable library of domain-level clinical information items and group, e.g. heart rate, blood pressure observation, Apgar, data items relating to recording a diagnosis, an order, etc. Known as *archetypes* in openEHR [6]; Detailed Clinical Models (DCM) by ISO	• openEHR ADL2 (ISO 13606-2)
Clinical data-sets: use-specific (e.g. local and/or application specific) data-sets constructed from archetypes. Known as templates in openEHR	• openEHR ADL2 (ISO 13606-2) • openEHR ADL 1.4 templates
Terminology value-sets: value-sets defining allowable values for specific data items within forms based on clinical data-sets, and in other contexts (e.g. documents and messages);	Various forms defined by HL7 FHIR, OMG CTS2, SNOMED International
Plans: structured plans consisting of Tasks, wait states, Events, decision points, used to represent work following a common structure, e.g. a birth procedure guideline; ante-natal care; chemotherapy administration Plans generally connect to decision logic rules to generate values determining control flow at decision points	• openEHR Task Planning; • OMG BPMN/CMMN (BPM + version; emerging)
Decision Logic: modules expressing clinical logic, rules, etc.; enables representation of scoring algorithms, diagnostic guidelines; treatment selection; risk analysis	• openEHR Decision Language (emerging). • OMG DMN (BPM + version; emerging). • HL7 CQL (development). • Others: Arden, GLIF, Gello
Portable Queries: queries based on clinical data-groups, data-sets	• openEHR AQL • HL7 CQL
Business Rules: condition/action rules (aka Event-Condition-Action or ECA rules) that are not specifically clinical logic or connected to clinical variables are often generically required in an environment, and are usually implemented with generic rule engine technology and languages	• ILOG, SWRL, KRL etc. • openEHR, OMG DMN/FEEL

We can visualise the knowledge-driven platform and related elements as shown in Fig. 5 as follows:

One crucial aspect of this platform concept may easily be missed: there is **one library of knowledge models for the whole platform**, that is, for all components, such as EHR, demographics, scheduling, orders, and so on. This means that components that may be used in the platform not only obey the basic type system (core information models), API conventions, and design, but are driven from a common set of domain definitions. Domain modelling thus becomes an independent activity, both for platform procurer and implementer alike.

The knowledge-driven platform thus presents the possibility of a major rearrangement of industry effort, with the possibility of realising far greater efficiency than today, where all products independently define their own version of needed semantics, inevitably in inconsistent ways.

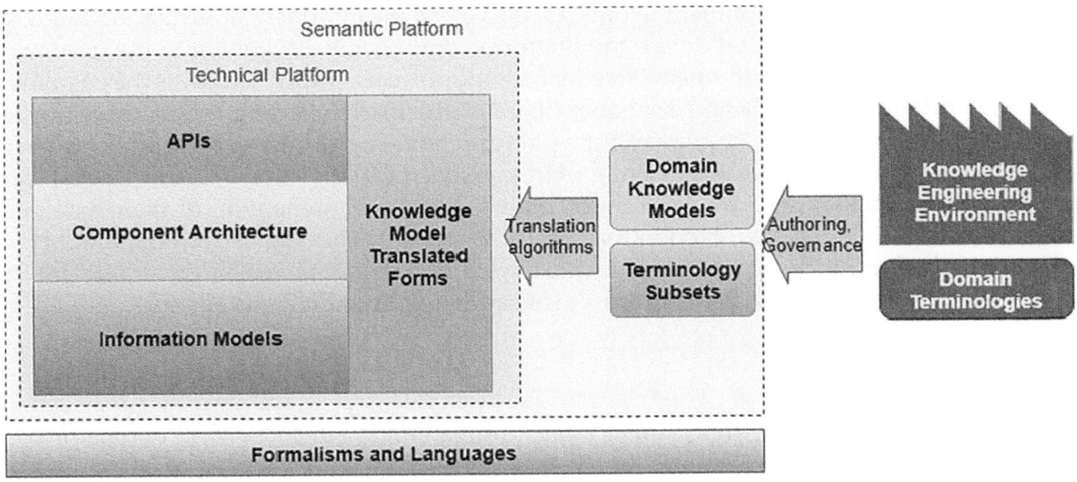

FIG. 5 Knowledge-driven platform and related elements.

The diagrams shown here are of course abstract visualisations of any platform; in practice, a particular platform is usually visualised in terms of its component architecture and related knowledge artefacts.

Platform evolution

A *realistic platform is always a living entity* rather than something that can be developed to completion. Hence, the set of services and components at any point in time represents a snapshot on the development journey of the platform itself. The core information models and particularly the formalisms need to be stable early on, in order to enable the development of higher-level components. The development and subsequent deployment of a platform are therefore an *incremental process.* A key aspect of its evolution is to maintain internal consistency and coherence, such that every new element added creates no semantic gaps or technical problems.

One simple route to improvement is the addition of new components over time: any component originally developed as a client of the platform may later become part of the platform, if it has a published interface that follows the established rules, and can be made coherent with the core type system, etc., and other assumptions of the platform.

Existing components may also evolve in the usual fashion with new functions added to APIs in a backwards-compatible way with previous versions.

Again, a major benefit of knowledge separation is that domain definitions are *consumed* by platform components rather than hidden inside them.

What does the platform mean for business?

A successful platform can be specified and maintained by the industry it serves. If it does this well, the result will be a **platform economy of plug-and-play vendors** competing on application quality rather than data lock-in. This will of course take time and will be different to the current health IT environment in one crucial aspect: the interfaces, information models, and knowledge definitions that define the platform components are known to the customer organisation and used as requirements to which vendors must comply. This puts the customer and user community in the driving seat operationally and financially. Fig. 6 represents the interactions between multiple services and the platform enabling the establishment of an open platform-based economic ecosystem using platform specifications as conformance points for procurement purposes. It means customers retain control and ownership of their data and that clinical experts can be involved in the specification and steady-state development of the system into the future. The ability to reuse existing components enables far greater flexibility and represents economy of scale. There is a growing list of openEHR suppliers.

Taking back control

From a business point of view, *a knowledge-driven platform represents a set of conformance criteria held by the procurement side,* rather than a secret architecture hidden inside products on the vendor side. It makes the buyer *the owner of the solution,* and the industry a supplier of well-defined components rather than total solutions. It also makes the procurement side (or a dedicated contracted entity) the *master integrator* in the overall IT environment, since it, not a vendor, is now the owner of the interfaces between components/products. This makes *integration a controlled and managed activity,* rather than an ad hoc one of endlessly trying to connect black-box products with ill-matched message standards. Since the onus is on vendors to implement the required component interfaces, integration is a far more tractable activity than it is today.

A platform specification can be used by procurement to:

- Purchase component implementations *incrementally* rather than in a big-bang fashion;
- As a *basis for component-level formal conformance* within tender documents or other modes of procurement of components;
- *Escape from vendor lock-in,* since any offering that conforms to a particular API and information model (such as Clinical Data Repository or Terminology Service) could

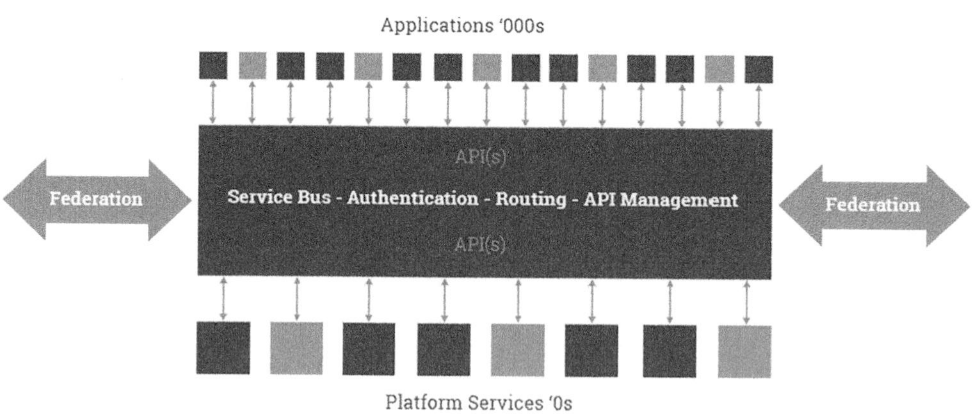

FIG. 6 Platform services.

be provided by more than one vendor, and a non-performing vendor (for example, one whose support is poor) could be replaced by another whose product has the same conformance; and

- Enable *data ownership by the provider* and/or data subject, not product vendors.

In addition to the above advantages that accrue with any well-defined technical platform, a knowledge-driven platform brings further value, including the following:

- *Domain semantics are no longer locked inside vendor products*, not even at component or UI form level, and are instead developed externally to the IT effort;
- *Domain experts and users become direct actors in the development* of final products through their efforts in creating the knowledge artefacts defining data, process, and forms of the final delivered solutions;
- Artefacts for a given domain entity (i.e. data group and process definition) can be *defined once for a country, and often the entire domain*, rather than being repeatedly redefined in competing, irreconcilable ways by both vendors and standards profilers; and
- *Vendors can make use of domain expertise* (e.g. via professional organisations), reducing their own costs, and expending effort on other aspects like performance and UX (user interfaces).

The big picture

The use of a *platform entails the emergence of a 'platform economy'* whereby small and large vendors can build components *according to the platform* to offer into a common market. This differs from the mega-suite approach to health IT which effectively locks all other vendors out of IT provision. A platform economy therefore requires *high confidence in the platform definition*, both in terms of technical quality and in terms of governance and management by its specifiers.

Current mega-technology companies such as Facebook, Amazon, Apple, Google, Microsoft, and Uber have adopted a platform business model able to offer multiple products and services globally. They have integrated e-commerce, cloud computing, logistics, consumer electronics, social media, gaming, and finance into their own unique ecosystems, to a greater or lesser extent, based on a customer focus. These companies have provided incentives and partnered with others to create new efficiencies [7] and demonstrated a myriad of advantages given the exponential growth of their platform users. This significant change where users are the producers and the platform owner has become the facilitator of value creation. Leveraging digital ecosystems requires collaboration between multiple stakeholders, including adjusting, supporting, and complementing numerous joint efforts and the integration of multiple ecosystems, where each is well positioned to meet the specific local needs. Chun et al. [7] note what's needed for an evolving ecosystem is a holistic approach and a design with a market gain focus creating value for all participants. Jacobides [8] noted that:

> Digital ecosystems can mimic organic organisms by growing in unpredictable directions, depending on where they find the most nourishment. Containment is necessary through strong frameworks to ensure that the broader societal implications receive appropriate consideration.

This observation is particularly relevant to the health industry given its size and complexity where the benefits of these transforming developments are expected to have enormous efficacy. These digital platform ecosystem developments are driving changes in conventional governance and trust models that are redefining traditional boundaries between public-private sectors, citizens and governments, and regulators and those being regulated.

A scholarly literature review of digital platform ecosystems [9] found that this concept has been viewed from multiple perspectives, market, business, or economic based, technical or socio-technical. An examination of its three structural elements, activities, actors, and architectures, reveals numerous interdependencies. These authors defined a digital platform ecosystem as:

> ... a platform owner that implements governance mechanisms to facilitate value-creating mechanisms on a digital platform between the platform owner and an ecosystem of autonomous complementors and consumers [9].

Digital platform ecosystems may be considered to have three essential, yet differing attributes [9]:

- Platform ownership
- Value-creating mechanisms
- Autonomy of complementors

Similar change is now emerging for the health industry as described in the previous chapter. The openEHR Foundation provides well-governed openly available platform specifications, now adopted and in use by multiple providers. Its value-creating mechanism is the Clinical Knowledge Manager, and its complementors represent the tooling made available to develop well governed formalisms which are evidence based, developed,and reviewed by its global multidisciplinary community.

Getting there

The mixed development economy

In the typical health IT environment of today, the adoption of a platform approach means introducing a new mode of development that will initially be used to add new components, typically at least a true patient-centric health record. However, platform-based development will not instantly replace the current mode of development and/or maintenance for many systems for some time, during which such replacements may be made carefully and prudently. Consequently, platform-based development will entail new kinds of activity and management in addition to extant modes of development and management. *It will therefore impose a cost in the short to medium term,* since existing activities will not in general disappear overnight.

Adopting a platform approach sustainably is therefore a question of managing two competing work-effort curves over time:

- The reducing work effort relating to legacy systems;
- The increasing work effort relating to new/modified platform-based systems.

The promise of the platform approach is that development of equivalent functionality under the new approach will be *substantially more efficient* and thus cheaper and quicker, than under the existing mode of operation.

A reality check

It is essential that an adopting organisation understand that substantial savings are only likely to be realised in the medium to longer term, and also only at a medium to large scale, rather than in small independent projects. This is because paradigm change involves systemic change, and it takes time to be normalised and internalised into any working environment. Applying it on a small scale will not generally create visible savings.

Following the above, it is *strongly recommended that the organisation focus initially on building the skills,* e.g. via the establishment of a platform-based 'lab' within the enterprise, without expectation of quick savings compared to current effort. This is not to say that no early gains can be made, just that the sum total of upskilling cost versus savings on the 'old mode' of development should not be expected to go negative (i.e. for the platform to start to pay for itself) for some time. It also has to be recognised that where major products exist, e.g. large-scale EMR, PACS, Lab solutions, or similar, and expected to stay, that development and maintenance activities within those products will follow their own structures for some time.

Initial steps

Initial steps towards these goals are mainly aimed at orienting key staff to the new concepts, technology and planning need for the approach, and include the following:

- Initial *training* for key staff who will be the core advocates and planners;
- Joining technology *communities* including openEHR, SNOMED, IHE, and HL7 as needed; and
- Setting up a *platform technology lab* to act as an internal familiarisation and training site, using, e.g., components like openEHR EhrBase and/or commercial tools, archetype, and template design tools, to demonstrate 'full-cycle' model-to-app development.

This level of orientation imposes only a modest cost, but will position the organisation to make its own detailed evaluation of both methodology and specific elements of technology, in order to determine feasibility, risks, and planning for the future. *It also provides the organisation a logical point at which to re-evaluate the strategy if it so needs.*

Moving towards, a stable target state involves the establishment of the three development streams/offices plus relevant oversight board/office, training group, etc. Extensive use can be made of the existing clinical and technical community members whose deep knowledge of the clinical and ICT environment will be invaluable to the success of the project. It is strongly recommended that some specific clinical needs, e.g. COVID, ICU care, or any other priority, be used to drive initial development, rather than a big-bang approach.

Early projects

In order not to disrupt the existing ICT environment in unexpected ways, deployments based on the new technology need to be carefully chosen. Projects that can add value with minimal disruption might include the following.

Secondary use: Reporting

Value can be added through reporting capabilities that are not currently available. These would be developed as applications running over an initial version of the patient-centric EHR, itself provisioned from existing data sources. This category of project involves only reading and provides a good path to familiarity with the new platform CDR (Clinical Data Repository) (openEHR), querying language, clinical models, and application development tools.

Regional summary EHR

A next possible step is to build up the content import from existing systems to the platform CDR to be sufficiently comprehensive for clinical use. This typically includes the following content:

- Basic demographics
- Managed Lists:
  - Master problems/diagnoses list
  - Current medications and other orders
  - Allergies and interactions
  - Vaccination history
- Family history
- Social situation
- Recent labs
- Past procedures
- Recent encounter/admission history

This summary content can be provided via new application screen(s) and dashboard to all care settings in the region, augmenting any locally available EMR. Initially, this may be particularly useful in, e.g. social care, mental health, out-of-hours encounters, and emergency services. In the simplest form, the summary EHR is read-only, but provides a starting point for centralising the managed lists (medications, problems, allergies) to the new patient-centric EHR from current disparate sources.

Adding intelligence: Diagnostic/risk assessment guidelines

The addition of diagnostic guidelines can be undertaken to add significant value via prevention, without harming existing clinical care delivery. Nearly 500 computable score- and scale-based GDL2 (Guideline Definition Language) guidelines are available, along with tooling [10]. These guidelines are based on archetyped data and would be implemented by using the relevant archetypes within dedicated templates to extract data from legacy sources, and populate the patient-centric EHR within the platform CDR. They could be run by a monitor that detects patients with certain risk profiles (e.g. high BP, smoker, and age) and executes relevant guidelines, or manually, as part of routine visits. The output of such guidelines is generally recommendations for certain medications or treatment pathways. This function is also read-only with respect to the back-end EMR systems.

Aggregating data for high-risk patients

One of the value-adds of the platform CDR is to act as a concentrator for data sources, including existing EMRs (GP, hospital EMR, etc.), wearable devices, and patient-provided data to build a new high-quality EHR for patients with certain key diagnoses. The HiGHmed project [11] is taking this approach with high-risk cardiology patients.

Medical research

Medical research projects constitute another category of function that only requires read-only access to the existing EMR systems, but needs intelligent querying over a high-quality longitudinal record. Oncology typically provides opportunities for outcome studies based on historical patient data, i.e. diagnoses, treatment pathways (particularly specific chemotherapy regimes), and outcomes over the subsequent years.

New functionality: Care management

A major new area of value available to most healthcare organisations is automation and computation, not only for routine performance monitoring and reporting to external entities but also at the process level of care, including the following:

- Computable care plans;
- Computable care pathways and therapeutic guidelines;
- Care coordination for team-based work, e.g. acute stroke and sepsis; and
- Long-term plan management, e.g. diabetic patient monitoring.

The openEHR approach to this area adds new components (e.g. Task Planning and Decision Logic engine) but does not assume an openEHR CDR—it is designed to have bindings to non-openEHR source systems as well. This is a longer term project, as the technology is still emerging.

Replacing a legacy EMR

Replacing a legacy EMR is not a universal goal, but for some organisations it is a necessity, usually due to technological obsolescence. Where it is embarked upon, it will undoubtedly be the most challenging project in an overall platform adoption programme. The key to success is likely to be in accordance with an implementation of a plan based on the following:

- Implement summary EHR, reporting, etc. as above, so that the technical capability to extract patient data into the platform CDR is already established.
- Establish a clinical dashboard(s) running over the platform CDR, to act as a 'container' for new apps.
- Break down all the system screens/functions to a well-defined set of clinical components (probably task-oriented rather than data-oriented) which are candidates for 'sub-applications' in a new implementation.
- Move any clinical intelligence/reporting functions over to new apps running over the platform CDR.

- Potentially re-engineer some forms into more coherent functional components that resemble eventual apps that would replace them in the platform CDR environment.
- Progressively implement the components as apps within a clinical dashboard environment running over the platform CDR.

For some time, it will be essential to limit the scope and impact of re-engineering the existing EMR. This can be done by various methods:

- Implementing initially for specific limited categories of patient, e.g. aged care, prison population, sports medicine, and ambulance;
- Implementing in a limited geography; and
- Deploying only to 'friendly' clinics or carers who have already been brought into the platform technology and knowledge engineering environment, and are motivated to bring the new technology to fruition.

Once a limited group or category of carers has successfully moved to the EMR replacement, experience will have been developed that indicates which particular strategies to follow for wider roll-outs.

Application development group

The Application Development Group is a group dedicated to quickly producing the usually major subset of relatively uncomplicated applications that are in widespread and frequent use across the user base. The mode of development is no- and low-programming, i.e. applications mainly or partially generated via dedicated tools that enable smart forms to be built from relevant knowledge artefacts (at least templates and terminology value-sets).

The responsibilities will include the following:

- Analysis of initial new application proposals;
- Creation and analysis of proposals to convert existing applications (or screens, etc.) to being platform-based;
- Establishment of UI/UX internal standards for the organisation;
- Form design, including re-engineering from forms in older applications;
- Determining requirements for new and modified knowledge artefacts based on application requirements; and
- Application usability testing.

Application quality testing prior to deployment will normally be performed in conjunction with the Platform IT group, based on clinical safety guidelines/standards, and on established platform technical compliance criteria.

Organisational structure for platform development

Below is a rough outline of *dedicated new or re-purposed* organisational units or functions that will need to be established, visualised in Fig. 7 as follows:

Note: the organisation unit names used here are for convenience in this document; the actual names and organisational unit types will of course vary from enterprise to enterprise.

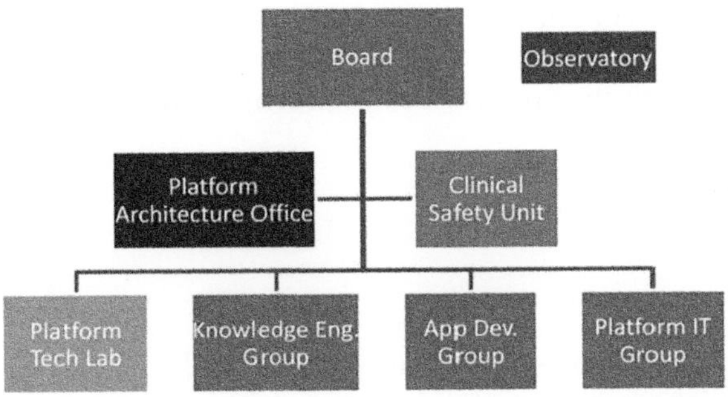

FIG. 7 Platform dedicated new or *Re*-purposed organisational structure.

Units for *deployment/delivery* and *change management* are not shown here, but are of course essential for the overall functioning of the organisation; they might be subordinate to the Board shown here, or to a higher-level managerial entity, depending on the current organisational structure and culture. Change management is covered in another chapter.

Project management is also not shown as a separate unit here, since it is assumed to be an organisational wide function. However, there may be a 'projects' group, depending on the style of development undertaken.

Generally speaking, deployment, change management, and project management would apply to the entirety of ICT work within the organisation, including both 'new' platform-based engineering and legacy maintenance, development, integration work, data, and knowledge governance.

Board-oversight unit

This unit is a core strategic group which needs to be established as the top-level coordinator and strategic decision-making entity for platform-based engineering and includes senior representatives of other organisational units mentioned here. The initial form of the Board (or differently named oversight group) will be informal and minimal, or else part of an existing oversight/management unit. It needs to undertake the following functions:

- Initial advocacy;
- Development of detailed proposals for introduction of new paradigm and technologies;
- Initial organisational set-up (i.e. of skills lab and other units); and
- Coordinating major organisational units as they are established, i.e. core IT, application development, and knowledge engineering.

This unit might also be part of a larger 'digital health board' whose scope is the totality of health IT in the organisation.

Platform architecture office

The Platform Architecture Office (however named) will be responsible for:

- developing and maintaining the *architecture of the platform* as used within the enterprise;
- determining how to *obtain and deploy* implementations of key platform components;

- *planning*, software engineering estimations, impact analysis, etc. of the platform as an emerging technology within the enterprise; and
- *tracking relevant standards* and other external activities relevant to platform evolution.

It may also participate in the development of a new platform specification (e.g. a new API) within the relevant external platform community.

Platform technology lab

This unit is a permanent technical group that enables the technology to be learned and tested in a dedicated space. The early establishment of a platform tech lab within the organisation will act as a location for developing organisational familiarity with the platform technology. One way to do this is to set up a dedicated hosting space where instances of the relevant technology can be set up, and technical staff—both IT professionals and candidate future application-builders—can learn the new technology and methodology. The deployed platform instance can also be used to test the technology in various ways, with the aim of refining the methodologies to be used within the organisation.

As time goes on, the tech lab would potentially take on the following functions:

- Train IT staff on the back-end aspects of the technology, possibly with a semi-formal or formal internal certification;
- Train analysts/staff with intermediate IT skills as low-/no-programming application developers;
- Deploy at least one platform implementation in a permanent test and demonstrate capacity;
- Potentially deploy and test various vendor platform implementations;
- Build demonstration applications;
- Undertake performance and volume testing on particular platform components;
- Test proof-of-concept technical ideas;
- Provide an environment for visiting experts to provide demonstrations, training, etc.; and
- Develop and document the organisation's version of the platform and methodology.

Knowledge engineering group

The Knowledge Engineering Group (however locally named) is a permanent organisational unit whose job is to select, validate, customise, and develop knowledge models in response to organisational needs. Its organisational functions will include the following:

- Training of new internal recruits;
- Creating regional network/outreach in order to find interested individuals from various kinds of clinical settings, medical specialties, etc.;
- Running training provided by internal and external experts;
- Determining local methodology, particularly with respect to reverse-engineering existing enterprise 'knowledge', language translation and other localisation needs;
- Planning the delivery schedule; and

- Acting as the coordinator and representative of the organisation on relevant national and international bodies concerned with knowledge artefact development, e.g. national digital health board, SNOMED International, national or international professional associations and colleges.

In terms of substantive deliverables, its work includes the following:

- Identifying, analysing, and capturing existing 'knowledge' embedded in the current IT, applications, etc., i.e. 'mining wisdom from the goldmine';
- Producing clinical content models (including translation);
- Identifying terminology-related resources for use by the organisation, including terminologies, subsets, and versions;
- Determining how to use terminology resources within the organisation; and
- Sourcing and producing clinical logic modules, including computable guidelines based on published guidelines in use in the organisation.

Platform IT group

This group is the unit consisting of IT professionals undertaking complex IT development activities, including any non-trivial programming-based development, tool building, performance testing, and application quality assurance.

Responsibilities will include the following:

- *System critical attributes*: managing performance, volumetrics, scaling, etc. of the platform deployment;
- *Operations:* hosting, virtualisation, and optimisation;
- *Security*: of applications, authentication, and back-end systems;
- *Back-end architecture*: custom needs of the back-end, e.g. special kinds of searching and customisation of generic platform components;
- *Complex services and applications*: any non-trivial application containing significant logic, not already provided by externally sourced components, e.g. bookings management; and
- *Systems Integration*: development and validation of integrations between platform components and external systems and data partners.

Clinical safety group

This unit is responsible for sourcing and/or developing standards and methods used to assure the clinical safety of deployed applications and other services, e.g. integration components. Such standards are designed to:

- Ensure fidelity of data rendering, i.e. that what is on the screen is a correct rendering or visualisation of the relevant data in the back-end;
- Ensure fidelity of data transformation, i.e. that in any scenario where data from one source (e.g. lab result messages) are converted to another form (e.g. a native platform format), no semantic errors are introduced into the data by the conversion.

The work of this group will be used to inform quality and safety testing of applications and other platform components.

Observatory

A special *monitoring unit* termed the Observatory here should be established towards the beginnings of routine operations of the platform-based development function. Its remit is to establish metrics and means for measuring effects of the technology and related (procedural, etc.) changes in the operational environment. Monitoring may occur for different types of outcome, to determine if:

- Clinician workload is reducing;
- ICT development efficiency is increasing; and
- Clinical outcomes are improving (e.g. due to the use of scores and guidelines).

This unit should be staffed in such a way as to be independent of development units and might be part of an existing organisational unit that undertakes such a function (typically one measuring at least clinical quality) or a clinical governance unit.

Conclusion

This chapter has described the knowledge-driven platform in technical, economic, and change management terms. The platform viewpoint sets the granularity of procurement and development to be the component rather than the product and empowers the procuring side by making it the owner of the overall platform architecture to which vendor offerings must conform.

The advent of formalised domain knowledge, expressed as computable models as the way to define most of the platform semantics, including clinical data, workflow, and decision logic, provides a second level of empowerment to procurement, since it is now clinical professionals who do this work.

As a consequence, the economics and management of work effort are substantially modified by the knowledge-driven platform, such that two new activities—knowledge engineering and low-code application development—replace much of the work effort (mainly product and integration and standards adaptation) in a non-platform environment. Professional IT effort is now limited to managing and developing the platform, tools, and development of high-value, complex applications.

Moving to a platform paradigm is not a simple task, and the introduction of new modes of development has to be done carefully over time to avoid endangering current systems availability and also to allow new modes of working to be absorbed by both clinical professionals and the supporting IT organisation.

References

[1] Intermountain-Healthcare, OpenCEM CEDAR, 2021, Available from: http://www.opencem.org/#/.

[2] T. Beale, Archetypes: constraint-based domain models for future-proof information systems, in: K. Baclawski, H. Kilov (Eds.), Eleventh OOPSLA Workshop on Behavioral Semantics: Serving the Customer, Northeastern University, Boston, 2002.

[3] openEHR-specifications, REST API specifictions (ITS-REST) Component—Latest Release 1.02, 2021, Available from: https://specifications.openehr.org/releases/ITS-REST/latest.

[4] openEHR-Specifications, openEHR Architecture Overview, 2020, Available from: https://specifications.openehr.org/releases/BASE/Latest/architecture_overview.html.

[5] openEHR, CDS, Guidelines and Planning Overview—Figure 9, 2021, Available from: https://specifications-test.openehr.org/releases/PROC/latest/overview.html.

[6] openEHR/CKM, Clinical Knowledge Manager, 2019, [cited 2019 21 February]. Available from: https://ckm.openehr.org/ckm/.

[7] V. Chung, M. Dietz, I. Rab, Z. Townsend, Ecosystem 2.0: Climbing to the next level: McKinsey-Digital, 2020, Available from: https://www.mckinsey.com/business-functions/mckinsey-digital/our-insights/ecosystem-2-point-0-climbing-to-the-next-level.

[8] M.G. Jacobides, A. Sundararajan, M. Van Alstyne, Platforms and Ecosystems: Eabling the Digital Economy, Briefing Paper: World Economic Forum, 2019, Available from: http://www3.weforum.org/docs/WEF_Digital_Platforms_and_Ecosystems_2019.pdf.

[9] A. Hein, M. Schreieck, T. Riasanow, D.S. Setzke, M. Wiesche, M. Böhm, H. Krcmar, Digital platform ecosystems, Electron. Mark. 30 (2020) 87–98.

[10] openEHR. Guideline Definition Langage (GDL). Available from: https://specifications.openehr.org/releases/CDS/latest/GDL.html.

[11] B. Haarbrandt, B. Schreiweis, S. Rey, U. Sax, S. Scheithauer, O. Rienhoff, P. Knaup-Gregori, U. Bavendiek, C. Dieterich, B. Brors, I. Kraus, C.M. Thoms, D. Jäger, V. Ellenrieder, B. Bergh, R. Yahyapour, R. Eils, H. Consortium, M. Marschollek, HiGHmed—an open platform approach to enhance care and research across institutional boundaries, Methods Inf. Med. 57 (2018) e66–e81.

CHAPTER

7

Are professional guidelines and regulatory standards fit for purpose?

Evelyn Hovenga[a,b,c]

[a]Digital Health, Australian Catholic University, Melbourne, Australia [b]eHealth Education Pty Ltd, Melbourne, VIC, Australia [c]Global eHealth Collaborative (GeHCo), Melbourne, Australia

Introduction

The previous chapter identified the diversity of technical standards development organisations in terms of their areas of focus, technical disciplines, investment commitments, stakeholder engagement, and motivation. Similar, if not greater, degrees of diversity are discernible amongst the many organisations and entities that have taken on a standards or guideline development role. Many of these are health discipline/clinical specialty specific; others such as entities concerned with health quality and safety are more encompassing in terms of scope. In every case, there are a multiple of authoritative national or global entities independently developing standards and guidelines for the same purpose, often resulting in overlapping or conflicting evidence-based requirements.

There are many types of standards, as well as guidelines and regulations, that collectively provide a desired set of infrastructure standards. These additional types of standards include occupational health and safety standards, health professional standards, patient safety standards, public health, and health reporting standards, which collectively contribute to any nation's health system infrastructure. Many of these are developed to uniquely apply to individual countries. All of these need to be considered regarding relevancy when developing a digital health ecosystem as there needs to be the capacity to monitor and evaluate adoption and compliance.

There is no shared opinion on the role and need for standards in health care, other than an acknowledged patient data sharing requirement. The world has a desperate need for an agreed vision of a transformed health system made possible by the best possible use of available digital technologies and knowledge, well suited for our new digital era.

For any given context, information system standards address issues of order and compatibility in the design, development, implementation, and operation of information systems and information technology [1]. Digital health ecosystems need to be able to provide evidence and new knowledge from which to generate new professional practice standards or update current standards and guidelines. These also need to be integrated for use within the health

Roadmap to Successful Digital Health Ecosystems
https://doi.org/10.1016/B978-0-12-823413-6.00017-3

system generally, preferably in a manner that provides ease of access for users, where and when required and with an ability to update the evidence, based on real-time operational data.

> Clinical practice guidelines are statements that include recommendations intended to optimize patient care, that are informed by a systematic review of evidence and an assessment of the benefits and harms of alternative care options [2].

Data and terminology standards reflect the health language in use and form the basis for representing these guidelines. Health data standards are intricately linked to health informatics standards that apply to system interoperability. How can guidelines and regulatory standards be integrated into digital health ecosystems for ease of use? How can health data best be collected and used to provide the necessary evidence required for the development of practice and regulatory standards or guidelines?

The previous chapter identified the diversity of technical standards development organisations in terms of their areas of focus, technical disciplines, investment commitments, stakeholder engagement, and motivation. Similar, if not greater, degrees of diversity are discernible amongst the many organisations and entities that have taken on a standards or guideline development role. Many of these are health discipline/clinical specialty specific; others such as entities concerned with health quality and safety are more encompassing in terms of scope. In every case, there are a multiple of authoritative national or global entities independently developing standards and guidelines for the same purpose, often resulting in overlapping or conflicting evidence-based requirements.

Data, terminology and their technical integration within digital health ecosystems, are explained in some detail in chapters to follow. Here we examine who are the primary developers of these standards and guidelines, what is their purpose, what are the foundations for these developments, how are practice guidelines represented, and do their presentation formats enable digital integration for ease of use?

Valuing infrastructure standards and guidelines

Successful digital health ecosystems need to be the result of adopting a set of key data, technical, regulatory, and governance standards that collectively specify a 'fit for purpose'. These types of standards, regulatory and legislative requirements, plus the adoption of any number of guidelines, provide a viable information platform for digital health technology commercialisation and implementation. Standards need to be evidence-based and suitably developed to meet global and national health, welfare, and digital health requirements. Widespread adoption of any type of standard has the potential to provide a cost-effective central point of agreement that reduces system development and information-exchange maintenance costs.

It is also useful to be aware of the most relevant or influential global and national organisations providing guidelines and principles that may be used to inform national standards as well as regulatory and legislative mandates. Global entities as discussed in the previous chapters need to work cooperatively with their members, many of whom are country jurisdictions. Members need to agree to adopt common rules and, in some instances, empower an institution, such as the World Health Organisation (WHO), or an internationally recognised

Centers for Disease Control (CDC) to oversee their implementation and compliance. This requires global investments, mutual trust, and sound governance to be effective.

The availability of infrastructure standards, guidelines, and common rules assists health-care delivery providers to establish their own policies and procedures, identify legislative or regulatory requirements, or make use of these for the development of system acquisition and procurement guidelines. A perceived need to comply with key standards may provide context or be incorporated in Request for proposal (RFP) or in Request for Tender (RFT).

The adoption of common standards or guidelines is valuable for vendors as this provides consistency, enabling software developers to streamline their responses and software application developments to meet the needs of multiple potential customers. There are many standards or guidelines developed to ensure staff and patient safety or to meet health service provider accreditation requirements. They provide consistency within any health ecosystem, improving system cost-effectiveness. Collectively, these standards and guidelines also inform policy initiatives and funding allocations, as well as providing investment opportunities for innovation and economic growth.

Key standards and guideline developments

All standards and guideline development activities must be guided by contributions that are ideally evidence based and provided by those with the right expertise. Experts are those individuals who have acquired a great deal of knowledge and/or ability associated with a specific knowledge domain via research, education, and/or practical experience. Such experts have preferably gained their extensive knowledge and practical experience by their application in multiple cultural and disciplinary settings, gaining international community recognition and endorsement.

In 2000, an informal collaboration of people with an interest in addressing the shortcomings of evidence-based grading systems in health care was established as a working group [3]. This group developed a common sensible and transparent approach to grading quality (or certainty) of evidence and strength of recommendations with input from many international organisations. This Grading of Recommendations Assessment, Development and Evaluation (GRADE) approach is now considered the standard to be applied for guideline development.

Current world reality

There are different types of international and national guidelines and health or clinical standards development organisations (SDOs) that provide this service. They include government-funded NGOs (nongovernmental organisations), and other entities that have emerged over time. Such organisations emerged in response to an identification of need resulting from research projects and/or by international professional associations covering numerous disciplines and clinical specialties. Most have developed their own guidelines for the development of standards and guidelines. Many guidelines are based on expert opinion, consensus statements, and appropriate use, rather than on evidence obtained from reliable, valid, and reproducible research.

The collection of evidence used for the development of clinical guidelines tends to be from randomised clinical trials and from any number of central data repositories, including disease registries and hospital discharge data. For example, a number of large databases were made available for researchers such as the MIMIC database established in 2003, containing de-identified patient records. This has been expanded since then to include physiological and clinical data [4]. Many others, identified by the MIT Critical Data Consortium [5], are limited in scope, often obscuring many confounding variables, due to a lack of significant clinical data required to answer a wide range of research questions. These types of databases continue to have many research limitations due to incomplete, and poor quality data, and the inconsistent organisation of data sets. In the UK, the NHS Digital has adopted the mission to harness the power of information and technology to improve health and care to overcome these obstacles [6]. Many large health data sets are freely available via the NHS web portal.

The Institute of Medicine (IOM) published guidelines to assess trustworthiness in 2011 [2]; however, a later review of a random selection of guidelines extracted from the US National Guideline Clearinghouse the following year found that fewer than half of these satisfied more than 50% of IOM's standards [7]. Most guidelines tend not to include any guidance on how to present their guidelines in computable formats, a necessary component for the preparation of their widespread adoption and use. Guidelines need to be made available in a manner best suited to a fully digitised health ecosystem.

The rapid expansion of new knowledge makes it impossible for clinicians to stay up-to-date in the absence of timely digital support. Ease of access to relevant evidence available for clinical decision-making, at the point of care, tends to determine the frequency of such decisions being made based on trustworthy evidence, rather than on practice experience alone. Many additional factors relevant to individual patients, including patient values and preferences, also need to be considered to ensure guideline applicability. In addition, clinicians need to be able to assess the trustworthiness, scientific validity, and strength of the evidence base used for guideline developments. This requires transparency.

The research infrastructure prevailing in 2016 was considered to be inefficient [8]. This may have improved to some extent by 2021. Randomised controlled trials were identified as having limitations as they are known to be costly, labour-intensive, and slow, with results that are seldom generalisable. This still appears to be the case. Participation in the global clinical trial industry does have an economic advantage that serves as an incentive to maintain the status quo. For example, the conduct of clinical trials in Australia attracts sizeable investment from international companies. This exposes Australian clinical staff to world-leading healthcare practices. Through both commercial and public investments, clinical trials in Australia support thousands of jobs within the research and healthcare systems including doctors, nurses, researchers, administrators, and many more across a vibrant ecosystem [9].

The extensive implementation of EMR/EHR systems in recent years was expected to open the door to a massive amount of previously untapped data. This led to the acknowledgement of the potential power of big data in knowledge creation and the potential value of secondary analysis of EMR/EHR data. The merging or pooling of data from multiple EMR/EHR systems is highly desirable as this provides larger and more diverse study populations. The latter is entirely dependent upon EMR/EHR structures (system architectures) and data collection governance, quality, and consistency between data sets stored in these systems. The high degree of complexity associated with large data sets stored in legacy EMR/EHR systems has resulted in an inability to make use of traditional data processing methods [5]. Current data silos continue the underuse of data collected, due to the inaccessibility of a vast amount of data and/or numerous data inconsistencies that have an impact on meaning. These factors explain the limitations encountered when collecting evidence to support guideline developments [10–13]. Data use possibilities are determined by the degree of success following digital health transformation initiatives undertaken.

From a digital ecosystem road map perspective, there is a need to identify required standard foundational infrastructures that are able to accommodate the many and varied health-related issues that any national health system needs to be able to address, including any essential global considerations such as public health. This principle is not different from that adopted by, for example, the car manufacturing industry who identified that various combinations of standard parts were able to produce different models.

Guideline development and provider organisations

There are, far too many, health-related guideline development and provider organisations to identify and explore their efficacy here. Our primary focus is to examine the most prevalent and globally influential entities for the purpose of identifying desirable structural transformations to be considered, to improve the enablement of national health systems, to optimise their functionality within this digital era.

International Council for Harmonisation (ICH)

The establishment of the International Council for Harmonisation of Technical Requirements for Pharmaceuticals for Human Use (ICH) in 1990 was in recognition of historical tragedies that associated with the prescribing of thalidomide. It has evolved to better respond to increasingly global pharmaceutical developments and a growing number of regulatory authorities [14]. The ICH, funded by annual membership fees, has recognised the urgent need to rationalise and harmonise regulations to address rising costs of health care, an escalation of research and development costs, and the need for minimum delays in making new treatments available to patients in need. ICH has developed and maintains a number of quality, safety, efficacy, and multidisciplinary guidelines.

The latter refers to the use of their medical terminology (MeDRA), the Common Technical Document (CTD), and the development of Electronic Standards for the Transfer of Regulatory Information (ESTRI). Another widely used guideline is based on the principles of Good Clinical Practice that originated as a result of the World Medical Association's Declaration of Helsinki first published in 1964. The Good Clinical Research Practice for trials on pharmaceutical products publication was first published in 1995, to assist national regulatory authorities, sponsors, investigators, and ethics committees. Such guidelines are made available via the WHO website. Adoption of any of these guidelines can be voluntary, the result of an official directive or a mandatory requirement via legislation.

World Health Organisation (WHO)

The WHO makes use of its guideline review committee for the development of global guidelines that can impact health policies or clinical interventions. Guidelines on numerous health-related topics are made freely available by WHO. This includes a guideline developed by specialised expertise to support digital health companies, known as the WHO's classification of digital health interventions [15] and a guideline on digital interventions for health system strengthening [16].

Clinical Knowledge Manager (CKM)

This online web-based Clinical Knowledge Manager (CKM) [17] originated in 2004, from an identified need to develop a repository for clinical models (archetypes) described by the openEHR specifications and now described in the ISO13606-2:2019 Health Informatics—electronic health record communication-Part 2: Archetype interchange specification. Over the years, this freely accessible online application, along with its structured governance, has evolved to become a transparent, evidence-based online repository of clinical knowledge made available in a computable form.

This web-based application has been designed to facilitate online model development for any specific clinical topic, by its large and growing number of global multidisciplinary contributors. This repository supports knowledge governance within and between the health enterprises. It is an application enabling online collaborative development, the management and publishing of a wide range of clinical knowledge resources, including archetypes, templates, and terminology subsets. Templates represent clinical data sets specific to a 'use case' where the data elements making up each data (value) set are able to be linked to a standard terminology, thus supporting automation.

The openEHR's information model is native (integrated) into the system and used as a reference for the clinical models. This is a standardised structure and content model for clinical information that can be used in any healthcare environment. The information is held in an open repository and developed by cloud-based collaboration between healthcare and IT professionals around the world seeking to represent all concepts that may be required for use in an EHR. Users select from the maximal model the level of information needed for their needs—the use of a standard model means that all pieces of that model can be shared and understood and clinical decision support can be developed and shared.

National entities with global influence

Cochrane, a British international organisation, was formed in 1993 to organise medical research findings based on systematic reviews of all relevant randomised controlled healthcare trials [18]. Its library provides accessible, credible information to support informed decision-making based on reviews undertaking in accordance with its review methodology.

A number of national government-funded research entities provide access to numerous guidelines via a dedicated web-based portal, for example, one provided by the National Health and Medical Research Council in Australia **(NHMRC)** [19]. Others are provided by National Medical Associations/Colleges such as Canada's **CPG Infobase** [20] or the

American Medical Association (**AMA**) who partners with over thirty-five medical societies and government agencies, to provide quick evidence-based reference tools for physicians via its **GuidelineCentral** website [21]. The US Agency for Healthcare Research and Quality (**AHRQ**)'s guidelines are now provided by the **ECRI Guidelines Trust** (Transparency and Rigor Using Standards or Trustworthiness) [22].

The Emergency Care Research Institute (ECRI) was founded in 1968, became the AHRQ-designated evidence-based practice centre in 1997, and rebranded to be known as the ECRI in 2007. A publicly available repository of clinical guidelines is managed by its ECRI Guidelines Trust developed in 2018. The ECRI officially affiliated with the Institute for Safe Medication Practices (ISMP) [23] in 2020. This created one of the largest healthcare quality and safety entities in the world. The ECRI makes use of the AHRQ's National Guideline Clearinghouse Extent Adherence to Trustworthy Standards (NEATS) instrument [24] to develop TRUST Scorecards for all its guidelines. A summary of these guideline and regulatory providers is presented in Table 1. This is by no means an exclusive list; it serves as an example of what is being made available by internationally recognised entities.

Adherence to such guidelines is frequently used as a required quality assurance and patient safety management performance measure. These guidelines represent valuable resources for

TABLE 1 Guideline developers and providers: some well-known examples.

Global entity	Area of focus	Relationships	Values/outputs
GRADE [3]	Evidence grading and evaluation	Numerous international GRADE centres and networks	• Grading of Recommendations, Assessment, Development and Evaluation (GRADE) approach • Publications
ICH [14]—International Council for Harmonisation of Technical Requirements for Pharmaceuticals for Human Use	• Global pharmaceutical developments • Drive harmonisation activities worldwide • Maintain associated MeDRA, CTD, and ESTRI standards • Training material development and delivery	• National regulatory authorities, sponsors, investigators, and ethics committees • Pharmaceutical and related industries • Health service providers	• Quality, safety, efficacy, and multidisciplinary guidelines • Principles of Good Clinical Practice • Good Clinical Research Practice for trials on pharmaceutical products • Electronic Standards for the Transfer of Regulatory Information (ESTRI) • Medical terminology (MeDRA), • Common Technical Document (CTD)
WHO [16,25–28]—World Health Organisation	Global guidelines on numerous health-related topics that can impact health policies or clinical interventions	Guideline Review Committee Digital Health Companies	• Classification of digital health interventions • Digital interventions for health system strengthening • ICD, Family of Classifications • International Health Regulations

Continued

TABLE 1 Guideline developers and providers: some well-known examples—cont'd

Global entity	Area of focus	Relationships	Values/outputs
CKM—Clinical Knowledge Manager	On-line, freely available, evidence-based, clinical knowledge associated with all specialties, at every level of granularity, in computable formats	• ISO TC215 • Strong and growing multidisciplinary global community of practice • Clinical Model Editorial Group • openEHR Foundation	• Multidisciplinary collaboration • Strong governance • Sharable specifications of clinical information • Transparency • Evidence based • Support safe healthcare computing
Cochrane [18] and its global community	Organise medical research findings based on systematic reviews of all relevant randomised controlled healthcare trials	Healthcare researchers and providers	• Cochrane systematic review research methodology • Library—a collection of databases containing systematic reviews of high-quality, relevant, and up-to-date synthesised research evidence
ECRI Guidelines Trust [22]	Repository of clinical guidelines	AHRQ—U.S. Agency for Healthcare Research and Quality [24,29] ISMP—Institute for Safe Medication Practices [23]	• National Guideline Clearinghouse Extent Adherence to Trustworthy Standards (NEATS) instrument • TRUST Scorecards for all its guidelines
NHMRC—National Health and Medical Research Council in Australia [19]	Promote health, prevent harm, encourage best practice, and reduce waste	University-based researchers and other research centres	Australian Clinical Practice Guidelines Portal

decision support software developers and for use within the health sector generally. Such guidelines must be evidence-based, developed by a panel of experts, governed, and maintained by a professionally recognised organisation. Their development is most commonly funded by governments and/or professional communities of practice, such as professional colleges/societies. Ideally, guidelines are made available in computable formats to enable optimum knowledge management processes to be adopted. This includes their use for decision support systems and for the development of artificial intelligence algorithms. This requirement is described in greater detail in other chapters to follow.

Professional organisations/associations' knowledge and digital health contributions

There are hundreds of health professional or industry associations, each with their own disciplinary focus and body of knowledge (BOK). Many national professional organisations have international links. They are known as associations, societies, institutes, or colleges, mostly operating as not-for-profit, although many also operate as commercial enterprises to benefit their members and to influence others. Some are powerful lobby groups able to influence government policy, resource allocation, regulations, professional registration requirements, educational standards, and public opinion. Activities include monitoring and influencing legislation, practice evaluation, quality assurance, service, or educational accreditation; identifying, examining, and taking action on issues of significance to their members, professional development; and delivering specialised education; and developing clinical guidelines, policies, and codes of ethical practice.

World Organisation of Family Doctors (WONCA)

Founded in 1972, this not-for-profit world organisation of family doctors' membership consists of 118 member organisations plus over 800 individual general practitioners and family physicians in 131 countries and territories, with a collective membership of around 500,000 family doctors servicing more than 90%of the world's population [30]. WONCA works closely with WHO. The provision of a forum for the exchange of knowledge and information is one of its membership services. It represents its members, policies, and educational, research and service provision activities to other world organisations and forums concerned with health and medical care. It has published numerous policy statements reflecting its mission that is to improve the quality of life of the peoples of the world through defining and promoting its values. These include respect for universal human rights, gender equity, and fostering high standards of care in general practice and family medicine.

World Medical Association (WMA)

Medical associations from 31 countries formed the World Medical association in 1946 and replaced the Association Professionnelle Internationale des Medecins which had been established in 1926 with a membership of 23 countries which had suspended its operations during World War II [31]. WMA now has 114 National Medical Association members. Its mission is to serve humanity by endeavouring to achieve the highest international standards in medical education, medical science, medical art and medical ethics, and health care for all people in the world. It has official observer status with WHO enabling WMA to submit policy statements.

The WMA has a strong medical ethics focus; its medical ethics manual is available in many languages. The WMA has published many statements, some are relevant to digital health, such as on Healthcare Information for All, Augmented Intelligence in Medical Care. Its declarations are available as a handbook; these guide ethical professional behaviour relative to the following topics:

- Code of Medical Ethics
- The Physician's Pledge
- Ethical principles for medical research involving human subjects
- Torture and other cruel, inhuman or degrading treatment or punishment in relation to detention and imprisonment
- On hunger strikers
- The rights of the patient

- Child health
- Professionally led regulation
- Ethical consideration regarding the use of genetics in medicine
- Euthanasia and physician-assisted suicide
- Law and ethics
- Guidelines for continuous quality improvement in health care.

World Health Professions Alliance (WHPA)

The WMA is a member of the World Health Professions Alliance (WHPA), together with the FDI World Dental Federation, the International Pharmaceutical Federation (FIP), the International Council of Nurses (ICN), and World Physiotherapy, formed in 1999 to develop and strengthen health professional's contribution to the health of the patients and communities they serve [32]. WHPA speaks for more than 34 million health professionals worldwide represented by more than 600 national member organisations. This organisation was established to be the key point for global access to healthcare professionals within the five disciplines represented.

A major deadly serious issue of concern is the estimated global shortage of 18 million health workers by 2030. This is expected to lead to poorer health outcomes and health worker well-being. This has been exacerbated by the recent pandemic. The WHPA is actively promoting the need for positive practice environments, enabling health settings to support excellence and decent work conditions with the power to attract and retain staff, provide quality patient care, and strengthen the health sector as a whole. It has produced downloadable campaign booklets for health employers, managers, and professionals as well as for national organisations representing healthcare professionals.

International Council of Nurses (ICN)

This federation of more than 130 national nurses associations today was established in 1899. It now represents more than 20 million nurses worldwide [33]. Nurses have also organised themselves according to clinical specialties, a number of which are international affiliates of the ICN. In addition, the ICN has a number of networks and a couple of foundations each representing a specific nursing specialty or specific purpose. The ICN works with various partners on campaigns of relevance to nurses and publishes a number of guidelines associated with regulation and education, endorsements, joint statements, and position statements on issues of concern or interest for the nursing profession. It provides access to a number of resources and manages a number of projects, including eHealth and the International Classification of Nursing Practice (ICNP) terminology, recognised within the WHO family of classifications. Guidelines are made available for the ICNP translation, catalogue development, and technical implementation. The ICN is an affiliate member of ISO TC215.

Health specialty professional organisations

A number of clinical specialty professional organisations or colleges actively participate in clinical research. There are more than 70 clinical specialties, each with their own practice guidelines and protocols based on their specific body of knowledge. To become a clinical specialist in any specialty, one needs to complete a postgraduate educational programme consisting of theory and supervised practice. Most of these programmes are developed, delivered, and assessed by senior college fellows and lead to the recognition of fellowship by the relevant professional organisation. Some of these programmes may need to be accredited by a national council or equivalent for successful graduates to be authorised to practice in that specialty. Most if not all clinical specialty entities develop clinical practice protocols and guidelines pertaining to their specialty and the types of patients they apply to.

The **American Academy of Family Physicians** (AAFP) [34] develops evidence-based clinical practice guidelines (CPGs) that serve as a framework for clinical decisions made, supporting best practices. They have developed a manual that summarises the processes used by the AAFP to produce their high-quality guidelines that adhere to the standards and principles as published by the Institute of Medicine [2] and the **Council on Medical Specialty Societies** (CMSS) [35]. This US-based CMSS began in 1965 and now have 45 specialty societies as members, collectively representing more than 800,000 medical clinicians. The CMSS acquired the **National Quality Registry Network** from the (NQRN) American Medical Association (AMA)'s Physician Consortium for Performance Improvement (PCPI) in July 2020. CMSS intends to integrate NQRN services and focus on registry work by building more expertise in measurement science and data standards.

National medical associations have been established in numerous countries; many of these provide practice guidelines. For example, the American Medical Association (AMA) has established Guideline Central [21], the **Canadian Medical Association** (CMA) has Joule [20]. There are also many specialty societies; most of them are national although many of these are affiliated internationally and have global influence. Wolters Kluwer hosts the **'UpToDate' website** [36] used by more than 1.9 million clinicians in 190 + countries. Its clinical evidence resource, covering at least 25 clinical specialties, is provided by thousands of world-renowned physician authors and reviewers. A rigorous editorial process is used to synthesise the most recent medical information into trusted, evidence-based recommendations. These are then made available to provide a clinical decision support service. Access to this clinical knowledge mostly presented in PDF formats is made available via embedded links within various large EHR/EMR software applications. Such guidelines should also be made available in computable formats for ease of use by the decision support applications themselves.

Every professional association has and governs its own body of knowledge (BOK). Even though most BOKs are not formally published as such, they reflect that organisation's scope of professional knowledge. Unfortunately, very few of these professional organisations are actively ensuring that their members' digital information and knowledge management needs are met and incorporated in the global or their national digital health agenda. Their highly specialised clinical knowledge and skills are needed to successfully drive digital health support for patient-centred care delivery, workforce capacity building, and the sustainability of all associated healthcare providers. This requires active knowledge conversion into digital formats, the appointment of data stewards and knowledge governance.

The **Data Management Association** (DAMA) represents the global data management community, has a well-documented and up-to-date body of knowledge (DMBOK2), provides an educational service, and certifies its practitioners [37]. In the 2019 foreword to DMBOK2, Zachman noted that this had been in the making for nearly 30 years and that today's focus must be on 'the meaning and purpose of data that are central to the creation of value and wealth'. It is not simply about improving productivity by reducing costs.

International Medical Informatics Association (IMIA)

Many countries have established Health or Medical Informatics Associations, which in turn are members of the International Medical Informatics Association (IMIA) [38]. In 1967, IMIA was a Technical Committee of the International Federation for Information Processing (IFIP) and evolved from a Special Interest Group to its current status in 1979. IMIA continues to maintain its relationship with IFIP as an affiliate organisation, has close ties with WHO, with the International Federation of Health Information Management (IFHIMA), and is recognised as a Liaison A category organisation in cooperation with ISO. IMIA has a growing number of Working and Special Interest Groups representing communities who share common interests in a particular digital health area of focus and regional groups that collectively cover the globe. It established the **International Academy for Health Sciences Informatics** (IAHSI) in 2017. Its fellows are internationally recognised leaders able to stimulate and guide future direction and advise governmental and nongovernmental organisations on health sciences informatics and digital health. IMIA has developed a Code of Ethics for Health Information Professionals together with an accompanying handbook.

Health information and Management Systems Society, Inc. (HIMSS)

The HIMSS was founded in 1961, at the Georgia Institute of Technology as the Hospital Management Systems Society; it affiliated with the American Hospital Association in 1966 and became HIMMS in 1986 to serve healthcare information systems professionals. HIMSS became independent in 1993 and merged with the Center for Healthcare Information Management (CHIM) in 2001 and has enjoyed a growing membership ever since [39]. HIMSS is a nonprofit organisation based in Chicago, USA. It is well supported financially by prominent players in the digital health market and private and public hospitals. HIMSS has strategically been expanding globally in line with the market expansions of its industry members, by providing conferences with extensive trade expositions, frequently in competition with national health informatics professional associations, with a desire to collaborate and gain access to local networks.

HIMMS is viewed by many as an advisor and thought leader serving global health information and technology communities across North America, Europe, the United Kingdom, the Middle East, and Asia Pacific. A popular interoperability measurement tool used internationally is the Electronic Medical Record Adoption Model (EMRAM), which defines seven levels of interoperability, and was first released in 2005; definitions of the evaluation criteria used for certification purposes were last updated in 2018 [40]. EMRAM is described as a 'progressively sophisticated road map that enables quality, safety, and operational efficiency for healthcare providers'. Hospitals implementing EMR systems invite HIMSS Analytics to make use of this tool to indicate the level of digital functionality achieved following implementation, Stage 7 validation being their ultimate goal.

HIMMS enjoys strong industry support and has a number of collaborations, such as the Alliance for Nursing Informatics, the Center for Information Technology Leadership, and IHE. It has initiated, for example, the establishment of HIMMS Analytics, the HIMMS Foundation, providing financial assistance to students in healthcare IT fields and other philanthropic activities. The HIMSS Foundation includes the Institute for eHealth Policy. Other initiatives include the HIMMS Electronic Health Record (Trade) Association (EHRA) and the Health Technology Alliance (HTA). HIMMS supports the Technology Informatics Guiding Education Reform (TIGER) initiative offering tools and resources for learners and educators.

Personal Connected Healthcare Alliance (PCHA)

The PCH Alliance is a membership-based HIMSS Innovation Company works to advance patient-/consumer-centred health, wellness, and disease prevention by accelerating technical, business, and social strategies necessary to advance personal connected health [41]. It publishes Continua Design Guidelines recognised by the ITU as the International standard for safe, secure, and reliable exchange of data to and from personal health devices.

Patient-centred global entities

Patient involvement and engagement in digital health transformation activities are important. The **International Alliance of Patients' Organisations** (IAPO) works closely with ISQua mentioned previously as guideline providers. IAPO's vision is to see patients at the centre of health care throughout the world [42]. They have a strong focus on patient safety based on a global commitment to recognise patient safety as a key health priority, as evidenced by the WHA7.6 Resolution 'Global Action on Patient Safety' passed on May 28, 2019, at the World Health Assembly. Digital health has not featured in their policy work to date, yet system errors resulting from any digital transformation activity pose significant patient safety risks. IAPO has established numerous cross-sector alliances and works collaboratively with like-minded individuals towards patient-centred care. IOPA works closely with the **Health Technology Assessment International** (HTAi) society that brings together organisations working to assess and approve new medicines, devices, and treatments.

Public health

Public health has been at the forefront due the COVID-19 pandemic that exposed many shortcomings. The science underpinning this health discipline is concerned with disease prevention, increasing life longevity and promoting healthy lifestyles through efforts and choices made by individuals, communities, and society as a whole, including public and private organisations. It deals with the population health as identified by any community's health burden determined by the incidence of infectious and noncommunicable diseases. It is multidisciplinary as it incorporates epidemiology, biostatistics, and health services. It has a strong interest in successful digital health ecosystems as it needs accurate, timely, and comprehensive information to guide its activities and interventions.

There are many nationally based professional public health and public health-related associations. In 1967, thirty-two (32) of these established the World Federation of Public Health

Associations (WFPHA). By 2020, this federation was composed of 115 association members representing around 5 million public health professionals. Its mission [43] is:

> to promote and protect global public health. It does this throughout the world by supporting the establishment and organizational development of public health associations and societies of public health, through facilitating and supporting the exchange of information, knowledge and the transfer of skills and resources, and through promoting and undertaking advocacy for public policies, programs and practices that will result in a healthy and productive world.

This federation is an accredited NGO in official relations with the WHO and accredited to the United Nations Environment Programme, holding consultation status with the United Nations Economic and Social Council and observer status with Codex Alimentarius of the Food and Agriculture Organisation of the United Nations. Its 2020 World Congress was their first ever with a digital health focus.

It initiated the development of a Global Charter for the Public's Health at its 69th World Health Assembly. This Global Charter brings together the best of all the existing models to provide a comprehensive, clear, and flexible framework that can be applied on a national or a global scale [44]. WFPHA joined the July 2020 launch of the Global Sustainable Health Equity Movement, supported by the UN and WHO leadership. This movement is about gathering networks and people from all regions, cultures, and professions around the ethical principle of the universal right to health.

Centers for Disease Control (CDC) and prevention

The US CDC was established in 1946 as the Communicable Disease Center to control Malaria [45]. This later became part of the US Public Health Service that included an epidemiology division. An epidemic intelligence service was established in 1951. Since then, it has dealt with smallpox, poliomyelitis, rabies, Asian flu, anthrax, hospital-acquired staphylococcus, tuberculosis, food-based salmonella, Hong Kong flu, legionnaire's disease, HIV and AIDS, Lyme disease, West Nile virus, the Ebola virus, the H5N1 avian flu, viral agents of gastroenteritis, severe acute respiratory syndrome (SARS), bovine spongiform encephalopathy (BSE), chemical terrorism, human monkeypox, H1N1 influenza virus, Middle East respiratory syndrome coronavirus (MERS-CoV), Zika virus, and COVID-19 virus. It conducted national health surveys, managed immunisation programmes for common childhood diseases such as rubella, diphtheria, and measles.

It has established a number of new divisions, institutes, and programmes including a chronic disease division in 1975, the international health programme office in 1981, the agency for toxic substances and disease registry in 1983, a National Center for Injury Prevention and Control in 1992, and the National Center on Birth Defects and Developmental Disabilities in 2001. The CDC worked with PAHO (Pan American Health Organisation) to develop a safe water system in 1990 and developed emerging infections programmes in 1995 as a prevention strategy for the United States. It has an Emergency Operations Center to assist in any public health emergency anywhere in the world. The CDC undertakes numerous additional public health-related activities. Its mission is to prevent disease, injuries, and disability and to encourage healthy lifestyles through all stages of life. It is about informing and educating the general public about the science of public health by demonstrating its relevance and impact and to make the case for its commitment to participate in global solutions to international public health challenges via its museum.

This snapshot of CDC's activities provides the reader with a brief overview of the complexity and need for continuing vigilance to identifying emerging public health issues at their earliest time possible to ensure a timely, appropriate and rapid response to minimise risk by maximising opportunities for prevention and containment. The US CDC's strategic framework is about securing global health, ensuring domestic preparedness, the elimination of disease and ending epidemics. This is supported by its foundational scientific work in virology and epidemiology along with its deep knowledge gained from more than 70 years of experience responding to and preventing disease outbreaks.

Any nation's CDC needs adequate resources to undertake this work, including timely and comprehensive information about the health of any nation's population; to determine the need for new investigations, programmes, travel advice, or rapid response initiatives; to optimise global and domestic preparedness; to stop the spread of contagions and to deal with biological, chemical, and radiation/nuclear terrorism or disasters or vector-borne diseases. Pandemic management in particular is highly dependent on successful national digital health ecosystems and an ability to share information in a timely manner.

Patient safety, audits and governance

In 2018, WHO and the UK Government established the **Global Patient Safety Collaborative** (GPSC) [46] with the aim to improve the safety of every country's health system. This is seen as an important component of healthcare delivery and is essential for achieving Universal Health Coverage (UHC). Its work will focus on leadership, education, training, and research. The GPSC will consist of a central, ever-evolving knowledge resource, managed by WHO headquarters, based on member contributions. Many of these country members have their own safety and quality organisations. For example, Australia has an independent statutory authority, known as the **Australian Commission on Safety and Quality in Health Care** (ACSQHC), accountable to the Australian Parliament and the Minister for Health. This authority develops, publishes, and maintains a number of standards, user guides, and guidelines [47]. Its primary purpose is to lead and coordinate national improvements in the safety and quality of health care.

Governance is about steering or directing any concept or entity in the desired direction by a governing body who has the authority to do so. Governing bodies have the authority to govern all aspects related to their role and areas of responsibility. Collectively, these enable the creation of a successful digital health ecosystem. Good governance is about avoiding fragmentation, resource wastage, duplication of efforts, nonstrategic investments and noncompliance with a key set of standards [48]. Governance is also about ensuring patient safety by auditing electronic data processing and undertaking risk management strategies. Digital health-governing bodies need to establish processes that comply with accountability, transparency, participatory, rules of law, respect, and confidentiality principles. Governance processes need to be equitable, efficient, inclusive, and ensure that the desired outcomes are achievable.

Healthcare accreditation services

Accreditation is the independent evaluation of conformity assessment processes, against recognised standards to ensure competence and degree of compliance. It is about providing governments, procurers, and consumer confidence in certifications provided, regarding the calibration of test results and inspection reports. Authoritative accreditation bodies are established for the purpose of ensuring conformity to enhance acceptance of products and services across borders.

A variety of international guidelines for quality and accreditation standards are available [49–52]. A legal framework for the provision of accreditation services across Europe became effective in 2008. The International Laboratory Accreditation Cooperation (ILAC) and the International Accreditation Forum (IAF) collectively determine the nature of accreditation globally. Their role is to confirm integrity. Accreditation then provides the basis for the acceptance of products and assurance to its markets and supporting international trade [53]. This is relevant for all surgical supplies and equipment used by healthcare workers in any healthcare organisation.

International Laboratory Accreditation Cooperation (ILAC)

ILAC began in 1977 and became a formal cooperation in 1996 [54]. It operates in accordance with the ISO/IEC 17011:2017 standard and is involved in the accreditation of conformity assessment bodies including calibration laboratories, testing laboratories, medical testing laboratories, inspection bodies, and proficiency testing providers based on ISO standards applicable to each of these entities. Its Mutual Recognition Arrangement (MRA) provides a credible and technically robust framework; it came into effect in 2001. One hundred accreditation bodies from over 100 economies have signed this MRA. This acts as an internationally recognised 'stamp of approval' to demonstrate compliance against agreed standards and requirements to minimise risk. It was extended in 2020 to cover the accreditation of reference material producers. ILAC works with a number of International partners and provides information to regulators. The use of the MRA minimises duplication, helps regulators meet their own legislated responsibilities, and realises the free-trade goal of 'accredited once, accepted everywhere' objective.

International Accreditation Forum (IAF)

Accreditation arrangements in the fields of management systems, products and services, personnel and other similar programmes of conformity assessments are managed by the IAF [55]. The IAF was formed in 1993 to cooperate with organisations operating similar accreditation programmes by exchanging information and procedures. Its membership consists of 66 accreditation bodies, 18 association members, and 6 regional cooperations. IAF works in partnership with ILAC to reach out to the World Trade Organisation and Government bodies to gain greater visibility and recognition of the value of accredited conformity assessment. The IAF also operates in accordance with the ISO/IEC 17011:2017 conformity assessment standards as well as the ISO/IEC 27006:2015 AMD1:2020 Information Technology—Security techniques—Requirements for bodies providing audit and certification of information security management systems. This is also relevant to digital health. Its role is to:

1. Ensure that its accreditation body members only accredit those that are competent to do the work they undertake and not subject to conflicts of interest.
2. Establish mutual multilateral recognition arrangements between its accreditation member bodies to reduce business and customer risks by ensuring an accreditation certificate may be relied on anywhere in the world and contribute to the freedom of world trade by eliminating technical barriers.

International Society for Quality in Health Care (ISQua)

The International Society for Quality in Health Care (ISQua) is another global, member based, not for profit entity that develops a number of guidelines, with a focus on patient safety and quality [56]. It has an extensive network of healthcare professionals across a large majority of countries. Its members manage similar nationally based organisations, each of which develops their own evidence-based guidelines. Its mission is:

> to inspire and drive improvement in quality and safety of health care worldwide through education and knowledge sharing, external evaluation, supporting health systems and connecting people through global networks.

Most countries have established an authority to develop patient safety and quality standards that form the foundation for healthcare organisations' accreditation and certification services. Examples are the Joint Commission in the Unites States [57], the Australian Commission on Safety and Quality in Health Care [58,59], and Accreditation Canada [60]. In the United Kingdom, the Trent Accreditation Scheme (TAS) that was National Health Service (NHS) based, withdrew from all hospitals' accreditation-related activities in 2010. In response a group of acknowledged clinicians and other experts actively working within the NHS and private medical practice, established the Quality Healthcare Advice, providing a surveying and accreditation service for hospitals and clinics located anywhere in the world [61]. Clinical pathology, imaging services, and physiological accreditation services are provided in the UK by a longstanding government-funded calibration and measurement service that continues to grow. This organisation became the United Kingdom Accreditation Service (UKAS) and was appointed as the National Accreditation Body by the Accreditation Regulations 2009 in July 2016 [53].

The Australian Council on Healthcare Standards (ACHS) manages the EQuIP programme that is part of its hospital and day surgery accreditation programme and publishes clinical indicator reports. EQuIP has mandatory criteria used for performance measurement and includes four criteria for Information Management:

1. Health records management systems support the collection of information and meet the consumer/patient and organisational needs.
2. Corporate records management systems support the collection of information and meet the organisation's needs.
3. Data and information are collected, stored, and used for strategic, operational, and service improvement purposes.
4. The organisation has an integrated approach to the planning, use, and management of information and communication technology [62].

Table 2 provides a summary of accreditation service providers described previously.

TABLE 2 Accreditation service providers: global examples.

Global entity	Area of focus	Relationships	Values/outputs
ILAC [54]	• Conformity assessment • Laboratory systems • Reference material production • Health and social care • Promote international trade	IAF • Recognised Regional Cooperation Bodies • Accredited conformity assessment bodies ISO/IEC	• Mutual Recognition Arrangement (MRA) • Safe food and clean drinking water • Harmonised laboratory, inspection body, and proficiency testing provider accreditation practices • ILC MRA recognition
IAF [55]	• Management systems incl. Security • Products, services, personnel and similar programmes • Develop guidance, rules and procedures based on standards to satisfy regulatory or government requirements	ILAC ISO • Other relevant international bodies • Regional groups of accreditation bodies • Industry groups	• Certified once, accepted everywhere • Impartiality, probity, and competence • Harmonised accreditation procedures • Use of recognised standards • Authoritative oversight • Support international trade • Remove technical barriers
ISQua [56]	Transform healthcare quality and safety through global communities with a special focus on lower-middle-income countries (LMIC) by providing expertise	• The International Alliance of Patients' Organisations (IAPO) • Strategic partnerships	• Inspire and empower people to advocate for improvement in the quality and safety of healthcare worldwide • Patient-centred care • Universal health coverage

Regulations/legislation

All regulatory and legal standards adoption requirements have a patient/citizen safety, security, and/or funding focus. Legal standards determines data acquisition and reporting policies for each health sector. This includes clinical data as this provides valuable information about health service demand. Medication management covers both patient safety and health service delivery costs. This is also the case for health equipment, including medical and patient monitoring devices, robotics, and surgical supplies. Every country or region has its own regulatory infrastructure. There is not a standard business process to undertake or manage these regulatory functions.

Most health informatics and clinical standards focus not only on patient safety, but also on patient well-being, continuity of care, patient outcomes, clinical decision-making, improved workflow, and new knowledge acquisition (research and advanced data analytics). Cybersecurity and data privacy need to be in harmony with any prevailing national cybersecurity/data privacy management process, although there may need to be special requirements to meet specific health system needs. All of these are essential foundational infrastructure requirements to be considered when transforming any country's health system to a digital health ecosystem. There is considerable diversity regarding how this multitude of regulatory requirements is managed from both legislative and regulatory perspectives, as these relate directly to each country's governing structures.

European Union

The European Union (EU) has multiple regulatory frameworks; each framework is associated with EU legislation and addresses a particular area, such as electronic communications or to ensure the safety and efficacy of medical devices and facilitate patients' access to devices in the European market.

Regulatory rules for medicinal products for human use and veterinary products are dealt with the European Medicines Agency (EMA). Its mission is to foster scientific excellence in the evaluation and supervision of medicines, for the benefit of public and animal health. It does this by publishing clear and impartial information about medicines and their approved uses, including public versions of scientific assessment reports and summaries written in lay language. The EMA is the hub of the European medicines network comprising 40 national regulatory authorities guaranteeing a constant exchange and flow of information regarding scientific assessment of medicinal products in the EU [63]. The authorisation of medicines, clinical trials, or medical devices is undertaken elsewhere. All EU legislation concerning medicines is developed by the European Commission and the European Parliament together with the Council of the European Union to adopt it.

US Health Insurance Portability and Accountability Act (HIPAA)

This federal act enacted in 1996 was created primarily to modernise healthcare information flow by stipulating how personally identifiable information, maintained by healthcare delivery and healthcare insurance providers, should be protected from fraud and theft and to address healthcare insurance limitations. Every health information system implemented in the USA must be HIPPA compliant; it covers five areas:

1. Healthcare access, portability, and renewability;
2. Preventing healthcare fraud and abuse, administrative simplification, and medical liability reform;
3. Tax-related health provisions governing medical savings accounts;
4. Application and enforcement of group health insurance requirements;
5. Revenue offset governing tax reductions for employers.

Every country has its own set of legislator requirements covering these topics. These are associated with each nation's government structures, including health data ownership, privacy, and security legislative requirements.

Food and Drug Administration (FDA)

The US FDA is responsible for regulating food, drugs, medical devices, radiation-emitting products, vaccines, blood and biologics, animal and veterinary, cosmetics and tobacco products, as the means to protect public health. Regulations are designed to ensure safety, efficacy, and security of these products. This includes the regulatory responsibility of their manufacturing, marketing, and distribution. The FDA advances public health by assisting the speed of innovations designed to make medical products more effective, safer, and more affordable, and it plays a significant role in counterterrorism capability. Its scope is broad, and its responsibilities are closely related to several other government agencies.

Its work must be undertaken in accordance with its legislative authority [64]. Action is taken in response to any one of many different factors such as new drugs or technologies or information regarding existing issues, concerns that have arisen from accidents or other societal problems, recommendations or petitions or requests from other agencies or interest groups, etc. There has been considerable discussion regarding the need to regulate and approve medical devices or clinical decision support systems. If so, what type of devices or systems? Its work is informed by the results of the regulator's assessments and monitoring activities. Many countries, such as Australia, have their own regulator, although some make use of regulations developed by other internationally respected regulators such as the FDA.

Therapeutic Goods Administration (TGA)

The TGA is the Australian regulatory authority for all therapeutic goods and is another international respected regulator [65]. It is involved in many activities with overseas organisations and regulatory agencies in Canada, Europe, New Zealand, Singapore, the United Kingdom, and the US FDA. It also has links with a number of other international organisations and activities. This includes a consortium of like-minded regulatory authorities established in 2007 to promote greater regulatory collaboration and alignment of requirements and an International Coalition of Medicines Regulatory Authorities (ICMRA) that aims to identify potential synergies and other similar collaborative arrangements. Their collective work not only protects public safety but their decisions also influence world trade as well as drug or medical device availability, over-the-counter medicines, medicine-shortage substitution notices, and prescription medicines registration approvals.

Discussion

All of the organisations aforementioned provide support for a global service industry. The main drivers for sustainable value cocreation need to be able to pinpoint the main elements of its service ecosystem [66]. Service theories adopt a holistic view to service management and identify a 'service ecosystem' as the most appropriate organisational model to support the emergence of value. This chapter has identified key entities that collectively represent these main drivers for the health service industry. Every entity provides a valuable service, although the primary obstacle to the full realisation of 'value' within the global health service industry is this complex multistakeholder, largely noncollaborative high-level infrastructure environment.

Every global data, knowledge, guideline and regulatory provider, and data user supporting this global service industry needs to be clear and transparent about its own data needs and avoid duplication. Communicating their data needs to other users enables them to consider automation and integration for sustainable value cocreation within a digital health ecosystem. The many existing digital health networks need to be viewed as collaborative value cocreating networks. All users need to be able to apply their specific talents to transform their service management approaches to produce mutual benefits for all.

These global or national organisational entities need to make use of data collected operationally anywhere within any nation's health system (secondary use). Such ability eliminates the need to collect such data separately as a distinct process to demonstrate compliance or how guidelines were used in practice and their outcomes. Within this digital era, it is possible to eliminate the need for any specific statistical or performance measurement data-/information-gathering programme. Global and national guideline and regulatory providers make use of data, information, and knowledge from a variety of sources that form the evidence base used for guideline development and regulatory requirements.

This analysis of key globally and nationally influencing organisations has exposed fragmentation, overlaps, diverse business, and governance processes making up an infrastructure that is not well suited to efficiently and effectively serve its customers and product users within a digital era. There is a need for every entity within the health industry to refocus and accelerate their digital investments in response to the need to support the delivery of patient-centred healthcare services in any location within the context of interactive global, national, regional and organisational digital health ecosystems. This requires an examination of desirable data, information, and knowledge flows to determine current and ideal data supply chains, consider the use of new data and advanced analytics, and improve business and service delivery operations. The next step is to plan for the selective modernisation of existing technology capabilities, to boost the speed of collaborative and innovative developments, and to increase organisational agility to deliver its products and services in a more timely and cost-effective manner.

The types of organisations listed here were identified as key providers of data and knowledge in the form of standards, guidelines, and regulatory requirements, developed to support the health service industry. Collectively, they form part of the foundational global infrastructure. The greater their successful transformation to best suit the digital era, the greater the ability of others to transform to benefit the health industry, and especially the global population.

Every organisation needs to develop its own road map, detailing a well-coordinated and detailed plan for what needs to be done, by whom and when from the leadership level down to the front line along with securing necessary investments. It needs to be based on their value propositions, assessed feasibility, and have an associated implementation plan.

Successful digital ecosystems need to be powered by modern software development methods and delivery capabilities and have robust and federated data and knowledge governance support to enable broad and continuous use of data by every frontline user whilst enriching the data over time throughout its supply chain within any national health system. This requires investing in people as well as in technology. Leaders need to adopt a digital thinking strategy, be data-driven, aggressively challenge the status quo, cultivate existing talent, foster innovation, and allow for and learn from failures.

The challenges faced by every organisational entity are to transform their own operations first. In the case of health data, knowledge, and regulatory providers identified here, they need to consider emerging customer/user needs, supply dynamics, and standards best suited for the digital era. This requires strong leadership and talent, a change in culture and ways of working for many. There needs to be a clear vision and implementation plan. Every entity needs to develop its own practical staged plan for each of many interdependent parameters. This requires all interconnected functionality within its digital ecosystem to be considered and investments that ensure the availability of sufficient resources that are also able to overcome technological bottlenecks whilst making the most of current assets including the informal organisational opinion leaders, who can be an effective 'neural network' to bring about change.

Patient-centred service delivery requires us to imagine patient journeys throughout the health system within a lifetime context, whilst providing for all necessary safety requirements. There is a need to leverage what already is available, plus the promotion of vendor independence to maximise flexibility and expand the data lifespan. This requires many self-steering, high-performing teams supported by a stable backbone of infrastructure support services.

There needs to be a staged approach within an agreed big picture vision context, initially focusing on the immediate local needs, identifying critical behavioural pain points in need of change. Once these are addressed, such local networks can be extended with the aim of working towards end-to-end solutions nationally and globally along an example of lines as shown in Table 3. This does require an agreed standard platform, well-governed data standards, a clinical content framework, and a high-level reference architecture everyone can work with

TABLE 3 Key value levers and their relationship with three digital health transformation options [67–69].

Key value levers	Growing core local network	Expanding networked services	Building end-to-end solutions	Value outcome examples
Streamlining patient-centred care	√	√	√	• Patient-centred real-time Information flow • Optimum service coordination
Operational efficiency	√	√	√	• Timely scheduling • Optimum resource management
New online specialist services		√	√	• Telehealth connecting with specialists beyond local location
Rapid response capability	√	√	√	• Timely information enabling immediate action needs • Actionable intelligence
Automated reporting			√	• Timely reporting to support funding • Performance reports • Accreditation/regulatory compliance reports
Advanced analytics/robotics			√	• Big data analytics, intelligence • Identification of best practice
Real-time research contributions		√	√	• Clinical trial data management

simultaneously, enabling progressive growth towards a well-connected and functioning global digital ecosystem.

Adoption of the 'enter data once and use many times' principle requires a strong focus of what data needs to be collected for what purpose and by whom. Such needs drive optimum system/network design, guided by an agreed clinical content framework. The following chapters focus on data and the various methods used to represent concepts.

Conclusion

The adoption or nonadoption of health informatics standards can impact safety of health information systems, yet these requirements are rarely incorporated into legislation or identified within regulatory requirements. This may be due to a lack of awareness amongst policy decision-makers or to a software vendors' lobby wishing to protect current financially lucrative, business models in place, or both. It is highly desirable to have an international cooperative responsible for the accreditation of clinical guidelines and/or clinical software based on an agreed set of standards similar to that of the ILAC MRA and IAF to overcome the existing fragmentation, duplication, and technology shortcomings of digital health.

There is a need to continually measure digital-channel performance, such as standards compliance, platform adoption, use of guidelines, data sharing and use, social and digital media use, search engine results, communication flows, and the development and use of apps, during any transformation, to facilitate quick adoptions accordingly as new learnings become apparent. Hundreds of clinical and operational decisions are made daily by any clinical provider. It is critical that timely accurate and complete data and information is made available at each decision-making point and that these data are processed in a manner that enables effective resource management, forecasting, and planning activities. Modelling such processes for every individual as well as for the most common mix of patient journey encounters or disease transmission is necessary to support planning for any digital transformation.

An inventory of core models that support all service delivery and associated business operational processes can mobilise the work required from clinical and functional leaders enabling them to identify priorities and the order within which digital transformation functions need to be implemented. This includes the need to define a programme focused on data/information/knowledge acquisition, exchange, processing, and advanced analytics, in concert with the most desirable local, national, or global desired data supply chains.

Successful execution requires an investment and development investment agendas based on anticipated new demand levels and the reinvestment of freed-up resources into patient-centred digital solutions and critical decision support systems at all levels within any ecosystem. It is critical for every player to consider their product/service supply chain including their own dependencies, as well as those who are dependent on their products and services. Some of the savings realised need to be dedicated to the selective modernisation of all digital technologies and software development tooling. The cost structures associated with hardware and software refresh cycles need to be well-balanced so as not to impede continuous improvements. A cloud-based platform plus a continuous software integration and delivery strategy is a very cost-effective method supporting rapid deployment of any digital transformation strategy. It is vital to also upskill the entire workforce.

References

[1] PAHO, in: H.D.F. Marin, R.J. Rodrigues, B.C. Delaney (Eds.), Building standard-based nursing information systems, Pan American Health Organisation (PAHO), Washington, DC, 2001.
[2] IOM, Clinical Practice Guidelines We Can Trust, National Academies Press, 2011.
[3] GRADE. Grading of Recommendations Assessment, Development and Evalution Working Group, Available from: https://www.gradeworkinggroup.org/.
[4] A.E.W. Johnson, T.J. Pollard, L. Shen, L.-W.H. Lehman, M. Feng, M. Ghassemi, B. Moody, P. Szolovits, L.A. Celi, R.G. Mark, MIMIC-III, a freely accessible critical care database, Sci. Data 3 (2016) 160035.
[5] L.A. Celi, P. Charlton, M.M. Ghassemi, A. Johnson, et al., in: MIT-Critical-Data (Ed.), Secondary Analysis of Electronic Health Records, Springer, Switzerland, 2016.
[6] NHS-Digital, National Information and Technology Partner to the Health and Social Care System, 2020, Available from: https://digital.nhs.uk/about-nhs-digital.
[7] J. Kung, R.R. Miller, P.A. Mackowiak, Failure of clinical practice guidelines to meet institute of medicine standards: two more decades of little, if any, progress, Arch. Intern. Med. 172 (2012) 1628–1633.
[8] MIT. Critical Data: Clinical Data Consortium, Available from: https://criticaldata.mit.edu/.
[9] MTPConnect, Clinical Trials in Australia: The Economic Profile and Competitive Advantage of the Sector, MedTech and Pharma Growth Centre, 2017.
[10] G. De Moor, M. Sundgren, D. Kalra, A. Schmidt, M. Dugas, B. Claerhout, T. Karakoyun, C. Ohmann, P.-Y. Lastic, N. Ammour, R. Kush, D. Dupont, M. Cuggia, C. Daniel, G. Thienpont, P. Coorevits, Using electronic health records for clinical research: the case of the EHR4CR project, J. Biomed. Inform. 53 (2015) 162–173.
[11] A.L. Nobles, K. Vilankar, H. Wu, L.E. Barnes, Evaluation of data quality of multisite electronic health record data for secondary analysis, in: 2015 IEEE International Conference on Big Data (Big Data), 2015, pp. 2612–2620.
[12] S.M. Meystre, C. Lovis, T. Bürkle, G. Tognola, A. Budrionis, C.U. Lehmann, Clinical data reuse or secondary use: current status and potential future Progress, Yearb. Med. Inform. 26 (2017) 38–52.
[13] T.G.R. Macieira, T.C.M. Chianca, M.B. Smith, Y. Yao, J. Bian, D.J. Wilkie, K. Dunn Lopez, G.M. Keenan, Secondary use of standardized nursing care data for advancing nursing science and practice: a systematic review, J. Am. Med. Inform. Assoc. (2019), https://doi.org/10.1093/jamia/ocz086.
[14] ICH, International Council for Harmonisation Geneva, 2021, Available from: https://www.ich.org/.
[15] WHO, Classification of Digital Health Interventions v1.0: A Shared Language to Describe the Uses of Digital Technology for Health, 2018, Available from: https://www.who.int/reproductivehealth/publications/mhealth/classification-digital-health-interventions/en/.
[16] WHO, Guideline: Recommendations on Digital Interventions for Health System Strengthening, World Health Organisation, Geneva, 2019. [cited 2019 18 April]. Licence: CCBY-NC-SA3.0IGO: Available from: https://www.who.int/reproductivehealth/publications/digital-interventions-health-system-strengthening/en/.
[17] openEHR/CKM, Clinical Knowledge Manager, 2019, [cited 2019 21 February]. Available from: https://ckm.openehr.org/ckm/#.
[18] Cochrane, Cochrane International Network of Researchers, 2018, [cited 2018 8 October]. Available from: https://www.cochrane.org/.
[19] NHMRC, Australian Clinical Practice Guidelines, National Health and Medical Research Council, 2021. Available from: https://www.clinicalguidelines.gov.au/.
[20] Joule-CMA, CPG Infobase: Clinical Practice Guidelines, Canadian Medical Association (CMA), 2021. Available from: https://joulecma.ca/cpg/homepage.
[21] AMA. GuidelineCentral: American Medical Association; Available from: https://www.guidelinecentral.com/about/.
[22] ECRI. ECRI Guidelines Trust (Transparency and Rigor Using Standards of Trustworthiness). Available from: https://guidelines.ecri.org/.
[23] ISMP. Institute for Safe Medication Practices (an ECRI Affiliate); Available from: https://www.ismp.org/.
[24] AHRQ-NEATS, The Agency for Healthcare Research and Quality's National Guideline Clearinghouse Extent Adherence to Trustworthy Standards (NEATS) Instrument, Available from: The Agency for Healthcare Research and Quality's National Guideline Clearinghouse Extent Adherence to Trustworthy Standards (NEATS) Instrument, 2021.
[25] WHO, International Classification of Diseases (ICD), World Health Organisation, 2018. [cited 2018 18 August 2018]. Available from: http://www.who.int/classifications/icd/en/.

[26] WHO, Family of International Classifications Geneva, 2018, [cited 2018 8 September]. Available from: http://www.who.int/classifications/en/.

[27] WHO, International Classification of Health Interventions (ICHI), World Health Organisation, Geneva, 2018. [cited 2019 2 December]. Available from: https://www.who.int/classifications/ichi/en/.

[28] WHO, International Health Regulations (2005), third ed., World Health Organisation, 2016. Available from: https://www.who.int/ihr/publications/9789241580496/en/.

[29] AHRQ, Guidelines and Measures, 2018, [cited 2018 13 September]. Available from: https://www.ahrq.gov/gam/index.html.

[30] WONCA. World Organization of National Colleges, Academies and Academic Associations of General Practitioners/Family Physicians, Available from: https://www.globalfamilydoctor.com/.

[31] WMA. World Medical Association, Available from: https://www.wma.net/.

[32] WHPA. World Health Professions Alliance, Available from: http://www.whpa.org/.

[33] ICN. International Council of Nurses, Available from: https://www.icn.ch/.

[34] AAFP. American Academy of Family Physicians, Available from: https://www.aafp.org/.

[35] CMSS. Council of Medical Specialty Societies, Available from: https://cmss.org/.

[36] UpToDate. Clinical Decision Support Resource: Wolters Kluwer, Available from: https://www.uptodate.com/home.

[37] DAMA. Data Management Association, Available from: https://dama.org/content/body-knowledge.

[38] IMIA. International Medical Informatics Association, Available from: https://imia-medinfo.org.

[39] OHSU. HIMSS (Health Information and Management Systems Society), Available from: https://www.clinfowiki.org/wiki/index.php/HIMSS.

[40] HIMSS. Electronic Medical Record Adoption Model (EMRAM)-HIMSS Analytics: Health Information Management Systems Society, Available from: https://www.himssanalytics.org/news/emram-criteria-changes-what-you-need-know.

[41] PCH-Alliance. Personal Connected Health Alliance: HIMSS, Available from: https://www.pchalliance.org/.

[42] IAPO. International Alliance of Patients' Organizations, Available from: https://www.iapo.org.uk.

[43] WFPHA. World Federation of Public Health Associations, Available from: https://www.wfpha.org/about-wfpha.

[44] WFPHA, Improving Health Globally, World Federation of Public Health Associations, 2018. Available from: https://www.wfpha.org/images/Improving_Health_Globally_2018.pdf.

[45] CDC. Centers for Disease Control and Prevention, Available from: https://www.cdc.gov/.

[46] WHO-GPSC, Global Patient Safety Collaborative Briefing Note, World Health Organization, 2018. Available from: https://www.who.int/patientsafety/partnerships/GPS-collaborative/en/.

[47] ACSQHC. Australian Commission on Safety and Quality in Health Care-Standards, Available from: https://www.safetyandquality.gov.au/standards.

[48] A. Marcelo, D. Medeiros, K. Ramesh, S. Roth, P. Wyatt, Transforming Health Systems Through Good Digital Health Governance, Asia Development Bank, 2018. [cited 2020 2 March]. Available from: https://www.adb.org/sites/default/files/publication/401976/sdwp-051-transforming-health-systems.pdf.

[49] T. Fortune, E. O'Connor, B. Donaldson, Guidance on Designing Healthcare External Evaluation Programmes Including Accreditation, International Accreditation Programme, International Society for Quality in Health Care (ISQua), 2015.

[50] E.-H.W. Kluge, Health information professionals in a global eHealth world: ethical and legal arguments for the international certification and accreditation of health information professionals, Int. J. Med. Inform. 97 (2016) 261–265.

[51] S. O'Connor, P. Hanlon, C.A. O'Donnell, S. Garcia, J. Glanville, F.S. Mair, Understanding factors affecting patient and public engagement and recruitment to digital health interventions: a systematic review of qualitative studies, BMC Med. Inform. Decis. Mak. 16 (2016) 1–15.

[52] WHO, Quality and Accreditation in Health Care Services: A Global Review, World Health Organisation, Geneva, 2003, p. 225.

[53] UKAS. United Kingdom Accreditation Service London: United Kingdom Accreditation Service; [cited 2019 1 June]. Available from: https://www.ukas.com/about/about-accreditation/.

[54] ILAC. International Laboratory Accreditation Co-operation, Available from: https://ilac.org/.

[55] IAF. International Accreditation Forum, Available from: https://www.iaf.nu/.

[56] ISQua. International Society for Quality in Health Care, [cited 2019 23 March]. Available from: https://www.isqua.org.

[57] JCR. The Joint Commission, Standards, [cited 2019 1 June]. Available from: https://www.jointcommission.org/standards_information/standards.aspx.
[58] ACSQHC, National Safety and Quality Health Service Standards, Australian Commission on Safety and Quality in Health Care, 2017.
[59] ACSQHC, Overview of the Clinical Care Standards, 2019, [cited 2019 20 March]. Available from: https://www.safetyandquality.gov.au/our-work/clinical-care-standards/overview-of-the-clinical-care-standards/.
[60] Accreditation-Canada. Health and Social Services Standards, [cited 2016 1 June]. Available from: https://accreditation.ca/intl-en/standards/.
[61] QHA. QHA Trent Accreditation United Kingdom, [cited 2019 1 June]. Available from: https://www.qha-trent.co.uk/.
[62] ACHS, EQuIP National Table—Standards, Australian Council on Healthcare Standards (ACHS), 2017.
[63] European-Commission, Medicinal Products—The EU Legal Framework, 2020, [cited 2020 16 August]. Available from: https://ec.europa.eu/health/human-use_en.
[64] FDA. Federal Drug Agency, Available from: https://www.fda.gov/regulatory-information/fda-rules-and-regulations.
[65] TGA. Therapeutic Goods Administration, Available from: https://www.tga.gov.au/.
[66] A. Botti, A. Monda, Sustainable Value Co-Creation and Digital Health: The Case of Trentino eHealth Ecosystem, 2020, Available from: https://www.mdpi.com/2071-1050/12/13/5263.
[67] M. Dietz, H. Khan, I. Rab, How Do Companies Create Value from Digital Ecosystems?, McKinsey & Company, 2020.
[68] J. Sengupta, V. HV, M. Dietz, X. Ji, L. Xiao, L. Li, How the Best Companies Create Value From Their Ecosystems, McKinsey & Company, 2019. Available from: https://www.mckinsey.com/~/media/mckinsey/industries/financial%20services/our%20insights/how%20the%20best%20companies%20create%20value%20from%20their%20ecosystems/how-the-best-companies-create-value-from-their-ecosystems-final.pdf.
[69] S. Singhai, B. Kayyali, R. Levin, Z. Greenberg, The Next Wave of Healthcare Innovation: The Evolution of Ecosystems, McKinsey & Company: Healthcare Systems and Services, 2020. Updated June 23, 2020. Available from: https://www.mckinsey.com/industries/healthcare-systems-and-services.

CHAPTER

8

Health data standards' limitations

Evelyn Hovenga[a,b,c] and Heather Grain[b,c,d]

[a]Digital Health, Australian Catholic University, Melbourne, VIC, Australia [b]eHealth Education Pty Ltd, Melbourne, VIC, Australia [c]Global eHealth Collaborative (GeHCo), Melbourne, VIC, Australia [d]LG Informatics Pty Ltd, Melbourne, VIC, Australia

Digital health ecosystem linkages

The use of data standards began in 1893 when the International Statistical Institute adopted the International Statistical Classification of Diseases (ICD) to record causes of death. This was extended to include morbidity in 1948 when this approach was adopted by all nations [1]. ICD is a historical example of a concept representation standard for a code system. This classification system is one of a number within the World Health Organization (WHO)'s Family of Classifications in common use [2]. The use of classification systems can provide some consistency, although this is dependent upon the classification rule interpretation. Some classification rules associated with individual systems in use are more complex or better formalised than others.

Numerous additional terminologies have been developed since that time, primarily to support statistical reporting. One national example of the number of classification systems in use is listed by Australia's Independent Hospital Pricing Authority (IHPA) [3]. Many of these types of classification systems are country specific. Collectively, these traditional terminologies tend not to provide the necessary richness required to optimally support today's healthcare service delivery directions and associated information needs when supported by digital health. People can usually manage to interpret small inconsistencies regarding the language used to communicate with others, but computers cannot, hence the need to adopt a widespread use of data standards. The link between the complexity of the health and clinical language used, associated data standards and digital technologies, is not well understood. As a result, available digital technologies are not able to be used optimally.

Today's digital health technical and data standards are created to identify best practice, content, structure, procedure and skill for design, implementation, use, sharing and governance of digital health information systems, and any nation's organisational or disciplinary health data assets.

Concepts are mental constructs, units of thought, or units of knowledge created by a unique combination of characteristics. Health concepts can be represented as terminologies that may be structured as standardised code systems, based on their corresponding classification rules, or through data specification within concept models.

https://doi.org/10.1016/B978-0-12-823413-6.00015-X

From a global population health perspective, desired improvements for health data standards are dependent upon the adoption of data and system integration strategies by health professionals, healthcare systems, and public health organisations. The optimisation of system connectivity requires us to overcome many existing interdisciplinary knowledge gaps prevalent among numerous health executives, terminology experts, and health information system users. All stakeholders need to be able to recognise that there are different requirements for concept representation in different parts of the healthcare system based on numerous diverse health data 'use cases'.

This requires a strong policy focus on merging the many silos of health practice realities, which are dependent upon the global adoption of agreed sets of technical and terminology standards.

At a high level, there is a need for health data design experts to adopt a well-defined standard public health taxonomy to reflect the desired outcome of a reporting strategy that enables the collection of comparable evidence from which we can learn best practice. Collecting comparable evidence requires the widespread and consistent use of data standards as defined by, for example, a public health ontology or logical information reference model [4].

Data need to be collected in a way that best suits the data source collectors and other users to suit all possible use cases throughout the data supply chain. There needs to be a number of data aggregation strategies for higher level reporting that are consistently designed and applied throughout the data supply chain. Public health data relies on the integration of data collected by all healthcare providers as described in the following:

> Linkage of programs, activities, and information among health care professionals, organizations, and systems and public health (including community-based organizations), to promote overall effectiveness and efficiency of the health system, improve the health and well-being of populations, and protect populations from health threats [5].

This description applies to any digital health ecosystem. Every concept included in the description already mentioned needs to have its own set of metadata standards to maximise usability across the many data supply chains from point of care and service delivery to global population health.

This chapter focuses on the need for information integration and sharing, in particular clinical information, as this provides the foundation for all health-related activities and supportive business processes. The pandemic has sown the seeds for change. This digital era is about unleashing exponential improvements towards health for all, sustainably. Our digital revolution requires all data, information, and knowledge to be machine- and human-readable and a large-scale system wide reformation. Who is or should be leading the transformation of existing health data standards to concept representation standards better suited to optimise global use and benefits of available technologies that define this digital era? Some activities towards this goal are in progress.

Leading health data standardisation and interoperability

In 2012, the World Health Organization (WHO), in its capacity as the United Nations agency for health, held an international forum with a view to support its Member States in their efforts to implement health data standards [6]. Participants represented standards development organizations (SDOs), UN agencies, academic and research institutions, implementing partners, donor organisations and subject matter experts in addition to representation from 54 countries, and WHO technical programmes. Its focus was on seven thematic areas, although three of these are in fact subthemes:

1. The eHealth and health data contextualisation
2. Country perspectives on health data standards
3. Impediments to health data standards access
4. In-country policy and governance mechanisms for health data standards adoption
   a. Innovative funding models for health data standards access and adoption
   b. Human capacity for health data standards implementation and maintenance
   c. Role of development partners in supporting Member States.

Key concluding forum outcomes reflecting relevance to all focus themes were listed as follows:

1. To achieve both semantic and syntactic interoperability, various categories of standards must be implemented as part of strengthening the overall eHealth systems and services at national and subnational levels.
2. It is essential to have national policies for eHealth and health information technology (HIT) standardisation.
3. The need for a global support mechanism to assist Member States, especially low- and middle-income countries (LMICs) to adopt standards at all levels of eHealth systems and services.
4. A competency-based workforce is essential for successful implementation of standards at national and subnational levels.
5. Funding needs to be part of a national eHealth strategy in order to sustain the implementation of standards.
6. The importance of national governments engaging academic institutions and health-related nongovernmental organisations to formalise specialised training programmes for existing healthcare professionals on standardisation and eHealth systems.
7. Funds currently provided to paper-based data collection need to be redirected to finance standards-based eHealth systems.

This was followed in 2014 by a Joint Inter-Ministerial Policy Dialogue on eHealth Standardization and a second forum on this topic [7]. Its aim was to facilitate further dialogue on the need for policy and governance mechanisms for the adoption of health data standards and to draft a policy and governance framework for full adoption by its Member States. Its concluding outcomes are represented in the following recommendations relating to data standards, indicating that eHealth policy for standardisation and interoperability in a national context should:

- Be embedded in a long-term national health plan providing continuity with long-term committed investments and, where relevant, embedded in an eGovernment plan.
- Be patient-centred, emphasising service quality, equity, patient outcomes, patient safety, and population outcomes.
- Be based on mutual trust and understanding, and genuine collaboration between all stakeholders from both private and public sectors.

- Support an evidence base for the socio-economic benefits of eHealth and encompass user utility and outreach programmes.
- Build capacity from country and ministry level down to that of frontline health workers.
- Ensure good governance, balancing top-down and bottom-up approaches, encompassing: equity and accessibility, legality, user rights in line with human rights, privacy, responsibility, and accountability to citizens and to the state. Compatibility of technologies, efficiency, open dialogue, and a shared vision on the use of data are necessary for implementation.
- Support competency-based education and capacity building in health informatics, with standardised curricula and measurable learning objectives at national and subnational levels.

The ability to action these is dependent on:

1. a sound understanding by national leaders of the science associated with health data and its relationship with the need to apply those principles in a manner that best suits this digital age,
2. the availability of data modelling, design, and development workforce skills, and
3. a willingness to collaborate to implement quality solutions to achieve desired outcomes.

During the 1990s, the US National Library of Medicine developed a Unified Medical Language System (UMLS) to facilitate the development of computer systems that behave as if they 'understand' the meaning of the language of biomedicine and health. This now consists of three sources of knowledge [1] listed and described later and demonstrates the complexity in need of simplification to better meet today's digital needs. These three knowledge sources are:

1. Metathesaurus based on 150+ terminologies and vocabularies, concepts, relationships, and attributes; this clusters synonymous terms from different terminologies by meaning into a single concept under a concept unique identifier.
2. Semantic network consists of semantic types or categories, such as Disease or Syndrome or Clinical Drug, and semantic relationships between semantic types; and links between categories used to help interpret meaning by indicating, for example, causes or threats.
3. Lexical resources that include the SPECIALIST lexicon, containing many words from the biomedical domain, and lexical tools, including a normalizer, a word index generator, and a lexical variant generator. Together these allow users to develop Natural Language Processing programs.

This chapter covers the many challenges associated with the adoption and implementation of these recommendations as they relate to health data standards. Recommendations made relative to technical issues and workforce capacity building are covered in other chapters.

Digital health data science

The digital era requires ICT systems to be more closely integrated into the real world of operational processes. This requires technology support for advanced knowledge-level and business process focused interoperability between all principals acting in digital health ecosystems, including persons, organisations, devices, applications, components, or objects necessary to achieve common health service delivery objectives. These represent a number of different interoperability levels as described in Table 1. The instances described refer to the types of

TABLE 1 Interoperability levels relative to instances [8].

Information perspective		Organisation perspective
Interoperability level	**Instances**	**Interoperability level**
Technical	Technical plug and play, signal and protocol compatibility	Lightweight interactions
Structural	Simple transactions, envelopes (EDI and HL7)	Data sharing
Syntactic	Messages and documents with the inclusion of agreed terms, vocabulary	Information sharing
Semantic	Advanced messages with common standard information models and terminologies	Knowledge sharing at IT concept level able to be processed by receiving computer
Organisation/service	Common operational process	Knowledge sharing at operational concept level. Agreed service function-level cooperation
Knowledge based	Multidomain processes, ontology focused	Knowledge sharing at ontology domain level. Cross-domain cooperation
Skill based	Individual engagement in multiple knowledge domains	Knowledge sharing in individual (patient) context. Moderated end-user collaboration

Wording modified from ISO 23903:2020.

interoperability schema in use from an information perspective. The organisation-perspective column makes a reference to the actual functionality achieved relative to each level of interoperability. The technologies referred to here were explored in some detail in the previous chapters.

Health-related data represents the most valuable asset within the health industry as a whole. It is the foundation for all our communication and provides content for all health information systems and devices used by many for decision-making purposes. Any patient's health journey starts with a first encounter where the healthcare responder needs to collect information from the patient about symptoms, situations (incidents that prompted the encounter), and their medical history. The healthcare provider then makes observations, such as vital signs, and makes decisions on how best to proceed based on all available information at that time. New information is added to such patient records and clinical knowledge is applied throughout such journeys. Each of these factual information exchanges represents numerous health-related concepts.

We need to have a sound understanding of the many and varied components of this content and the context within which this is described to extract meaning. This requires us to decompose the complexity of health information and available knowledge, in order to enable such knowledge to be digitised in a manner that is well suited for the 'use case' (fit-for-purpose) and leverage the benefits of computer processing.

Terminology is the term usually used to indicate a computable terminology, one that can be referenced by computers and used to support processing and decision-making. This terminology provides a link between the formal computable reference terminology, which is more complex than any other type of terminology and not necessarily something that clinical staff would actively use in the workplace.

BOX 1

Definitions of common terminology-related concepts

Vocabulary	A set of terms belonging to one special language or culture [8]
Reference terminology	"A reference terminology is a terminology designed to uniquely represent concepts. It does this by listing the concepts and specifying their structure, relationships and, if present, their systematic and formal definitions. It normally contains a unique identifier, a rubric and may contain reference to alternate terms to the preferred term (which may be conceptualized as an interface term (interface terminology) and it may contain maps or pointers to aggregate terminology." [8]
Interface terminology	"A maintained set of unique identified terms designed to be compatible with the natural language of the user." [8]
Aggregate terminology	"A grouping of similar concepts, for particular purposes, using relationships that may be hierarchical and/or uni- or multi-dimensional" [8]
Clinical terminology	"the component of health language used at the point of care for the purpose of clinical management of subject(s) of care" [8]

The reference terminology describes the unique concepts of health care, and the preferred (canonical, natural unique representation) term used to describe that concept, as well as all the other terms used to describe the concept, including the terms used in different languages. Local expressions can be used to describe a concept, or specific definitions can be selected to be used in local or domain-specific instances (the interface terminology). It is common for a terminology to include the reference elements and the interface components together in one system. It is also common for the reference terminology to provide mechanisms for grouping or aggregating ideas together for reporting or decision support use [9]. Box 1 provides the definitions of five common terminology-related concepts.

Data interoperability

The art of generating data-driven solutions through the comprehension of complex real-world health problems within a well-connected and well-integrated digital health ecosystem is dependent upon quality data representation at the source. This requires the adoption of the representation theory that posits that:

> the essential purpose of an information system (IS) is to provide a faithful representation of some focal real-world phenomena, thereby assisting its users to track states and state changes (events) in the phenomena it represents [10].

One most important and useful function of health data is for healthcare professionals to capture information in a precise, standardised, and reproducible form, represented in a manner able to achieve semantic data interoperability. Data interoperability can be defined as follows:

- The ability to, transfer data to and use data in, any conforming system such that the original semantics (meaning) of the data are retained irrespective of its point of access [11].
- HIMSS defines semantic interoperability as it "*provides for common underlying models and codification of the data including the use of data elements with standardized definitions from publicly available value sets and coding vocabularies, providing shared understanding and meaning to the user*" [12].
- Various health informatics standards define data interoperability as follows [13]:
  - Ability for data shared by systems to be understood at the level of fully defined domain concepts.
  - Capability of two or more systems to communicate and exchange information, and for each system to be able to interpret the meaning of received information and to use it seamlessly with other data held by that system.
  - Ability of computer systems supporting healthcare practice and management to correctly and consistently interpret the information being exchanged.

Data management principles that collectively make up the data science discipline include:

- Data collection,
- storage,
- integration,
- analysis,
- inference,
- advanced analytics, including artificial intelligence (AI),
- communication,
- ethics, and
- their relationships with computer processing protocols, including
  - information system design and
  - health data exchange schemas.

Digital health technical experts and health professional knowledge managers need to be familiar with this science. It is important to create synergies between computationally oriented information technology (IT) professionals, system architects, computer scientists, biomedical, and health informaticians, as each bring their own unique, yet complementary perspective to any digital health project.

Digital health data science requires interdisciplinary expertise and multidisciplinary teamwork to be effective. Neither IT professionals nor computer scientists are trained to develop or apply terminologies to clinical and research data [14]. This is a major knowledge gap reflecting the poor understanding and importance of links between computing technologies and data science.

Over the past several decades, it has increasingly become apparent that digital health applications need to make use of controlled health and clinical terminologies. Differences between such terminologies in use by multiple systems continue to be a major obstacle to optimum automation [15]. For example, one systematic literature review identified the following barriers to clinical decision support development:

- a lack of incentives,
- lack of standardised clinical terminology in use,
- outdated legacy electronic health records (EHRs),
- lack of transferability of clinical decision support (CDS) logic from one system to another,
- lack of experts needed to translate medical knowledge into a CDS knowledge base, and
- the low computer literacy of the end user [16, 17].

In this digital era with a desire to provide patient-centred care, healthcare system connectivity needs to be achieved within highly distributed environments. Patient-centred digital data, information, and knowledge transfers need to be independent of time and resource location. Such transfers also need to be in accordance with individually tailored diagnosis, care, and treatment protocols to meet real-world requirements. This requires:

- Every healthcare organisation's operational management to be virtually supported,
- Interactions and processes to be dynamically created and performed, and
- All relevant disciplines need to be integrated to provide a ubiquitous health service delivery ecosystem [18].

Supporting digital technologies need to be open, scalable, flexible, business process driven, trustworthy, user accepted, and user friendly. This requires compliance with a specific set of standards, semantically harmonised (mapped) terminologies at the required level of granularity, and computable knowledge representations.

Most national Health IT initiatives have demonstrated a significant knowledge gap regarding the link between the use of terminologies and information system design. There needs to be a greater appreciation of the significance of the ability to represent the semantics of a domain within an information system as shown in Fig. 1. Where these ontological domain concepts are not completely and accurately represented and reflected in the information system design, its usefulness is likely to be severely undermined as is its ability to connect with any number of distributed systems. This best explains the many current health IT shortcomings. The adoption of ontological foundations for information system design is critical, as such systems need to be able to represent the real world [19].

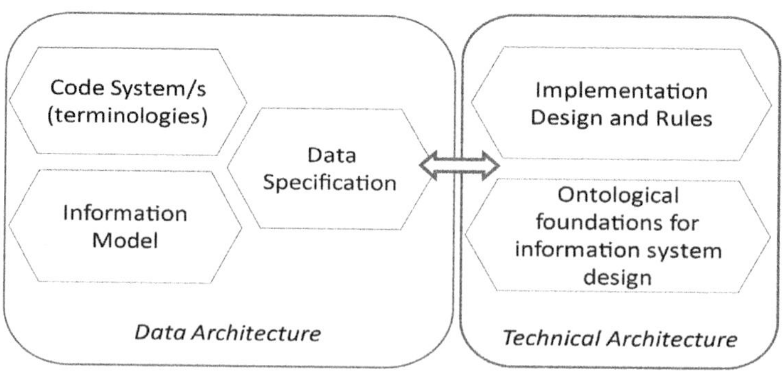

FIG. 1 Representing domain semantics within an information system.

A previous chapter identified numerous organisations engaged in the development and/ or provision of guidelines. All clinical guidelines are based on knowledge. Knowledge is derived from data and information that collectively provide content. Such content can be presented electronically in various formats such as a decision-support algorithm or a PDF document and made searchable by the use of metadata standards as well as query languages. Retrievability is dependent upon the storage format, database design, or data architecture adopted. Content that makes up each guideline relates to multiple knowledge domains, where each domain reflects its scope of knowledge. For example, the Glasgow Coma Scale is a knowledge component of both neurology (design focus) and nursing (user focus) knowledge domains. Digital knowledge management is covered in more detail in a following chapter.

Collectively, these concepts make up a content framework best viewed from a health data science perspective with a focus on the data itself, as these form the foundation for not only decision-making purposes, but also for the generation of health statistics, and big data, facilitating research and the use of advanced analytics, including artificial intelligence (AI).

From a clinical perspective, atomic-level data must be accurate at all times, throughout data life cycles and use, along the entire data supply chain. This requires health data managers to fully understand how data semantics are retained at every level of granularity suited to each individual use case, within digital ecosystems.

Data can be grouped into data sets or categorised according to classification rules. The need for the adoption of data standards and the avoidance of ambiguity is well known. The degree of expressivity and formalism determines the maximum possible level of interoperability associated with any data exchange between information systems. This capacity is dependent upon system design. Poor design may result in data collection duplication or a change in meaning between point of collection and secondary use. This in turn determines any system's data processing automation capacity.

System developers have traditionally created ad hoc sets of controlled terms for use in their applications in the recognition of the need for controlled vocabularies. As a result information in one system cannot be recognised or used by another. Despite the greater adoption of standard terminology codes, these data continue to be inconsistent between vendors as well as between organisations using the same vendor's system. This may be considered beneficial from a vendor perspective as it makes data migration from one vendor system to another more difficult and costly.

Terminology standards are seldom integrated into information systems in a manner that enables the consistent capture of standard data at various stages of the healthcare process. This has resulted in a heavy reliance upon data mapping within organisations and networks, and an increasing adoption of 'platforms' where data from multiple systems are stored to improve the desired enablement of data sharing and secondary data use [20, 21]. This reality concerning traditional 'platforms' also impedes research, as this requires solid data governance rules and sustainable data management infrastructures. A *knowledge-driven platform* is one that defines the technical architecture as shown in Fig. 1, whose domain semantics are *explicitly* defined by knowledge artefacts rather than being implicitly expressed in software and database schemas. The difference between traditional platforms and a knowledge-driven platform is explained in detail in a chapter to follow.

Clinical content structures

Any clinical content knowledge domain is made up of data and data relationships enabling the provision of new information. For example, Fig. 2 shows a mind map of blood pressure showing that this knowledge concept relates to a patient's positional state, an event, a number of data elements and data collection protocols. Above the mind map, it shows data constraints associated with the two actual measurements. This data formalism enables a computer system to ensure consistent data entry, including data associated with the context of the blood pressure reading. This knowledge is essential for the clinical interpretation of the recorded measures. Greater detail about these types of knowledge models is provided in the chapter to follow.

Data may be represented in a variety of ways, by text, standard terms, different types of terminologies, images or signals or sounds, and they may be presented for computer processing purposes in any of the formalism examples (data structures) listed in Fig. 2.

Clinical content consists of any patient-related clinical information and specific clinical knowledge made available as a resource that needs to be applied to that information for clinical decisions to be made. Clinical knowledge covers many knowledge domains at multiple levels of granularity (detail).

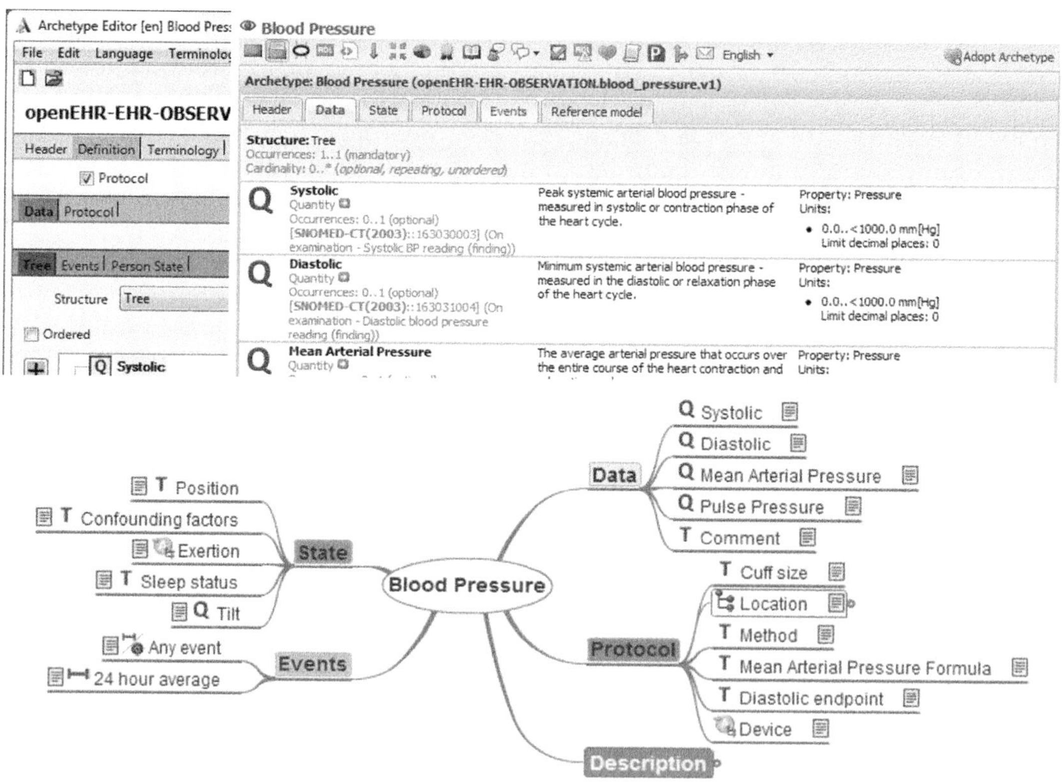

FIG. 2 Blood pressure mind map—a formalism example [22].

Patient information represents another data domain as does clinical reasoning. Collectively these domains represent a major knowledge component of the health data science discipline. Currently, there is not one terminology or data dictionary able to represent all clinical content at every level of granularity able to serve multiple users for a variety of different purposes. For example, a 'lung condition' is a high-level concept, whereas a 'bacterial lung infection' has greater granularity as this concept provides more detailed information. Drilling down further would describe the type of bacteria. The level of granularity of any concept determines possible use.

Clinical content can be represented as terms described in a data dictionary, as a terminology, a coding system, a reference terminology, or as a reference information model or an ontology. Such representations integrate knowledge from linguistical, philosophical, informatics or health science perspectives. The context in which an item of clinical information is represented can dramatically affect its meaning or semantic value.

Digital environments need information and knowledge representations to be formalised in the required level of detail to be effective. Data representations stored in patient records to suit clinical decision-making for individuals differ from data requirements to suit operational care delivery management, although the first 'use case' informs the other. Representing the former clinical use case needs to be far more detailed than the data representation of the latter use case.

Healthcare terminological systems were originally developed as a way of organising health data and information (clinical content) for the enablement of efficient high-level data management and statistical processing to suit policy development, administrative and operational management purposes. Such traditional terminological systems represent a set of designations within a healthcare domain that, where appropriate, have associated rules, relationships, and definitions.

Ontologies and software engineering relationships

The Object Management Group (OMG) SDO has developed a specification for a set of enabling capabilities for model-driven architecture-based software engineering, an ontology definition metamodel [23]. This provides the formal grounding for representation, management, interoperability, and application of business semantics. It is applicable to knowledge representation, conceptual modelling, formal taxonomy development, and ontology definition and enables the use of a variety of enterprise models as starting points for ontology development. A number of different ontologies are used for IT/IS analysis, design, and development. The most prominent one is the BWW (Bunge-Wand-Weber) ontology, widely regarded as a major contribution to the theory and practice of information technology/information systems (IT/IS) and conceptual modelling [24].

There are many types of ontologies ranging from those least formal and expressive (natural language), to highly formalised with a high level of expressivity (formal ontology). The many different ontology domains used to describe clinical content vary considerably across these ranges. Many originated well before the advent of computing technologies and were designed to suit a specific purpose. Each ontology's degree of formality and degree of expressivity determines how computers are best able to reliably and accurately process such data representations.

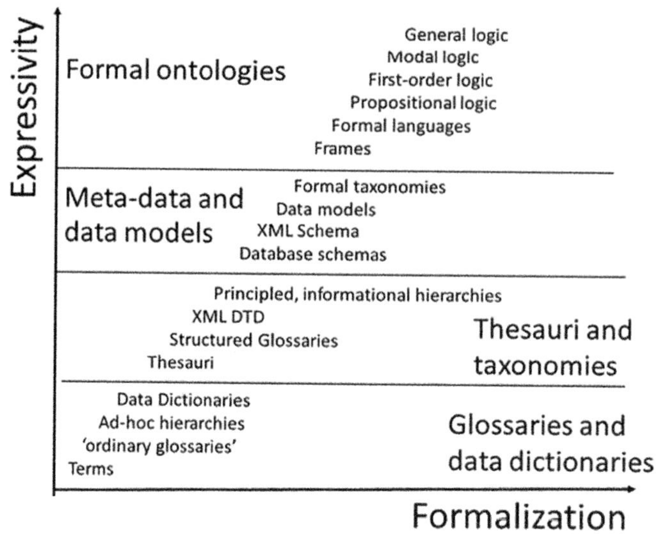

FIG. 3 Degree of formalism and expressivity for each group of formalism. *Reproduced with permission from B. Blobel, Ontology driven health information systems architectures enable pHealth for empowered patients, Int. J. Med. Inform. 80 (2011) e17–e25.*

Domain ontologies, also referred to as terminologies, may be grouped into four distinct types of formalisms [18]:

1. Glossaries and data dictionaries,
2. Thesauri and taxonomies,
3. Metadata and data models, and
4. Formal ontologies.

The hierarchy in terms of the degree of formalisation and expressivity together with examples within each of these groups of formalisms is shown in Fig. 3. There are differences in terms of actual degrees of expressivity and formalism between examples of each type listed in this hierarchy. For example, data dictionaries will vary considerably in terms of their level of expressivity.

Each type of formalism determines how each data set is being structured. This in turn determines the level of granularity, how data are being represented and presented for computer processing purposes. This in turn determines potential usage in a digital environment. Each of these groups of formalisms make up clinical content representations reflecting any clinical content knowledge domain that collectively makes up the clinical content structures.

Clinical content formalisms and expressivity

An examination of the many standardised clinical content representations in use today was undertaken to identify and describe the foundations from which to develop a supportive framework for a digital transformation. Many terminologies have historical characteristics to be considered when determining, not only their suitability for this digital era, but also how best to continue to use them during the transformation process where appropriate.

Terminologies used as data standards

Every terminology represents a knowledge domain or ontology. Terminologies can be differentiated by the different ways that concepts (data) are represented, they may be:

- standardised,
- specified through a code system,
- possibly further specified through the rules used to determine the codes, and
- by the application of their codes in a given data specification use case.

A clinical content framework, underpinned by clinical data, information, and knowledge, consists of many intersecting domain ontologies that collectively need to represent a high-level overview of the most common domain ontologies, data dictionaries, and terminologies in widespread use. There is also a need to consider the many interdependencies between various domain ontologies where concepts (terms) included in one or many different domains are required to undertake a measurement that describes a concept in another domain.

For example, the 'burden of disease' concept is determined by the gap between current 'population health' and the optimal state where all people retain full life expectancy without suffering major ill-health. Again determining 'population health' requires the use of an additional set of data standards. Such interdependencies are present, in all high-level secondary use data based on any prior source data aggregations that occur, in any data supply chain to serve many users at all levels within any national healthcare system.

A following chapter has detailed the currently available viewable knowledge representations presented in the form of guidelines, developed and maintained by numerous authorities, research institutes and professional organisations. These, mostly text-based, guidelines are least formal or expressive; they represent the starting point from which to determine the current status of available clinical content representations and may be used for natural language processing. Adopted data structures associated with any knowledge domain determine the ease of searching and retrieving data. This in turn determines how computers are able to process retrieved data. Examples will be identified from the least (more or less) to the most expressive formalism, for each of the types of standards as identified in Fig. 3.

Glossaries and data dictionaries

Many countries and organisational entities have developed their own sets of data dictionaries and guides, or data glossaries to provide information about standardised terminology and data elements used by them for data collection and reporting purposes, such as incident or workforce or asset data and for registries.

Data dictionaries are known as information and technology (IT) documents storing business and/or technical terms; they represent a traditional data management requirement. They vary in terms of the amount of detail provided for each listed term.

Glossaries tend to be developed to represent specific knowledge domains. They are key deliverables for data governance purposes and should include not only terms and definitions, but also data owners and data stewards for each term. This assists with the management of data quality within any healthcare organisation. Any organisational analysis of data use is likely to identify a number of differing definitions for the same term and conversely a number of different terms with the same definition. Once such occurrences are known, it is possible to agree on standard definitions for glossary inclusion to represent individual concepts.

Multiple standards development organizations (SDOs) define concepts included in each standard as an upfront glossary in each standard. This evolution of glossaries has resulted in standards products where, increasingly, terms are defined and/or used in different ways. ISO has developed a technical specification that provides metadata details and requirements for inclusion and construction of quality terms and definitions in health informatics glossaries [25].

Glossary examples are:

- The ISO TC215 committee recognised the requirement for a single international health informatics glossary. The Joint Initiative Council of Health Informatics Standards Development Organizations (JIC) now has a shared online Standards Knowledge Management Tool (SKMT) to assist in finding and managing standards, documents, products, terms, and definitions [13].
- The HL7 International has its own glossary of terms [26].
- The CDISC provides a glossary of terms used in clinical research, particularly eClinical investigations sponsored by the pharmaceutical industry or a federal agency. This glossary includes links to other glossaries of relevance.
- McMaster University has developed a glossary of terms used to describe health evidence for local use [27].
- The WHO published its first version of a Health Promotion Glossary in 1986. This was updated in 1998 and in 2006 when Smith et al. [28] recommended the addition of new terms. This glossary was written to facilitate understanding, communication, and cooperation among those engaged in health promotion activities locally, regionally or nationally. Regular updates are necessary to ensure there is consensus regarding meanings and to accommodate new practice developments, such as the use of 'burden of disease' and 'social determinants of health' concepts.
- The Centers for Disease Control and Prevention provides online access to a number of glossaries for specific knowledge domains [29].
- Australia's national MyHealth Record has its own glossary [30].
- The US Office of the National Coordinator (ONC) for Health Information Technology (Health-IT) has adopted its first version of the United States Core Data for Interoperability (USCDI) [31] as a standard in the ONC Cures Act Final Rule. Its USCDI sets a foundation for broader sharing of electronic health information to support patient care. It is a required part of the new application programming interface (API) certification criterion and replaces the Common Clinical Data Set (CCDS).

Thesauri

A thesaurus structures its terms hierarchically and adds nonhierarchical relationships between concepts and other properties to each concept, in the form of synonyms. This makes it fairly simple to maintain. The generic ISO 25964 standard [32] covers all aspects associated with the development of a thesaurus and includes a data model and an XML schema for data exchange. Part 2 of this standard was developed to encourage high-quality information retrieval across networked resources indexed with different vocabularies. This standard explains how to set up mappings between the concepts in such vocabularies and other forms of complementary use, and is compatible with the W3C's Semantic Web Simple Knowledge Organization System (SKOS) standard.

Mapping terms between health taxonomies need to be compliant with the Health Informatics ISO 21564 standard [33] to ensure that the desired quality is retained for every health data map.

A health thesaurus provides a long list of concepts that may be organised within concept schemes to assist consistency of use and subject retrieval. Concept schemes include diagnosis, facilities, population characteristics, and organisations. Within each of these schemes, additional hierarchies exist. For example, the diagnosis scheme includes diagnostic procedures, prognosis, and risk assessment. Many of these types of concepts represent national system infrastructure standards where clinical content is related and used for statistical reporting or performance measurement or for funding purposes. For example, the Australian Health Thesaurus [34] catalogues health-related medical and human services terms (concepts). These are organised into categories to improve discoverability of content by connecting search terms and concepts from multiple sources.

What is metadata?

When data contained in data dictionaries and glossaries are described in terms of meaning, concept representation, use, and misuse, their descriptions are referred to as 'metadata'. There is a need to produce high-quality metadata to improve consistency in data recording where required. The use of metadata allows the resources thus described to be understood by both machines and people, a necessary function for promoting interoperability and resulting in minimal loss of content or data functionality.

Metadata represent facts, concepts, or instructions that can provide meaning and context to data by describing how data are captured and the business rules associated with collecting data. This assists in the interpretation of data so that data can be better managed, shared, and used for subsequent data analysis. Metadata sets can be established to reflect, for example, regulatory guidelines or as an aid to enhance searchability and retrieval or to serve as a conceptual map to suit any specific purpose. Not everyone defines metadata in a consistent manner.

Definitions for what metadata is have been developed by many standards organisations—the definitions in the following are all intended to mean the same thing but are described in different ways. Sometimes even by the same organisation or even in the same standard [35]. Box 2 lists five different metadata definitions.

The metadata repository requirements standard [36] describes requirements for the collections of metadata about data elements and their containing models and data sets in a healthcare environment. It focuses on the description of data that is persisted (not likely to be modified) in healthcare systems. Its adoption promotes interoperability between metadata registries and provides opportunities for the development of tooling that consumes metadata for the generation or the parameterisation of code. These are highly desirable features for a well-functioning digital environment.

BOX 2

Five examples of differing metadata definitions included in some standards

Definition	Standards organisation	Context
Data that define object class and property for the information collected	ISO and IEC 11179-3 CEN	Software development
Data that define and describe other data	ISO and IEC 11179-3 Standards Australia	General
Information stored in the data dictionary that describes the content of a document	ISO TR 2221:2006 Health Informatics—Good principles and practices for a clinical data warehouse	General
Data describing context, content, and structure of records and their management through time	ISO 15489-2001	Health records
Data that describe other data, particularly XML tags characterising attributes of values in data fields, such as version, unique identifier, and maps	Integrating the Healthcare Enterprise	Software development

Metadata identifiers, structures, and uses

Metadata can be identified by the use of 'tags', as in many instances such 'tags' refer to entire data sets contained within each 'tagged' metadata category. In one instance, it was recommended to "facilitate the rapid mapping of existing semantic taxonomies into tagged data elements, while continuing to encourage the longer-term harmonization of these taxonomies by vendors and other stakeholders" [37].

The US National Information Standards Organization (NISO) identified three main types of metadata:

1. descriptive used for discovery and identification,
2. structural indicating how compound objects are put together such as pages ordered to form chapters, and
3. administrative for the purpose of managing a resource consisting of two subsets, rights (i.e. IP) management and preservation metadata [38].

Metadata schemes (schema) are designed for a specific purpose and need to be digitally identifiable. These principles are also applied to concept modelling where individual data elements represent content value. Metadata can be encoded in any definable syntax. For metadata to be most effective, it needs to meet the FAIR (findable, accessible, interoperable, and reusable) principles as defined in Box 3.

Guidelines for metadata development are detailed in a handbook published by Standards Australia [40]. This requires the development and use of a data management and stewardship plan, the adoption of an 'open-platform' and vendor-neutral data repositories to suit each unique digital health ecosystem as described in various other chapters.

One Australian health example of metadata is freely accessible via the metadata online registry (METeOR). Its structured repository is based on the International Organization for Standardization and the International Electrotechnical Commission 11179 Metadata Registries (ISO/IEC 11179) [41]. This ISO/IEC standard was updated in 2019 to better reflect the needs of the health industry [36].

The United States Health Information Knowledgebase (USHIK) is also based on this core ISO standard's architecture and is accessible via the Agency for Healthcare Research and Quality (AHRQ) Website [42]. This provides an online, publicly accessible registry and repository of healthcare-related data, metadata, and standards to support the health information technology initiatives of the HHS (Health and Human Services) Secretary and the AHRQ Director. This is made available as a resource for the US HHS Office of the National Coordinator for Health IT (ONC)'s Standards and Interoperability initiatives and other federal agencies' programmes. Metadata standards are also used to support a systematic approach for clinical and/or operational research.

Classifications

Classifications are designed to group similar things together—to classify them, rather than intending to represent meaning. The structure of a classification system consists of groups identified according to one or more attributes. A classification system does not necessarily have a relational hierarchy but can do so. Many are not structured to be a formal taxonomy, although this can apply to some classifications. The structure adopted indicates the potential for ambiguity and degree of expressivity and formalism.

The health industry globally makes use of numerous classification systems for pricing, funding, reporting, and research purposes. Classification systems are developed for the purpose of grouping data for counting and analytical purposes; they enable categories to be coded for ease of computer processing purposes. Many of these systems were developed prior to the digital revolution. As a result, many are not well suited to this digital era. Their maintenance is equally complex and time-consuming. All terminologies require regular updating based on a change request system; this terminology maintenance activity is responsible for generating an entire global industry in its own right for terminologists.

The most widely recognised classification is the International Classification of Diseases (ICD), developed and maintained by the WHO as one classification system within its Family of Classifications in common use [2]. The use of classification systems can provide some consistency, although this is dependent upon the classification rule interpretation. For example, we are aware that there are many inconsistencies between those responsible for ICD-10 coding.

The implementation and interpretation of classification rules applicable to this classification system vary between jurisdictions. These variations can compromise coding outcomes

BOX 3

The FAIR guiding principles [39]

To be findable

F1	(meta)data are assigned a globally unique and persistent identifier.
F2	Data are described with rich metadata (defined by R1 later).
F3	Metadata clearly and explicitly include the identifier of the data it describes.
F4	(meta)data are registered or indexed in a searchable resource.

To be accessible

A1	(meta)data are retrievable by their identifier using a standardised communications protocol.
A1.1	The protocol is open, free, and universally implementable.
A1.2	The protocol allows for an authentication and authorisation procedure, where necessary.
A2	Metadata are accessible, even when the data are no longer available.

To be interoperable

I1	(meta)data use a formal, accessible, shared, and broadly applicable language for knowledge representation.
I2	(meta)data use vocabularies that follow FAIR principles.
I3	(meta)data include qualified references to other (meta)data.

To be reusable

R1	(meta)data are richly described with a plurality of accurate and relevant attributes.
R1.1	(meta)data are released with a clear and accessible data usage license.
R1.2	(meta)data are associated with detailed provenance.
R1.3	(meta)data meet domain-relevant community standards.

and their subsequent comparative use. This is especially relevant for those who make use of these codes for research purposes. Updates to the classification or the rules that govern its implementation, including identification of changes and how the old and new versions relate to each other is another issue that needs to be managed. There have been instances when ICD updates have changed the meaning of codes, and new terms may be added in ways that can change the meanings of other terms (e.g. 'not elsewhere classified' terms) adding to issues to be considered by users of clinical data repositories [43].

The International Classification of Nursing Practice (ICNP) [44], a recognised member of the WHO Family of Classifications, was developed by the International Council of Nurses (ICN) as a seven-axes, compositional terminology. Terms from multiple axes can be combined to form a nursing diagnosis, intervention, or outcome via the pre- or postcoordination of terms from two different axes. It identifies relationships among concepts and vocabularies, enabling its use as a reference terminology for a number of other nursing terminologies. The ICN have recently entered an agreement with SNOMED International for them to continue the maintenance and integration of this terminology with SNOMED-CT described in the formal ontology section to follow.

Some classification rules associated with individual systems in use are more complex or better formalised than others. One national example of the number of classification systems in use is listed by Australia's Independent Hospital Pricing Authority (IHPA) [3]. Many of these types of classification systems are country-specific.

Categorical structures

Classification systems are organised by defining categories, or entity types, that are shared by all the individual instances in existence in the present, past, and future. Such categorical structures consist of a minimal set of domain constraints for representing healthcare terminological systems' entities (terms or codes), in a precise subject field (special knowledge) to achieve a precise goal (intended use).

Domain constraints are the rules that prescribe the set of representations of relations that are valid to specialise a category in a certain domain. ISO TC215 has developed, or has work in progress to develop, a number of standards for categorical structures, such as:

- for terminological systems of surgical procedures,
- of representation for evaluation of clinical guidelines of Traditional Chinese Medicine (TCM),
- for representation of TCM clinical decision support system,
- for representation of nursing diagnosis, nursing actions and nurse sensitive outcomes in terminological systems, and
- categorical structure and data elements for the identification and exchange of immunisation data.

Formal taxonomies

A formal taxonomy has a hierarchical structure providing a relationship map to support consistent indexing or tagging and retrieval. It can also be viewed as a top-down navigation structure in a user interface. It is a system for naming and organising things within any domain. One example is the Medical Subject Headings (MeSH) terms developed and used to access publications indexed by the US National Library of Medicine (NLM) [45]. Taxonomies are often developed to enable qualitative research or to improve access to web-based content or to define complex domains such as any number of public health domains of interest.

Data/information models and value sets

Conceptual data models represent a structured organisation of data elements, showing how these relate to each other and their real-world properties in a graphical form. They are abstract formalisations or artefacts representing individual concepts. Data models are used to define information requirements and associated rules to support the development of information systems and applications. They are fundamental to effective health data exchange.

Every data model is a single view of the world, often from a single perspective like looking out a window at the world. Every view from every window is different even if the world outside is the same. From a standards perspective, the issue is to identify the view needed to solve a specific problem and how each view connects with every other view.

Every type of data model needs to include a mechanism to 'bind' data standards to every data element included in the model. Data elements may be defined by metadata or by any chosen formal or informal terminology used to identify 'value sets'. Value sets are uniquely identifiable sets of valid data representations. They are used to represent all possible values of the data element making use of the most suitable terminology code system or standard terminology, such as SNOMED-CT or LOINC or ICNP codes or any other standard terminology that best represents the relevant health or clinical domain. They specify the allowable content for each coded element. Some data elements within a data model can be expressed by any one of multiple code systems based on equivalency.

The US National Library of Medicine established a Value Set Authority Centre (VSAC), a repository and authoring tool for public value sets, including HL7 C-CDA Value Sets, in 2012 [46]. The VSAC repository includes downloadable access to all official versions of value sets specified by the Centers for Medicare & Medicaid Services and electronic Clinical Quality Measures (eCQMs). The repository structure determines the ease of search and retrieval.

Conceptual data models are particularly important for determining database design. The logical and physical versions of these models are required for database development and implementation purposes. There are a number of common data models (CDM) in use, such as those developed by national regulators, for example, the Federal Drug Administration (FDA)'s Sentinel System [47], or the US National Patient-Centered Clinical Research Network (PCORNet) [48] or the Observational Health Data Sciences and Informatics (OHDSI) common data models [49]. These CDMs define database structures for sets of healthcare concepts and create the opportunity for the interchange of those clinical concepts as standardised data.

The structure and metadata definitions contained in a data model heavily influence which data elements can be stored, how data values should be interpreted, and how easily these data can be queried and extracted from a database. A common feature of all data models is their leveraging of standard terminologies and their associated coding systems applicable to the data elements within each model. Unfortunately, most of the common data element and clinical information modelling efforts to date have not been primarily driven by the clinical care use case. Instead, these efforts were driven by research, public health, or surveillance requirements [50]. Data models in use are not necessarily based on a grounded conceptual modelling theory that determines their degree of expressiveness and formalism.

Types of data models

Some types of data models, which may also be referred to as semantic artefacts, are developed specifically for the purposes of health data exchange using HL7 messaging protocols between systems within a network or ecosystem. Exemplar specifications of this type are:

- the HL7 International SDOs Clinical Document Architecture (CDA), [51]
- a subsequent version of C-CDA (template),

- Clinical Information Modeling Initiative (CIMI) models, [52] and
- Fast Health Interoperability Resources (FHIR) [53].

The C-CDA development methodology focuses on the transfer of documents. CIMI models can be used as physical representations for content within its logical model framework for clinical content. FHIR models are usually referred to as FHIR profiles.

FHIR profiles group data structure definitions and corresponding value sets defined for a common purpose suitable for specific use cases and workflows. They do not have an underlying logical model. The inclusion of metadata into FHIR profiles provides substantial optionality resulting in numerous inconsistencies between users. These types of models frequently do not provide the desired degree of expressivity.

Standardisation of the clinical data model content itself, in a manner that preserves semantic meaning and is natively computable, is still lacking. This 2018 study, undertaken by the Duke Clinical Research Institute [50], found that attempts to extract information from documentation systems via FHIR and other translation-based approaches to populate registries, for example, have simply not succeeded to date. Although all HL7 data models are freely available, they are not stored in easy to find formats for clinicians to select for review, in terms of clinical suitability. Some are made available as a formal standard.

Data model design influences their ease of usability. They vary in terms of degrees of expressivity and formalism. For example, the operational data models made available by the Clinical Data Interchange Standards Consortium (CDISC) specify the information that needs to be shared among different software systems at three different stages: during the study set-up, operation, and analysis [54].

Each HL7 data model type has its own associated technical development and implementation methodology, designed exclusively for transferring clinical and administrative data or documents between healthcare providers. Their value sets are based on the fundamental definitions found in the HL7 v.3 Core Principles document and the functionality as described in the OMG's Common Terminology Services 2 (CTS2) specification [55, 56].

All HL7-related data models are very different to openEHR archetypes. The latter are data models designed to be platform, terminology, language, and location independent. This means that the abstraction of data concepts from common usage to higher level data concepts, as applicable across all healthcare systems, is possible irrespective of technology or terminology in use [57]. These formal models of domain concepts control data structures and the data content. They are stored in the Clinical Knowledge Manager (CKM) database [58], to represent a freely available single governed and curated collection of shared implementable clinical information models, built collaboratively by a global multidisciplinary community. The evidence used for the development of each data model reflects best clinical practice or expert knowledge and is made transparent.

The openEHR approach is unique in that an archetype is strategically designed using the ontological modelling approach. Archetypes represent notional maximal data sets of relevant data elements, the universal use case, using a use-case agnostic mindset to model data requirements supporting all imaginable uses. This minimises the number of archetypes required to be able to represent all possible clinical uses and enables the use of data at the finest level of granularity. Its design method not only provides the best possible level of expressivity and formalism but also streamlines model storage and retrieval methods for ease of use.

Data models facilitate the translation of knowledge into computable formats. As such, they need to convey knowledge 'truths' (epistemology) and reference world reality (ontology). This requires data models to:

- make use of a rigorous modelling methodology based on an underlying theoretically grounded conceptual framework,
- adopt a participatory approach,
- make use of a sound evidence base able to demonstrate real-world applicability and usability, and
- be trustworthy and transparent.

The availability of openly available, governed universal data models, facilitates the implementation of EHR systems and interconnected applications. For example, Demski et al. [59] used the openEHR model-driven engineering methodology for its application development across multiple applications. This software development methodology simplifies the design process by promoting communication between developers. These authors demonstrated that open data models can be regarded as enablers for interconnecting the fragmented and highly diverse stand-alone applications in healthcare.

Fig. 4 shows the relationships between the reference information model, archetypes (data model), templates, and terminologies (representing value sets in templates to suit specific use cases).

Universal data models are reusable, as they provide the foundation for the development of any number of models to suit individual 'use cases'. This significantly reduces the time and development costs as demonstrated in this case study where total archetype reuse of use-case agnostic archetypes (universal data models) ranged from 40% to 90% across unanticipated, novel modelling challenges and multilingual deployments [60]. Given the availability of multiple types of data models, it was pertinent to be able to assess the data model quality dimensions. These were identified and validated by Moody and Shanks as presented in Table 2.

Kahn et al. [62] noted that the most visible component of any information system or application is the user interface as this defines which parts of the data model can be 'seen' and

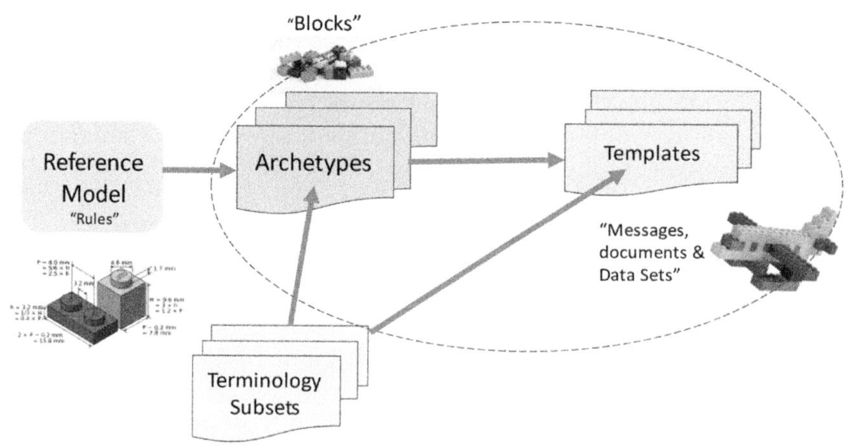

FIG. 4 OpenEHR infostructure.

TABLE 2 Data model quality factors [61].

Correctness	Technically sound, conforms to the rules of data modelling technique adopted
Completeness	Contains all user requirements
Integrity	Business rules definitions (integrity constraints)
Simplicity	Contains the minimum possible entities and relationships
Flexibility	Ease with which the data model can cope with business and/or regulatory change
Integration	Consistency with organisational data
Understandability	Ease with which concepts and structures can be understood
Implementability	Data model can easily be implemented within project time, budget and technology constraints

'manipulated' by the user. This determines access capabilities that can make a flexible data model appear to be very limited to the user, or vice versa. The simplest type of data models make use of the entity-relationship diagram format. Data models represent entities (concepts) which in turn need to be organised to inform the database design or data extraction requirements to suit a specific research study or meet reporting requirements.

Data model presentations and tooling

Some of the data models identified previously, such as the CDA and C-CDA, are in fact documents or standard templates. These models differ significantly from openEHR templates that are able to meet any specified user requirements to suit any niche application. openEHR templates are created to represent a specific data set, comprised of one or more archetypes that are aggregated, configured, and constrained to precisely match the data set requirements for a specific clinical use case, health domain, profession, or geographical location.

The number of archetypes used in a template reflects the required scope of content and/or the level of detail. Some simple templates representing a laboratory test report may comprise only a single archetype, while others representing a complex document or consultation note, may be comprised of attributes selected from upward of 50 archetypes—there is no upper limit. This dual-level modelling approach consists of first defining and standardising archetypes, which is followed by combining and constraining them to create clinical templates. This is a unique openEHR data modelling approach explained in more detail in a chapter to follow.

Data models can be presented in a variety of formats such as mind maps, diagrams using any familiar modelling tool such as UML, or computer-aided software engineering (CASE) tools or in computational formats such as Extensible Markup Language (XML) or Java or the Archetype Definition Language (ADL) or JSON. These are computer languages that are platform independent enabling the development of applications that can be used by a variety of computer types or devices using associated operating systems, without having to modify the data models' code in any way for many years. Platform independence means that the applications are not limited by the capacity or idiosyncrasies of the computing devices on which they are used enabling developers to focus on application design.

A number of clinical information modelling tools (CIMTs) are now available. Nine were evaluated according to performance, using evaluation metrics that were the result of a

TABLE 3 List of domains covered by functional requirements for CIMTs.

• Data types and specifications	• Clinician involvement
• Support for testing and validation process	• Searching capabilities
• Clinical information models (CIM) metadata	• Terminology and ontology binding process
• Support for CIM evolution and specialisation	• Semantic relationships
• Collaboration	• Communication with terminology servers

consensus of experts regarding essential requirements for CIMTs for which tool capabilities were described [63]. It is important to note that every tool has one of a number of supported specifications, including reference information models. Evaluation metrics were based on the following domains covered by functional requirements for CIMTs as presented in Table 3. Data models and their associated reference information model need to be able to meet these functional requirements, as well as co-inside with each use case, where they need to interact, to best achieve the desired degree of functionality as described and coded within the data models.

Reference information models

Health-related information systems need to make use of many data models which in turn need to be organised in a manner that facilitates retrieval. Such organised structures are informed by a reference information model developed to suit any one specific health-related domain. There may be hierarchies of data models showing the concept represented in an organised manner that may be in accordance with an agreed reference information model (RIM) that, for example, represents an agreed high-level health record structure as is the case of the openEHR RIM.

The openEHR's information model is native (integrated) into the system and used as a reference for the clinical models. In this instance, the health record-related documents or reports may be structured by combining one or more data models to suit specific use cases or small vignettes that indicate the system's needed functionality. Each of these use cases may be represented in the form of a template providing high specificity to perfectly meet the user's needs within resource limitations. This is only possible where the RIM's focus, e.g. health record, is separated from the system's technical reference architecture. The openEHR RIM has been designed to minimise the need for software and schema updates. Benefits of this separation are explained in a chapter to follow.

Information models define relationships between managed objects such as data models. Data modelling is a critical first step taken once requirements of analytics and reporting have been defined. Data models need to be aligned with what drives and motivates and are the goals of intended data users. The ability to share data, as defined within data models, between information systems, is dependent upon agreed reference information models as well as data standards adopted.

A great variety of reference information models have been developed for use within the health industry. For example, the HL7 International SDO makes use of its RIM as the high-level foundation for organising its messaging standards, including the HL7 Fast Health Information Resources (FHIR) or the use of its Clinical Information Modeling Initiative (CIMI) data models. This RIM does not focus exclusively on clinical data or health records, its focus is on the use of a messaging methodology. Therefore, the openEHR RIM differs significantly from the HL7 RIM resulting in very different types of data models even though the content of many will overlap and need to be used in harmonious ways within any healthcare organisation's enterprise information system and related ecosystem.

There are many other health-related RIMs, each developed to suit a specific purpose and covering a specific domain. Most Electronic Health Record (EHR) systems or Clinical Data Repositories in hospitals in use today, rely on proprietary reference information models. This includes the IBM's Unified Data Model for Healthcare that supports evolving standards, regulations and use cases within the healthcare industry [64]. There is no apparent consensus regarding a standard patient information model or a standard health/medical record structure, other than traditional paper-based medical records. This fragmented reality is making effective patient-centred data sharing extremely difficult to achieve. The impact of these data model and RIM relationships is neither well studied nor published [62].

Standards development organizations (SDOs), data standards and data management

A previous chapter identified a number of terminologies, including classification systems, developed and maintained by internationally recognised standards development organizations (SDOs). These data standards need to be examined within the context of their degree of expressivity and formalism as this determines, for example, their applicability within data models and their general usability within a digital era. For example:

1. GS1 manages a number of taxonomies for its supply chain domain which may be represented in the form of barcodes, QR codes, data matrix, and radiofrequency identification (RFID).
2. The Clinical Data Interchange Standards Consortium (CDISC) develops and maintains data standards, including taxonomies of relevance to clinical trials which vary in terms of their degrees of expressivity and formalism.
3. The best known SDOs that only manage and maintain data standards are SNOMED International and the Regenstrief Institute. These SDOs have been working collaboratively.
4. ISO TC215 WG3 has developed a semantic content framework, as shown in Fig. 5, which is used to identify areas where standards work is needed. This has led to the development of standards associated with data specification for repositories, governance, and visualisation in user interfaces.

Data types

Data types refer to the formats used for data storage to enable distinct types of data or range or values used to express a variable or function, to be stored and processed. Data types

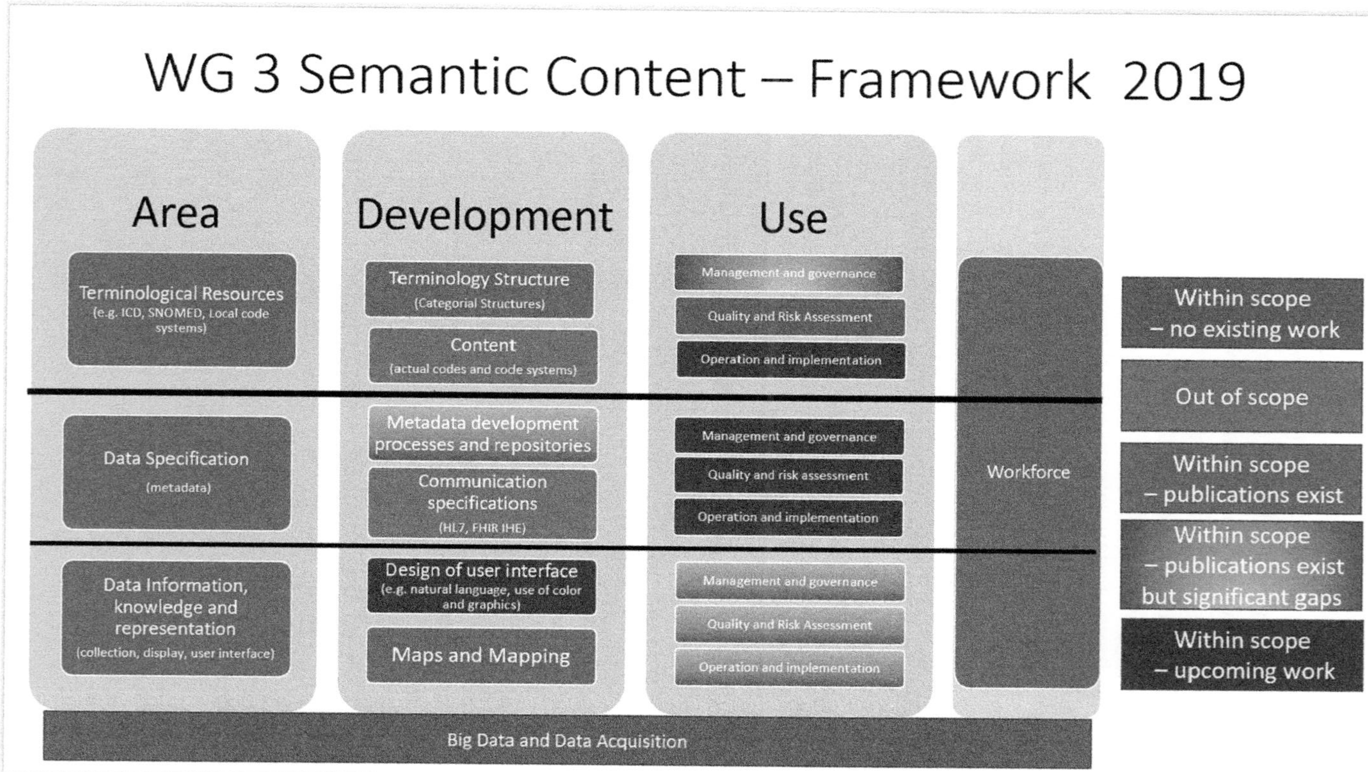

FIG. 5 ISO TC215 WG3 semantic content framework 2019.

define the operations that can be done on the data and the meaning of the data. Data-type format includes integers, characters, strings, and floating-point numbers. Almost all programming languages explicitly include the notion of data type, though different languages may use different terminology.

The terminology of most relevance and unique to the openEHR specifications is its own set of data types that act as a higher level reference. Other SDOs, such as HL7 International, ISO/IEC and CEN/TS, have developed their own sets of data types. Many of these are the same but there remain a number of discrepancies that hinder the reliable exchange of information in the absence of their harmonisation.

The ISO 21090:2011 Health Informatics—harmonised data types for information interchange—took many years to develop. This standard was underpinned by ISO/IEC11404, ISO13606 and CEN/TS 14796 offering a practical and useful contribution to the internal design of health information systems. Its primary intent is for this standard to be used when defining external interfaces or messages to support communication between them.

Data management within and among local and distributed information systems' environments is not only dependent upon the adoption of data standards at whatever level of expressivity or formalism. Reliable data management is also dependent upon the use of reference models and frameworks, definition of data domains, data types, data structures and their associated semantics as well as languages, services and protocols for persistent storage, concurrent access, concurrent update, and data interchange.

Data reference models

The United States (US) Government has adopted a Federal Enterprise Architecture of which its Data Reference Model (DRM) is one of five reference models that make up that architecture. Its primary purpose is to enable information sharing and reuse across the US Federal Government via the standard description and discovery of common data and the promotion of uniform data management practices. This reference data model consists of multiple sets of metadata. Such an arrangement does not exist for the US Health Industry.

Many countries have adopted similar government, focused enterprise architectures, few if any have applied these principles specifically to their national health ecosystem. These generic reference architecture principles apply to the latest ISO standard's interoperability and integration reference architecture [65]. This standard was developed to facilitate the deployment of existing standards and systems and to accommodate ongoing organisational, methodological, and technological paradigm changes in health and social care that is transforming health systems enabling them to support personalised, preventive, predictive, and participative precision medicine (P5).

Formal ontologies

An ontology is a formal representation of knowledge within a domain. Such a representation may be a hierarchically arranged set of unique terms known as concepts, their attributes, and the semantic relationships between those concepts [17]. Ontologies organise domain knowledge into structures to represent a set of concept definitions that may be implemented in software of some kind for knowledge engineering purposes. The use of ontologies supports effective data retrieval and different sorts of reasoning. Ontologies are also created to foster consistency for the purpose of more effective integration of clinical and scientific data [66]. Ontologies need to represent reality.

Ontological models are based on various formal logical theories such as frames, description logics, and formal logics. Ontological modelling embraces the notion of inheritance and uses conceptual abstraction to define upper-level or base-level classes making use of parent and child classes. An ontology is a particular type of knowledge representation. Its design uses formal logic enabling coding and computer processing.

Most clinical terminologies and classification terminologies in use are informal ontologies which do not enable decision support and are not inherently interoperable [67]. Table 4 provides a synthesis and summary of formal ontological modelling concepts. The coarser levels may serve as reference ontologies.

Ontologies can be created specifically for human use, such as decision support or for computer use such as intelligent agents/bots. They need to be both human and computer interpretable. Ontologies can be developed as high-level models of reality or of some specific clinical context at low levels or at various levels of abstraction in between these boundaries. They can be used as reference models to assist with mapping one ontology to another when the semantics are not naturally interoperable. Formal, well-defined ontologies are extensible and reusable. Adoption of a coherent framework is expected to improve the capacity for accurate clinical content data sharing and automation.

For example, Jorm et al. developed a public health ontology [4] to improve the ability to compare public health activity across jurisdictions and countries. This public health ontology may also ascertain differences in what is considered to be public health activity. They made use of open-source Protégé software [68] and its published procedures to construct such an ontology, which was initially based on a number of existing terminologies. This was then further refined following consultations with other public health experts. The Protégé software supports current Semantic Web standards.

TABLE 4 Characteristics of ontologies at varying levels of granularity [67].

Granularity	Knowledge type	Number of concepts	Level of abstraction	Scope of problem domain	Changeability	Comments
Coarsest	Intentional[a]	Few	0	Universal	Unchanging	Timeless
Coarse	↑	↑	1	Discipline specific	Unlikely to change	Longitudinal stability
Medium	↕	↕	2	Specialisation specific	May change with new knowledge	Medium to long term stability
Fine	↓	↓	3	Problem context specific	May change with change in context	Short to medium term stability
Finest	Extensional[b]	Many	4	Location & implementation specific	Fluid, likely to change with needs & practices	May describe this as 'agile', high flexibility can be changed as needed.

[a]Intentional knowledge in this context refers to specific instances of things that exist.
[b]Extensional knowledge refers to the collection of all the objects falling under it.

Terminologies and ontological structures

The increasing recognition that terminologies need to be ontologically structured to maximise their value in a digital world, led to the establishment of a grass-roots effort in 2002 to create a suite of interoperable reference ontologies in the biomedical domain. The Berkeley Bioinformatics Open-source Projects (BBOP) [69] focuses on the development, use, and integration of ontologies into biological data analysis. It hosts the OBO Foundry [70, 71] listing an expanding family of ontologies designed to be interoperable and logically well formed so as to incorporate accurate representations of biological (including clinical) reality.

Ontologies not only have a hierarchical structure but also include axioms, restrictions, and relations in their structure. For example, the Systematized Nomenclature of Medicine—Clinical Terms (SNOMED CT) terminology is structured as a polyhierarchical ontology with the capacity to employ rule-based or 'intentional' value sets that leverage the relationships within the ontology [72]. This is the most comprehensive clinical terminology available. SNOMED CT enables the concise identification of included and excluded subtypes of a clinical condition by referring to the SNOMED CT's hierarchical 'is a' supertype-subtype (parent-child) relationships. Such relationships can closely match clinicians' thinking about, for example, clinical phenotypes and the subtypes of conditions they wish to be included or excluded.

This terminology is able to continually add new concepts to ultimately represent all clinical concepts. SNOMED International [72] maintains a published customer guide that details how to prepare a content change request submission. A number of terminologies have been mapped to SNOMED-CT, e.g. LOINC [73], ICNP, Clinical Care Classification (CCC) [74].

Clinical content: An ontological framework

When clinical content is computerised, there is a need to be able to find, extract, and process data using predetermined rules and algorithms. This requires structure and consistency to be effective for widespread use. We need to have a good understanding of the many ontological intersections, including data and technology, to ensure we have the best possible structured design. A clinical content domain ontology may be viewed as an 'upper' or 'foundation' ontology representing commonly shared relations and objects applicable across the many clinical knowledge domain ontologies, as these intersect with each other.

Few health (clinical) professional entities have developed an ontological structure that represents their specific knowledge domain. For example, the US National Cancer Institute (NCI) has developed an integrated biomedical informatics infrastructure, now known as the cancer Biomedical Informatics Grid (caBIG) to support collaboration within the cancer research community [75]. This work was based on an evaluation of four standard terminologies; the gene ontology, the NCI Thesaurus, the Common Terminology Criteria for Adverse Events, and the laboratory portion of LOINC.

The NIH makes use of their data standards metadata repository, which is ISO11179 compliant, for system developers and users to describe data and systems as commonly understood concepts and definitions. Their aim was to implement terminological and semantic consistency in caBIG and to develop an open-source, open-access information network for secure data exchange on cancer research [76].

This development work demonstrates the complexity associated with terminology development and selection. This work has resulted in a set of 105 criteria to be considered relative to the terminology's structure (data model and concept orientation), content (domain coverage), documentation (intended use and organising principles) and editorial process (changes, updates and additions), which were used to evaluate individual terminologies for semantic consistency. The use of this evaluation method identified many shortcomings and limitations of the LOINC terminology and demonstrated that this is a sound method for evaluating any existing terminology in terms of its suitability, 'is it fit for purpose?' Data accuracy represents just one criterion; it is a basic terminology requirement not only for research, but also for all clinical decision-making.

Data supply chain

Effective use of health data is integral to the efficient functioning of any nation's health ecosystem. Increased electronic data sharing, across public and private health sectors, can facilitate greater leveraging of technology to improve health outcomes and system integrity and increase administrative efficiency. This requires the adoption and maintenance of an agreed set of data standards and health data governance. Data standards need to be layered, for example, data standards representing a professional knowledge domain need to be international, with national approaches to best suit local language, legislative, and reporting needs.

Fig. 6 provides a big picture overview of the data supply chain and the many different administrative and clinical uses over time, ranging from individual patient care by episode with

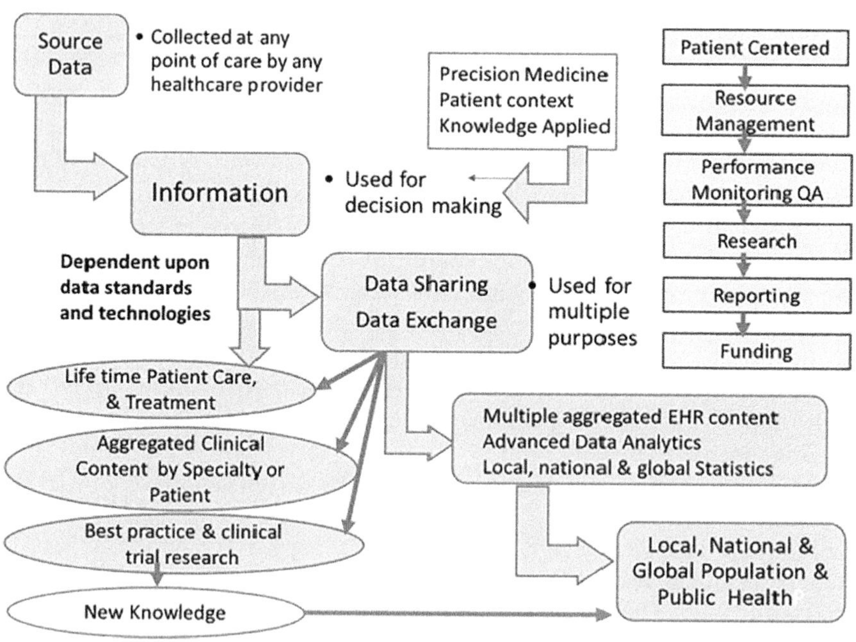

FIG. 6 Overview of data supply chain.

contributions from numerous providers, for any patient health journey and over a lifetime, to the use of aggregated data for multiple secondary use cases.

The data generated provide the foundation for decision-making at every level within any healthcare system for multiple purposes. Central to all is the need for data sharing, data aggregation, data linkage, and data exchange.

The generation of new knowledge necessary for continuous improvements, both clinically and administratively, is dependent upon:

- the source data collected by individual and organisational healthcare providers,
- data standards used consistently by everyone,
- the technologies adopted for all health data exchange,
- agreed data sharing/linkage protocols to suit a variety of different purposes, and
- supportive legislation/regulatory controls.

It is important to consider the desired information result for every type of use case, as this determines the source data required.

Standard value sets, Application Programming Interfaces (APIs), reference sets, algorithms, natural language processing, machine learning, and health data exchange schema, all link to standard terminology codes. This needs governance to maintain patient safety/data accuracy. Streamlining the data supply chain in a manner that maximises the potential for automation at various levels also needs to consider the best point (location and by whom/role) for data source collection and or processing.

Data governance and coordination

A 2017 inquiry into the benefits and costs of options for increasing availability of and improving the use of public and private sector data by individuals and organisations concluded that there is a need for a statutory office holder, a National Data Custodian, to guide and monitor data use [77]. Legislation in the form of a Data Sharing and Release Act is especially necessary where this refers to health data. National data custodians need to be supported by local data stewards who are responsible for managing data governance processes within any healthcare organisation, supported by cybersecurity expertise. Data governance needs to extend to technical health data exchange.

A National Health Data Custodian, together with a small advisory board, needs to provide the necessary leadership and ongoing updating of a new data framework, including the integration of specialty data sets, managing the broader ethical considerations around data use and the implementation of business rules relative to the technologies in use.

Those occupying these positions need to have an appropriate mix of health data science and technology knowledge and skills, and be provided with a strong supportive infrastructure. Functions would include the accreditation of processes and capabilities of a suite of Accredited Release Authorities (ARA) (independent entities) and publish guidance on data use for other organisationally based data custodians. The ARAs represent sectorial hubs of expertise tasked with developing a national health/clinical data management framework necessary to improve the curation and quality of all data sets, value sets, and terminologies in use.

Without the leadership and collaboration across the global clinical communities, governments, regulatory bodies, and health IT stakeholders, neither a standard for common clinical data elements nor a process for governing these has emerged. Professional societies representing specialised clinical expertise, need to consider taking on such knowledge and data governance roles for their discipline. Such expertise is required to:

- maintain practice guidelines,
- mandate data sharing needs between providers,
- best suit the types of patients they manage per episode or long term, and
- ensure continuity of care for every patient throughout that individual's health journey.

Disciplinary experts are best placed to determine the degree of data granularity, data standards, and value sets required to best represent their specific knowledge domain and to evaluate best practices. This proposal has similarities with the US NLM Value Set Authority Centre (VSAC) [46] mentioned previously which is similar to Australia's MeTeOR initiative that began in 1991 with its first National Health Information Agreement between the Australian, and State and TerritoryGovernments [41].

Current issues and limitations

An environmental scan of health terminologies and vocabularies identified numerous issues and limitations associated with their governance, coordination, maintenance, access, dissemination, and adoption [1]. A major Australian impediment to effective health/clinical data use was considered to be distrust around data use. This can only be improved through greater data accuracy.

This chapter overview and evaluation of the range of health terminologies in use, relative to their degrees of expressivity and formalism, have exposed many of the continuing complexities in need of reformation. Many of these issues were already identified and documented from a US perspective in 2006 [78], little progress is evident.

This overview of health data standards in use has also highlighted many variable levels of expressivity and formalisms even among a variety of ontologies, such as metadata or data models. These variations have significant implications for health information systems and applications and their enablement to interoperate semantically.

There are no generally adopted practices to ensure consistency and rigour in system development, no maintenance, and no requirements or funding for rigorous efficacy and safety evaluation. There is no widespread agreement for the adoption of a standardised representation model or standardised contribution model, and very little attention is given to the metrics of quality and life-critical system practices [1].

Our greatest impediment to achieving semantic interoperability is the lack of a shared information model followed by the lack of an agreed metadata framework. This gap makes it impossible to map concepts that are not the same. There are many terminologies, and many overlaps between some clinical domains, such as nursing and medicine. Some terminologies are open source; many attract royalty payments or have commercial limitations. Redundant health terminologies (often used in niche developments covering one domain, e.g. nursing) present a barrier to interoperability. Seeking to apply data standards once data have been collected is too late. There are no singular solutions; these issues need to be resolved as part of a layered process by linking data at the point of collection to metadata and an information model.

Semantic continuity continues to be an issue with some data standards, for example, the reissue of old codes for new terms. There is no uniform measure or test of quality control across sources to ensure that data standards/codes conform to the official standard. Fragmentation will continue if standard terminologies are not openly available. Open international science will continue to be compromised as a result. Data standards represent a key component of a foundational global infrastructure that must be publicly funded and openly available to support global population health and minimise the cost of managing national disease burdens.

There is a need for industry and a national/global adoption of best practice health data exchange and data standards, for their maintenance, support, and to enable comparative evaluations to be undertaken, within each comparable level within a digital health ecosystem. The current trend is the adoption of SNOMED-CT, as this is the most comprehensive and ontologically structured terminology able to represent clinical concepts.

Access to this terminology is limited to financial members of SNOMED International. This terminology is currently unable to represent all clinical and additional content necessary for every use case. This is likely to change over time, although this requires an understanding of how to best engage with the process for requesting changes. There is minimum capability to implement such requests at this time.

SNOMED International is not operating as an SDO using consensus. Consensus of priorities for improvement should be more open and transparent. Its governance structure does not enable users to have much influence. The use of this terminology is augmented by the adoption of LOINC, which is open-sourced, and other terminologies that have been mapped to SNOMED-CT.

There are few national or global authorities driving terminology standards development, selection, maintenance, adoption, or compliance, other than national authorities with the responsibility for the management of statistical data collections and global reporting. Maintenance timeliness and version controls are important particularly when these terminology standards are used for devices and procedures. Every standard terminology has its own governance, coordinating mechanisms, and change request processes, depending on ownership. Each named standard considered for use, needs conformance testing, uniform implementation guides, and other support to ensure up-to-date information on system capability and performance is available for users.

There is an urgent need to develop comprehensive, cross-standard governance, and coordination mechanisms, globally and nationally, as well as public and private collaboration, to overcome existing fragmentation, siloed efforts, political, intellectual property (IP), and private interest conflicts with public interests. We need to create integrated, safe, and effective ways of representing and exchanging detailed patient clinical data, despite potentially nonviable markets. This requires a well-understood health terminology vision to unify efforts and guide adoption [1].

Data sharing examples

Historically, data repositories have represented concepts as a single point of data collection within the data supply chain. This needs to change to enable access to data representations at all points along the data supply chain and aggregation of data across repositories.

Only then is reporting and communication automation using data collected operationally, enabled. Digital repositories containing basic raw data (unprocessed computer data, such as database or spreadsheet content) are viewed as critical and valuable components of the United States' research infrastructure [79].

Various research groups are part of an international community making use of a number of open-access computing platforms for digital data repositories that were the result of a consortium of representatives from each of the US National Institute of Health (NIH) institutes' Biomedical Information Science and Technology initiative that began in 2000 [80]. Its aim was to make optimal use of computer science and technology and to address problems in biology and medicine by fostering new basic understandings, collaborations, and transdisciplinary initiatives between computational and biomedical sciences.

The Informatics for Integrating Biology and the Bedside (i2b2) and Observational Medical Outcomes Partnership (OMOP) platforms are used to, for example, map data from EHR systems to a common data model. Metadata repositories need to standardise data not only at source but also at aggregation and points of processing throughout the entire data supply chain. Harmonised data can then be analysed locally and converted to a shared aggregate form for further analysis and visualisations to note regional differences and global commonalities [81].

This process was used during 2020 by an international consortium of 96 hospitals across five countries to address critical clinical and epidemiological questions about coronavirus disease (COVID-19). This study methodology was designed to address a major data sharing gap that continues to limit optimum use of data stored in numerous data silos. Data widely available in electronic health records (EHRs) are not effectively shared across clinical sites, or with public health agencies or with policymakers.

The critical importance of data repositories to adopt accepted principles for data sharing and reuse has been noted by this 'open-source, open data' community. This community has recognised that EHR data users need to have access to metrics and methods of assessments that cover the full range of repositories, from raw data to complex databases that have advanced query and analytical tools, for situations where databases are explicitly linked or where researchers need to integrate data from different sources. This requirement is yet to be achieved.

Discussion

It is clear from this overview that data provide the foundation for all health- and health industry-related activities. The current fragmentation and other limitations identified are compromising progress towards sustainable digital health ecosystems. The selection of the most suitable terminology standards and artefacts to be adopted, to best represent clinical content and related concepts, cannot be undertaken in isolation of the many related interdependencies, including system architectures and data repository structures.

The ontological data modelling approach was found to be the most comprehensive with its ability to develop artefacts linked to standard value sets, with the highest degrees of expressivity and formalism available today. Once widely adopted, many stakeholders will benefit exponentially as a result of greater semantic interoperability between systems that enables increased automation throughout the ecosystem. This approach is discussed in greater detail in other chapters.

There are many stakeholders, numerous knowledge domains, many points for data collection, and extraction of data standards regarding existing related entities. In addition, the technology platforms in use or to be adopted specifically to enable data sharing need to be considered along with software engineering principles, data stewardship, governance, and legislation. Advanced data analytical methods to be adopted are dependent upon access to prime data, most of which is contained in and associated with electronic health records in the first instance. Later chapters expand on these concepts in greater detail.

There needs to be strong committed global leadership regarding the streamlining of all data standards used for global population health and disease burden reporting, in a manner that will enable automation from national routine data collections. Global clinical leadership is necessary by specialty to optimise the potential for timely knowledge sharing, reducing the time gap between obtaining research results and their use for real-world practice. Clinical data collections at every point of care need to be able to contribute to the data needs of any clinical trial to not only significantly reduce these research costs, but also increase sample size and to make use of real-world data, where cohorts can be controlled by the use of data analytical methods.

These significant reformations may well take generations to achieve due to their complexity and the many stakeholders who need to collaborate. There needs to be an agreed vision for all leaders at every level to follow. The many high-level entities discussed in the previous chapters all have a role to play. They are in influential positions to drive subsequent complementary developments at national levels. Knowing the desired outcome is essential for local- and middle-level entities to simultaneously take steps to drive the necessary reformations accordingly, so that over time top-down and bottom-up approaches can be harmonised globally. It requires a strong collaborative open approach and significant investments to be successful.

> Yesterday's data architecture can't meet today's need for speed, flexibility, and innovation. The key to a successful upgrade—and significant potential rewards—is agility [82].

Conclusion

The exploration and overview of available data standards have identified many limitations that explain why today's health systems are struggling to efficiently provide personalised, preventive, predictive, and participative precision medicine (P5). A foundational requirement for health system reformation is to develop a high-level clinical content data framework, to enable a better evaluation of our very fragmented health systems from which to develop a roadmap for digital health system reformation. Such a framework needs to begin with an agreed high-level record structure designed to meet the variety of patient-centred clinical functions. Such structures do exist for paper-based medical records. Record structures that best meet all functional needs in this digital era need to significantly differ from those paper-based records. Its design needs to focus on maximising the potential of all available digital technologies.

The way forward requires us to build consensus regarding desired outcomes along with the adoption of an agreed roadmap. This needs to include the adoption of more comprehensive and coordinated planning, resourcing, research, and evaluation to guide legislative changes, governance, adoption, maintenance, dissemination, and use of terminology standards.

The creation of a shared, public library of standard-based, vendor-neutral clinical information models, linked to standard value sets developed for potential reuse across a diverse range of health data sets, effectively represents a strategically designed ecosystem of data. This is foundational for health and health system improvement. Global archetype developments enable value set standardisation and more streamlined concept organisation, while accommodating existing terminologies in use.

References

[1] S. Roy, V. Aukd, Health Terminologies and Vocabularies Environmental Scan: National Committee on Vital and Health Statistics—Subcommittee on Standards, 2018, Available from: https://ncvhs.hhs.gov/wp-content/uploads/2018/10/Report-Health-Terminologies-and-Vocabularies-Environmental-Scan.pdf.

[2] WHO, Family of International Classifications Geneva. Available from: http://www.who.int/classifications/en/. (Cited 8 September 2018).

[3] IHPA, Independent Hospital Pricing Authority—Classifications. Available from: https://www.ihpa.gov.au/what-we-do/classifications.

[4] L. Jorm, S. Gruszin, T. Churches, A multidimensional classification of public health activity in Australia, Aust. New Zealand Health Policy 6 (2009) 9.

[5] D.S. Porterfield, L. Rojas-Smith, M. Lewis, L.A. McCormack, T.J. Hoerger, D.J. Holden, A Taxonomy of Integration Interventions Between Health Care and Public Health (Internet), RTI Press, Research Triangle Park (NC), 2015. Available from: https://www.ncbi.nlm.nih.gov/books/NBK532450.

[6] WHO, Forum on Health Data Standardization and Interoperability, World Health Organisation, Geneva, 2012.

[7] WHO, Joint Inter-Ministerial Policy Dialogue on eHealth Standardization and Second WHO Forum on eHealth Standardization and Interoperability, World Health Organization, Geneva, 2014. Available from: https://www.who.int/ehealth/events/standardization_forum_2014/en/.

[8] AS5021: The Language of Health Concept Representation, Standards Australia, IT/14, Sydney, 2005.

[9] H. Grain, Clinical terminology, Stud. Health Technol. Inform. 151 (2010) 70–83.

[10] J. Recker, M. Indulska, P.F. Green, A. Burton-Jones, R. Weber, Information systems as representations: a review of the theory and evidence, J. Assoc. Inf. Syst. 20 (2019) 735–786.

[11] R. Qamar, A. Rector, Semantic issues in integrating data from different models to achieve data interoperability, Stud. Health Technol. Inform. 129 (2007) 674–678.

[12] R. Blake, T. Shaw, A. Blake, U. Hübner, R. Kaye, S. Schug, J. Thye, EU*US eHealth Works to Improve Global Workforce Development—White Paper, 2017.

[13] JIGSH, Joint Initiative for Global Standards Harmonization (JIGSH) Health Informatics Document Registry and Glossary, Standards Knowledge Management Tool (SKMT) (web page). Available from: http://www.skmtglossary.org/.

[14] E.V. Bernstam, W.R. Hersh, S.B. Johnson, C.G. Chute, H. Nguyen, I. Sim, M. Nahm, M.G. Weiner, P. Miller, R.P. DiLaura, M. Overcash, H.P. Lehmann, D. Eichmann, B.D. Athey, R.H. Scheuermann, N. Anderson, J. Starren, P.A. Harris, J.W. Smith, E. Barbour, J.C. Silverstein, D.A. Krusch, R. Nagarajan, M.J. Becich, C.B.I.K.F. Committee, Synergies and distinctions between computational disciplines in biomedical research: perspective from the Clinical and Translational Science Award programs, Acad. Med. 84 (2009) 964–970.

[15] J.J. Cimino, Desiderata for controlled medical vocabularies in the twenty-first century, Methods Inf. Med. 37 (1998) 394–403.

[16] B. Middleton, D.F. Sittig, A. Wright, Clinical decision support: a 25 year retrospective and a 25 year vision, Yearb. Med. Inform. Suppl. 1 (2016) S103–S116.

[17] P.I. Dissanayake, T.K. Colicchio, J.J. Cimino, Using clinical reasoning ontologies to make smarter clinical decision support systems: a systematic review and data synthesis, J. Am. Med. Inform. Assoc. 27 (2019) 159–174.

[18] B. Blobel, Ontology driven health information systems architectures enable pHealth for empowered patients, Int. J. Med. Inform. 80 (2011) e17–e25.

[19] Y. Wand, R. Weber, An ontological model of an information system, IEEE Trans. Softw. Eng. 16 (1990) 1282–1292.

[20] S. Janaswamy, R.D. Kent, Semantic interoperability and data mapping in EHR systems, in: 2016 IEEE 6th International Conference on Advanced Computing (IACC), 2016, pp. 117–122.

[21] S. Cha, A. Abusharekh, S.S. Abidi, Towards a 'Big' health data analytics platform, in: 2015 IEEE First International Conference on Big Data Computing Service and Applications, 2015, pp. 233–241.

[22] CKM, openEHR Clinical Knowledge Manager. Available from: https://www.openehr.org/ckm/. (Cited 26 September 2018).

[23] OMG, Ontology Definition Metamodel Specifications: Object Management Group, 2014, Available from: https://www.omg.org/spec/ODM/About-ODM/.

[24] R. Lukyanenko, A journey to BSO: evaluating earlier and more recent ideas of Mario Bunge as a foundation for information systems and software development, in: S. Nurcan, I. Reinhartz-Berger, P. Soffer, J. Zdravkovic (Eds.), Enterprise, Business-Process and Information Systems Modeling. BPMDS 2020, EMMSAD 2020. Lecture Notes in Business Information Processing, 387, Springer, Cham, 2020, pp. 345–358.

[25] ISO/TS-17439. Health Informatics: Development of Terms and Definitions for Health Informatics Glossaries, 2014. Available from: https://www.iso.org/standard/59744.html.

[26] HL7-International, Health Level Seven International Glossary of Terms, 2018, Available from: http://www.hl7.org/documentcenter/public/calendarofevents/FirstTime/Glossary%20of%20terms.pdf.

[27] McMaster-University, Health Evidence Glossary. Available from: https://www.healthevidence.org/glossary.aspx.

[28] B.J. Smith, K.C. Tang, D. Nutbeam, WHO health promotion glossary: new terms, Health Promot. Int. 21 (2006) 340–345.

[29] CDC, Centers for Disease Control and Prevention. Available from: https://www.cdc.gov/.

[30] ADHA, My Health Record Glossary, Australian Government, Australian Digital Health Agency. Available from: https://www.myhealthrecord.gov.au/glossary.

[31] ONC, United States Core Data for Interoperability (USCDI), 2020, Available from: https://www.healthit.gov/isa/united-states-core-data-interoperability-uscdi.

[32] ISO-25964-1. Information and Documentation—Thesauri and Interoperability With Other Vocabularies—Part 1: Thesauri for Information Retrieval, 2011. Available from: https://www.iso.org/standard/53657.html. https://www.niso.org/schemas/iso25964.

[33] ISO/TS_21564, Health Informatics—Terminology Resource Map Quality Measures (MapQual), International Organisation for Standardisation, 2019. Available from: https://www.iso.org/standard/71088.html. (Cited 6 December 2019).

[34] HealthDirect-Australia, Australian Health Thesaurus. Available from: https://about.healthdirect.gov.au/australian-health-thesaurus.

[35] JIC, Joint Initiative Council for Global Health Informatics Standardisation. Available from: http://www.jointinitiativecouncil.org/.

[36] ISO/TS21526:2019. Health Informatics: Metadata Repository Requirements (MetaRep). Available from: https://www.iso.org/standard/71041.html.

[37] IOM, C. Grossman, B. Powers, J. McGinnis, Digital Infrastructure for the Learning Health System: The Foundation for Continuous Improvement in Health and Health Care: Workshop Series, Institute of Medicine, Washington, DC, 2011.

[38] NISO, Understanding Metadata, National Information Standards Organisation, 2004. Available from: https://web.archive.org/web/20141107022958/http://www.niso.org/publications/press/UnderstandingMetadata.pdf. (Cited 5 September 2018).

[39] M.D. Wilkinson, M. Dumontier, I.J. Aalbersberg, G. Appleton, M. Axton, A. Baak, N. Blomberg, J.-W. Boiten, L.B. da Silva Santos, P.E. Bourne, J. Bouwman, A.J. Brookes, T. Clark, M. Crosas, I. Dillo, O. Dumon, S. Edmunds, C.T. Evelo, R. Finkers, A. Gonzalez-Beltran, A.J.G. Gray, P. Groth, C. Goble, J.S. Grethe, J. Heringa, P.A.C. 't Hoen, R. Hooft, T. Kuhn, R. Kok, J. Kok, S.J. Lusher, M.E. Martone, A. Mons, A.L. Packer, B. Persson, P. Rocca-Serra, M. Roos, R. van Schaik, S.-A. Sansone, E. Schultes, T. Sengstag, T. Slater, G. Strawn, M.A. Swertz, M. Thompson, J. van der Lei, E. van Mulligen, J. Velterop, A. Waagmeester, P. Wittenburg, K. Wolstencroft, J. Zhao, B. Mons, The FAIR guiding principles for scientific data management and stewardship, Sci. Data 3 (2016) 160018.

[40] SA, HB 291-2007 Health Informatics—Guide to Data Development in Health, Standards Australia, 2007. Available from: http://www.e-health.standards.org.au/Home/Publications.aspx. (Cited 3 May 2019).

[41] AIHW, Metadata Online Registry (METeOR). Available from: http://meteor.aihw.gov.au/content/index.phtml/itemId/181162. (Cited 5 September 2018).

[42] USHIK, United States Health Information Knowledgebase Registry and Repository of Healthcare-Related Metadata, Specifications and Standards, Agency for Healthcare Research and Quality, USA. Available from: https://www.ahrq.gov/cpi/about/otherwebsites/ushik.ahrq.gov/index.html and https://ushik.ahrq.gov/mdr/portals.

[43] J.J. Cimino, L. Remennick, Adapting a Clinical Data Repository to ICD-10-CM through the use of a Terminology Repository, AMIA Annu. Symp. Proc. (2014) 405–413.

[44] ICN, International Classification for Nursing Practice (ICNP®), 2018, Available from: http://www.icn.ch/what-we-do/international-classification-for-nursing-practice-icnpr/. (Cited 21 August 2018).

[45] MeSH, Medical Subject Headings Browser, US National Library of Medicine. Available from: https://www.nlm.nih.gov/mesh/mbinfo.html. (Cited 22 February 2019).

[46] NIH-VSAC, Value Set Authority Center, US National Library of Medicine. Available from: https://vsac.nlm.nih.gov/.

[47] Sentinel, FDA Sentinel System, Sentinel Initiative and Sentinel Operations Center. Available from: https://www.sentinelinitiative.org/.

[48] PCORNet, The National Patient-Centered Clinical Research Network. Available from: https://pcornet.org/data-driven-common-model/.

[49] OHDSI, Observational Health Data Sciences and Informatics—Data Standardisation. Available from: https://www.ohdsi.org/data-standardization/. (Cited 18 March 2019).

[50] Duke-Clinical-Research-Institute, Improving Healthcare Interoperability—Project Report, 2018, Available from: https://dcri.org/wp-content/uploads/2018/12/Pew-Report-v2018DEC08FINAL3.pdf.

[51] H. International, HL7 Version 3 Clinical Document Architecture (CDA) Release 2. Available from: http://www.hl7.org/implement/standards/product_brief.cfm?product_id=7 (Cited 26 September 2018).

[52] HL7-International, Clinical Information Modeling Initiative (CIMI). Available from: https://www.hl7.org/Special/Committees/cimi/index.cfm.

[53] HL7-International, Fast Healthcare Interoperability Resources Specification (FHIR). Available from: https://www.hl7.org/implement/standards/product_brief.cfm?product_id=343.

[54] CDISC, Clinical Data Interchange Standards Consortium. Available from: https://www.cdisc.org/. (Cited 23 May 2019).

[55] OMG, Object Management Group. Available from: https://www.omg.org/.

[56] HL7, Health Level Seven International. Available from: http://www.hl7.org/. (Cited 6 February 2019).

[57] M. Madsen, Knowledge and information modeling, in: E. Hovenga, M. Kidd, S. Garde, C. Hullin Lucay Cossio (Eds.), Health Informatics: An Overview, IOS Press, Amsterdam, 2010.

[58] openEHR/CKM, Clinical Knowledge Manager. Available from: https://ckm.openehr.org/ckm/#. (Cited 21 February 2019).

[59] H. Demski, S. Garde, C. Hildebrand, Open data models for smart health interconnected applications: the example of openEHR, BMC Med. Inform. Decis. Mak. 16 (2016) 137.

[60] H. Leslie, openEHR archetype use and reuse within multilingual clinical data sets: case study, J. Med. Internet Res. 22 (2020) e23361.

[61] D.L. Moody, G.G. Shanks, Improving the quality of data models: empirical validation of a quality management framework, Inf. Syst. 28 (2003) 619–650.

[62] M.G. Kahn, D. Batson, L.M. Schilling, Data model considerations for clinical effectiveness researchers, Med. Care 50 (Suppl) (2012) S60–S67.

[63] A. Moreno-Conde, T. Austin, J. Moreno-Conde, C.L. Parra-Calderón, D. Kalra, Evaluation of clinical information modeling tools, J. Am. Med. Inform. Assoc. 23 (2016) 1127–1135.

[64] IBM, Unified Data Model for Healthcare V9.4 Supports Evolving Standards, Regulations, and Use Cases Within Healthcare Industry, 2018, Available from: https://www-01.ibm.com/common/ssi/rep_ca/0/897/ENUS218-560/ENUS218-560.PDF.

[65] ISO-DIS23903: Health Informatics—Interoperability and Integration Reference Architecture-Model and Framework, International Organization for Standards (ISO), Geneva, 2020. Available from: https://www.iso.org/standard/77337.html.

[66] B. Smith, W. Ceusters, Ontological realism: a methodology for coordinated evolution of scientific ontologies, Appl. Ontol. 5 (2010) 139–188.

[67] M. Madsen, Health care ontologies: knowledge models for record sharing and decision support, in: E. Hovenga, M. Kidd, S. Garde, C. Hullin Lucay Cossio (Eds.), Health Informatics: An Overview, IOS Press, Amsterdam, 2010.

[68] Protege. A Free, Open-Source Ontology Editor and Framework for Building Intelligent Systems, Stanford University. Available from: https://protege.stanford.edu/.

[69] BBOP, Berkeley Bioinformatics Open-source Projects (BBOP). Available from: http://www.berkeleybop.org/index.html.

[70] AHRQ, Methods for effective health care, in: R.E. Gliklich, N.A. Dreyer, M.B. Leavy (Eds.), Registries for Evaluating Patient Outcomes: A User's Guide, Agency for Healthcare Research and Quality (US), Rockville, MD, 2014.

[71] B. Smith, M. Ashburner, C. Rosse, J. Bard, W. Bug, W. Ceusters, L.J. Goldberg, K. Eilbeck, A. Ireland, C.J. Mungall, O.B.I. Consortium, N. Leontis, P. Rocca-Serra, A. Ruttenberg, S.-A. Sansone, R.H. Scheuermann, N. Shah, P.L. Whetzel, S. Lewis, The OBO Foundry: coordinated evolution of ontologies to support biomedical data integration, Nat. Biotechnol. 25 (2007) 1251–1255.

[72] SNOMED-International, Leading Healthcare Terminology, Worldwide. Available from: http://www.snomed.org/.

[73] LOINC, Logical Observation Identifiers Names and Codes from Regenstrief, Regenstrief. Available from: https://loinc.org/. (Cited 25 February 2019).

[74] Sabacare, Clinical Care Classification (CCC™) System, 2018, Available from: https://www.sabacare.com/.

[75] J.J. Cimino, T.F. Hayamizu, O. Bodenreider, B. Davis, G.A. Stafford, M. Ringwald, The caBIG terminology review process, J. Biomed. Inform. 42 (2009) 571–580.

[76] NIH-NCI, cancer Biomedical Informatics Grid™ (caBIG™), National Institute of Health, National Cancer Institute. Available from: https://biospecimens.cancer.gov/relatedinitiatives/overview/caBig.asp.

[77] Productivity-Commission, Data Availability and Use, Overview and Recommendations Report No. 82 Canberra, 2017, Available from: https://www.pc.gov.au/inquiries/completed/data-access/report. (Cited 5 June 2019).

[78] AMIA-&-AHIMA, Healthcare Terminologies and Classifications: An Action Agenda for the United States—White paper, American Medical Informatics Association and the American Health Information Management Association, 2006. Available from: https://perspectives.ahima.org/healthcare-terminologies-and-classifications-an-action-agenda-for-the-united-states/.

[79] Big-Data-Interagency-Working-Group, Measuring the Impact of Digital Repositories, National Science and Technology Council, 2018. Available from: https://www.nitrd.gov/pubs/BD-IWG-Digital-Repository-Workshop-Report-2018.pdf.

[80] i2b2, tranSMART Foundation, An Open-Source-Open-Data Community. Available from: https://i2b-2transmart.org/.

[81] G.A. Brat, G.M. Weber, N. Gehlenborg, P. Avillach, N.P. Palmer, L. Chiovato, J. Cimino, L.R. Waitman, G.S. Omenn, A. Malovini, J.H. Moore, B.K. Beaulieu-Jones, V. Tibollo, S.N. Murphy, S.L. Yi, M.S. Keller, R. Bellazzi, D.A. Hanauer, A. Serret-Larmande, A. Gutierrez-Sacristan, J.J. Holmes, D.S. Bell, K.D. Mandl, R.W. Follett, J.G. Klann, D.A. Murad, L. Scudeller, M. Bucalo, K. Kirchoff, J. Craig, J. Obeid, V. Jouhet, R. Griffier, S. Cossin, B. Moal, L.P. Patel, A. Bellasi, H.U. Prokosch, D. Kraska, P. Sliz, A.L.M. Tan, K.Y. Ngiam, A. Zambelli, D.L. Mowery, E. Schiver, B. Devkota, R.L. Bradford, M. Daniar, C. Daniel, V. Benoit, R. Bey, N. Paris, P. Serre, N. Orlova, J. Dubiel, M. Hilka, A.S. Jannot, S. Breant, J. Leblanc, N. Griffon, A. Burgun, M. Bernaux, A. Sandrin, E. Salamanca, S. Cormont, T. Ganslandt, T. Gradinger, J. Champ, M. Boeker, P. Martel, L. Esteve, A. Gramfort, O. Grisel, D. Leprovost, T. Moreau, G. Varoquaux, J.-J. Vie, D. Wassermann, A. Mensch, C. Caucheteux, C. Haverkamp, G. Lemaitre, S. Bosari, I.D. Krantz, A. South, T. Cai, I.S. Kohane, International electronic health record-derived COVID-19 clinical course profiles: the 4CE consortium, npj Digit. Med. 3 (2020) 109.

[82] A. Castro, J. Machado, M. Roggendorf, H. Soller, How to Build a Data Architecture to Drive Innovation—Today and Tomorrow, McKinsey Technology, 2021. Available from: https://www.mckinsey.com/business-functions/mckinsey-digital/our-insights/how-to-build-a-data-architecture-to-drive-innovation-today-and-tomorrow.

CHAPTER

9

Quality data, design, implementation, and governance

Evelyn Hovenga[a,b,c] *and Heather Grain*[b,c,d]

[a]Digital Health, Australian Catholic University, Melbourne, VIC, Australia [b]eHealth Education Pty Ltd, Melbourne, VIC, Australia [c]Global eHealth Collaborative (GeHCo), Melbourne, VIC, Australia [d]LG Informatics Pty Ltd, Melbourne, VIC, Australia

Health data

So far in previous chapters, we have identified numerous issues associated with data capture and use impacting on data quality, value, and use in this digital world. These issues may be summarised as:

- Poor data design due to lack of recognition of the continuum of data, resulting in the lack of consistency and collectability of data;
- Lack of consistent representation of data which:
  - increases the costs associated with information exchange and sharing, and
  - reduces the ability to analyse and maximise the value of the data;
- Lack of recognition of the nature of computer-based information requirements needed to inform data design and support use and semantic interoperability of that data and the need for longevity.

Health data represent the primary asset for any healthcare organisation, research group, and governance structure within any nation's health system. The appropriate application of all data use requirements has the potential to reduce the burden and cost of data capture and collection in the long term, as well as improve the quality and use of the data collected supporting any nation's health system sustainability.

Known barriers that need to be overcome are:

- Data silos due to a lack of standardisation,
- Lack of understanding and skills to enable the data supply chain needs to be recognised and meet these requirements at the data source,

Roadmap to Successful Digital Health Ecosystems
https://doi.org/10.1016/B978-0-12-823413-6.00013-6

- Lack of skills in data design, leading to replication of historical methods which can now be done more efficiently than in the past,
- Leaving data design to the software vendor, where the supply chain issues may not be obvious and the use case environment poorly understood,
- Lack of sufficient or appropriately applied data quality measures across the health data continuum.

There is an urgent need to transform how health data is currently represented, collected, aggregated, linked, exchanged, processed, used, and reused which requires us to explore:

- Data specification and design principles including:
  - Concept representation principles in data design—choice and currency,
  - Data design protocols—information models, maps [1], and technical issues
  - Data granularity and the relationship to a maximal dataset,
  - Avoidance of silos of data—'collect-once-use-often principle',
  - Ongoing governance, accountability, and reporting,
- Relationship to global data specifications and national extension,
- Data collection and collectability design,
- Management of data design—vendor management
  - Data standardisation and its relationship to workflow, clinical decision-making, and the data supply chain,
- Data design quality measures—strategic needs and identification of pathways to improvement,
- Openness, transparency of software's ability to implement and apply standardised information models, concept representation, and data definitions.

The previous chapter identified numerous ways that data are structured and standardised. We noted overlaps and possible gaps existing in health data standards. Ontology-based structures were identified as representing structures that relate to an information model. Such ontology domain models can accommodate new additions and are most suitable to support the semantic interoperability required in any digital health ecosystem.

Legacy information systems are based on old designs that may limit the development of new functionalities. Linking new systems requires the relationships between the older data and new data to be considered when systems are upgraded or when implementing any health data exchange schema. Such data mapping needs to be undertaken in a manner that retains the original meaning to ensure safe and optimum future use.

This chapter addresses how users can identify their issues and establish the desired level of data quality for their use case and ensure that quality is maintained.

Healthcare data needs to be viewed as a valuable asset as it:

- Has intrinsic value,
- Requires significant cost, in money and other resources, to collect, store, and make it available for use,
- Forms the basis for core clinical decisions, core day-to-day business operations, population health management, and health service funding mechanisms.

Health data supply chain

Current approaches include increasing use of data maps to aggregate and standardise data after the collection of that data at the source. This approach results in a lack of consistency, clarity, retention of precise meaning, and provenance. We currently lack a clear vision of data within any healthcare organisation and the quality of that data. As a result, software vendors, healthcare organisations, care providers, and patients do not know what is required for good information governance in any healthcare setting.

These data-related problems occur across the continuum of information in healthcare and the use of healthcare information systems referred to as the data supply chain. The standards required must be produced in a coordinated fashion by the relevant organisations and be agreed nationally and/or globally. The use of universal data formalisms, such as the openEHR archetypes, can go a long way towards resolving this issue. This is due to the requirement of considering all possible use cases for any concept being modelled, throughout its supply chain, during the knowledge modelling process.

The creation of mechanisms to support secondary data use requires widespread adoption of digital thinking so that the unknown is more easily discoverable. Getting any new implementation structure right allows for easier data collection, potential automation, and transformation throughout the journey of the data supply chain. Data analytics is dependent upon the quality of the data collected.

Fig. 1 represents multiple data sources from the collection at individual organisations through to high-level supply chain indicators.

Fig. 2 shows the data cycle from the original source data through aggregation and data sharing, resulting in analysis and identification of new knowledge which is returned to users to inform further practice. This includes the use of AI (artificial intelligence) and machine learning. There are issues of data quality throughout this cycle.

Fig. 3 shows the relationships between the different elements of data and the pathway from a simple concept through to general intelligence. A concept is the representation of an idea or knowledge object. These ideas may be expressed in many ways, verbally or in writing, including in the patient's health records as well as in treatment plans, orders, and referral letters. When describing concepts in healthcare, considerable detail is required to ensure clarity and to enable others to make use of these concepts. We need to be able to electronically represent all health concepts and retain their meaning when transferring these between systems. Data, information, knowledge, and intelligence (wisdom) ascend hierarchically.

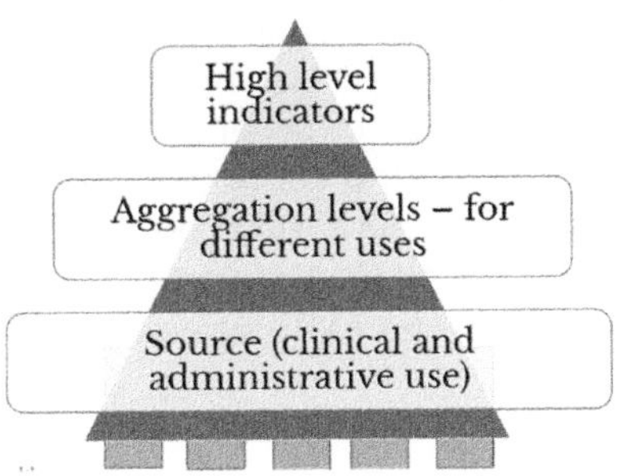

FIG. 1 High-level data supply chain.

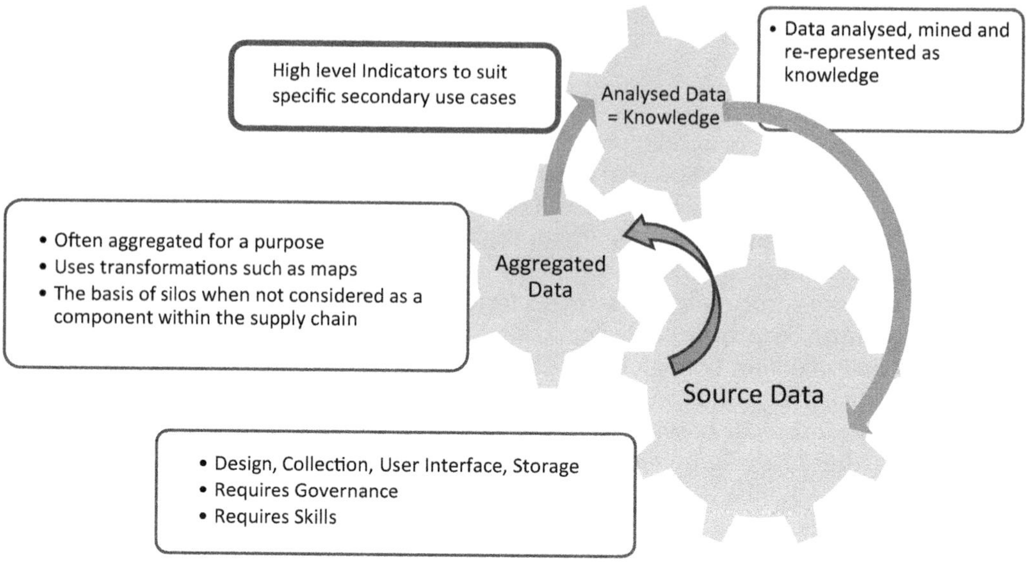

FIG. 2 The data cycle.

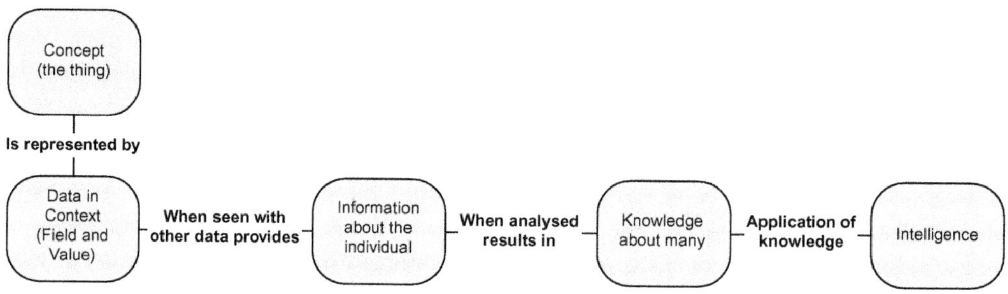

FIG. 3 The pathway from idea to intelligence.

What is quality data?

Data needs to be actively managed at the executive level as a strategic resource of any healthcare organisation [2]. Such data management also applies to specific clinical disciplines as each speciality has its own unique knowledge domain. We need to ensure data quality at the time of data collection relative to potential secondary use requirements.

For example, morbidity reporting around the world usually includes information about the smoking status of the individual. Where these data are collected and represented in a standardised way, such as using an openEHR archetype, then such reportable data can be collected by automatic transposition from the EHR repository to the morbidity data collection. This achieves two objectives:

- Reduced morbidity data collection costs
- Improved data quality through consistent data collection

This approach needs those who design data collections, such as morbidity data, to have new skills to develop standard algorithms for data extraction to leverage the design and requirements of the patient care data system. Similar examples of collecting data to represent 'times' and 'dates' can reduce manual data collection and calculation, resulting in the more effective use of data.

Health data asset management is greatly assisted by the adoption of an enterprise-level information model to provide a common, consistent view of data across the enterprise and of their relationships within, and external to, the enterprise. Such models, including the use of diagrams, a business data glossary, technical data dictionaries, specifications, and grouping frameworks for datasets and electronic records (e.g. related to confidentiality) are extremely valuable as a means of communication between those who understand the data flow and application developers.

These representations of data use throughout healthcare ecosystems and patient-centred EHRs (the data collection mechanism) need to be considered vs the proliferation of disparate, vendor-specific models which reduce interoperability. Such information models support the data supply chain by reducing the interoperability complexity through a shared source data design structure.

Here is a simple example from the business world.

- The concept of interest paid by financial institutions or processes has a consistent meaning throughout the world; it is represented by 3 pieces of data:
  - Interest rate—represented as a percentage.
  - Interest period—the time frame to which this percentage of interest applies.
  - Interest object—the amount of money to which the interest rate is applied each period.
  - Interest charge—the result of the multiplication of the interest object by the interest rate.
  - Date of interest calculation—the date at which the interest charge is calculated.

There are many other data items which can be relevant to interest but these basic items (a selection of all the items you could apply) are internationally standardised. This is one reason why computerisation in banking is much easier than in healthcare—healthcare has not yet matured to an agreed representation of our data.

Data quality design needs to focus on data requirements, including content specification, representation and information exchange, data design quality measurement, data collectability, and ongoing data quality assurance management. The scope of data quality design includes:

- Health record structure (data concepts),
- Concept representation,
- Implementation of these standards for the transmission, translation, and transformation, when moving data between platforms,
- Consideration of natural language processing and transformation between data standards,
- Quality measures associated with the implementation of information models, concept representation, and data exchange to determine conformance to the defined requirements.

Data access and risk management

In 2017, Australia's Productivity Commission published its review of data availability and use and found [3], amongst many other findings, that there was a 'lack of trust by both data custodians and users in existing data access processes and protections', and there were 'numerous hurdles to sharing and releasing data, choking the use and value of Australia's data'. It noted that: 'Marginal changes to existing structures and legislation will not suffice'.

Its many recommended reforms aimed at moving from a system based on risk aversion and avoidance to one based on transparency and confidence in data processes, treating data as an asset and not a threat. It noted that:

> Significant change is needed for Australia's open government agenda and the rights of consumers to data to catch up with achievements in competing economies. Continuing denials of access to health data for data use as is required throughout the data supply chain within any digital health ecosystem represent numerous lost opportunities, including lives saved.

One of the most relevant recommendations for the reform was the need for 'a new Data Sharing and Release Act and a National Data Custodian to guide and monitor new access and use arrangements, including proactively managing risks and broader ethical considerations around data use'. This recommendation was provided to address the continuing restrictions to access and use data as specified within a variety of national and state legislation and other program-specific policies. Health data was found to exemplify the problem identified. This highlighted the need for suitable national digital health-supportive infrastructures. The Office of the National Data Commissioner has now been established by the Australian Government [4].

Data sharing and linkage is not without risk which needs to be managed. This is a well-developed area with a mature body of knowledge and practice. Commonly used methods of assessing risk are based on standards now regarded as the internationally recognised benchmark that provides guidance, principles, framework, and processes to manage risks in an organisation [5]. This standard applies equally to data governance processes as the lack of governance increases the risk of error in patient documentation or decision support algorithms, compromising patient safety and increasing the risk of adverse events. This framework is shown in Fig. 4.

Big data

Connected devices and information systems are now generating huge amounts of data that may be grouped in a variety of ways. The value of these data is directly associated with the quality of the original data and our ability to access and analyse such data to find new patterns, new meanings, new data relationships, new knowledge, and intelligence. The use of data held in records generated in real time in clinical systems has the potential to reduce the time it takes to gain knowledge.

To meet these demands for usable information presented and analysed immediately, it is essential to consider the ability to handle 'big' health data in terms of volume, heterogeneity, numerous variables, and real-time data generation. Big data represents 'high-volume,

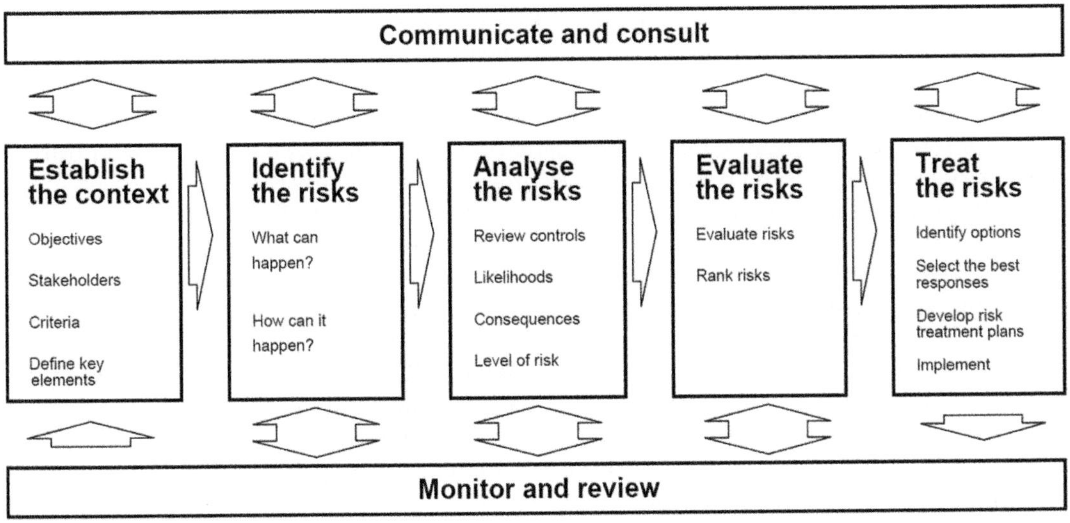

FIG. 4 AS/NZS 4360 risk assessment approach [6].

high-velocity, and high-variety information assets that demand cost-effective, innovative forms of information processing for enhanced insight and decision-making' [7]. An ability to analyse and mine big data provides policy and decision makers with new insights into various aspects of:

- work or information flow,
- operational business patterns and trends,
- safer and more effective healthcare, and
- how to achieve greater efficiencies and safer and more effective healthcare.

There is less need to collect specific information to answer any question or for formal research purposes following the adoption of open knowledge-based platforms within well-designed ecosystem architectures.

Access to and analysis of big data has the potential to transform how we live and work. This requires our current leaders and decision makers to fully understand and appreciate the value of all digital health ecosystem fundamentals required and act accordingly. Only then are we enabled to make the best possible use of all electronic data generated and stored.

Quality management responsibilities

Our new digital era requires us to be more diligent about the data collected, processed, and used within the healthcare system, as that data is no longer on a single sheet of paper, used often for a single purpose and after that mostly used as a record. Data quality management responsibilities apply to all source data as well as secondary data and knowledge although processes may need to differ.

Today's data must suit multiple purposes. This requires the adoption of data governance processes by every stakeholder within a digital health ecosystem making use of the health data. It is good practice to collect data in a consistent manner, recognising the need for greater and lesser data specificity in different healthcare settings. If data is collected using standardised information models, concept representation, and data specifications, then information exchange standards can be simpler and be implemented at less cost.

Quality measurement includes ways to measure the quality of all data components from data design, collection, storage, and transition through the supply chain, including data mapping activities undertaken for any information exchange purpose. Sharma et al. [8] noted that for example, where risk prediction models can be integrated, then misaligned clinical terminology is likely to limit its implementation and use.

Quality management also includes the management of knowledge provided for decision-making, such as that in clinical practice, but also the management of the acquisition of knowledge from data (data mining). For example, clinicians should not have to go to a website or open a separate application and manually complete data fields to access a clinical guideline to assist their decision-making process, as this tends to be viewed as a significant access barrier [8]. The data triggering the perceived potential benefit of accessing any evidence-based practice guideline should be readily available within the EHR, and the system needs to support the provision of a streamlined workflow and enhance the quality of care delivered. Another related program design feature is the need to assess points of data collection, associated data collection mechanisms adopted, and data collectability.

Data quality management/assessment

Data quality management operational principles need to focus on safe practice first, efficiency second, and be pragmatic for any use case as a last resort only. The safe practice focus also applies to open communication of data specifications and requirements, such as metadata repositories [9], metrics for maturity measurement, and people skills needed for quality management and assessments. These practices are essential prerequisites for the successful implementation of systems.

The use of assessment conformity ensures success. This requires the development of a conformity assessment framework based on overarching requirements for information systems, including health record structures within which data are captured. There needs to be a robust set of information criteria, i.e. testable attributes and measures, for the identification of maturity of implementation beyond the definition of specific data elements.

Data quality needs to be assured prior to use at any level within the data supply chain. Data processing requires users to check for accurate values and conformance to applicable standards. For example, variables such as blood pressure, age, cholesterol level, etc., need to be within reasonable limits. Such constraints are embedded when clinical models for such concepts are used. Errors may occur during any data entry, archiving, data exchange between systems, and interpretation processes. Data inaccuracy can be costly as it may result in any clinical adverse event, including misdiagnosis and any number of erroneous decisions [10]. The need for data cleansing, data validation, data verification, data formatting, or updating prior to use can be minimised through appropriate system design.

Diagnosis and procedure codes are frequently used to group patients for a variety of research, performance monitoring, or reporting purposes. The quality of those codes is, to a large extent, dependent upon the medical record content as well as the clinical coders' knowledge and skills [11]. Well-designed systems based on a knowledge-based platform can eliminate these errors by automating the coding of at least 80% of all inpatient records. High-quality coded data also form the foundation for casemix-based payment systems.

The ability to automate coding not only improves the accuracy but also results in significant cost savings. Today's diagnosis and procedure coding errors are known to be prevalent. In addition, every jurisdiction tends to have its own unique interpretation of coding rules. Mitigation of these errors is being addressed by the appointment of auditors to check individual organisations and coders. Only one effort has been made to share and improve coder training processes in order to improve coding quality and consistency; this has resulted in the development of a cloud-based simulated coding training tool, now known as eHRoL (electronic health record online learning) [12]. This is made available to anyone under a licensing agreement.

The quality of record content can in part be addressed by appointing data stewards who understand and are willing to assume the responsibilities of effective stewardship in order to earn and retain the support of all data users, including patients and the public. They need to be appropriately educated to collect, maintain, aggregate, analyse, and use health data in accordance with applicable coding or grouping rules, laws, regulations, and policy requirements [13].

Generally endorsed, health data use principles that need to guide data stewards include:

- Access to and use of health data should be viewed as a public good. Data should be available and 'fit for use', with proper security, for appropriate purposes not only for but also beyond direct patient care.
- Health data must be as consistent, comparable, timely, accurate, accessible, complete, and as reliable as possible. Users must be able to track the degree to which the data have attained these attributes. Understanding the context and provenance of the data is also critical in determining their 'fitness for use' [13].

Data quality and access management suitable for a digital health ecosystem need to be incorporated within overall health ecosystem governance processes, from local data collection through information sharing between service providers to data collections, at all levels of healthcare. This is not the domain of software vendors—responsibility goes far beyond any individual product or process.

Metadata facilitates data integration

The support structure needed to enable and support quality data and information in healthcare includes the management and creation of metadata to facilitate integration. Metadata needs to be relevant and kept up to date. The metadata also needs to go beyond the representation of a single item of data at a point in time but to define the method/s of aggregation to be used. The implementation of centralised metadata repositories and updating metadata files to align with standards used by others enhances metadata management and supports publication and data exchange. The use of metadata standards supports the ease of finding data quickly.

Metadata includes the specification of the data element and the codes used to represent that data element. When used to suit computer data entry, such metadata needs to not only have an agreed definition but also needs to describe any other features, such as use, characteristics, and rules that need to be applied, such as the entry of this information is required or this field may be left blank. Such data specifications assist system developers and users to clearly understand what data needs to be collected, the level of detail, the way it is to be collected, and how it may be used.

Metadata has a standard specified structure and needs to be governed to ensure consistent use and interpretation of all data collected and used. This requirement is vital to facilitate information sharing amongst primary and secondary users of the data and to ensure comparability and consistency of the data collected and produced from the collection. An additional new requirement that relates to EHRs is the need for the maintenance of meaning over time. Changes to either elements or code systems require a rationale and metadata for all modifications made. Another important consideration is that many different datasets have overlapping data elements within single collections. Such duplications should be avoided as there should be only one data standard for every data element representing a specific concept within any ontological knowledge domain.

Each data element within any dataset needs to be defined, as do the codes used to represent different meanings within that data element, to ensure all users apply the same meaning. Within the health industry, there is an increasing need for lots of 'little data', that is atomic-level data to fit numerous clinical use cases. Little data gets aggregated, grouped, linked, transferred, and processed to meet the needs of further use cases along the data supply chain. It is critical that their meaning is retained at all times.

Value-based healthcare requires the identification of data elements that best represent what is being valued to enable these values to be measured. These form part of any agreed data value set. These may be organisationally, nationally, or internationally and/or clinically focused and need to be governed to ensure reliable data usage to suit any use case. Many jurisdictions are now recognising the need for health information governance and have implemented policies, procedures, and rules accordingly, although these tend to focus on high-level secondary data use where metadata is most relevant, rather than clinical data use.

Data governance needs

Our new digital era requires us to be more diligent about data collected, processed, and used within the healthcare system, recognising that this data will be used for many purposes. This requires the adoption of data governance processes by every organisational entity making use of health data and people able to lead and manage these processes. It is good practice to collect data in a consistent manner, recognising the need for greater and lesser data specificity in different healthcare settings. If data is collected using standardised information models, concept representation, and data specifications, information exchange standards can be simpler and implemented at less cost.

Governance is about steering or directing any concept or entity in the desired direction by an entity (governing body) that has the authority to do so. In addition to this general governance concept, it is important when transferring data between information systems within the healthcare industry or any digital health ecosystem to also manage trust and reliability so that clinicians (and other data users) can confidently make use of the information received.

Our challenge is to develop a standard way of documenting any concept (clinical, demographic, environmental, etc.) that needs to be considered by any decision maker (clinician in any speciality, manager, policy developer, etc.) in the health industry. Such data standards need to suit multiple purposes, including obtaining new information or knowledge discovery. Each data standard may be included in any one of many different datasets where each dataset was developed to meet a specific need. Data governance is about ensuring that everyone engaged in facilitating any type of data collection makes use of the data standards as agreed by all relevant stakeholders. For example, data collected for national reporting purposes must be defined in a standardised way to facilitate automation. There should be NO data collected just for national reporting purposes. Data collected for patient care might be transformed into data required for national reporting and the method of transformation must be clearly defined, preferably by a repeatable consistent algorithm.

The adoption of a standard information model representing any knowledge domain needs to be part of the overall governance implementation strategy that forms the basis for the development of a governance program roadmap. This includes enterprise-level conceptual, logical, and physical models of all data assets, including standard value sets, that need to be managed by the organisation within the context of their healthcare ecosystem EHR model.

Conceptual models represent meaning (truth), logical models depict generic-shared relationships amongst resources, activities, outputs, outcomes, and impact—a chain of cause and effects leading to the desired outcome—whereas physical models represent their technical implementations. The logical model positions the concept in the platform's architecture. A healthcare ecosystem with an EHR model at the centre rejects the proliferation of disparate vendor-specific models which reduce the ability to fully interoperate.

The governance program's design needs to consider each knowledge domain's content and how this is best represented by standard terminologies. These apply to all required concept levels of granularity, including genomics, and how these standard terminologies are made use of within any data exchange schema. Also, within the design scope is the need to consider natural language processing and transformation between data standards, including quality mapping of data where different terminologies are used to describe the same concept.

Data governance links to attaining value

Any health information governance program needs to be aligned with the organisations' general and digital health strategic directions, its overall future vision, mission, and performance evaluation goals. Desired outcome measures need to be able to represent the value for the healthcare provider, its human resources delivering and supporting services provided, all its care recipients and their associated significant others, including the local community and overall population health. Health data needs to include links between significant social determinants of health data, such as clean water, shelter, food security, and access to health services and internet access by geographical location.

Large healthcare facilities collaborate with research institutions to undertake many types of investigative studies to learn more about, for example cause and effect of various commonly occurring diseases, referred to as non-communicable diseases (NCD). These activities consume a very large share of any health system's delivery costs by contributing significantly to any country's overall health burden, productivity, and growth capacity. Many healthcare facilities also collaborate in the conduct of clinical trials another costly endeavour.

All clinical and patient-related data collected and processed need to be governed to ensure consistency for all collected data, provided by multiple healthcare facilities, to benefit individuals and numerous others within any digital health ecosystem. Research outcomes derived from using accurate 'big data' made available in a timely manner represent great value.

Governing and maintaining coherent datasets over time, as may be required to attain the desired value, is extremely difficult in the current environment. Data collection and processing requirements for every desired value need to be based on a coherent data design that enables the complete dataset to be viewed to identify any inconsistencies or divergence so that these can be managed in a manner so as not to compromise value. There is a need to intentionally design data semantics to optimise value, quality, accuracy, and the ability to reuse data for multiple subsequent use cases. In every project, we need to ask if the data is informatics-informed? Are the semantics fit for purpose?

The case study in Box 1 demonstrates the adverse impact of the continuing use of traditional data collection methods. This supports the argument that investing in a digital health transformation is cost-effective, adds value, and is necessary to optimise the use of today's technologies.

BOX 1

Case Study: Data Coherence and Oversight

One research institution set up a large prospective study, intending to investigate the effects of lifestyle factors on seven common cancers using a combination of questionnaires and physical assessments. At the time of engagement, the study had been running for over 20 years and was still accumulating data. In that time, they had had 4 data managers, and the questionnaires had been 'tweaked' several times. The data dictionary was managed in 83 Excel spreadsheets. Each spreadsheet represented a discrete data collection, a single concept for a single cancer, representing a range of clinical, lifestyle, and social factors including past medical history, medications, diet and nutrient intake, physical exertion, smoking and alcohol consumption, sexual health, and physical measurements.

While the data collections progressed, the data dictionary had evolved. Despite best endeavours, no one had adequate oversight of the contents of the 83 spreadsheets. After all, it really is an almost impossible task when the semantics for each dataset were represented in 10 or more discrete spreadsheets per cancer. Each data collection contained the same concepts but not necessarily consistent questions across each cancer type. The cumulative effect of each 'minor' questionnaire

Continued

BOX 1

Case Study: Data Coherence and Oversight—cont'd

tweak over decades was almost impossible to measure, mainly because of the distribution across 83 discrete spreadsheets. The overall coherence and integrity of the dataset were not well understood.

Asked to redesign the dataset with strong semantic consistency, each of the spreadsheets was carefully analysed and reaggregated by concepts, rather than cancer. In this way, all tobacco-related questions were aggregated together. Systematically, all questions related to each concept were grouped across all cancers—one for diet, another for smoking, etc. And, this is where the project got interesting!

The researchers were shocked to discover the range and divergence of questions that had occurred for each concept. They stopped the data accumulation and worked to normalise the questions again. The consequences for the data collection were significant, both in terms of the data already collected and in the new data model used for future data collections.

If we focus on smoking as an example, typical questions found across all data dictionaries included:

- Number of cigarettes currently smoked each day
- Average number of cigarettes per day

Embedded assumptions in questions are essential to identify and understand their consequences. Note how the first question asks for a number, implying but not stating that this is a typical number, and the second asks for an average. Both are specifying cigarettes, but if the respondent smokes other forms of tobacco or uses tobacco in ways other than smoking, these options are not available.

- Ever smoked 7 cigarettes a week?—true/false
- Currently smokes 7 cigarettes a week—true/false

Rightly or wrongly, seven cigarettes a week seems to have been used as a benchmark to establish a significant smoking behaviour in this study. The meaning and intent of 'ever' are unclear. An answer of 'true' could be interpreted in many ways: a former smoker; a one-off smoking episode decades ago; someone who currently smokes 7 cigarettes every Friday night; even someone who smoked 7 cigarettes yesterday. If the interpretation of any answer is ambiguous or unclear, then the question should be reconsidered and potentially redesigned.

- Smoked 7 cigarettes a week for at least 1 year—true/false
- Smoked cigarettes over the last 5 years—true/false

It is common to see durations of time included in questionnaires; however, there needs to be a clearly defined reason for each duration or consistency between all durations.

- Stopped smoking for at least 1 year between starting and stopping—duration (units of year, month, week, day, hours, minutes, and seconds)
- Stopped smoking for at least 1 year between starting and stopping—true/false

The implied purpose of this question is not clear, but likely to be about cessation-related behaviour. However, different data dictionaries contained the same questions but used different data types to record the answer, making data aggregation or inferencing impossible.

Data governance strategic principles

Data governance policies and coherent data designs need to be based on the following strategic principles:

- Identify the governance boundaries or scope, e.g. within a single entity or across multiple entities.
- Reflect the needs of the data user community, this will evolve and change over time.
- The entity's priorities, where entities can represent any jurisdiction, healthcare facility or clinical discipline. Priorities reflect its value propositions.
- Once value propositions are defined, datasets can be defined relative to data standards to be complied with, and the most appropriate data collection points for all required data sources and data collection methods can be identified. Appropriate data designs making use of well-defined data standards are best able to attain a value.
- Incorporate data/information at rest and when in motion, irrespective of where it originates, and the processes used in each state. Governance must consider the data not only from its original source but also though to use cases outside the entity's own boundaries. This means that any data-collecting entity must be aware of the format in which those data are required for information exchange. Such entities need to adopt design governance that ensures that the data representation needs of the receiving entity are met. This minimises the use of data transformations such as maps which may introduce errors and are costly to maintain.
- Apply these principles to all data, irrespective of collection format, storage, and use. E.g. paper, born-digital or reproduced data.
- A focus on data sources is important as many data points can be derived from the automation of data collected previously. For example, the date of birth stored in a system should be used to define age anywhere else where this fits a use case.
- Data linking and aggregation to link datasets and standardise how data may be aggregated (e.g. health regions, age bands, and minimum cell size) to facilitate comparison and use.
- Apply a shared data specification and concept representation approach such as the use of global standards for all aspects of information governance wherever possible. This principle supports the leverage of best practice and shared community development of standards and specifications, rather than allowing software vendors to dictate data representations. Such vendor specification is usually highly beneficial to the vendor but less so to the healthcare system, as it makes data less compatible across the system and forces information mapping within any digital health ecosystem.
- Electronic data retrieval from the existing operational systems is a preferred strategy as this method is best able to result in consistency.
- Compliance with standards that confirm and support the collection of quality data. It is essential that the data are designed to be collectable and of value to those who collect the data. Asking for data not available at the point of collection or of no use to those at the point of collection does not result in quality information.
- Engage with stakeholders to ensure collectability. Remember that when data collected is of no value to the collector it may not be accurate. Patient care data must have the highest governance priority to maintain patient safety.

- Broad user engagement is essential to develop and sustain trustworthy and trusted data.
- Data governance priorities need to be based on the frequency of use, as some data are used for many use cases, others only sporadically.
- Processes for managing the full life cycle of data and records for their use locally and with trusted partners for further value generation.
- Data collection and preparation to support the collection, parsing, cleansing, remediation, and storage of data to enable it to be linkable and fit for use in analysis and care.
- Records and data maintenance in technology systems, which may include storage, encryption (in motion and at rest), retention, archiving, logging, destruction, recovery, and secure access mechanisms.
- Focus on specific issues in need of resolution and keep the scope manageable. Expand data governance projects gradually as other members of the workforce become familiar with the policies, procedures, and the data stewardship role.
- Use a roadmap that offers a mechanism to understand, plan, and evaluate governance requirements and processes considered essential for quality data. Here, the entity's cultural capacity to adapt and adopt new approaches for data and information management needs to be considered.
- Enable workforce capacity building in all areas of information governance. It is desirable to employ independently certified health information managers to manage these governance processes.

Data governance and stewardship

Health data, information, knowledge, and wisdom (intelligence) (DIKW) represent the key strategic assets that need to be managed locally, nationally, and globally within any health ecosystem. Previous chapters have described DIKW characteristics and current uses as well as highlighting the many quality gaps. It is widely known that the more structured the format of data entry, the easier it is to find the piece of information you want, provided the person who originally captured that information had put the right stuff in the right place. Where data are captured by any device, then its data retrieval capacity is dependent upon the built-in software and its capacity to interoperate with other systems.

Information governance applies to the whole lifecycle of data, from definition, through collection in different systems for different purposes, aggregation of data for statistics and reporting, analytics, and finally destruction. Information governance must also consider the data life cycle and ensure that there is sufficient capacity and capability in the health data ecosystem to carry out operations and controls necessary to achieve the desired outcomes. Entities within any digital health ecosystem need to consider their place in the data supply chain. Everyone needs to consider their own requirements and conformance to standards for information governance across all areas within the digital health ecosystem.

Data governance management provides the necessary leadership and strategic directions for the organisational data governance program's scope. Data stewards have the responsibility to manage and oversee the local operational implementation of its data governance policies, procedures, and rules associated with data collection and usage to ensure the quality of the organisation's data assets maintained throughout each data element's life cycle.

Data governance has four components defined by KPMG [14] as follows:

1. Data stewardship is accountable for the management of data assets; they are the caretakers of the enterprise data assets, ensuring the quality, accuracy, and security of data.
2. Data ownership—responsible for the creation of the data and the enforcement of enterprise business rules by defining data use within the organisation.
3. Data policies document the enterprise-adopted rules guiding the management of its data assets including authentication and access rights to data, legal, and regulatory compliance requirements.
4. Data standards document the precise criteria, specifications, and rules for the definition, creation, storage, and usage of data within an organisation.

Using data within a data governance environment requires a new approach to system architecture that starts with the data itself—what is its value and how can we best capitalise on that? Data governance includes the specification of data in order to support the required consistency and comparability. There is an urgency to implement data governance to ensure the integrity of healthcare information used. Governance processes need to ensure DIKW accuracy, relevance, validity, completeness, collectability, ability to represent flavours of null, consistent transformation through the data supply chain, and timely availability in the most appropriate format for every user.

This digital era provides the necessary capacity to optimise the benefits of DIKW governance. We are now required to collect, retrieve, and use health record information in electronic formats. This gives great flexibility, as data can be retrieved, displayed, and processed in different places and in different formats to that in which it was collected enabling greater use throughout its supply chain. To achieve this, electronically maintained records require greater structure and consistency than that of their paper-based predecessors. These structures are represented on the screens for both data entry and as reporting purposes, as generated by the record's repository-associated systems and applications. Although information may be presented in different ways for different purposes in an electronic record, there is still a need for consistent screen locations for certain types of information.

Data governance scope

To achieve this, the information about the individual and the knowledge must be represented in a compatible and consistent manner. That also enables the application of knowledge gained from individual patient records, through recognition of triggers based upon that knowledge, such as in clinical decision support systems. Data governance refers to the policies, procedures, and rules that need to be complied with operationally, to facilitate this. These need to define how to:

- determine and enforce rules regarding how best to relate disparate data to commonly understood concepts,
- identify how to systematically integrate health data assets to produce high-quality information that is safe to use for clinical decision-making, and
- ensure that data is useful, up to date, and relevant for policy or management decision-making or for the development of algorithms for artificial intelligence use and for other applications.

The scope of data governance includes the specification of individual elements of data, the relationship to other data (information) about the individual, and the way that data is represented. In health informatics, we not only refer to information and data but also to 'concepts'.

Increasing the availability of data and its use boosts innovation and competition and enables the sharing of associated benefits and costs. Effective use of data is increasingly integral to the efficient functioning of the economy. Improved availability of reliable data, combined with the tools to use it, is creating new economic opportunities. The increasing availability of data can facilitate the development of new products and services, enhance consumer and business outcomes, better inform decision-making and policy development, and facilitate greater efficiency and innovation in the economy. Greater use of data requires open data policies. This increases the transparency and accountability of governing processes.

Data governance needs to be viewed as a foundational requirement for any successful digital health transformation activities. In its absence, we cannot fully capitalise on advanced analytics or meet personalised healthcare delivery and population health management needs.

Data/information governance operational principles

1. Preserve data provenance with traceability to the point of origin, be it human, device, or software, including:
   - who captured (human, device, or software),
   - what was captured (data content and context, including individual to whom the data pertained),
   - when captured (date/time), including relationships such as retrospective (past), concurrent (now), future (planned),
   - where captured (physical location, network address, device ID),
   - why captured (purpose, rationale),
   - the custodian and/or owner of the data,
   - alterations made (ensure fidelity to source) e.g., transformation, map quality requirements, and aggregation algorithms,
   - support non-destructive updates,
   - points of data verification over time,
   - completeness (or non-completeness markers associated with data content),
   - relationships between key data components, such as problems, symptoms, diagnoses, and actions taken.
2. Preserve relationships between data and individual identity of patients, clinicians, scribes, including identity assurance/verification actions.
3. Retain data source through the supply chain. As information moves through different systems and use cases, carry the original content alongside transformed content, retaining caveats, such as the persistence of consented use associated with the data.
4. Specify organisational accountability.
5. Consider the data supply chain and the utility of data throughout that chain to minimise the need for data transformation.
6. Apply a cohesive approach to governance of data, analytics, and data technologies.
7. Evaluate and enable skill development to support information governance and quality.

8. Ensure sufficient capacity and capability in the health data ecosystem to carry out operations and controls necessary to achieve the desired outcomes.
9. Specify vendor accountability.

Effective vendor management has the potential to improve data and information quality and safety and reduce costs and the need for extensive data mapping. There needs to be a mechanism to manage an ongoing operation of a cohesive, enterprise-wide health data and information governance program in an ethical and effective manner. This needs to align with the entity's business strategies, ensure that stakeholders attain and retain buy-in, orchestrate, change management when required, and manage risks.

Clinical governance and data reliance

Clinical governance is a well-established concept that was first initiated in 1995. Standards have been developed for clinical governance which are based on the recognition that leaders of any health service organisation have the responsibility to the community they serve to ensure that appropriate guidance is given to those working in the organisation. This requires them to engage in continuous improvement of the safety and the quality of services delivered by ensuring that their services are patient-centred, personalised, safe, and effective. Clinical audits need to be undertaken regularly to review clinical performance against agreed standards of practice. The need to focus on the evidence of clinical and operational performance requires the assembling and critical evaluation of large datasets. The provision of meaningful results is dependent upon the accuracy of the original data and how that data has been aggregated, linked, and analysed to measure outcomes (value).

Given that there are many and varied data sources stored in numerous data silos, there are numerous limitations to the acquisition of large useful datasets. Quality management, the potential use of decision support systems, and clinical governance activities are severely compromised as a result. Decision support systems, including the generation of alerts, need to be designed and managed within a comprehensive framework of intrinsic and extrinsic determinants of safe and effective care to avoid significant errors [15]. Building blocks differ for every decision point but generally a decision support system needs:

- An appropriate clinical terminology and classification [16],
- A basic set of evidence-based rules for clinical decisions, with an appropriate ontology, that can be written into a computer-readable form [17],
- Structured (coded) data from the electronic health record (EHR) [18] to construct trigger events that feed data from the EHR,
- A decision support engine that applies the rules to the EHR data and to display the guidance such that it achieves clinical, management, or research purposes [15].

National governments, other jurisdictions, and healthcare service providers have usually developed their own clinical governance framework using fairly standard headings reflecting principles, expectations, roles, different stakeholders, their area of responsibility, and required actions with a strong focus on leadership, culture, teamwork, and partnerships. The use of robust and accurate data may be mentioned, but the need to include clinical data and knowledge governance is rarely if ever mentioned. Addressing this requires policymakers to establish regulations to standardise EHR integration tools, such as the national adoption of an open knowledge platform and vendor/technology neutral clinical data repositories. These types of quality management activities need to be included in any organisational data and clinical governance framework.

The continuing reliance on disparate tools, inconsistent value measures, the lack of processes to ensure quality data collection and processing, inconsistent adoption of data standards, poor understanding of data assets and ownership, an absence of data stewardship, and an apparent lack of awareness regarding the importance of data governance reflects the existence of widespread low digital health and data governance maturity levels across the healthcare industries.

For example, the Australian clinical governance standard makes no mention of clinical data or knowledge governance. One action statement refers to 'uses clinical and other data collections to support risk assessments' [19]. There is one section on healthcare records requiring health service organisations to have record systems that are made available to clinicians at the point of care. Healthcare service providers are required to:

- Support the workforce to maintain accurate and complete healthcare records,
- Comply with security and privacy regulations,
- Support systematic audit of clinical information,
- Integrate multiple information systems, where they are used.

There are associated standards, guidelines, and policies applicable to healthcare record documentation, which primarily apply to paper-based records. They tend not to be written to best suit the digital era or with a focus on the need for data and knowledge governance. Yet, both clinical and related data and available clinical knowledge represent the key assets of any healthcare service provider. Data represent the essential resources needed to undertake any audit, evaluation, or performance-monitoring activity. The lack of focus on clinical data and knowledge governance may explain why most healthcare provider organisations appear to be unable to optimise their data assets. The health workforce needs to adopt digital thinking!

The Australian Digital Health Agency has published a clinical governance framework that focusses on the end-user experience and is designed to effectively manage risk, promote a shared responsibility for safe and high-quality care, and enable continuous improvement [20]. One of this agency's core functions is to:

> Develop, implement and operate comprehensive and effective clinical governance, using a whole of system approach, to ensure clinical safety in the delivery of the national digital health work program.

Clinical risks in digital health are identified by this agency and based on themes seen in reported incidents associated with the national MyHealth Record initiative, they manage as:

- Misidentification—(this issue is discussed in another chapter);
- Loss of information—electronic document upload failure;
- Incorrect information—incorrect documents uploaded;
- Human error—clinical staff may use systems in good faith, however, inadvertently not detect that they are contributing to inaccuracies in clinical content;
- Expired authentication certificates.

Again, this framework makes no mention of the need to govern clinical data and knowledge. Addressing this requires policymakers to establish regulations to standardise EHR integration tools, such as the national adoption of an open knowledge platform and vendor/technology neutral clinical data repositories. These types of quality management activities need to be included in any organisational data and clinical governance framework.

Professional clinical knowledge leadership and governance

Previous chapters identified some of the international professional societies/associations that need to provide leadership regarding their own knowledge domain governance, including digital representation formats. Other chapters covered data standards, formalisms, and knowledge management; here, the focus is on data, knowledge, and their collection and representation governance.

Every professional health discipline has its own body of knowledge that needs to be governed in accordance with a shared data use vision. Professional bodies develop the knowledge and skills within their discipline and provide leadership and environments for their members to learn through conferences, publications, etc. Here, we discuss the need for a new type of clinical professional engagement, one that supports the representations of their discipline and its knowledge electronically in systems for patient care as well as for knowledge acquisition. This includes data in all its states:

- Data needed for patient care and how that data needs to be collected.
- How that data can be aggregated to support the discovery of new knowledge.
- How to present knowledge back into the clinical care process to improve the speed of knowledge uptake in the clinical practice domain.

This is a major leadership function, representing a key responsibility for professional bodies of all types in healthcare today. One for which most are ill-prepared and for which they do not appear to have the skills. It is also important to realise that administrative engagement is also necessary.

Every clinical speciality has its own group or set of types of care recipients (patients/clients) for whom best practice diagnosis, treatment, and care options are made available. These protocols are based on evidence obtained from expert practice and research. They form the foundation for clinical practice guidelines used to inform software developers of decision support and artificial intelligence applications. Each clinical knowledge domain also provides the necessary content used for educational, specialist certification, and practice assessment.

Digital health ecosystems with EHRs at the centre need clinicians to drive how their knowledge is best represented digitally in computable formats. This requires the adoption of standard terminologies and formalisms with the appropriate degree of expressivity that best suits each use case, including value sets to be adopted for a range of use cases applicable to each clinical knowledge domain. Every knowledge concept has a definition and a terminology code to represent it. These need to be adopted and governed as a standard to ensure consistent use. Many high-level clinical concepts can be de-composed into many sub-concepts, each with its own definition and code such as an Apgar score.[a]

Strong disciplinary clinical leadership for knowledge domain governance is not yet seen globally or nationally in any significant way, other than ad hoc arrangements to suit a specific use case, although the adoption of these ideas is gathering momentum. There is an urgent need to convince current clinical leaders in every clinical specialisation to take on the responsibility of their own knowledge domain's digital transformation.

Patient-centred care demands interdisciplinary collaboration, with consistent terminology standards adopted to enable that collaboration throughout digital environments. Such activities have the potential to be far reaching, benefit many stakeholders, and contribute to health system financial sustainability. Clinical data and knowledge represent a key strategic asset that needs to be governed by clinicians in the first instance; this includes governance of clinical formalisms which need to be evidence based. Clinical knowledge forms the digital health foundation.

For example, every type of patient has his/her own set of commonly occurring, clinically relevant characteristics, plus commonly used treatment/care protocols, and key information needs to accompany them throughout his/her healthcare journey to ensure continuity of care. Each of the above metadata concepts can be represented by one or more standard terminologies/ontologies and every knowledge concept can be represented by a computable formalism. Only clinical specialists, familiar with treating/caring for their groups of patients and their own specific knowledge domain, are able to define this. Such definitions influence data transfer and system connectivity requirements that best suit these types of patients.

This information also needs to be readily available when legislation is being drafted to regulate data access or data linkage options or to specify reporting needs to suit public health or funding options. There are many interdependencies where every stakeholder needs to make a significant contribution based on their own knowledge domain and operational needs.

Health ecosystem governance frameworks

Nations govern their own healthcare industry to a greater or lesser degree. A nation's government or sovereign ruler or their principal representatives provides the authoritative directions or restraints that determine what can, should, or must be done and by whom. Governance is based on delegated executive power. The WHO defines governance [21] as:

> Ensuring strategic policy frameworks exist and are combined with effective oversight, coalition building, the provision of appropriate regulations and incentives, attention to system design, and accountability.

Governance of the health system is the least well-understood aspect of health systems [22]. Many governance elements or indicators used to assess health systems' governance reflect values such as equity. One could argue that sovereignty and/or politics provide the means

[a] a measure of a baby's condition after birth.

by which the governance process in any nation operates. The WHO has identified that the generation and strategic use of information, intelligence, and research on health and health systems is an integral part of the leadership and governance function [4]. Indeed, leadership/governance, associated with information and research, healthcare financing, the health workforce, medical products and technologies, and service delivery, constitutes any health system's building blocks in WHO's health system framework; all of which need to be governed.

There is a trickle-down effect along with a pyramid of control, as various governance processes can be nationwide, regional, community, organisational, clinical domain-specific within an organisation, or entity-specific. Such processes rely on legislation, policies, regulatory, and professional, moral, cultural, and/or ethical standards adopted, based on what we value as a nation. A nation's governance infrastructure determines not only how its health system is governed but also its data, information, knowledge, and wisdom (DIKW) assets, environmental health influences, its health system infrastructure, administrative and management processes, as well as which and how health services are funded and in what manner made available to whom and under what conditions [23,24].

Such high-level infrastructures determine any nation's health system strategic directions, including digital health-supportive infrastructures, such as the determination to mandate the adoption of a knowledge-based platform, concept representation, terminology/coding, and any other standards. Any national digital health ecosystem's architecture needs to be designed accordingly. One essential supportive infrastructure is a national entity able to provide the necessary leadership for the management of data governance throughout the digital heath ecosystem. The traditional approach to data governance from a national perspective has been the identification of minimum datasets which are to be reported to the government for a specific purpose, when originally created technology had no better alternative. This process has generated the data silos from which all data collection systems around the world still suffer.

However, the technology has changed, and the national infrastructure needs to reconsider how they define and use data. This has the potential to avoid further creation of data silos and maximise the value of all data collected, enabling every stakeholder to make the best possible use of national technology/vendor-neutral, cloud-based data collections. A governance framework, as presented in Table 1, may be used to guide the development of a strategic governance plan.

Data governance policies and processes

Every healthcare facility needs to develop policies and processes for:

- managing the full life cycle of data and records for their use locally and with trusted partners for further value generation,
- data collection and preparation to support the collection, parsing, cleansing, remediation, and storage of data to enable it to be linkable and fit for use in analysis and care,
- data linking and aggregation to link datasets and standardise how data may be aggregated (e.g. health regions, age bands, and minimum cell size) to facilitate comparison and use,
- records and data maintenance in technology systems, which may include storage, encryption (in motion and at rest), retention, archiving, logging, destruction, recovery, and secure access mechanisms,

TABLE 1 Overview of a health information governance framework.

	Governance	Design	Use	Implementation and Support
Data supply chain governance (all use cases)	Vision	Content Specification Representation (code systems) Exchange (messaging)	Privacy Access Ethics Display	Metadata standards Standards for use cases
	Strategy Including a standardised information model	Quality measurement and assurance	Ownership Custodianship	Quality metrics and maturity models
	Vendor management	Global governance with national extension		Vendor accountability
	Roadmap	Collection/ collectability	Audit	Engagement and change management
	Program management Policies Processes (life cycle, consent, privacy, security, maintenance)	Management (including risk management)	Analytics (AI)	Workforce Skill requirements Skills for roles Education consistency
	Reuse			

- defining the risk tolerance level of the organisation, processes to assess individual risks, and decision rights to act on the analysis including:
  - consideration of negative risks (e.g. harms resulting from breaches),
  - positive risks (e.g. failure to realise benefits from sharing data and information), mitigation strategies to maintain those risks within the organisation's tolerance level,
  - consideration of ethics (e.g. intended use and unintended consequences),
  - both health data and information security risks (i.e. protection against deliberate threats),
  - safety risks (protection against unintended threats—e.g. loss of information or data from lack of backups, unsupported file formats, encrypted files with lost password),
  - explicit measures of the likelihood and impact of pre- and post-mitigation activities,
  - retention and destruction of data.
- Compliance with legal and statutory privacy rule requirements including reference to legislative authority and intent, governance and accountability structures, consent management, compliance, training, and incident management (identification, notification, and remediation). Alignment with social license to collect, store, distribute, and use data and information. It is important to consider the impact of risk appetite on decisions of access and privacy management.
- Standards that can be used to assess vendor conformance to data standards and specifications including the production of maturity metrics and conformance assessment schedules (timing) and contractual requirements. Such requirements could include contractual obligations to meet specific standards e.g. standards for map quality

assurance or use of standardised information models (such as those in the clinical knowledge manager).
- Data/information internal and external access and sharing based on purpose, e.g. care, health system use, for-profit, monetisation, across various types of organisations, e.g. public sector, researchers, innovators, private sector, open data, aligned with any existing mandates and/or social license, e.g. open data vs formal data requests. This needs to include requirements for de-identification and anonymisation. Application of advanced analytics practices such as machine learning, artificial intelligence, and predictive analytics, as well as testing for the generation of valuable insights to the organisation needs special attention including the considerations for ethics in the interpretability, bias, and the impact of generated insights.

Every healthcare facility needs to consider developing a framework for the review of compliance with legislation, organisation policies, strategic, and operational decisions, including data standards, quality assessment processes, and privacy policy. This may include practices for spot checks, remediation, communication, and active learning from data policy compliance issues encountered. May also include interaction with regulatory or international bodies for periodic or mandatory reporting (e.g. privacy commissioners, OECD).

Data management needs to consider all possible data corruption or misrepresentation opportunities throughout the data supply chain and potential associated risks impacting reuse. Use cases, data displays, and who needs access to data must be considered within the context of access control management strategies, policies, regulations, legislation, and ethical use. This includes the identification of data ownership and custodianship.

The roadmap needs to consider how these requirements are best audited as well as data suitability and limitations for advanced analytics. Implementation and support are dependent upon the availability of metadata standards and catalogues of standards as they apply to the many and varied use cases. The quality management plan needs to be based on the desired outcomes and associated value sets to be adopted and includes assessment and measurement protocols to be followed.

The OECD has published the following key elements of a good governance framework for personal health data [25]:

1. The health information system supports the monitoring and improvement of healthcare quality and system performance, as well as research innovations for better healthcare and outcomes.
2. The processing and the secondary use of data for public health, research, and statistical purposes are permitted, subject to safeguards specified in the legislative framework for data protection.
3. The public are consulted and informed about the collection and processing of personal health data.
4. A certification/accreditation process for the processing of health data for research and statistics is implemented.
5. The project approval process is fair and transparent, and decision-making is supported by an independent, multidisciplinary project review body.
6. Best practices in data de-identification are applied to protect patient data privacy.

7. Best practices in data security and management are applied to reduce re-identification and breach risks.
8. Governance mechanisms are periodically reviewed at an international level to maximise societal benefits and minimise societal risks as new data sources and new technologies are introduced.

This framework applies to every organisational data user throughout the data supply chain. Data clarity is applicable for every data collection point where the data originates, as well as to all information shared. Data governance applies to their definitions through collection via one or more systems or devices, data aggregation processes adopted to suit statistical analysis, performance monitoring, routine reporting, advanced analytics making use of big data, and finally archival storage or destruction.

Benefits of adopting network and cross-sector collaboration

Today's technologies are able to capture and harness data in a manner that would enable significant improvements in care outcomes to be realised while at the same time reduce costs and waste. We are at a critical turning point for value in healthcare [26]. The Global Coalition for Value in Healthcare, a collaboration between the World Economic Forum and leading healthcare stakeholders to continue promoting and driving global health system transformation, was launched at Davos 2019. This focusses on policies and regulations, payment/funding systems, tools for data analytics, care delivery models, and informatics. The latter is essential for the achievement of the former four areas of focus. Informatics facilitates the capture, analysis and sharing of health outcomes, and relevant data for each population segment. The transformation of our digital health ecosystem provides the foundation that enables all other desired health transformations.

The major limitation is the continuing existence and further establishment of data silos and the limited consistent use of data standards. Collectively, this prevents useful data aggregation the leveraging of these potential benefits [27]. The adoption of standards has enabled the utility, scalability, and global growth of the Internet and the Global System for Mobile (GSM) communication. The World Economic Forum has demonstrated the value and benefits attained to the world at large as a result. There is an urgent need to leverage the power of global health data by improving its accessibility, utility, and scalability as this is considered to have a profound positive impact on medical research, drug pricing, clinical decision-making, patient empowerment, and health service outcome improvements. Even small locally focussed digital transformations to optimise its digital health ecosystem is expected to yield very significant benefits for many.

The focus of this text is to contribute to the creation of a vision for a global health informatics architecture designed to increase value and ensure the sustainability of any nation's health system. There is a strong focus on the need to adopt the best available set of standards. The exposure of gaps and overlaps in the standard landscape should assist with identifying how best to progress towards the coordination and endorsement of standards that fit closely with this global vision. Health data represent any health systems' key asset that needs to be governed and managed by those best prepared to do so. The desired benefits of adopting network and cross-sector collaboration can only be realised with strong national and local leadership and the adoption of the most suitable suite of technologies that collectively makes up the digital health ecosystem. This requires a strong commitment to global digital transformation. We need to think big and start small by adopting both top-down and bottom-up approaches.

Transforming to optimise the value of this digital era

A willingness to share and collaborate is not inherent in some healthcare systems. The lack of trust by both data custodians and users in existing data access processes and protections and numerous hurdles to sharing and releasing data are choking the use and the value of health data. In some countries, the ownership of data represents fiscal control, used as an incentive for patients to return to the service where their information is held and thereby those services retain patient patronage. Where organisations use this fiscal model, sharing and collaboration are actively discouraged. Incremental changes in the data management framework to date have failed to deliver a culture of making data available for widespread use. The range and volume of datasets now held in the public or private sector, that could potentially be made more widely available, and the associated opportunities are monumental. Culture shift requires both persistent changes of attitude and permission to demonstrate that [3].

Marginal changes to existing structures and legislation will not suffice. Adding more ad hoc adjustments to existing structures and legislation will not suffice and will continue to be very costly. Fundamental and systematic changes are needed to the way governments, businesses, and individuals handle data. Reforms need to aim at moving from a system based on risk aversion and avoidance to one based on transparency and confidence in data processes, treating data as an asset and not a threat [3]. Improving trust community-wide needs to be a key objective.

Change management

Transforming current practices and behaviours to better suit ongoing digital health development, efficient and safe information system design, and implementation requires the adoption of a well-designed enterprise-wide change management structure. First and foremost is a requirement for the right type of leadership, a review of existing organisational structures, lines of authority, and the appointment of suitably qualified people to take on these new data management and governance roles.

The required data management and governance activities described above are not well known as is evident from the data/informatics relationship knowledge gaps we have identified. There is a need to define competency requirements for skills to undertake these specific tasks safely and effectively including:

- Competency requirements associated with specific roles. Competency covers all levels of learning and skill development from knowing what to do, through being able to do it, up to evaluation and strategy development associated with a topic or task and leadership.

- Educational engagement and sustainability approaches aligned with target audiences (employees, management, board) on key topics related to their area of responsibility and their use of health data and information. This includes all components of the information governance framework.
- Adopt a well-planned approach, effort, and timetable for meeting requirements for information governance capacity to achieve its mandate for information governance across all areas of governance scope, such as management, analytics, security, and privacy.
- Consider assessment of workforce supply and demand, career structures to support sustainability, and succession planning.
- Identify upcoming talent and give them the space to innovate and explore.

Managing change requires strong leadership from those with the right skills, attitudes, and talent, including the ability to make use of inductive reasoning, ability to work flexibly with unfamiliar information and find solutions, emotional intelligence, and strong people skills. These attributes are largely determined at birth; they represent key precursors to success and are difficult to observe during interview processes, but can be tested for. Someone with the right knowledge and skills but who does not possess these personality traits is highly likely to fail. Change management processes to be adopted for these workforces, people, culture, and human resource management and changing needs are provided in greater detail in the workforce chapter.

The literature has identified a number of common barriers to change when instituted to achieve a digital transformation. These tend to be the result of a lack of the following factors [28]:

- Awareness and meaning amongst many stakeholders which often leads to the non-acceptance of sharing data for example.
- Confidence and trust regarding privacy, confidentiality, and security protection.
- Interoperability enabling the free flow of data at the technical and semantic levels.
- A clear legal framework related to trust regarding data access management and data use.
- Staff training to ensure correct data collection and use in day-to-day health professional practice, including for medical and nursing students engaged in workplace practice.
- Meaningful and integrated solutions leading to behavioural changes.
- Innovation in funding models.

As part of any change management planning process, one needs to consider all of the above and include strategies to address these factors. In all instances, it is important to consider all possible anticipated and possible unanticipated consequences of proposed changes as well as the positive or negative severity of their impact when developing the plan for the change process.

Any organisational change needs to be well planned, adopt a holistic organisational perspective, and have a clear vision regarding the outcomes to be achieved [29]. Change needs to benefit every individual employee in order to be successful. Successful change management requires an intimate understanding of human behaviours and work and information flow processes. Automation should result in the availability of timely more complete information, ease of use, staff time-savings, as well as improved patient flows and outcomes to provide value in return of investments made [30].

Part of the planning process must ensure that everyone has an opportunity to ask questions and to alert implementers to possible issues requiring resolution. Everyone needs to fully understand what is being proposed. Leaders cannot assume that others have an equal appreciation of the need or a desire for a change as they do. The cultural landscape at every level and within every department plays an important role. This needs to be well understood as it influences what people value, their perceptions, and behaviours. Leaders need to make use of this knowledge to enable them to influence and drive the change process in the desired direction [31].

Conclusion

Health data quality, its management, governance, and use are fundamental to everything associated with digital health ecosystems. Every ecosystem entity has a role to play to bring about a major transformation from current practices, which are no longer fit for purpose, to alignment with today's available digital technologies to benefit us all. Many other components in need of transformation are dependent on the successful adoption of information models and agreed data formalisms and standards. The next chapter explores how this can contribute to the transformative requirements for guideline and knowledge management in a digital world.

References

[1] ISO/TS_21564, Health Informatics—Terminology Resource Map Quality Measures (MapQual), Intenational Organisation for Standardisation, 2019. [cited 2019 6 December]. Available from: Health Informatics — Terminology resource map quality measures (MapQual).

[2] K. Andronis, K. Moysey, Data governance for health care providers, Stud. Health Technol. Inform. 193 (2013) 299–315.

[3] Productivity-Commission, Data Availability and Use, Overview and Recommendations Report No.82 Canberra, 2017, [cited 2019 5 June]. Available from: https://www.pc.gov.au/inquiries/completed/data-access/report.

[4] ONDC, Office of the National Data Commissioner, Australian Government, 2021. Available from: https://www.datacommissioner.gov.au/.

[5] ISO-131000, Risk Management: SafetyCulture, 2018, Available from: https://safetyculture.com/checklists/iso-31000-risk-management/.

[6] AS/NZS-4360, Risk Management: Standards Australia, 1999, Available from: http://www.epsonet.eu/mediapool/72/723588/data/2017/AS_NZS_4360-1999_Risk_management.pdf.

[7] Gartner's-IT-Glossary. Available from: https://www.gartner.com/en/information-technology/glossary/big-data.

[8] V. Sharma, I. Ali, S. van der Veer, G. Martin, J. Ainsworth, T. Augustine, Adoption of clinical risk prediction tools is limited by a lack of integration with electronic health records, BMJ Health Care Inform. 28 (2021), e100253.

[9] ISO/TS21526, Health Informatics: Metadata Repository Requirements (MetaRep), 2019, Available from: https://www.iso.org/standard/71041.html.

[10] J. Nahar, T. Imam, K.S. Tickle, D. Garcia-Alonso, Issues of data governance associated with data mining in medical research: experiences from an empirical study, Stud. Health Technol. Inform. 193 (2013) 332–361.

[11] B. Reid, Casemix systems and their applications, Stud. Health Technol. Inform. 193 (2013) 316–331.

[12] eHRoL, Electronic Health Record Online Learning, Global eHealth Collaborative (GeHCo), 2021. Available from: https://gehco.org/ehrol/.

[13] G. Hripcsak, M. Bloomrosen, P. Flatelybrennan, C. Chute, J. Cimino, D. Detmer, M. Edmunds, P. Embi, M. Goldstein, W. Hammond, G. Keenan, S. Labkoff, S. Murphy, C. Safran, S. Speedie, H. Strasberg, F. Temple, A. Wilcox, Health data use, stewardship, and governance: ongoing gaps and challenges: a report from AMIA's 2012 Health Policy Meeting, J. Am. Med. Inform. Assoc. 21 (2013).

[14] KPMG, Data Governance Driving Value in Health, 2018, Available from: https://home.kpmg/xx/en/home/insights/2018/06/data-governance-driving-value-in-health.html.

[15] S.T. Liaw, Clinical decision support systems: data quality management and governance, Stud. Health Technol. Inform. 193 (2013) 362–369.

[16] A. Kumar, P. Ciccarese, B. Smith, M. Piazza, Context-based task ontologies for clinical guidelines, Stud. Health Technol. Inform. 102 (2004) 81–94.

[17] D.M. Pisanelli, A. Gangemi, If ontology is the solution, what is the problem? Stud. Health Technol. Inform. 102 (2004) 1–19.

[18] S.E. Bleeker, G. Derksen-Lubsen, A.M. van Ginneken, J. van der Lei, H.A. Moll, Structured data entry for narrative data in a broad specialty: patient history and physical examination in pediatrics, BMC Med. Inform. Decis. Mak. 6 (2006) 29.

[19] ACSQHC, Australian Commission on Safety and Quality in Health Care-Standards, 2021, Available from: https://www.safetyandquality.gov.au/standards.

[20] ADHA, Clinical Governance Framework v2.5, 2019, Available from: https://www.myhealthrecord.gov.au/sites/default/files/clinical_governance_framework_v2.5.pdf?v=1566786544.

[21] WHO, Everybody's Business: Strengthening Health Systems to Improve Health Outcomes, 2007, [cited 2020 3 march]. Available from: https://www.who.int/healthsystems/strategy/everybodys_business.pdf.

[22] S. Siddiqi, T.I. Masud, S. Nishtar, D.H. Peters, B. Sabri, K.M. Bile, M.A. Jama, Framework for assessing governance of the health system in developing countries: gateway to good governance, Health Policy 90 (2009) 13–25.

[23] E.J. Hovenga, H. Grain, Health data and data governance, Stud. Health Technol. Inform. 193 (2013) 67–92.

[24] E.J.S. Hovenga, National healthcare systems and the need for health information governance, in: E.J.S. Hovenga, H. Grain (Eds.), Health Information Governance in a Digital Environment, IOS Press, Amsterdam, 2013.

[25] OECD, Health Data Governance: Privacy, Monitoring and Research—Policy Brief, 2015.

[26] World-Economic-Forum, Value in Healthcare: Accelerating the Pace of Health System Transformation, 2018, Available from: http://www3.weforum.org/docs/WEF_Value_in_Healthcare_report_2018.pdf.

[27] World-Economic-Forum, How to Unleash the Enormous Power of Global Healthcare Data, 2018, Available from: https://www.weforum.org/agenda/2018/12/global-healthcare-data-is-a-vast-untapped-resource-until-now/.

[28] ECHAlliance, The Digital Health Society Declaration, 2019, [cited 2019 18 February]. Available from: https://echalliance.com/page/DHSdeclaration.

[29] AIIM, in: Hyland (Ed.), State of Information Management: Are Businesses Digitally Transforming or Stuck in Neutral? AIIM Industry Watch, Ohio, 2017.

[30] PWC, Strategy and Ten Guiding Principles of Change Management, 2004, [cited 2019 14 February]. Available from: https://www.strategyand.pwc.com/media/file/Strategyand_Ten-Guiding-Principles-of-Change-Management.pdf.

[31] E. Hovenga, C. Lowe, Measuring Capacity to Care Using Nursing Data, Elsevier, Academic Press, 2020.

CHAPTER

10

Guideline and knowledge management in a digital world

Evelyn Hovenga[a,b,c]

[a]Digital Health, Australian Catholic University, Melbourne, VIC, Australia [b]eHealth Education Pty Ltd, Melbourne, VIC, Australia [c]Global eHealth Collaborative (GeHCo), Melbourne, VIC, Australia

Language and communication: The data, information, knowledge, and wisdom (DIKW) continuum

The health industry relies on the use of language to communicate between those in need of any healthcare service and any healthcare service provider. Such communication needs to be able to clearly represent any number of concepts or ideas or things in the form of data. Previous chapters have indicated that data represent a key foundational asset in constant use to benefit all associated with any aspect of health service provision. Electronic use of health data requires health data to be able to be manipulated or processed to produce new information. The application of existing knowledge to such information may result in the acquisition of new knowledge as shown in Fig. 1. Wisdom and knowledge are applied to any new information generated via data analytics to create new knowledge, a process that is traditionally undertaken by people but increasing so by computers (machine learning) for artificial intelligence (AI) purposes.

Information is technically not the same as data. Information is defined as:

> data that are interpreted, organized and structured [1].

For example: *Mary Smith is 4 years old*. This is information that can be applied to an individual instance. It is something that is a fact about a given person or circumstance.

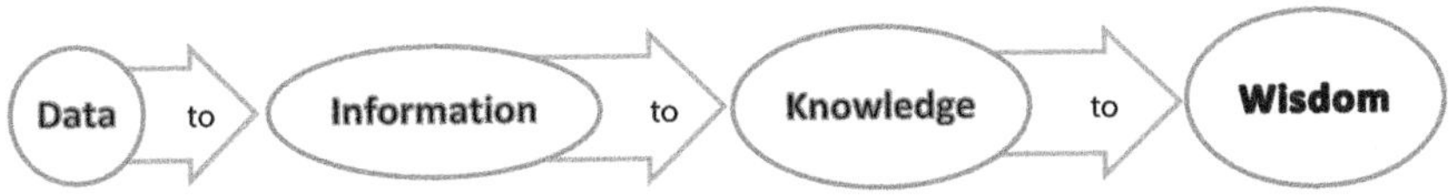

FIG. 1 DIKW hierarchy.

Roadmap to Successful Digital Health Ecosystems
https://doi.org/10.1016/B978-0-12-823413-6.00012-4

Level	Example
Data	4
Data element	Age at admission '4'
Data in a person's record (information about Mary)	Mary's record shows her 'age at admission' to be '4'

The concept of encoded data organised as 'information' is the common feature of genetics, biotechnology, language, communications, mathematics, electronics, computing, and robotics.

Data need to be represented in ways that enable computers to accomplish a great variety of processing tasks, such as aggregating, linking, identifying patterns, trends, or relationships, using any one of a number of different computer languages. Electronic health data can be represented either as natural language symbols or as a structured terminology in addition to images, video or audio types of data.

The language used in health care is complex and multifaceted, more so than most language domains. In fact, looseness in the usage of words such as terminologies and classifications tends to confuse discussion and debate. It has been estimated that between 500,000 and 45 million different concepts are needed to adequately describe concepts like conditions of patients and populations, actions in health care and related concepts, such as biomedical molecules, genes, organisms, technical methods, and social concepts. Simple coding schema cannot adequately represent these concepts [1], nor can this number of data points be adequately processed using legacy systems as explained in Chapter 6. Knowledge is acquired from the formation of an understanding of various sets of relationships between any number of concepts represented in a variety of ways. Individual knowledge acquisition is dependent upon an individual's ability to identify such relationships to extract greater meaning.

Types of data

Atomic source data, from which new knowledge can be discovered, tested, and established, need to be structured to enable such data to reliably support computer searches, analytics to aid discovery, testing ideas or validity or any automated use case. This requirement has resulted in the adoption of 'data types' as used by computing languages. There are different types of data types. For example, those that refer to groups of data such as identifier data, demographic data, and clinical data, or those that support programming. These latter types of data represent distinct values, characterised by properties of those values, and by operations on those values as these relate to any given domain. Other data types consist of three main features: a value space, a set of properties, and a set of characterising operations [2]. These types of data types are critical for enabling data exchange (interoperability). For example, an integer (data type) has a value of 6. Integers can be processed, e.g. multiplication or subtraction.

There are many specialised health data types to represent unique clinical concepts that need to be able to be represented in computable formats in EHRs. Numeric data attributes include whole numbers (integers) or numbers written with fractional components such as

2.75. These in turn may represent a currency or any other type of contextual attribute such as a volume or length measure. The same applies to numbers associated with pathology test results where each type of test has its own representative attributes or numbers used to, for example, represent a date where the data type adopted determines if this is represented as day/month/year or year/month/day.

HL7 standards make use of its own set of data types, and ISO/IEC has its own set, as does openEHR. Where different data type standards are in use, then the necessary matching algorithm(s) need to be programmed to enable data exchange between systems. The adoption of a standard set of data types improves interoperability, maximises productivity, and reduces the opportunity for error.

As multiple domains apply to any digital health ecosystem, it was necessary to harmonise multiple sets of data types, as a foundational step towards enabling electronic information exchange, and to publish the results as a new health informatics standard. This ISO standard [2] publishes a set of data type definitions for representing and exchanging basic concepts commonly encountered in healthcare environments. It specifies a collection of healthcare-related data types by declaring their semantics, providing Unified Modelling Language (UML) definitions, and specifying Extensible Markup Language (XML)-based representations of these data types. Grouping, classifying, drawing images, and modelling relationships are traditional methods used to describe and represent viewed reality. Artificial intelligence techniques emulate and automate such cognitive processes, hence the need to represent these concepts in computer-processable formats.

Knowledge representation

Knowledge representation approaches or formalisms include a variety of computer science techniques of using symbols or models or statistical inferences. Such approaches include the representation of knowledge by using terms from controlled terminologies or numbers. Ideally, formal multidimensional concept (knowledge) representation systems are developed to enable the computer processing of this complexity. Such systems need to result from highly interdisciplinary approaches by integrating knowledge from linguistics, philosophy, informatics, and the health sciences to simulate perceived world reality. Atomic source data used to formulate new knowledge need to have a high degree of accuracy to be truly meaningful. This is especially important when using big data derived from many individual records and data repositories.

Many terminology structures in code systems in current use are the result of historical foundations. Many concepts are defined by multiple terminologies. Many terminologies are no longer fit for purpose in this digital era. Terminology development began with the use of classification in the days of Plato and Aristotle, and the first medical terminology began in the 16th century for the London Bills of Mortality [3]. This resulted in the identification of patterns of mortality that provided the foundation for the population-based epidemiology discipline.

Cimino [4] noted that 'Medical knowledge representation has traditionally focused more on the structure than the symbols' and that a 'knowledge-based terminology can help us get raw patient data' by the right names' and set us on the path to knowledge, to:

- Gain a better understanding of our patients;
- Access knowledge relevant to the care of our patients;
- Enable the use of smart systems to apply knowledge to the care of our patients;
- Discover new knowledge from health data.

Optimum electronic health information management is dependent upon the use of a major infrastructure designed to enable the development of electronic data collection, storage and retrieval systems, able to support electronic communication of health information and knowledge. Terminologies represent a component within the field of health concept representation. This knowledge domain includes rules and relationships used to interpret clinical terms within confined contexts. The context within which an item of clinical information is represented can dramatically affect its meaning or semantic value.

The many ways health concepts can be represented indicate the variety of formats in use based on historical workplace-related use, to provide source data and influence meaning, at the first point of collection representing the start of the data supply chain referred to in the previous chapter. Facts can be represented by any natural language, where its grammar influences meaning. Facts can also be represented by the use of a standard terminology, consisting of any standard set of symbols or words used to describe concepts, processes, and objects pertaining to a specific domain or field of study. Meanings of facts are best represented by a standardised ontology. Each of these representations has any number of attributes that determine the best electronic formats that need to be used for computer processing purposes. Such processing includes data retrieval, data linkage, and sharing to arrive at new information to which prior knowledge can be applied to generate new knowledge to be used for decision-making purposes.

Clinical practice requires its practitioners to learn from and interpret the patient information made available and apply their existing knowledge to arrive at new information or knowledge including a diagnosis, treatment, and care options. Knowledge is the end result of learning from things or experiences, and the wisdom that may follow is the result of gaining an understanding of why things happen. This requires us to consider any number of possible relationships and potential impacts. We then have a choice on how we wish to make use of new knowledge and/or wisdom gained.

One issue that the digital health discipline can address is the abundance of existing knowledge that no one person can possibly remember. We've now arrived at an era where all relevant existing knowledge needs to be readily accessible with maximum usability in a timely manner. The Venn diagram presented in Fig. 2 demonstrates the general scope of health concepts and the many ways these can be represented illustrating the overlapping nature and complexity associated with health concept representation [1].

This chapter focuses on the representation of existing knowledge in computable, that is, executable formats, as this is a key foundational requirement for the development of decision support and artificial intelligence systems. In this digital era, every custodian for any one of the many discipline-specific knowledge domains has a responsibility to ensure that existing knowledge is available to its members in an online accessible computable format. To this end, the University of Minnesota initiated an Annual Nursing Knowledge: Big Data Science Conference in 2013 [5] with the objective to:

> develop a roadmap for achieving "sharable and comparable" nursing data and to ensure the timely adoption of big data methodologies across all of nursing's domains.

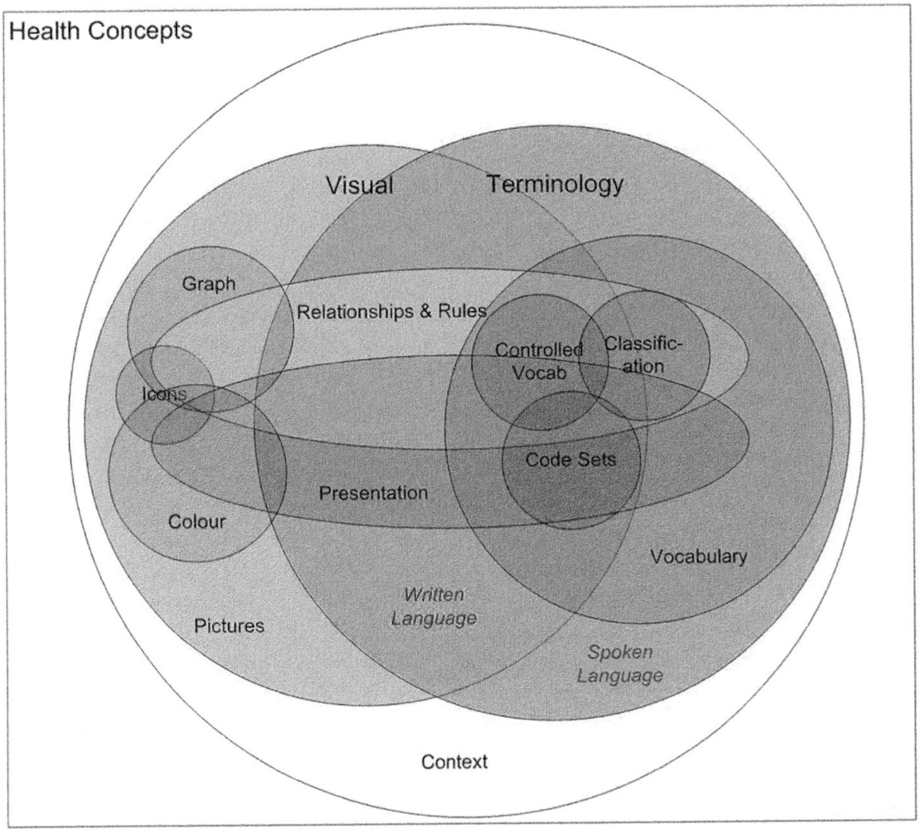

FIG. 2 Pictorial representation of major elements on health concept representation.

As demonstrated in a previous chapter, health professional colleges are well overdue in initiating similar activities to ensure their unique body of knowledge is kept up to date and readily accessible to its members in formats to best suit this digital era. Knowledge represents various relationships between data, which may be seen as patterns or sequences or trends no matter how randomly data become available. From an information systems perspective, these observations of real-world activity are used to model data flows or develop data processing rules based on observed relationships. Databases are structured to enable data retrieval and processing to meet functional requirements. Knowledge bases also need to be structured as a set of concepts, together with associated rules that determine how these concepts can be meaningfully arranged. Such structures or models are referred to as ontologies [6]. Knowledge representation schema has evolved to arrive at the best available computable formalisms with a high degree of expressivity over many years, to maximise the potential for knowledge computability.

Types of knowledge and knowledge management theories

Every organisation is dependent on the availability and use of knowledge. Knowledge is contextual, relevant, and actionable. Knowledge is a prerequisite to problem-solving. There are different types of knowledge to reflect knowing what, how, or why. Knowledge can be presented in a number of different ways to enable knowledge to be communicated from one entity to another. Once received, this knowledge needs to be interpretable by the receiver (person, organisation, or computer) to gain understanding, and it needs to be able to build on and expand, deliver, or store existing knowledge. Knowledge is the result of explorative research, observation, and experience. The use of available knowledge is particularly valuable for the delivery and management of healthcare services. Tacit and explicit knowledge combined represents the intellectual capital and core competencies of any organisation.

Tacit knowledge is cumulative and refers to subjective, cognitive insights, personal acumen, know-how, trade secrets, skill sets, and knowledge gained by individuals through experiential (on-the-job) learning. This type of knowledge includes procedural knowledge. It is about available expertise and high-level skills difficult to codify, and it may include or be based on explicit knowledge. It reflects situational knowledge acquired by individuals. Organisational cultures influence knowledge sharing opportunities, as culture represents a pattern of shared basic assumptions and values. Individual knowledge and skill acquisition is dependent on available learning opportunities and strategies adopted to facilitate learning. The content and format used for medical training are very regulated making the management of tacit knowledge less of a challenge in health care [7]. These same principles tend not to apply to other health disciplines.

Explicit knowledge refers to evidence-based, objective, rational, and technical knowledge stored as data (facts), policies, procedures, legislation, or documents. Although complex, this type of knowledge is easier to codify and model into computable formalisms to enable electronic knowledge management, yet information systems need to be a good match with the prevailing organisational culture, to be successful. The use of explicit knowledge by all individuals working in the health industry can influence the further development of tacit knowledge. Health care has generated huge amounts of constantly changing explicit knowledge making its management a significant challenge [7].

Propositional knowledge, also known as descriptive or declarative knowledge, refers to well-known facts, such as the colour of an object, the description of a mammal or other known object such as car or house, or the fact that Susan knows her husband is a musician. This type of knowledge is explicit. Propositions refer to claims made about the world. The knower needs to have a strong indication that the proposition made represents the truth. It should not just be based on belief as truth needs to be justifiable. This type of knowledge refers to all scientific disciplinary known facts. It reflects an individual's understanding of the world around them. Procedural knowledge refers to knowing how to do something, that is being in possession of a skill.

Any knowledge management initiative is about a strategic commitment to improve an organisation's effectiveness (best possible outcomes by service type or product produced) and operational efficiencies. Healthcare organisations have a variety of different knowledge management needs, organisational, discipline-specific, and patient-centred, including the application of precision medicine. These variable knowledge management requirements add to the complexity associated with the need to integrate tacit and explicit knowledge in a manner that serves multiple purposes. Another additional dimension to be considered is knowledge ownership (intellectual property—IP) and associated conditions for sharing. Effective knowledge management is fundamental to achieving desired outcomes, and it has the power to prevent repeating errors of the past. A good knowledge management system needs to be dynamic and able to be updated as new knowledge becomes available.

Knowledge management tools

A great variety of knowledge management tools have emerged and been used over the years. Ancient civilisations and indigenous communities made use of storytelling to disseminate knowledge to others. Knowledge management requires available tacit knowledge to be identified, gathered, organised, refined, accurately represented, and distributed. Explicit knowledge has similar needs in addition to being stored in some format, to be found by those who need to know, retrievable, and presented in a usable manner. It is important to differentiate between human-readable computer-based knowledge, such a PDF documents or HTML text, and computer-executable knowledge stored in a format (computing language), developed from a human-readable format, to enable computer processing.

Knowledge sharing methods have evolved from these foundational principles for the vocational trades and academic teaching and learning professions. Successful knowledge management implementation requires the adoption and use of the right combination of methods, technologies, and tools. Its key desired requirement is to facilitate existing knowledge to flow from those who know to those who need to know, throughout any organisation within the context of recognising that knowledge evolves over time and grows during the learning process. This requires lifelong learning principles to be adopted by all, made so much easier due to digital transformations within the publishing industry and greater ease of access to government publications and other grey literature.

Knowledge management is multidisciplinary as it leverages various collections of available intellectual capital. It draws on various aspects from science, research, interpersonal communication, organisational learning, cognitive science, publishing, business process analysis, and information technologies' disciplines. Successful knowledge management requires it be represented in computable formats in addition to being dependent on motivation, incentives, and perceived potential benefits of value to both knowledge providers and recipients. Whichever knowledge management tools are adopted, they need to be integrated within existing systems in use and match the specific requirements associated with the type of knowledge to be disseminated.

For example, social media may be the tool of choice to disseminate tacit knowledge but not considered suitable for the distribution of explicit knowledge, such as details about COVID-19. Within the health industry, knowledge management implementation strategies need to reflect the best possible applied methodology suitable to support business and clinical healthcare service practices enabling them to meet desired outcomes. This means that any combination of knowledge management systems and technologies as listed in Table 1 can make up any organisational specific knowledge management system designed to share information internally and externally. Organisational policies and procedures influence these selections as these govern the way every stakeholder can interact with each other.

The list presented in Table 1 indicates that knowledge management strategies have evolved over time. Many of these types of knowledge management systems have become individual proprietary software products difficult to integrate with other systems. Some knowledge has been computerised in a variety of different ways by multiple application developers and is used extensively to support healthcare delivery where these systems are in use. Examples are embedded calculators, risk algorithms, clinical prompts, and drug interaction warnings.

Many of these are cottage industry products, including those supplied by large vendors, which have not been maintained or implemented correctly or have not been reformed to best suit today's digital era. This explains the continuing fragmentation causing many frustrated users, cost increases, and inconsistencies of information sharing. This reality has also resulted in serious algorithm-related incidents [7], such as data maps built for one purpose being used for other purposes and delivering inaccurate data. Organisations need to focus on maximising the use of available collaborative clinical and informatics experts ('talents') within a sound professionally governed national (and/or international) infrastructure. The healthcare workforce needs to be able to practice in accordance with an up-to-date and trustworthy common knowledge base and standards of care.

Digital health ecosystems should consider the inclusion of a knowledge map with reference to people, processes, locations, and technologies to make it easier for stakeholders/users to find required knowledge in a timely manner to improve efficiencies. Every organisation needs to develop their own knowledge management policies, including a code of conduct relative to the use of each available technology. The US Mayo Robert-D and Patricia E Kern Center for the Science of Health Care Delivery, Informatics and Knowledge Management team provides a nice example [8], as represented by one instance of their knowledge management programme developed to curate, store, and disseminate clinical knowledge [9].

Computable knowledge needs to be sustainable and its library must be trustworthy and known. Users need to be able to determine content currency. Wyatt [7] lists the following professional challenges associated with computable knowledge:

- Protection of intellectual property;
- Freedom from bias;
- Measuring and improving knowledge quality and currency;

TABLE 1 Examples of knowledge management systems, technologies, and methods.

• Electronic communication tools and networks	• Knowledge bases, FAQ
• Cloud computing	• Data lakes, hubs, warehouses
• Social software, incl. LinkedIn, ResearchGate	• Databases, big data, e.g. relational, object-oriented, technology-neutral
• WWW, browsers, and search engines	• Videoconferencing, teams, groups
• Retrieval tools, agents, filters	• Group support and task management systems, e.g. Asana, Slack, MS Teams
• Content repositories, incl. EHRs, Coroner reports, guidelines, protocols, standards	• Systems linking distributed systems
• Document management systems	• Decision support systems
• Scientific publications, publishers	• Artificial intelligence agents and apps
• Directories for expertise	• Helpdesk apps, call centres
• Open computing platforms	• Learning content management system (LMS)
• Incident reporting systems	

- Legal liability for suppliers and users of clinical knowledge bases and decision support systems;
- Procurement or quality assurance of computerised knowledge;
- Tracking the source of knowledge and how it has changed since the original version.

He listed the following technical challenges:

- Encoding knowledge to enable safe reuse in a variety of clinical systems;
- Indexing computerised knowledge for accurate recall;
- Locking the knowledge to protect its integrity;
- Preserving its meaning when used in systems with different clinical coding systems;
- Adjusting computerised knowledge to fit specific contexts;
- Archiving obsolete computerised knowledge to support later legal investigation.

One known scientifically proven solution best able to address these challenges is the adoption of a *knowledge-driven platform* by all entities within any identifiable digital health ecosystem. Such a platform is where any knowledge domain's semantics are *explicitly* defined by knowledge artefacts rather than being implicitly expressed in software and database schemas. Such a platform provides a common base, defined by its architecture, on which numerous higher-level components (applications) can be built. This is explained in some detail in another chapter.

Knowledge management barriers and benefits

A number of barriers that inhibited or limited informed decision-making were found in one literature review [10] that focused on knowledge management from an evidence-based practice perspective and found a number of barriers that inhibited or limited informed decision-making. These were challenges such as technical constraints, the lack of employee motivation to share knowledge, system unreliability, lack of senior management support, clinician reluctance to use ICT tools, lack of attention to results and use of evidence, lack of incentives for documentation and dissemination, inadequate awareness about knowledge management systems, investment costs, poor quality data/information, organisational culture, knowledge sharing inhibitions due to status inequities, lack of trust, lack of access to modern technologies and/or up-to-date research results, lack of applied research, lack of communication between policy-makers and researchers/scientists, and poor interface usability IT application access issues.

Emerging opportunities for improvement mentioned included advances in health information and communication technologies, clinical decision support systems, electronic health record systems, communities of practice, and advanced care planning [10]. This requires the adoption of a comprehensive knowledge management infrastructure framework embedded within a digital health ecosystem architecture. An additional key requirement is for organisations and professional disciplines to create a culture of sharing, collaborating, teamwork, innovation, knowledge sharing protocols, and a willingness to change current practices. This has been well-demonstrated by the international scientific collaboration during the COVID-19 pandemic. The adoption of knowledge management systems enables better decision-making, faster response times to key issues, evidence-based best practice sharing, innovation, significant cost savings, staff well-being, and productivity growth.

Digital health ecosystem frameworks need to be designed in a manner that best represents this complex knowledge ecosystem to support integrated knowledge management in a manner that best serves multiple purposes, at every point of decision-making in the form of 'knowledge on demand'. Such frameworks need to be based on:

- applicable knowledge sharing principles,
- virtual communication and knowledge flows within and between healthcare facilities and disciplines,
- evidence of practice,
- guideline and protocol dissemination,
- the availability and appropriate use of knowledge management tools, and
- associated IT-specific artefacts and knowledge codification methodologies as specified within an organisation's knowledge management framework.

Knowledge representation and management continue to present us with many challenges generating a lot of research activity.

Knowledge discovery and representation

Knowledge discovery is about using all available resources, including computer-based data and analytical power to enlist computing power to extract and process data from any number of structured and unstructured sources of data, information, and/or knowledge. This process is used to, for example, determine disease trends and risk factors, to identify treatment pathways relative to outcomes, or to make predictions or identify the probability of adverse events. It requires a sound understanding of available sources of knowledge, storage entities, formats, and potential channels of knowledge dissemination including computer processing and result presentation requirements. Knowledge discovery can consist of any process adopted to search for, find, and extract data from any source. The quality of the new knowledge gained is entirely dependent on the quality (accuracy and completeness) and presentation format of the source data used. Reliable digital health data represent the foundational elements of a continuous knowledge discovery or learning health system [11]. Explicit knowledge is easier to be represented in a number of formalisms (artefacts) than tacit knowledge. The latter requires direct contributions to be made by those who possess this knowledge.

Data analytical methods are then used to identify patterns or trends or new relationships to produce new meanings or understandings. Computational intelligence, data mining, machine learning, and natural language processing (NLP) represent some computer processes that may be used to discover new knowledge. NLP makes use of both syntactic and semantic language analysis. Machine learning enables computer algorithms to be improved automatically through processing experience, and it is a particularly useful process for knowledge discovery from big data. In this digital era, it is a requirement for both existing and new knowledge to be in both machine-readable and machine-interpretable formats to enable optimum and timely data use, irrespective of technologies used for data extraction and processing.

These computer processing techniques make use of any number of mathematical processes and algorithms suitably applied to available data sources, including data analytics, resulting in a variety of different outputs, including decision trees and variants thereof, rules of various types, clusters, and visualisations. Computer software-based analytics assists with the discovery of hidden patterns. The value of its findings is totally dependent upon the quality of the available data. Reliability of knowledge discovery outcomes is to a large extent also dependent on prevailing data governance, data structures, and data cleaning techniques adopted prior to the application of knowledge discovery techniques to enhance data quality. Data sources need to be appropriately aligned with their intended use [11]. Traditionally, health information systems in use worked independently without standardisation or integration between administrative and clinical areas of practice. Such data fragmentation within healthcare organisations limits successful health data exchange, as well data linkages, a situation that continues to be prevalent today.

An absence of well-connected digital health ecosystems designed to optimise interoperability and the widespread use of data governance are limiting the ability to discover new knowledge. One solution has been the evolution to the development and use of data hubs, metadata, and a greater use of cloud computing. Another evolving development is the establishment of artificial intelligence and knowledge discovery joint research laboratories. The interest in the use of 'big data' has become an entirely new industry for computer scientists although, specifically in the health industry, this requires a multidisciplinary approach to ensure that newly discovered knowledge and associated interpretations are of relevance to the applicable disciplinary science. A well-functioning knowledge management system needs to also enable new knowledge discovery.

One recent example is the European FAIR4Health Project, which aims to make research data Findable, Accessible, Interoperable, and Reusable (FAIR) by adopting a responsible approach for the use of healthcare data from publicly funded research results stored in digital health records [12]. This project enables the reuse of data in a secure, controlled ecosystem compliant with the European Commission's legal and ethical framework. This ecosystem consists of 17 partners from 11 different countries and will be in place for 36 months. This project is expected to have a strong impact on research results and on health and social care by accelerating scientific discovery, improving the strength and quality of the scientific evidence provided, and providing innovative digital health services. One of its partners is the European Federation for Medical Informatics (EFMI) [13].

Learning health systems

A growing desire to focus on value-based healthcare delivery was the precursor for adopting what are now known as 'learning health systems (LHS)'. These LHS are knowledge discovery systems embedded in daily practice, designed to capture and process operational data to report on and learn from financial, operational, and clinical performance as this relates to individual patients, departments, disease types, or the organisation as a whole. Such systems are used to support the best diagnosis and care decisions as well as to avoid providing too much, too little, or inappropriate care based on the best available evidence.

There is a need to adopt a standard framework to enable comparisons to be made. Such frameworks essentially identify various levels of metadata, such as core values, resource availability, processes, and outcomes to assist with the identification of suitable data(value) sets, data collection/extraction requirements, suitable relationships, meaningful structuring, and organising of source data relative to the entity of interest, such as patient and disease type. This type of study is no different than what was traditionally referred to as 'work study', industrial engineering, operational research, or business analysis. Another way of describing such frameworks is what was originally named by Toyota as 'Lean Six Sigma', a quality management methodology.

Such methodologies have collectively resulted in numerous useful tools, including many types of flow charting and modelling tools, to assist with the identification and presentation of numerous concepts and their relationships, at various levels of detail, to discover and enable the use of new knowledge gained. Learning health systems are now being promoted by its own journal, published to provide content pertaining to a new transdisciplinary science incorporating behavioural, social, and organisational science; cognitive, information, and computer science; industrial and systems engineering; and other areas of expertise. The LHS Journal began in 2017, to publish computable knowledge publications, technical reports, research reports, experience reports, policy analysis, briefs, and commentaries.

The LHS concept was originally conceived by the National Academy of Medicine in 2007, followed by a series of workshops evolving this concept over time [11,14,15]. LHS need to have the ability to 'continuously, routinely, and efficiently study and improve themselves' [16]. Friedman et al. described five observable features and can exist as a cybersocial system within any digital health ecosystem:

- Patient characteristics and experiences securely available as data to learn from.
- The immediate availability of best practice knowledge supporting health-related decisions.
- Continuous improvement through ongoing study.
- Significant level of automation within a suitable socio-technical infrastructure and economies of scale.
- An integral component of prevailing organisational culture accepted by all stakeholders.

Interestingly, these data to knowledge, knowledge to performance, and performance to new data, cyclical model for automation, reflect the vision nursing informatics leaders used to first identify desirable nursing informatics competencies at a Stockholm Workshop in 1987 [17,18]. We now have the technologies to realise this vision. This newly described LHS concept now informs those developing a framework for the creation of LHS to support the delivery of value-based health care [19,20]. New knowledge discovered within one ecosystem may also need to be transferrable to other digital health ecosystems, as demonstrated for the sharing of knowledge regarding the optimal treatments to manage those hospitalised during the recent COVID-19 pandemic. Successful knowledge transfer for any domain requires the adoption of the same frameworks and/or metadata supported by a collaborative and cooperative culture to achieve the necessary alignments.

Large-scale digital health ecosystem-centric LHS require new approaches to system conception, design, implementation, orchestration, evolution, certification, and governance [16]. Highly valued LHS are trusted and valued by all stakeholders, economically sustainable and governable, adaptable, self-improving, stable, certifiable and responsive and capable of engendering virtuous cycles of improvement. These types of systems need to be made use of by clinical guidelines and protocol standard developers, governors, and distributors, as described in a previous chapter.

Representing concepts using entities, objects, and agent formalisms

We have previously described how data can be represented. This chapter's introduction described the many ways health language can be represented. Data reflect concepts that relate to ideas or things. Data can be collected by any number of different types of devices. Entities, objects, agents, and models are containers describing data attributes and relationships. These containers represent concepts about which information is stored. Container structures determine data retrievability, usability, and the ability to transfer knowledge from one application to another. For example, entity relationship and Unified Modelling Language (UML) diagrams are models used to describe interrelated things of interest within a specific knowledge domain. Such models are composed of types of entities and include specified relationships that can exist between them.

Objects represent software entities to denote one instance that is different from another. They can represent a data structure, a function, or a method of value referenced by an identifier. Object-oriented programming makes use of variable objects that can be a combination of variables, functions, and data structures referring to a particular instance of a class such as a table or column. Key objects in a system are less likely to change over time than the data in a system. Objects are selective about what services and data are accessible to its users.

Domain-specific object models encapsulate the key concepts in a domain and the relationships between those concepts as shown in reference models representing a mature domain, such as a health record representing a consensus view from a representative community (clinicians). Agents encapsulate both data and behaviour, to exhibit a degree of autonomy when requested to perform a service, whilst meeting specified constraints. Agents are proxies to represent, for example, patients or service providers [21]. Their use enables greater degrees of automation. Any one of these approaches needs to make use of data standards and draw upon rich ontological support at both the syntactic and semantic levels.

Knowledge is dynamic, and it requires interpretation, via appropriate procedures, of information that is structured in a manner that reflects context. For example, a clinical decision support system (CDS) may need to be designed to determine if a patient's recorded blood pressure reading (information) is consistent with any normal reading or indicative of a particular problem (knowledge). This is a 'publication use case'. A CDS system can also be used to develop information.

Knowledge acquisition is the result of a sophisticated reasoning activity based on accurate and factual data. For example, data from patient encounters, such as demographics, diagnosis, treatments, and outcomes, can be analysed to determine patterns where a patient's treatment plan can be modified, based on knowledge derived from the system in use. Data must be comparable and consistently represented in order to provide knowledge and insights to inform future care decisions. This requires the adoption of a data standard to consistently represent a given concept or data element (content), to enable the sharing of the commonly associated meaning attached to that data standard's syntactic (structure) and semantic data (meaning) normalisation.

Health data exchange challenges

Health data exchange (interoperability approaches) is first and foremost transport-oriented, making use of the lowest machine-level exchange protocol. For data transfer, this approach may be data-oriented, service-oriented, or agent-oriented where each approach represents a different level of computer processing that needs to make use of some form of data normalisation. The same normalisation is also necessary for big data approaches to queries. However, most clinical data in the United States and elsewhere, even within institutions, are heterogeneous, which presents a major challenge for harmonisation and automation efforts [11,22,23]. For example, multiple systems are in use within general practice in Australia that make use of any one of three different drug-related warning databases.

Syntactic and semantic data normalisation requires collected data to be represented in accordance with a standard terminology (data standard). This is a precursor for health data exchange (interoperability) and can be achieved by:

(1) Collecting normalised data at the source. This requires appropriate system design and architectures and an understanding of the desired normal (meaningful) form of representation.

(2) Transforming and mapping data on the back end of data systems. This method is commonly used as a component of a health data exchange schema, which can be effective and suit some use cases, but is commonly associated with ambiguous meanings and other transformation difficulties. The result is that systems dependent upon this method are not able to fully represent clinical data [24–26].

(3) Adopting a hybrid approach where new systems normalise their data at the source, whilst established systems implement standard normalisation protocols to transform data from legacy systems. This method is becoming more popular as this improves consistency.

Machine-readable (formalisms), rather than human-readable (free text), standard representation is necessary for large-scale inferencing and secondary data use in accordance with its data supply chain. For knowledge to be managed, it needs to be organised and structured in some way in terms of content, processes, decision-making logic, and natural relationships between these conceptual requirements. This also requires a determination regarding how these concepts are best represented in terms of governed terminologies, ontologies, and routinely identifiable value sets. Collectively, these knowledge structures need to represent the health domain's 'truths'.

Such 'truths' can only be defined by multidisciplinary teams of experts who need to be able to identify and make use of available evidence. Experts have the necessary knowledge, attained through education and wide-ranging experience, to develop reliable and safe computable truth representation. These 'truths' are also known to be in need of constant updates and sustainable management, to enable the continuing accommodation and use of new knowledge acquisitions, although some high-level 'truths' remain fairly constant. Domain knowledge represented by formalisms needs to have version control measures and be governed to ensure their quality and timely availability are assured.

One method able to provide a normalised form for the exchange of clinical data is the use of clinical element models (CEMs) or openEHR archetypes [27–33]. Their structures (formalisms) retain a consistent meaning for data when they are exchanged between heterogeneous computer systems or when clinical data are referenced in decision support logic or other modalities of secondary use. They retain context and data provenance and make use of a knowledge-based digital health platform.

These formalisms enable the recording of data which is specific to an individual patient in a specific context, for example a patient's position and body location alongside his or her blood pressure reading to provide clinical meaning [11,34]. Well-documented knowledge is best presented into a standard formalism such as the Glasgow Coma Scale, or the Boston Bowel Preparation Scale and the many other examples shown in the openEHR CKM or as represented by HL7 CIMI or other models. The development of these models requires the use of both tacit and explicit multidisciplinary knowledge in order to include all possible contextual use cases. The Clinical Knowledge Manager's (CKM's) repository and use are presented in more detail in the chapter on clinical knowledge modelling.

Bodies of knowledge, ontologies, and frameworks

Complete sets of concepts, terms, and activities that make up any professional domain are known as 'bodies of knowledge' or BOKs. These are likely to be of the highest quality when defined and governed by the relevant knowledge organisations such as learned societies, groups of experts in the field or professional organisations, to relate to any profession's scope of practice or educational curricula designed to develop new generations of professionals or trade experts. These design and governance activities do require specific skills.

A formally documented body of knowledge is one that permits its use for purposes such as the development and accreditation of academically sound educational courses and programmes, certification of specialists or for professional licensing. It sets the standard for professional practice, endorsement, and accreditation criteria. It promotes the advancement of both the theory and practice for those who wish to specialise in any aspect contained within this domain. The body of knowledge needs to be underpinned by the scientific foundations for the domain [35]. Its objectives are to:

- Promote a consistent view of the digital health body of knowledge worldwide, including the core (what needs to be known by the health workforce as well as health informaticians and health IT specialists).
- Specify its scope and clarify its place with respect to other related disciplines and bodies of knowledge.
- Be publicly accessible.
- Enable the identification of role-specific competencies from which position descriptions can be developed and associated with essential skills, knowledge, and attributes identified, to suit the many different types of healthcare organisation.
- Provide a foundation for health informatics course and curriculum design, development, accreditation, and professional development programme endorsements.

Any BOK plays an important role in defining the knowledge required for any trade or profession or clinical specialty. There are many types of explicit knowledge, including knowledge used as a foundation for its application. For example, knowledge is applied for innovative developments, constructions, designs, new processes, new work methods, or new treatment options including practice guidelines. From a digital health perspective, every relevant health or clinical knowledge domain representing these types of knowledge is ideally represented in a form that enables their body of knowledge to be machine-readable when its application requires digital support. That is, the type of knowledge that needs to be usable for inclusion in decision support or artificial intelligence systems, automated machinery, or robotics must be machine-readable. These types of knowledge are embedded in any occupational or role-focused BOK and can be represented as ontology.

There are numerous BOK presentation formats, and some are easier to navigate than others. Their potential use determines the most desirable format to be used for presentation purposes. A well-structured framework representing a BOK makes these most usable. This is well-demonstrated by the global use and recognition of the Skills Framework for the Information Age (SFIA) Framework. The SFIA Foundation has listed the many bodies of knowledge of relevance to professional disciplines which are within the scope of skills and competencies included in their framework [36]. All of these are of relevance to a digital world, including health, although none of these relate to any of the health professions as such. Various components of these collective bodies of knowledge do apply to the health informatics or digital health professional disciplines. The SFIA Framework was designed to be applicable to any industry; hence, its definitions are written generically.

The SFIA list provides useful examples of how bodies of knowledge can be represented. Many of these also describe how they are used, by whom, and for what purpose. The list is organised in terms of high-level topics included in SFIA's Framework, which are as follows:

- Strategy and architecture
- Change and transformation
- Development and implementation
- Delivery and operation
- Skills and quality
- Relationships and engagement

The SFIA Framework has been designed for ease of use to suit multiple practical purposes including human resource management and education. The SFIA Foundation has many contributors who collectively maintain its relevance by publishing updates. It is now in its 8th edition. In the absence of such a well-structured framework, it is very difficult for users to make use of this knowledge in a consistent manner.

One of the BOKs listed in SFIA's strategy and architecture section was recently developed by the Open Group, a global consortium that enables the achievement of business objectives through technology standards. Their very comprehensive BOK was developed in accordance with its mission, which is to drive the creation of Boundaryless Information Flow, to suit any digital practitioner in any industry [37]. This philosophy should also be adopted to drive the establishment of successful digital health ecosystems. This BOK is considered to be of great relevance to the readers of this text.

The Open Group BOK is divided into four stages any digital practitioner may find themselves in, starting with 'individual/founder', followed by 'team', then 'team of teams', to ultimately arrive at the 'enduring enterprise' stage. Everyone needs to acquire sufficient digital knowledge and skills to support their chosen lifestyle; others need to further develop to enable them to apply their digital knowledge and skills within any work environment at all levels, from the most junior to the most senior executive level. This staging framework reflects any digital practitioner's knowledge acquisition and user journey, which needs to be replicated and used for the development of educational curricula designed to best prepare digital practitioners to suit any industry. This large and comprehensive document made use of many resources, all listed as references.

Another completely different example is the American Health Information Management Association's 2019 version of their body of knowledge. This is presented relative to the following topic areas which collectively make up their BOK [38]:

- Coding, classification, and reimbursement
- Confidentiality, privacy, and security
- Information governance and standards
- Health information technologies and processes
- Health informatics
- Healthcare leadership and innovation
- Consumer engagement and personal health information

Each of these topics contains numerous subtopics for which multiple supportive sources are listed including publications representing AHIMA-owned content complemented by links to public material. Their BOK encompasses the theory and practice of health information management. All content is indexed by metadata fields consisting of subject, author, title, work setting, specialty area, document type, publication type, and source.

This presentation is comprehensive but has not focused on presenting their foundational most up-to-date theory and practical knowledge in a usable framework within which to present their body of knowledge, and it is up to the individual to analyse the many resources in a manner of relevance to the person seeking this knowledge for their intended use. Many of the topics listed are equally applicable to other health professional BOKs. This demonstrates the difficulties encountered to define each profession's core BOK and their boundaries or scope of practice which tend to change over time. There are many overlaps.

The digital health (health informatics) BOK is huge as this integrates all core knowledge associated with both the ICT and health industries. The International Medical Informatics Association (IMIA) has undertaken such a development task that began with a 'think tank' of experts and resulted in the identification of 14 distinct topics representing a cognitive map of the health informatics discipline. This was followed up by the use of an extensive data extraction method that identified the most commonly used keywords published in the health informatics literature. This was followed by a consensus method to produce a final framework and knowledge base [39]. The resulting spreadsheet shows 14 themes, each with numerous subthemes, was endorsed by IMIA, and complements its educational guidelines [40]. These themes are follows:

(1) Computer Science for Health Informatics (health IT),
(2) Health and social care processes,
(3) Health (care) records,
(4) Health and social care industry,
(5) Health informatics standards,

(6) Knowledge domains and knowledge discovery,
(7) Legal and ethical,
(8) People in organisations,
(9) Politics and policy,
(10) Technologies for health,
(11) Terminology, classification, and grouping,
(12) Uses of clinical information,
(13) Using informatics to support clinical healthcare governance,
(14) Computer systems applications in health (toolkits).

Each of these has subcategories ranging in number from 9 to 28. From a knowledge governance perspective, it is important to separate core from additionally relevant knowledge that may need to be governed by a different professional group. Every health informatician needs to have some foundational knowledge about all of the above but will have specialised in any one or more of many possible subcategory knowledge domains.

Knowledge structures and frameworks

The examples above highlight the importance of the structural design adopted in terms of usability for any body of knowledge. Such designs are guided by the underpinning philosophy and purpose of projected use. These need to be explained in the introduction of any BOK as they determine the high-level categories or metadata used to decompose and organise the domain's knowledge. The structure or framework, its presentation, and usability determine the BOK's value. Developing and maintaining a BOK is time-consuming and resource-intensive, and it requires many contributions and a governance strategy. It is of great importance for knowledge transfer over time, from one generation to another.

Every knowledge domain can have multiple subdomains. This is certainly the case for the 'digital health' domain. Digital health as a term describes this knowledge domain at the highest level. It can be decomposed in numerous ways. The complete BOK is best decomposed into multiple knowledge domains as it requires the ability to process all knowledge at any level of granularity, including highly granular atomic levels consisting of a wide variety of data used to represent concepts and numerous relationships creating new meaningful concepts. Many of these knowledge domains are represented as 'terminologies' such as SNOMED CT or LOINC, described elsewhere in this text.

The SFIA Framework and the Open Group BOK focus on the digital aspects, and it is also necessary to be able to identify a structural framework that has a health focus. Given the depth and breadth of the digital health knowledge domain, such a framework needs to reflect all relevant knowledge domains based on both digital and health foci but presented in an integrated manner. This in essence is the digital health ecosystem knowledge domain that needs to be reflected by the architecture of any national or organisational digital health ecosystem. Its presentation needs to assist individuals to gain an understanding of how the many components of digital, health, and IT management knowledge domains fit together in this new digital world.

Semantic knowledge graphs—Data asset linkages

The data, information, knowledge, and wisdom continuum referred to previously has highlighted the critical importance of accurate source data. This is especially so if and when such data are used for any artificial intelligence (AI) or automated applications which make use of knowledge (intelligence) that may be represented or acquired by various means. Designing such applications requires an organisation's or discipline knowledge domain to be represented as artefacts understood by both humans and machines. Such representations include available knowledge assets, content, and data pertaining to people, places, documents, images, data in relational databases, etc., and how these are related to one another. This is achieved by the development of a metadata model, or 'ontology' that defines classes for the things, properties for the things, and relationships between the things also referred to as a knowledge graph.

Such collections of interlinked descriptions of entities, objects, events, or concepts representing the real world provide a framework for data integration, unification, analytics, and sharing. This is not an easy but very necessary task. Such a knowledge graph provides the foundations from which to establish and implement relevant policies and procedures required to link up a critical mass of any entity's knowledge assets. These principles apply to any digital health ecosystem. Gartner [41] explained that:

> The rising role of content and context for delivering insights with AI technologies, as well as recent knowledge graph offerings for AI applications, have pulled knowledge graphs to the surface.

Knowledge graphs provide descriptions of formal semantics that allows both people and computers to process these in an efficient and unambiguous manner. Such graphs or models represent how entity descriptions contribute to one another, and they form a network that provides context supporting interpretation. The recommended framework for data integration, unification, linking, and reuse is the Resource Description Framework (RDF), a standard developed by W3C to embed semantics [42].

Ontologies

Making a body of knowledge machine-readable requires its domain to be analysed and modelled in a manner that shows key knowledge concepts and their relationships. Such models formalise the meaning of that knowledge. Formalisms with a sufficient degree of expressivity to convey meaning represent a highly desirable requirement for machine processing that needs to make use of knowledge pertaining to any knowledge domain. Ontologies are only able to represent what is needed when they are applied to a known information model.

Similarly, the 'semantic web' makes use of a language for encoding knowledge on web pages to make this machine-readable for electronic agents searching for information. Knowledge concepts can be described or labelled with a term and/or given a code to reflect the knowledge domain's structure. An ontologically described knowledge domain needs to reflect mutual exclusivity at all levels; that is, every concept has a unique place within the structure, and there can be no overlaps. This is particularly important when making use of clinical knowledge. There can be no ambiguity or inconsistencies in terms of meaning.

Many existing bodies of knowledge are yet to be structured in a manner that makes them machine-readable and applicable for use in the digital age.

Ontologies need to be able to describe its knowledge domain at the lowest atomic level. Such structures are then viewed as vocabularies or terminologies that reflect a specific knowledge domain such as the globally used systematised nomenclature of medicine with clinical terms, known as SNOMED CT that is maintained and governed by SNOMED International [43]. This very extensive terminology can be used to define specific specialised clinical knowledge domains; however, as a consequence of doing so, concepts may be

- identified that are not yet included in SNOMED CT, and these then need to be considered for inclusion by the governing body, or
- be more appropriately represented in the information model, e.g. whether a condition is present, absent, family, or personal history.

Professional bodies may have accepted the responsibility for governing specialised clinical knowledge used for specialist training, assessment, and certification. They may also have adopted the responsibility for the development of clinical guidelines, treatment options, and protocols. However, few such professional bodies have kept up with digital health technology advances or recognised the need for their explicit knowledge to be made available in a machine-readable format. Many terminologies in use were developed prior to the digital age and are not necessarily ontologically sound, thereby creating inconsistencies between uses. Computers are binary, it is either right or wrong, and there is no in-between although there may be many possibilities.

Problem identification and reasoning ontologies

Digital health requires an ability to represent any kind of information and knowledge we possess about the world around us, in a form that a computer system can process in a manner that suits multiple purposes. Such computer processing includes reasoning and problem-solving. Knowledge can also be processed to manufacture objects or automate any type of operational work or administrative process. Knowledge representation and reasoning are fundamental requirements for artificial intelligence (AI). Knowledge-based or intelligent systems must be able to apply its knowledge to new situations, acquire knowledge from its interactions with the world, and infer new knowledge from its existing knowledge, as required.

Intelligent systems need to represent knowledge and understand its meaning, in order to make good use of that knowledge. Designers of intelligent (expert) systems may need to make use of human expert knowledge that is captured in the form of, for example, reasoning used to make decisions or to undertake real-world problem-solving. This requires the capture of strategies used by an expert who has solved similar problems in the past. Such knowledge is based on any number of experts' past experiential learning. Analysing existing knowledge is about capturing human thought processes to enable these to be represented in a way that facilitates computer processing.

We are constantly being confronted with new problems for which existing strategies are not available. In such instances, the problem-solving process requires the application of knowledge about relationships between entities and objects, and the properties associated with those objects. It may then be possible to apply that knowledge to define it as some form of logic. Knowledge reasoning requires us to make logical conclusions or predictions. Such outcomes can be the result of rule- or case-based inferencing, but first, we need to analyse existing knowledge within the scope of its relevant knowledge domain to be able to structure this in a meaningful ontological framework or model.

For example, the system security knowledge domain applicable to the connectivity of many stand-alone Internet of things (IoT) systems through the Internet was analysed to develop an ontology-based cybersecurity framework using knowledge reasoning for IoT [44], with the aim to be able to improve system security. This work is highly applicable to digital health, as individuals are increasingly making use of numerous devices that need to be able to connect to personal health records. Another example is based on the use of a clinical reasoning ontology [45].

The use of semantic technologies making use of explicit statements representing the meaning of data and data relationships provides other opportunities to bridge and interconnect data, content, and processes. Semantic technologies are able to be adaptable for the representation and reasoning over shared knowledge and ontologies, when used as a formal explicit specification of a shared conceptualisation of a domain of interest [46]. The requirement to make use of domain ontologies is included in the latest ISO standard [47] in response to the recognition that open and intelligent digital health ecosystems, best able to support personalised, patient-centred health systems, have a need to adopt a system-theoretical, ontology-based, policy-driven reference architecture model [48].

Decision theories and artificial intelligence

Individuals make decisions that indicate choices made. Choices are made based on knowledge, values, beliefs, desires, risk assessment, and attitudes. Decisions may be strategic, tactical, or operational, and they may need to include the identification of alternatives or the weighing of evidence. Every one of these types of decisions may be rational or intuitive or follow no specific process. Some decisions are reversible; others are not. Every decision consists of an act, an event or process, an outcome, and a pay-off (value). Acts may be based on available information or some other stimulus including motivation. Events or processes may be analytical, directive, conceptual, or behavioural. The reasoning or thinking processes adopted can be categorised in terms of decision theories which are based on how these many attributes cohere together. Reasoning processes may be supported by decision-making techniques or tools such as the use of a Strength, Weakness, Opportunity, and Threat (SWOT) diagram or a decision matrix. A variety of methodologies are available to assist with analysing decisions.

Decision theories may be defined as interdisciplinary approaches to arrive at the most advantageous decision within an uncertain environment pertaining to a specific domain. The normative decision theory models the most ideal decision for a given situation to result in the best possible outcome. The descriptive decision theory models a definitive result; for example, if act Y is taken, then X will be the outcome. Decision theories integrate the psychology, statistics, philosophy, and mathematics disciplines. Many are uniquely applicable to the domain of statistics and applied exclusively to data analytics and visualisations; others are uniquely applicable to psychology or any other disciplinary domain.

Knowledge-based decision support is a form of artificial intelligence (AI). AI is based on the results of studying human thought processes and then representing these processes via machines, including computers and robots. All of the above complexities associated with making decisions relate to intelligence which include the ability to [49]:

- learn and understand based on experience;
- make sense out of ambiguous or contradictory messages;
- respond quickly and successfully to any new situation;
- use reasoning to solve problems and direct conduct/activity;
- deal with perplexing situations;
- understand and infer rationally;
- apply knowledge (based on environmental observations, perceptions, and prior knowledge) to manipulate your environment;
- think and reason;
- recognise the relative importance of different elements within a situation.

Intelligence also includes visual perception, speech recognition, and language translation, functions now being undertaken by machines, although humans are far superior in interpreting their natural language in common use, in part based on concurrent body language. Conscious, emotional, and ethical responses remain unique components of human intelligence. Human reasoning is also influenced by their perceptions of space, time, and physical interactions within any real-world environment. Computers are increasingly capable of replicating these perceptual influencers.

Artificial intelligence represents a very large knowledge domain making use of psychology, philosophy, linguistics, electrical engineering, computer science, and management science knowledge domains, as its foundation. The integration of various components of these multiple disciplines has resulted in numerous new fields of study representing new knowledge domains, all of which come under the umbrella of AI. Many of these are applicable to digital health, including genetic algorithms, natural language processing, speech understanding, biometrics, robotics, deep learning, machine learning, expert systems, intelligent tutoring, game playing, and virtual reality.

Machine learning and other AI techniques are increasingly being applied to the unprecedented amounts of clinically relevant data now available for clinical and research use, including predictive analytics, precision medicine, and differential diagnosis. The use of ontologies has become essential to enable data extraction, coding, and the analysis of clinical information by machine learning algorithms [50]. Recent literature continues to demonstrate the added value of adopting ontology-based data integration approaches [51]. Artificial intelligence in this digital era requires precise descriptions and machine-readable representations of human intelligence.

AI also makes use of advanced analytics using big data. This may be problematic as its source data may not be suitably representative of the world's reality. As a consequence, various types of bias may become entrenched, or the use of some learning methods may result in dramatically different outcome accuracies. In health, it is critical to ensure that all source (little) data are accurate, prior to undertaking data analytics of any kind.

One form of representing human intelligence is via the use of algorithms, consisting of sets of unambiguous computer-executable instructions. Given the complexities associated with decision-making, it follows that algorithms vary greatly in terms of their levels of complexity. Many of the decision-making components are nonalgorithmic, in that they do not necessarily follow a logical step-by-step approach based on well-defined start and end points. A large component of AI concerns its ability to process symbolic, nonalgorithmic methods of problem-solving and its use of search mechanisms. One needs to question if, for example, self-driving cars are able to accurately interpret their physical location, or assess a pedestrian's intention as accurately as humans can. How does one assess any AI system's trustworthiness given the many complexities?

Ethical data use

Health data fragmentation, the many known inconsistencies, and poor use of data standards continue to severely compromise many AI developments due to its dependence on data extraction and processing methods using big data. Big data is the result of data sharing and data linkages, which are dependent upon compliance with regulatory and legal requirements. These determine applicable consent, privacy, and security frameworks, although these privacy regulations do not apply to health-related data generated and used outside clinical settings, including de-identified data and when used in social media. Data need to be managed throughout their life cycle to ensure its future potential use. There needs to be a clear differentiation between the need to share identifiable data to ensure continuity of care, and the sharing of de-identified data. Both routinely collected health data and the use of AI in health care need to be curated.

There is an urgent need to make better use of current systems and data to improve science and to innovate with AI techniques for electronic decision support and data analytics. This was stressed at the 2018 WHO-UNICEF Global Primary Health Care Conference in Astana, Kazakhstan. There is an urgent need to make better use of the data stored in many public health data sets, disease, death, and health services registries, genomic databases, and biobanks. This is considered an essential foundation from which to achieve Universal Health Coverage, better and fairer personalised, patient-centred health care for individuals and populations as a whole [52].

All health data need to be ethically collected, processed, and used throughout its supply chain. The International Medical Informatics Association has developed a code of ethics listing six fundamental ethical principles, seven general ethical principles of informatics, rules of ethical conduct for health informatics professional (HIP), and HIP duties towards healthcare professionals, institutions, employers, agencies, and society [53]. This document also describes duties that apply to HIPs themselves and to the standing of their profession. IMIA's Primary Care Informatics working group has developed the following five recommendations concerning the ethical use of electronic health record data and AI [52]:

(1) Ensure consent or formal processes to govern access and sharing throughout the data life cycle;

(2) Implement sustainable data creation/collection to build and maintain trust as well as permission;

(3) Pay attention to extract-transform-load processes as they may have risks that go unrecognised;
(4) Integrate data governance and data quality management to support clinical practice in integrated care delivery and systems;
(5) Recognise that additional processes are needed to address the ethical issues arising from AI in primary care.

Tools for knowledge processing

A variety of tools have been developed over the years to support all functions associated with knowledge discovery, management, and dissemination. Such 'knowware' and knowledge management platforms may be incorporated as subsets of content management software within, or connected with, any type of enterprise system used for internal and external knowledge sharing. These include learning management systems or customer/supplier relationship systems and knowledge bases. These tools or software applications organise information so that any of its users can quickly access and distribute knowledge content. Collectively, these tools need to provide the capacity to identify, capture, develop, share, and support effective use of available knowledge by supporting the following functions:

(1) Store and provide access to knowledge repositories—web pages, databases/repositories, information portals, and workflow management systems
(2) Search mechanisms to enable knowledge retrieval—browsers, search engines, agents or filters and computer languages, e.g. SQL and AQL
(3) Communicate knowledge and capture knowledge contained within any type of communication tool—email, encrypted messages, Wi-fi or Bluetooth document transfer mechanisms, videoconferencing, virtual reality, chat rooms, record online meetings, helpdesk applications, and call centres.
(4) Manage the transfer of knowledge amongst teams or groups of collaborators.
(5) Knowledge visualisation tools, statistical applications, and data analytical tools.
(6) Decision support tools that enable the discovery of patterns of knowledge from data, including reporting software.
(7) Big data storage and tools to enable fast processing of high variety and volumes of structured and unstructured data.

Any digital health ecosystem needs to explore which knowledge management tools are best able to interoperate to collectively provide the necessary support functions in a timely and accurate manner. Such tools need to be user-friendly to ensure widespread engagement and use to optimise safe care delivery and overall health service performance.

The adoption of evidence-based practice requires practitioners to be able to find and make use of available evidence. Electronic health records collectively contain all the necessary source data. The challenge is to store and organise these data in a manner that best facilitates data sharing and linkage in a way that is ethical, complies with privacy and security requirements, and meets the need of every potential user throughout its data supply chain as discussed in a previous chapter. There needs to be an agreed framework detailing the conditions that must be present for identifiable data sharing and de-identified data transfers and use. Once in place, it will be possible to optimise clinical knowledge management technologies that benefit all of us.

Many governments, private health service providers, professional societies, health research and educational organisations have established knowledge management toolkits or frameworks to inspire interest and use of available knowledge. Such frameworks need to focus on their specific essential and desirable knowledge management needs and then choose and implement the best suited set of tools. Ideally, we develop and adopt a global knowledge management framework to enable every health knowledge discipline to collaborate with others to maximise the new learnings to be realised from multidisciplinary knowledge sharing and collaboration.

Use of knowledge in applications

Healthcare organisations are increasingly looking to implement knowledge management applications and stimulate cultural change and innovation. Various jurisdictions have taken steps to make new knowledge accessible from the workplace. This began with data warehouses which are gradually being transformed to various cloud systems. In addition, applications supporting effective and timely communication between staff and contractors need to be considered first and foremost.

Many individual patients need to receive care from multiple providers in numerous locations. In the absence of secure communication systems, staff will make use of any application available to them, including WhatsApp or equivalent applications on their personal smartphones. Many new health professional graduates, familiar with today's personal digital support, find the technologies they need to use within the health system antiquated. As a consequence, they are innovative; the use of shadow systems is rife. That is, where official and enterprise approved systems and procedures fail to meet operational needs, then staff will develop their own shadow system to fill the gap.

The progression of knowledge management application use, relative to functions, tends to follow a well-defined pathway starting from fragmented point-of-care solutions. Data warehouses, the use of standard terminologies, and patient registries have been in use by most healthcare facilities and nations for many years. The adoption of enterprise systems has enabled automated internal reporting for many healthcare facilities. External reporting still requires a lot of manual data extraction and data entry, such as clinical coding of discharge summaries used for billing and funding purposes.

Improvements in this external reporting area alone would save billions of healthcare dollars and improve statistical accuracy. Widespread adoption of evidence-based practice, making use of the timely availability of such evidence at points of care, has the potential to significantly reduce waste, duplications, time delays, and unnecessary care variabilities, in addition to improving patient outcomes. Such evidence may be incorporated in any number of decision support systems or is used to inform AI applications [10].

National and global adoption of population health and clinical data standards has the potential to improve research, in terms of accurate representative data availability and volume and early adoption of research findings, minimise clinical risks, improve predictive and prescriptive analytics, and support personalised, patient-centred care.

Automated external reporting/sharing has the potential to greatly improve learning from others about new diseases, new treatment and care options, timely effective global population health, and pandemic management. The use of analytics unlocks data and identifies areas of opportunity and needs to be able to scale using big data over time. This latter functionality requires the adoption of an agreed open platform and health data exchange framework. A current limitation is the fact that every country collects these data differently, making meaningful comparisons problematic and new learnings uncertain.

Knowledge governance, validation, and certification

Every healthcare facility needs to manage and make use of knowledge from numerous knowledge domains. Every domain deserves its own governance and maintenance structure, and many of these activities need to be external to individual healthcare organisations and managed by those with the necessary authority to govern any given knowledge domain. Governance is about steering or directing any knowledge concept or entity in the desired direction.

Most current health data governance practices are haphazard. Many healthcare organisations do not have a formal data governance or data asset management strategy. Those with governing responsibilities must have experience in, and a thorough familiarity with the domain to be governed, technical and corporate governance knowledge, as well as negotiation and interpersonal skills. They need to be able to function well as team members and with key stakeholders. Health data governance roles require its incumbents to have health data science and health informatics knowledge and skills.

Authority to govern is to a large extent determined by knowledge ownership. Current and new knowledge is generated in vast quantities by many within the health industry. Much of this knowledge is the result of contributions by many, making it difficult to determine ownership. Governance processes associated with digitising knowledge are not straightforward as a result [54]. Many relevant knowledge resources are held in 'common'. That is, they are shared and not privately owned, yet such knowledge still needs to be governed to ensure continuing relevance and currency. Some clinical software, including devices, making use of AI or any decision support method, may be privately owned but will need to be regulated to ensure patient safety.

With the advent of the mass adoption of a collaborative and sharing culture, legal changes are needed as IP ownership, governance processes, and accountability have become fuzzy concepts in many instances. From a governance perspective, there is a need to be able to clearly differentiate between open, free, and proprietary software and knowledge resources used within the health industry. Ownership, licensing arrangements, and accountabilities differ between these and influence who are the greatest beneficiaries of their use. Knowledge acquisition is costly and time-consuming, and such investments need to be appropriately acknowledged and rewarded.

There is a gap between issues associated with copyright, IP ownership, and IP governance. In some instances, permission is required from the owner to make use of such resources (IP). In other instances, IP resources may have an identified need to govern its use in the interest of public safety and/or be made available in the public domain enabling anyone to make use of this knowledge resource freely.

This gap was filled by the establishment of a 'Creative Commons (CC)' in 2001. This legal framework has resulted in the public availability of creative common copyright licenses that enable free access for certain uses. There are six types of licenses which stipulate the permissions and conditions applicable to the use of a creative copyrighted work. Only one of these licenses permits commercial use [55]. Their work is continuing towards the development of a global infrastructure that enables and optimises knowledge sharing. Its sustainability is dependent upon its adoption of the most appropriate economic theory applicable to the management of common goods.

IP, ownership, and copyright

We are witnessing a major global change regarding resource ownership and use. This requires a reliable and democratic type of authority as that adopted by the openEHR Foundation. Its clinical models represent multidisciplinary, cross-sectoral knowledge, transparently contributed by its global community based on acknowledged evidence. These critical knowledge resources need to be governed in the interest of public safety [56]. This work continues to be self-governed by its community who have adopted a governance framework.

The need to maintain patient safety and its sustainability is what Ingram and Arikan identified as a 'wicked problem'. They identified the following six features of an open-source ecosystem contributing most towards a progressive solution.

(1) The promotion of effective and efficient innovation within software development communities. Clinical pioneers have traditionally had to build whole local infrastructures, and unnecessary efforts to 'reinvent the wheel' are still common.
(2) The pooling of development and maintenance costs for essential infrastructure.
(3) The enablement of the research interface. Discipline grows through sharing, testing, and reviewing methods.
(4) Increased awareness and understanding of the inner workings of systems to improve procurement.
(5) The integration of systems and services and continued efforts to combat fragmentation.
(6) The enablement of more patient-focused services that can be managed and regulated more effectively and transparently. The goal is to create a more trusting environment of professional practice, patient participation, and public engagement.

Validation

In addition to an ability to make use of available knowledge, we also have an obligation to ensure that this knowledge represents the best available 'truth', is accurate, and is representative. Whatever its format, it needs to authentically represent the best available knowledge resources. Validation refers to an assessment process adopted to verify that the representation of a knowledge concept may be regarded as evidence demonstrating the desired degree of accuracy.

Reliable sources of evidence can be provided in multiple forms, which need to be critically evaluated for trustworthiness before being accepted as evidence. Evidence should be the result of scientific studies. Many scientific publications now include a section detailing what is known before presenting what has been newly learned from the study conducted. Evidence consists of a combination of findings from empirical studies, data, facts, and figures gathered from any organisation or study domain, professional expert experiences and judgements as well as values and concerns of stakeholders.

The design of all health information systems and applications has made use of available knowledge. Such knowledge is usually collected by undertaking a business or workflow or communication flow analysis within any healthcare entity or ecosystem. Systems and applications intended for a global market need to consider any number of cultural and geographic variations to ensure their system can accommodate these.

Few systems are designed and structured based on evidence from comprehensive empirical studies, and one exception is the openEHR specifications [57,58]. Its first empirical studies were undertaken in Europe and funded by the European Commission. The Good Electronic Health Record (GEHR) Project began in 1992 and resulted in the first version of specifications for an electronic healthcare record architecture for Europe [59]. The openEHR Foundation and its global community have adopted rigour, engagement, and trust as its foundational principles guiding this work.

The health informatics discipline has evolved since the early computing days. Its knowledge domain is well-represented by numerous published and peer-reviewed evaluation studies of health informatics applications. Experiences from editors, reviewers, authors of systematic reviews, and readers of this scientific literature were studied. This has resulted in the publication of a list or set of STARE-HI principles relevant for properly describing health informatics evaluations [60]. This work was validated and enriched by comments from many HI experts and was endorsed by the International Medical Informatics Association (IMIA) in 2008. Compliance with these guidelines ensures that readers are able to assess and validate the quality of the evidence being presented. All technologies used to support the delivery of health care need to be proven to be safe, effective, and most appropriate compared with alternative solutions.

Certification

Many health professionals need to be registered to gain permission to practice or to be employed by a healthcare provider. Such registration is based on the successful completion of a validated nationally recognised and professionally endorsed educational and final assessment programme.

Health IT professionals may have successfully completed any one of many IT or health IT programmes, but official national registration is not a requirement to be employed in such roles. Professional endorsement in the form of a certification is made available for some by their professional society or by various commercial solution providers, such as Citrix or Microsoft or Oracle knowledge management. These types of organisations offer certification examinations designed to validate relevant, real-world, and in-demand skills required to successfully design, implement, and manage IT solutions consisting of their technologies. Certifications should relate to their domain's body of knowledge.

Use of global intellectual capital in this digital era

Global intellectual capital determines survival, renewal, and growth by focusing on what is important and by justifying investments in strategic knowledge management initiatives. The digital era has a strong focus on knowledge workers, representing our collective intellectual capital that needs to be recognised as a primary asset. Successful knowledge management outcomes need to demonstrate:

- Knowledge content quality, ease of retrieval, and sharing;
- Recognition and respect for intellectual ownership, no undesirable consequences, and tangible benefits;
- A reduction in the loss of intellectual capital within individual healthcare delivery organisations, within professional disciplines, governing bodies and jurisdictions resulting in economies of scale by:
  - reducing redundancy of knowledge-based activities;
  - increasing productivity, improving workflows, and quality performance;
  - having made knowledge available more quickly and easily to support multiple use cases throughout the data supply chain;
- perceived usefulness and increased user satisfaction by enabling greater personal and professional disciplinary development and empowerment;
- the generation of valuable innovations;
- having met strategic global, national, regional, and/or professional public health directions;
- the realisation of value resulting from the establishment of:
  - global, national, technical, and organisational supportive infrastructures based on a standard, flexible knowledge structure;
  - a collaborative and multidisciplinary knowledge-friendly culture of sharing;
  - a clear agreed purpose;
  - changed motivational practices;
  - multiple channels for knowledge transfer;
  - strong high-level leadership [61].

Privately owned and/or some open-source healthcare software using clinical and healthcare knowledge lacks transparency and is frequently ungoverned. This makes it extremely difficult to, for example, identify the cause of errors made, or of the numerous adverse events experienced by many recipients of healthcare services, some of which may be due to a system error. The quality of proprietary software suffers greatly from the secrecy of its internal workings. Medical software is not regulated. This places the burden of safe and effective use on its users, who as a general rule pay little attention to the software fundamentals [62]. This may be referred to as the 'black box' phenomenon. This is a major concern associated with any decision support or AI system, or embedded algorithms, as health professionals continue to have a legal responsibility and are accountable for the use of advice received and their actions. Clinical software is rapidly becoming a patient safety issue in clinical practice, yet this is not subject to regulation.

A collaborative sharing culture, authoritative knowledge governance, and appropriate timely use of up-to-date readily available knowledge have the potential to make significant improvements, not only in patient outcomes and population health, but also in more rapid new knowledge acquisition, knowledge availability and use, resulting in major cost reductions.

Widespread global adoption of a knowledge-based platform requires software developers to think differently. Traditional database schema no longer matters as application data will then be committed and accessed via relevant service APIs, rather than any direct access to a database. Queries are then model based rather than being based on concrete database schemas. This utopic global environment can only be achieved with support from global entities, governments, industry and health professionals driving the necessary transitional initiatives required, to secure a viable open-source framework and supportive community.

References

[1] AS5021, The Language of Health Concept Representation, Standards Australia, IT/14, Sydney, 2005.
[2] ISO-21090, Health Informatics—Harmonized Data Types for Information Interchange, ISO, 2011.
[3] C.G. Chute, Clinical classification and terminology: some history and current observations, J. Am. Med. Inform. Assoc. 7 (2000) 298–303.
[4] J.J. Cimino, From data to knowledge through concept-oriented terminologies: experience with the medical entities dictionary, J. Am. Med. Inform. Assoc. 7 (2000) 288–297.
[5] C.W. Delaney, C. Weaver, 2018 nursing knowledge big data science initiative, Comput. Inform. Nurs. 36 (2018) 473–474.
[6] E. Coiera, Guide to Health Informatics, CRC Press, 2015.
[7] J. Wyatt, P. Scott, Computable knowledge is the enemy of disease, BMJ Health Care Inform. 27 (2020), e100200.
[8] MayoClinic, D. Robert, E. Patricia, Kern Center for the Science of Health Care Delivery, Informatics and Knowledge Management, 2021, Available from: https://www.mayo.edu/research/centers-programs/robert-d-patricia-e-kern-center-science-health-care-delivery/research-activities/informatics-knowledge-management.
[9] J.L. Shellum, R.A. Nishimura, D.S. Milliner, C.M. Harper Jr., J.H. Noseworthy, Knowledge management in the era of digital medicine: a programmatic approach to optimize patient care in an academic medical center, Learn. Health Syst. 1 (2017), e10022.
[10] L. Shahmoradi, R. Safadari, W. Jimma, Knowledge management implementation and the tools utilized in healthcare for evidence-based decision making: a systematic review, Ethiop. J. Health Sci. 27 (2017) 541–558.
[11] IOM, Digital Data Improvement Priorities for Continuous Learning in Health and Health Care: Workshop Summary, The National Academies press, Washington, DC, 2013.
[12] EC_Cordis, Improving Health Research in EU Through FAIR Data, European Commission, 2021. Available from: https://cordis.europa.eu/project/id/824666.
[13] EFMI, European Federation for Medical Informatics, 2021, Available from: https://efmi.org/.
[14] IOM, The Learning Healthcare System: Workshop Summary, National Academies Press, 2007.
[15] IOM, C. Grossman, B. Powers, J. McGinnis, Digital Infrastructure for the Learning Health System: The Foundation for Continuous Improvement in Health and Health Care: Workshop Series, Institute of Medicine, Washington DC, 2011.
[16] C.P. Friedman, N.J. Allee, B.C. Delaney, A.J. Flynn, J.C. Silverstein, K. Sullivan, K.A. Young, The science of learning health systems: foundations for a new journal, Learn. Health Syst. 1 (2016).
[17] H. Peterson, U.E. Gerdin-Jelger, Preparing Nurses for Using Information Systems: Recommended Informatics Competencies, National League for Nursing, New York, 1988.

[18] E.J. Hovenga, Milestones of the IMIA-NI history and future directions, Stud. Health Technol. Inform. 146 (2009) 3–10.

[19] M. Menear, M.-A. Blanchette, O. Demers-Payette, D. Roy, A framework for value-creating learning health systems, Health Res. Policy Syst. 17 (2019).

[20] T. Noseworthy, Innovation in the Canadian health system, Healthc. Manage. Forum 34 (2021) 5–8.

[21] D.H. Jarvis, J.H. Jarvis, Interoperability, Stud. Health Technol. Inform. 151 (2010) 45–55.

[22] C.G. Chute, S.A. Beck, T.B. Fisk, D.N. Mohr, The Enterprise data Trust at Mayo Clinic: a semantically integrated warehouse of biomedical data, J. Am. Med. Inform. Assoc. 17 (2010) 131–135.

[23] B. Cohen, D.K. Vawdrey, J. Liu, D. Caplan, E.Y. Furuya, F.W. Mis, E. Larson, Challenges associated with using large data sets for quality assessment and research in clinical settings, policy, Policy Polit. Nurs. Pract. 16 (2015) 117–124.

[24] T. Beale, FHIR v openEHR—concreta, 2018, [cited 2019 21 February]. Available from: https://wolandscat.net/2018/10/10/fhir-v-openehr-concreta/#more-156.

[25] T. Beale, A FHIR Experience: Consistently Inconsistent, 2019, [cited 2019 8 May]. Available from: https://wolandscat.net/2019/05/05/a-fhir-experience-consistently-inconsistent/.

[26] T. Beale, A FHIR Experience—The Formalism, 2019, [cited 2019 9 May]. Available from: https://wolandscat.net/2019/05/08/a-fhir-experience-the-formalism/.

[27] CKM, openEHR Clinical Knowledge Manager, 2021, [cited 2018 26 September]. Available from: https://www.openehr.org/ckm/.

[28] K. Atalag, T. Beale, R. Chen, T. Gornik, S. Heard, I. McNicoll, openEHR-A Semantically-Enabled, Vendor-Independent Health Computing Platform—White Paper, 2017.

[29] H. Demski, S. Garde, C. Hildebrand, Open data models for smart health interconnected applications: the example of openEHR, BMC Med. Inform. Decis. Mak. 16 (2016) 137.

[30] T. Gornik, Application-Centric and Data-Centric Interoperability: openEHR, 2019, [cited 2019 19 March]. Available from: https://www.openehr.org/news_events/industry_news/260.

[31] S. Kropf, C. Chalopin, D. Lindner, K. Denecke, Domain modeling and application development of an archetype- and XML-based EHRS. Practical experiences and lessons learnt, Appl. Clin. Inform. 8 (2017) 660–679.

[32] L. Marco-Ruiz, D. Moner, J.A. Maldonado, N. Kolstrup, J.G. Bellika, Archetype-based data warehouse environment to enable the reuse of electronic health record data, Int. J. Med. Inform. 84 (2015) 702–714.

[33] OpenEHR, Archetype Query Language (AQL), 2021, [cited 2019 12 February]. Available from: https://specifications.openehr.org/releases/QUERY/latest/AQL.html.

[34] R.R. Freimuth, Q. Zhu, J. Pathak, C.G. Chute, Simplifying complex clinical element models to encourage adoption, AMIA Jt Summits Transl. Sci. Proc. 2014 (2014) 26–31.

[35] E. Hovenga, H. Grain, Learning, training and teaching of health informatics and its evidence for informaticians and clinical practice, in: E. Ammenwerth, M. Rigby (Eds.), Evidence-Based Health Informatics, Studies in Health Technology and Informatics, IOS Press, Amsterdam, 2016.

[36] SFIA-Foundation, List of Bodies of Knowledge, SFIA Foundation, 2021. Available from: https://sfia-online.org/en/tools-and-resources/bodies-of-knowledge/list-of-bodies-of-knowledge.

[37] OpenGroup, Digital Practitioner Body of Knowledge Standard, The OpenGroup, 2020. Available from: https://publications.opengroup.org/.

[38] AHIMA, Health Information Managers Body of Knowledge, American Health Information Management Association, 2021. Available from: http://bok.ahima.org/.

[39] G. Wright, The development of the IMIA knowledge base, S. Afr. J. Inf. Manag. 13 (2011) 455–458.

[40] IMIA, IMIA Endorsed Documents, International Medical Informatics Association, 2021. [cited 2019 22 February]. Available from: https://imia-medinfo.org/wp/imia-endorsed-documents/.

[41] S. Sicula, K. Brant, Hype Cycle for Artificial Intelligence, Gartner Research, 2018.

[42] Ontotext, Knowledge Hub Fundamentals, 2021, Available from: https://www.ontotext.com/knowledge-hub/fundamentals/.

[43] SNOMED-International, Leading Healthcare Terminology, Worldwide, 2021, Available from: http://www.snomed.org/.

[44] B.A. Mozzaquatro, C. Agostinho, D. Goncalves, J. Martins, R. Jardim-Goncalves, An ontology-based cybersecurity framework for the internet of things, Sensors (Basel, Switzerland) 18 (2018).

[45] P.I. Dissanayake, T.K. Colicchio, J.J. Cimino, Using clinical reasoning ontologies to make smarter clinical decision support systems: a systematic review and data synthesis, J. Am. Med. Inform. Assoc. 27 (2019) 159–174.

[46] R. Šendelj, I. Ognjanović, E. Ammenwerth, W. Hackl, Towards semantically enabled development of service-oriented architectures for integration of socio-medical data, in: 2016 5th Mediterranean Conference on Embedded Computing (MECO), 2016, pp. 436–440.

[47] ISO-23903, Health Informatics—Interoperability and Integration Reference Architecture-Model and Framework Geneva, International Organization for Standards (ISO), 2021. Available from: https://www.iso.org/standard/77337.html.

[48] B. Blobel, P. Ruotsalainen, D.M. Lopez, F. Oemig, Requirements and solutions for personalized health systems, Stud. Health Technol. Inform. 237 (2017) 3–21.

[49] E. Turban, J.E. Aronson, Decision Support Systems and Intelligent Systems, sixth ed., Prentice Hall, 2001.

[50] P.N. Robinson, M.A. Haendel, Ontologies, knowledge representation, and machine learning for translational research: recent contributions, Yearb. Med. Inform. 29 (2020) 159–162.

[51] F. Dhombres, J. Charlet, Design and use of semantic resources: findings from the section on knowledge representation and management of the 2020 International Medical Informatics Association Yearbook, Yearb. Med. Inform. 29 (2020) 163–168.

[52] S.-T. Liaw, H. Liyanage, C. Kuziemsky, A.L. Terry, R. Schreiber, J. Jonnagaddala, S. de Lusignan, Ethical use of electronic health record data and artificial intelligence: recommendations of the primary care informatics working Group of the International Medical Informatics Association, Yearb. Med. Inform. 29 (2020) 051–057.

[53] IMIA, Code of Ethics for Health Information Professionals, 2016, Available from: https://imia-medinfo.org/wp/wp-content/uploads/2019/08/IMIA-Code-of-Ethics-2016.pdf.

[54] E.J.S. Hovenga, National healthcare systems and the need for health information governance, in: E.J.S. Hovenga, H. Grain (Eds.), Health Information Governance in a Digital Environment, IOS Press, Amsterdam, 2013.

[55] CC, CreativeCommons-CC, 2021, Available from: https://creativecommons.org/.

[56] D. Curto-Millet, Sustainability and Governance in Developing Open Source Projects as Processes of In-Becoming: Technology Innovation Management Review, 2013, Available from: https://timreview.ca/article/649.

[57] D. Ingram, Origins of openEHR, 2002, [cited 2020 24 February]. Available from: https://www.openehr.org/about/origins.

[58] openEHR, An Open Domain-Driven Platform for Developing Flexible E-Health Systems, openEHR Foundation, 2021. [cited 2019 6 February]. Available from: https://www.openehr.org/.

[59] EC_Cordis, The Good European Health Record, European Commission, 2021. Available from: https://cordis.europa.eu/project/id/A2014.

[60] J. Talmon, E. Ammenwerth, J. Brender, N. de Keizer, P. Nykanen, M. Rigby, STARE-HI—statement on reporting of evaluation studies in health informatics, Int. J. Med. Inform. 78 (2009) 1–9.

[61] N.A. Ali, A. Tretiakov, D. Whiddett, I. Hunter, Knowledge management systems success in healthcare: leadership matters, Int. J. Med. Inform. 97 (2017) 331–340.

[62] A. Gropper, Open-source health care sofware, AMA J. Ethics (2011), https://doi.org/10.1001/virtualmentor.2011.13.9.stas1-1109.

CHAPTER

11

Modelling clinical knowledge

Heather Leslie[a,b]

[a]Atomica Informatics, Melbourne, VIC, Australia [b]Clinical Program Co-Lead openEHR International, London, United Kingdom

Introduction

The digital content in a health and social care ecosystem needs to accurately represent the information that clinicians need to record about each patient—their history, context, current status, and plans for future care. In addition, with appropriate consent and authorisation, it also needs to capture data that are suitable for related activities such as health information exchange, reporting, research, population health analysis, decision support, and the evolving sphere of artificial intelligence.

There is a foundational need to accurately represent health and related knowledge within a digital health ecosystem. Such knowledge is best communicated to information system developers through clinical models developed by clinicians and other domain experts. Clinical models, or health-related concept models, represent the meaning of terms and concepts, relationships between concepts, and data patterns, expressed as atomic data structures, suitable for implementation within digital health systems.

A common misconception about the scope of clinical modelling is that it is limited to constructing a clinical data set that can capture and record a specific data collection for a single purpose. Often, the source of information, data structure, and potential or pitfalls of secondary use are not considered; the only design thinking is limited to ensuring that anticipated data values can be recorded in some workable manner. In practice, most clinical modellers have a technical background rather than direct experience of health or clinical care.

The often unappreciated skill of high-quality clinical model design is the aptitude and skill to deconstruct and analyse a clinical domain expert's inherent knowledge and understanding. This chapter will focus on the principles of clinical modelling that will support the synthesis of this knowledge into a collection of consistent and coherent clinical models that accurately represent health-related data requirements and are suitable for use in a digital health environment. Within the context of creating a collaborative health data ecosystem, it is essential to consider principles related to model design, validation and verification as 'fit for use', and governance.

The decision to assemble a health data ecosystem is not trivial. The first requirement is building a community with a shared vision for the future of digital health, a willingness to

https://doi.org/10.1016/B978-0-12-823413-6.00016-1

defy prevailing health technology trends, and the courage to challenge the current status quo. The second requirement is creating a 'whole of health' domain information model as a prerequisite for data interoperability, as discussed in the previous chapters. The advantage of a standardised domain model for the health domain is to provide a stable and coordinated structure of all health-related information. It establishes a robust foundation to underpin hands-on, integrated patient care, data aggregation and analysis, reporting, task management, user interfaces, and knowledge-based activities such as decision support or personalised medicine. The domain model requires common clinical knowledge resources to be defined, shared, and reused, including clinical models, terminologies, data sets, guidelines, queries, clinical processes, and workflow [1].

The alternative is to maintain the current situation indefinitely—a collective lemming mindset that holds us hostage to perpetual incremental innovation and interminable patching of digital health with an unsustainable assortment of messages and documents. The resulting costs and negative impact on healthcare and health outcomes from ever-increasing technical debt and loss of opportunities are potentially enormous.

Is it feasible to develop an information model for all of health? Is it not hubris, an unwarranted arrogance or overinflated self-confidence to aim so high? Or could it be considered negligent or foolhardy to ignore it as a possibility? Of course, dominant groups will push to preserve both the status quo and their power base, conceivably discrediting moves towards change. Nevertheless, common sense would imply that standardising as much of the digital representation of health as possible would benefit all, even at the most generic level. The challenge lies in how far this hypothesis can be applied to the vast breadth and depth of health, primary and secondary use of data, and the dynamic evolution of health knowledge.

Clinical modellers have traditionally had computer engineering or system analysis backgrounds, mainly because of a lack of tooling that supported the direct involvement of grassroots clinicians. The inherent assumption is that the engineers and analysts can bridge the gap between clinical knowledge and software requirements, including ambiguities, uncertainties, and nuances that naturally occur in clinical practice—at least enough to 'translate' the knowledge into clinical models. Unfortunately, this assumption is often incorrect. Resulting clinical systems and data sets expose the mismatch or distortion that can occur when domain experts, with their intrinsic knowledge, try to communicate with software engineers and vice versa. Despite best endeavours, communication between each party is imperfect; foreign concepts and processes, honest errors, ambiguous semantics, and subtle misunderstandings can undermine the noblest of intentions.

The recognition of health informatics as a professional sphere has been rapidly growing in recent years, including the very specialised role of the clinician informatician. With their unique combination of clinical knowledge and health informatics, clinician informaticians have the ideal skill set to bridge the clinical and information technology domains; they are the ultimate clinical modellers.

A scarcity of direct clinical domain experts or health informatician input in designing and creating clinical models should trigger serious concern in those responsible for health technology policy or clinical system procurement. Anecdotally, providing a group of clinicians and a group of technicians with identical clinical requirements exposed significant differences in the resulting models. Not surprisingly, and on multiple occasions with different groups, clinicians developed a model closely aligned with their knowledge of clinical practice. In contrast, the technicians developed a substantially different model based on theoretical logic.

It follows that if we utilise clever tooling to remove the technical burden and ensure technical consistency and validity, then tech-savvy clinicians or clinician informaticians would make the best clinical models. Similarly, clinicians and other health domain experts are the experts who should determine if a clinical model is fit for use within clinical systems.

In an ideal world, all aspects of managing and controlling digital representations of their clinical knowledge, from model design to governance, should be managed and owned by expert clinicians, health domain experts, and clinician informaticians, supported by appropriate tooling. Imagine clinicians designing and governing their purpose-built digital domain knowledge—generalists and specialists, primary and secondary care, and those delivering direct patient care, doing research, or developing clinical decision support algorithms. This philosophical change is potentially a significant turning point in how we design and deliver digital health.

Clinical model design

The role of health informaticians in the leadership of health data design will continue to rise, as recognition of the need for clinicians to design and govern health data becomes the de facto approach. At the same time, the role of the software engineer shifts away from the domain of data design towards a tighter focus on the design and implementation of more intelligent data-driven systems.

The unfortunate reality is that there are no standard methods or approaches to clinical modelling. In the past, the design of clinical models and aggregate data sets have often been reactive to collecting requirements for a specific project or document—existing data sets simply mimicking an existing paper form or computer screen. Without a shared strategy for a data ecosystem, it would always be impossible to create an intentional and coordinated approach to standardising health data.

The philosophy underpinning digital health for many decades has been the exact opposite—with everyone creating data based on their best efforts and with best intentions, but fundamentally believing that 'their information model' gave them a competitive advantage. In this context, it is easy to understand why digital health has evolved in what appears to be a highly fragmented and haphazard way, application- and silo-centric.

There appear to be three main approaches to the current health data design:

(1) Ad hoc—this kind of informal, 'it seemed sensible at the time', modelling is probably the most common methodology for data design in use. This approach demonstrates humans simply doing the best they can, in the current conditions and with the knowledge they have at hand and locking data patterns into proprietary databases. Unfortunately, given the lack of transparency of closely held proprietary data models, it is impossible to gather evidence to measure the extent of actual data variance. In this context, it should be no surprise that when current systems attempt to share information, the only commonality between clinical systems is by accident rather than by design.

(2) Reactive—this kind of pragmatic modelling is primarily focusing on exchanging health information between clinical systems. The intent is to create data models that match or align with most of the typical content currently in clinical applications or data sets. This approach is usually technically led by engineers designing a solution in response to an identified current problem. The HL7 Fast Health Interoperability Resources (FHIR) [2] standard is a widespread example, designed as a specification for exchanging healthcare information electronically, primarily industry-driven and leveraging previous HL7 standards.

(3) Proactive—this kind of modelling is an anticipatory approach, primarily focusing on modelling clinical knowledge, capturing all of the things that clinicians and domain experts know and do in a digital representation, in addition to the existing content in clinical systems. The intent is to create a coherent health data ecosystem of shared clinical models, reusable across multiple clinical scenarios—effectively initiating a universal health language. One of the most successful examples of a coordinated clinical modelling effort has been within the openEHR International [3] Clinical Models Program [4], which has developed a methodology that puts clinicians and informaticians at the forefront of designing, peer-reviewing, and governing their clinical models. The first openEHR clinical models, known as archetypes, were built in 2004. As of June 2021, their community portal, known as the Clinical Knowledge Manager [5] (CKM), contained over 500 clinical models, known as archetypes, and over 2500 registered users from 101 countries.

If interoperability is the holy grail of healthcare, the ad hoc approach that has prevailed for the past 60 years clearly has not delivered the desired results. The FHIR standard grew out of ongoing technical frustration with health data exchange. It has vastly improved health information sharing, albeit limited to the primary administrative nature of the underlying FHIR resource model and generic approach to most clinical data observations. However, it is doubtful that interminable patching between systems is a sustainable strategy and unlikely that FHIR is scalable to incorporate all clinical data. Typically, any messaging approach requires mapping or transforming data in and out of both the source and receiving systems. This wicked messaging problem will persist until there is widespread adoption of standardised, or at least closely aligned, content in both clinical systems and messages.

Unfortunately, in practice, the intentional phases of data analysis, design, and modelling are often overlooked or bypassed because of the rush or the cost of creating software. As a result, there is often a misperception that intelligent data design and the deliberate reuse of common, standardised clinical models in a proactive health data ecosystem are a slower and more expensive course. In reality, there is always going to be some time lag between identifying the need for a new clinical model and its eventual publication as ready for use. When few clinical models were available for reuse in the early days, the process appeared to be glacially slow. However, over time, as the library of reusable clinical models has grown, their immediate availability is accelerating data set development, reducing modelling and implementation costs, enhancing health data quality, and increasing the potential for data interoperability.

So how do we design good data? How do we extract modelling patterns from the clinical knowledge held by domain experts? What level of detail should a model be? What data types to use? What is needed to support the consistency of modelling in a distributed environment? The solution is not trivial, nor feasible without being underpinned by scientific rigour. Many have tried with limited success, but usually only within a constrained scope such as a single organisation or application. There is no doubt that there is an 'art' to clinical modelling—a knack or an aptitude for problem- and pattern-solving that distils and decomposes clinical domain knowledge into discrete, atomic concepts and computable data structures. In addition, there is also a need for a component of scientific rigour, which is not usually recognised or understood.

From a clinical modelling perspective, the openEHR Reference Model (RM) is unique amongst the three approaches and is crucial for establishing a coherent data ecosystem. It specifies the foundation level of modelling by describing and defining the immutable technical rules and definitions that ensure that data created using any archetype is valid and implementable in an openEHR-enabled clinical system.

Technically, archetypes are constraints on the RM, representing the second level of modelling focused on expressing the clinical content. The RM ensures that any clinical data created using an openEHR archetype can be stored, retrieved, versioned, and queried consistently. Also, essential technical components are represented and recorded in the same way across all archetypes—for example, data types, subject of care, author, attestor, information source, participants, and critical date. Garde et al. [6] proposed the openEHR RM as providing a syntactic (data) level of interoperability. Modelling formalisms that omit an underlying RM must deliberately add each of the universal, nonclinical requirements to all of their logical models, with a substantial risk of human inconsistency or errors.

Clinician engagement has not been easy to achieve in traditional standards organisations. The expectation of technical competence excluded many clinicians from participation. However, the openEHR approach has always intentionally defocused on minimising technical barriers to entry by embedding and concealing the RM within modelling tools. As a result, the clinicians and domain experts are free to focus on representing their clinical knowledge in each archetype. The RM's ontological model represents the health/medical record knowledge domain, as shown in Fig. 1.

First described by Beale and Heard [7] in 2007, the Clinical Investigator Record (CIR) ontology (Fig. 1) is a unique part of the openEHR approach. They identified five main types of health information that needs to be captured or accessed in healthcare: historical observations or activities that have taken place; opinions or summaries about the patient's current

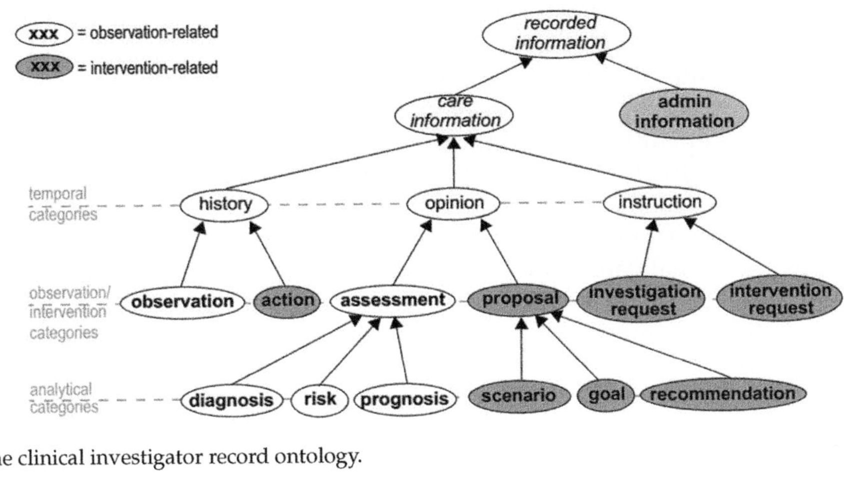

FIG. 1 The clinical investigator record ontology.

situation; instructions for future interventions or activities; and data about an administrative activity or event. The CIR ontology has been embedded within the openEHR RM as five stand-alone ENTRY archetype classes, which carry the primary attributes that support modelling for each different type of data:

(1) OBSERVATION—recording data created by an act of inspection or examination, measurement, questioning, or testing, including history-taking, examination findings, vital signs, test results, and answers to questionnaires. This class also supports recording complex timing, especially for measurements, such as point-in-time or interval measurements or a time series. In addition, it assists accurate interpretation of the data by recording simultaneous information about the state of the patient at the time of data collection;

(2) EVALUATION—opinions about the present situation and intent for the future, including summaries, statuses, goals, or recommendations;

(3) INSTRUCTION—orders for future investigations or interventions, ranging from prescriptions and laboratory or imaging tests through to referrals for care or community services;

(4) ACTION—recording the data relevant to documenting the progress of an INSTRUCTION, in compliance with the openEHR standard Instruction State Machine (ISM) [8], as shown in Fig. 2. The ISM supports an ideal pathway—what is intended (Planned and Scheduled), what is currently happening (Active), and what has been done (Completed). Temporary deviations to the ideal path include statuses for Postponed or Suspended, and permanent alternative end points include Cancelled or Aborted; and

(5) ADMIN ENTRY—nonclinical data that must be kept in the health record, such as admission details.

For example, medication-related data can be recorded using an order archetype [9] (using the INSTRUCTION class) or a medication management archetype [10] for tracking progress around activities such as dispensing and administration (using the ACTION class). A medication statement archetype [11] (using the OBSERVATION class) represents a snapshot view about medication use, especially in the context of a data extract or message exchange, aligning closely with the FHIR medication statement resource [12]. The medication screening questionnaire archetype [13] is another example of the use of the OBSERVATION class. In addition, a medication summary archetype [14] uses the EVALUATION class and summary-related attributes to cater for the situation where the pattern of use or cumulative dosage of a medication needs to be monitored to prevent possible adverse health impacts. A later refinement to the openEHR specification is the CLUSTER archetype class, representing concepts designed explicitly for broad reuse within the context of an ENTRY archetype, such as anatomical location or medical device.

Successful modelling patterns have been designed at many levels and for a variety of purposes. Many ask, 'What came first—the model or the pattern?' The answer is 'neither and both!' openEHR clinical modelling pattern development commenced in 2004. The relatively slow growth in momentum has allowed modellers to investigate, experiment, innovate, and iterate in response to feedback from domain experts and implementers. As a result, many original archetype patterns have been replaced as part of the learning journey. A new modelling pattern may take a significant investment of time to design, review, and test in implementations. However, as each new modelling pattern is confirmed, the experience

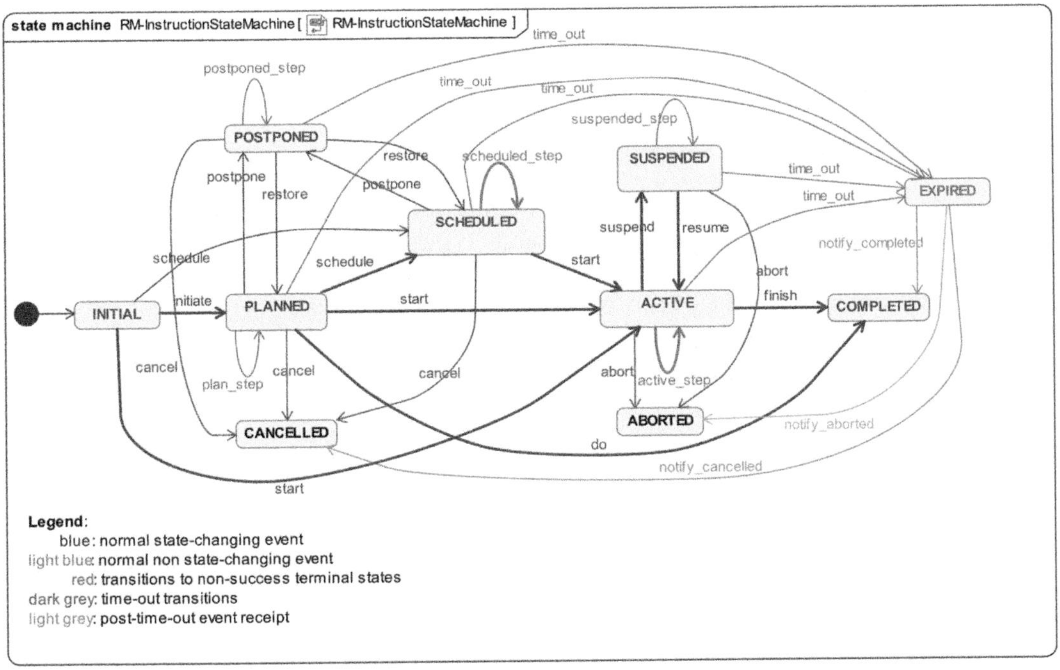

FIG. 2 The openEHR standard instruction state machine.

and insight are applied to the next. Taking advantage of the existing data patterns, new archetypes for related concepts can often be fast-tracked through the peer review process to implementation.

Simple modelling principles underpin the content of many archetypes, such as the common inclusion of a 'Clinical description' data element to capture either a clinical narrative about the concept or the legacy data that might only exist as text blobs. Similarly, most archetypes contain a generic 'Comment' field, effectively added as a placeholder for any additional narrative about the clinical concept that has not already been captured in other data fields.

The multitude of scores and scales is universally represented using the OBSERVATION class of archetypes, faithfully representing the original score or scale as much as possible. A family of Exclusion archetypes have been designed to represent safety-critical concepts—to allow clinicians to assert positive statements of exclusion, such as 'no known adverse reactions', rather than relying on negation.

More complex patterns have been designed to support the thinking patterns and mnemonics used by clinicians at the bedside. For example, the 'Symptom' archetype includes details about the date/time of onset, a description of the course, and identification of exacerbating or relieving factors. Systematically representing the fractal nature of any physical examination was particularly challenging. The requirements include capturing global findings about a patient as observed from the end of the bed, then zooming in to focus on examining their injured limb, and finally targeting the condition of a single blood vessel or nerve within a finger laceration. This task was further complicated because there is no universal method of physical examination—the solution needed to support all permutations and combinations of examination-related archetypes, no matter which clinician is conducting the examination or the clinical problem under investigation.

Shared model ecosystem

In parallel, the philosophy of data design within openEHR International is focused primarily on creating an ecosystem of shareable archetypes. The philosophical design intent of archetypes within the CKM tool is to create archetypes for a generic, universal use case. The goal is to create a maximal data set for each clinical concept, inclusive of relevant data points ranging from existing data in current systems to known and anticipated clinical or domain requirements. In practice, the aspirational maximum data set is often reduced to a practical, inclusive data set, albeit still intentionally designed for reuse across multiple clinical scenarios and future extensions towards the ideal maximal state. This perspective may present an opportunity to investigate the creation of a lingua franca, a universal health language, that might underpin future digital health developments.

Implicit in the successful development of a coherent ecosystem of archetypes is the presence of an underlying modelling strategy and methodology, based on a robust RM. Each openEHR archetype is a specification for a single, discrete clinical concept. Multiple archetypes are aggregated to represent a complex data set, with each archetype constrained to reveal only the data elements required for the specific use case.

For an ecosystem to be achievable, a coherent collection of archetypes needs to be designed that potentially covers all of health and care—a daunting task. But, following the notion of eating the elephant one bite at a time, the openEHR community focused initially on modelling core content with broad reuse across many health domains—for example, adverse reaction risk, problem/diagnosis, and medication order. With the development of clinical models for documentation of clinical care well under way, the natural next steps extend this approach to standardisation of clinical models representing the more abstracted and categorical data required for clinical trials, data registries, and research.

Construction of each new data set reuses existing core content where possible, identifies existing archetypes that required modification or extension, and recommends new ones for each novel concept. These new archetypes are, in turn, added to the public library of archetypes for reuse in the next project. Even in the earliest days of modelling, upward of 70% of a new data set reused preexisting archetypes. Thus, the initial collection grew into a library of increasingly refined and reusable archetypes after each completed project. In addition, openEHR modelling editors have slowly been untangling the 'art', or the knack, of modelling within a coherent ecosystem. Think of it as a complex multidimensional jigsaw, with editors devising how each archetyped concept sits within the ecosystem as a whole, such that there is no semantic overlap, nor gaps between, related archetypes. Local archetypes that are not universally reusable, such as models carrying content that is jurisdiction- or culture-specific, are also necessary within the ecosystem.

Outside of this ideal vision for a unified health data ecosystem and despite the RM, an established methodology and preexisting data patterns, anyone can create an archetype for any purpose. And in a local scenario or project, it can be enormously valuable. However, a problem arises when these are reused more broadly, and we find ourselves in a situation where models overlap, diverge, or are inconsistent. In reality, any archetype can be used in any modelling scenario, no matter how well or poorly designed it might be. However, anyone using 'wild' archetypes should also be mindful of potential risks to interoperability and data consistency—convenience may come with a parallel cost of data integration and technical debt.

Within the openEHR community, we have increasingly seen the benefits of the growing ecosystem of published clinical models. For example, in mid-2020, a case study [15] published details about the significant reuse of archetypes during the global scramble to rapidly develop clinical systems in response to the beginning of the COVID-19 pandemic. A Norwegian vendor developed an app for symptom screening in the earliest days, implemented in a hospital setting within 10 days. Similarly, data sets based on the World Health Organisation (WHO) guidance were quickly deployed within the UK National Health Service, and a COVID decision support system implemented within a Chinese hospital. Each of these three applications, deployed in three languages, was based on the same core published archetypes sourced from the same international CKM. In addition, in quick response to identifying new requirements for symptom, disease, and management screening, a flexible questionnaire pattern was designed, and a new family of screening-related archetypes [16] were created, ready for subsequent deployments. In a matter of weeks, archetype reuse rose from 40% in the initial implementations to 100%. These same questionnaire archetypes are now also being reused across other clinical scenarios, including data registries and additional patient screening settings.

Clinical model quality

How can the quality of a clinical model or modelling pattern be assessed? For clinical systems comprising concealed, proprietary data models, the best they can do is to have a 'trusted clinical advisory board' that endorse their data model. In reality, this is little more than a marketing manoeuvre and provides zero insight into the actual quality of the data.

Quality assurance is a critical factor in developing a successful health data ecosystem, but what does high quality look like in the context of an ecosystem of logical models and health data sets? The simplest benchmark is possibly the hardest to achieve—that it is deemed 'fit for purpose' and safe to implement in clinical systems. A quality assurance approach has two main components—the first is the assessment of the model as fit for purpose at a point in time and the second is a process of clinical knowledge governance that oversees and maintains it over time.

Kalra et al. [17] suggested that shared clinical models need to be trusted before we can expect wide adoption and use. He also identified a key challenge in designing a library of logical models is to ensure that archetypes are evidence-based or meet de facto agreed clinical needs, which may be established by consensus or considered to reflect the existing practice. Whilst this may reflect a baseline requirement for clinical health records, the logical models also need to support other activities, including data exchange, research, population health, reporting, and knowledge-enabled activities, such as decision support.

Content quality assessment

Within the standards development organisations, there have been some attempts to provide measures of clinical model quality. For example, ISO/TS 13972 [18] described general design principles for logical models and ISO/TS 18864 [19] represents the first attempt to define a set of high-level quality metrics for the evaluation of a framework of consistent metadata headings. However, the presence of a comprehensive metadata framework alone should not be considered a de facto measure of clinical model quality. Informative metadata descriptions are an essential starting point for quality assessment—ensuring that they are correct, precise, and appropriate for the concept and type of data. Yet, even this simplest level of quality assurance is poorly understood and commonly neglected.

There is no doubt that each clinical model should be compliant with quality design principles, such as those described in ISO/TS 13972. But vague and abstract principles are not adequate to assess quality. For example, the presence of a defined scope and purpose provides zero information about whether they are clinically relevant, semantically precise, and suitable for use. In addition, whilst we may all aim for comprehensive and accurate models, how can they be objectively assessed?

Common sense would suggest that each logical model within an ecosystem should:

- clearly define the concept that is at the heart of the model. Identifying a clinical concept, its intent and scope of use, especially positioning it within a coherent health data ecosystem, is possibly one of the most critical skills in clinical modelling.
- contain metadata that describes: the purpose, intended use, and misuse of the model; the unique factors that differentiate it from related models; and include references to evidence supporting the model design
- store technical attributes, including a unique ID and version
- document the people and organisations involved in the development of the model
- accurately reflect clinical and other domain requirements for data capture, storage, sharing, and querying,
- contain clinical content specified by precise semantics, especially definitions, descriptions, and constraints; and
- be coherent with modelling patterns used to represent similar or related concepts.

Within each clinical model, the content will vary enormously—for example, the concepts for a medication order, goal, or procedure are discrete and distinct. There is no universal objective or quantifiable measure of quality in this situation. As a result, each clinical model has equally unique assessment requirements, and the composition of the appropriate community of expert reviewers may vary significantly between models. Thus, clinical knowledge management is hugely complex and has potential clinical safety implications that should not be underestimated.

For example, to determine if a medication order model is fit to represent a digital prescription within a clinical system, the domain expert reviewers should include representatives from multiple professions: prescribers and dispensers, consumers, informaticians, terminologists, and software engineers. In addition, the clinical professionals should represent both the community and acute care settings, and input from a wide variety of jurisdictions and geographical locations will potentially support broad international prescribing requirements.

For a clinical model to be deemed appropriate for implementation, it needs to have integrated and merged five expert perspectives:

- Clinical and domain content—from clinicians, consumers, and other domain experts
- Data design (including correct data types, constraints, and classes), modelling pattern, positioning within the modelling ecosystem, and potential for reuse—by clinician informaticians
- Technical validity and readiness for implementation—by software engineers and health informaticians
- Terminology binding—by terminologists
- Translations—by native language speakers, especially flagging if there are cultural or language barriers to translation

There are very few examples of organisations or businesses applying these principles in practice. Online libraries of models may be displayed or shared in generic, public hosting services or private websites. HL7 International reviews FHIR artefacts, such as profiles, as part of the core specification ballot cycle approximately every 1.5–2 years or as implementation guides for specific use cases [20]. FHIR logical models are undergoing the same occasional and prolonged balloting process as their technical specifications. However, in the context of a complex and dynamic health knowledge domain, especially where clinical safety is paramount, it is not hard to make a case for clinical models to warrant an independent process outside of traditional standards development cycles. Clinical model development, review, and publication need to be adaptable and responsive to changes in healthcare, evolving clinical safety issues, or health emergencies such as a pandemic.

As early as 2005, Garde et al. [6] proposed the concept of clinical knowledge governance, defined as a set of processes that enabled the creation, development, organisation, sharing, dissemination, use, and continuous maintenance of openEHR archetypes, despite no governance tools being available at the time. Kalra [17] also identified four types of requirements as part of quality assessment in logical models—business, technical, clinical, and information governance. In 2017, Leslie and Bakke [21] described the International openEHR Clinical Models Program approach, established in 2008, as prioritising broad community participation, robust editorial accountability, and transparency at every level of governance. In practice, multiple levels of quality assurance are built into the openEHR clinical modelling methodology, embedded within the tooling to optimise consistency and compliance.

Archetypes are standardised and agreed, reusable formal models representing aspirational maximum data sets about a single clinical concept. Templates are formal specifications for use case-specific data sets representing forms, documents, messages, or screens created by combining and constraining archetypes. The openEHR RM, built into both archetype and template modelling tools, handles most business and technical requirements by defining the rules that support the structural organisation, classes, constraint patterns, and technical attributes of archetypes to ensure they are valid for implementation. As a result, clinical modellers only need to focus on expressing the clinical domain content as they build an archetype, and on data set requirements as they create a template.

The openEHR Clinical Knowledge Manager (CKM) tool is purpose-built to meet the operational requirements for the clinical and information governance of archetypes. The CKM has three primary functions—a library of freely available, public domain archetypes, an online collaboration portal for asynchronous review of content, and a comprehensive methodology and inbuilt processes for consistent artefact governance.

As per the requirements specified in ISO/TS 13972, each openEHR archetype comprises:

- Name of the concept represented—including a definition of the concept within the context of the archetype's class—for example, differentiating between a medication order (INSTRUCTION class) and medication management (ACTION class)
- Purpose of the archetype—in terms related to the intended data use
- Use of the archetype—usually focusing on explaining the design pattern intent and practical insights or examples about how to model the archetype within a template or how it can be used to support data capture within an implementation
- Misuse of the archetype—primarily identifying alternative model options better suited to record related clinical concepts.
- References—Usually to identify any source material or evidence used in the design phase, including original papers or other clinical models. Often there is no formal source material as the content reflects common clinical knowledge or practice.
- Details that identify the archetype—including a unique archetype ID, version, date of authoring, and build ID.
- Identification of all people involved in the quality assurance for the content of the archetype—comprising the primary author, plus all design contributors, content reviewers, translators, and translation reviewers, including their organisations, profession, expertise, and nationalities.
- Ownership or custodianship details about the archetype to provide current contact points and establish provenance.
- Licensing details—including copyright statement, Creative Commons license, and flagging any requirements for users to obtain third-party licenses or copyright permissions.

A checklist can be used to ensure that each of these metadata components is present and contains content; however, determining if the actual content is appropriate relies on human input, especially from subject matter experts. It is simply not possible to assess the quality of the clinical content by any simple objective means. There is no objective or authoritative knowledge base nor an existing universal library of health models that we can use as a source of truth.

In practice, it can be problematic if there is too much emphasis on using academic sources as the trustworthy authority as the archetype content represents the natural messiness of clinical practice, not the purity of theory. In many archetypes, theory and clinical practice are closely aligned, but not always. One of the most familiar examples of academic divergence is the measurement of blood pressure. The academic definitions usually describe the effect of the pressure of circulating blood against the walls of blood vessels. By comparison, the archetype definition reflects the reality of measurement in clinical practice: the local measurement of arterial blood pressure, which is a surrogate for arterial pressure in the systemic circulation [22].

Aspirational, high-quality content within an archetype is the result of juggling several competing tensions. The primary goal is for each archetype to accurately reflect a maximal collection of data elements describing current clinical knowledge and practice about a single clinical concept. Concurrent considerations support alignment with data in current and legacy clinical systems (where this is sensible) and standards, such as HL7 FHIR, to ensure that the archetype also supports existing data integration. Above all, the most significant responsibility of reviewers is to ensure that the content of each archetype is safe to use and suitable for implementation. As a result, the only way to assess the quality is by using a qualitative approach. Thus, each archetype needs to be evaluated on its own merits by a group of reviewers whose knowledge and skills complement the archetype's class, scope, intent, and content.

Traditional standards development organisations tend to operate through a democratic voting process of credentialled experts or nominated representatives. In contrast, the openEHR community has established a crowdsourced approach as a quality measure, applied to the focused area of reviewing, refining, and publishing archetyped content. Crowdsourcing of knowledge has been attempted in many domains with varying outcomes [23]. The online Clinical Knowledge Manager tool has proved central to the success of the openEHR crowdsourcing methodology, the coordinating hub for collaborative review and content publication. The CKM has been purpose-built to support an active crowdsourcing content review process, including:

- building and managing a community of practice,
- inviting appropriate cohorts of the reviewer community to participate in targeted reviews of archetypes,
- collate and organise the community contributions,
- support editors to facilitate the review contributions, providing individual feedback to each contributor;
- the reviewer community recommending next steps in the publication process for each; and
- ensuring transparency and accountability of all editorial decision-making.

Despite the library of clinical models being freely available, the CKM hosts a growing virtual community of more than 2500 people from 102 countries as of June 2021. Self-registration is the sole requirement for becoming an active participant in the inclusive community. This lack of formal credentialing or screening may be challenging to some involved in the traditional standards development organisations; however, this alternative strategy encourages input from the broadest range of participants and contributors who might otherwise not be able to participate. To become a reviewer requires only a name, an email address and a consent flag, indicating a willingness to be invited to content reviews. The optional additional recording of professional and clinical expertise allows editors to select and invite an appropriate cohort of reviewers to participate in each archetype review.

The openEHR reviewers intentionally volunteer their time, effort, and expertise to collaboratively refine and, eventually, achieve consensus about each archetype's content. No single person is considered an expert about all aspects of any archetype; each participant's contribution, no matter how large or small, inexperienced, or expert, introduces a unique perspective and augments the published archetype outcome. All feedback is encouraged, from the most trivial grammatical correction to questioning complex informatics representations. In addition, those who ask questions or challenge inherent assumptions demonstrate that the archetype needs, at the very least, further clarity and sometimes even a total redesign. In the international setting, consideration also needs to include concepts that are not globally understood, such as the lack of an equivalent concept in another language.

The CKM user interface and review processes have been specifically designed to ensure that nontechnical experts, such as grassroots clinicians, can participate equally alongside the technology savvy. Each CKM review follows a 'wizard' process that systematically guides the reviewer through the metadata and specific clinical content, allowing them to comment on each archetype component and respond to a standard series of questions. In addition, the content can be displayed in various formats, ranging from a mind map to the underlying Archetype Definition Language. Thus, it removes the need for clinicians to acquire additional technical skills or training to participate, and they can focus on the domain content under review.

Editors, usually clinician informaticians, are responsible for iterating the clinical content of each archetype from an initial draft state through one or more reviews towards a published end state. At the end of the typical 2-week review period, editors accept all contributions, respond to the feedback, iterate changes to the archetype, and decide on the next steps. The only mandatory part of a review is for each reviewer to record a publication recommendation—Accept, Minor revision, Major revision, Reject, or Abstain. If all recommendations are 'Accept', the consensus permits the archetype content to be published and the archetype is updated from a v0 draft to a v1 published lifecycle state. 'Minor revision' recommendations are treated as 'Accept' when the content changes are trivial, such as spelling corrections. If any 'Major revision' or 'Rejected' recommendations are recorded, further review rounds are required until consensus is achieved. Any subsequent content changes made to a published v1 archetype strictly follow the openEHR lifecycle model [24] and versioning rules [25].

There are no strict rubrics about the actual number of reviewers required to confirm the content of each archetype. However, the number of reviewers, the composition of the reviewer community, and the total number of completed reviews provide insight into the credibility of the review process. For example, an archetype review for a validated score or scale will ensure that the original intent has been accurately represented in the digital format. In this situation, the number of reviewers can be relatively small. Conversely, the review of an archetype that is a core concept but for which there is no broadly accepted clinical content, such as an adverse reaction risk or a diagnosis, requires a larger, more diverse cohort of reviewers and a more rigorous review process. In addition, if only administrators or consumers review a clinical archetype or only reviewers from one geographical region, publication of the archetype would be inappropriate.

In addition, the CKM tool supports:

- detailed artefact history and audit logs to document provenance, traceability, and any changes made to each clinical model;
- change requests for existing archetypes submitted by the community, on the identification of new requirements, ambiguity, or errors; and
- threaded community discussion.

Clinical knowledge governance

Clinical information modelling governance is a relatively novel and experimental challenge. Until recently, most of our only experience in governance of health data standards has been at the complete message or document data set level and as part of technical balloting cycles. Furthermore, there are few working examples of clinical knowledge governance—it is a phenomenally complex, poorly understood, and evolving field.

In the past decade of developing and evolving the Clinical Knowledge Manager, the most critical success factor for the openEHR Clinical Models Program has been developing a knowledge governance strategy. It facilitates simultaneous management of clinical content publication alongside technical versioning of artefacts, closely aligned with the semantic versioning [26]. Without this coordinated approach to governance, managing and maintaining archetypes would be a semantic and technical nightmare. Instead of impossible, governance becomes possible, and the rules embedded in the CKM tooling simplify the process. Possibly the most important realisation from the openEHR journey is the paradox that as more is understood about the nature and complexity of the clinical knowledge governance, the more is realised about the sheer magnitude of what remains to be understood and mastered.

Benefits of a 'whole of health' data ecosystem

Whilst the notion of a coherent health data ecosystem is only imaginary to some, the openEHR International community has been progressively advancing towards this outcome for well over a decade. It is a mammoth task, but, model by model, the clinical modelling programme is evolving a methodology, tooling, and data patterns that work in practice. This long, slow maturation process has allowed the openEHR community to research, test, and refine their clinical modelling methodology. With the relative luxury of time, the openEHR mantra of 'implement, implement, implement' has provided a robust foundation for the current library of archetypes and modelling work to continue well into the future.

The current library of archetypes avoids semantic overlap at the same time as minimising semantic gaps between concepts, always aiming for a coordinated and coherent collection of complementary clinical models as the basis for the fledgling health lingua franca. The initial archetypes developed were considered 'core'—the models represent the most common concepts that underpin everyday health data recording, with the most likely impact on saving a life.

For national programmes and jurisdictions, the ecosystem functions as a single, coherent national repository of clinical knowledge artefacts and a road map for future health data advancement, ideally including both clinical models and related terminology subsets. This results in:

- Improved data quality—a consistent representation of purpose-designed data, validated as fit for use by clinician and domain experts, to support shared EHR programmes, health information exchange, and secondary data use.
- Data independence—the data created using neutral, vendor-independent clinical models can be decoupled and stored separately to any single application, messaging, implementation, or technology trend.
- Improved data longevity—data captured and stored independent of any single application can be safeguarded and endured, despite application migrations or evolving technology trends. Neutral, nonproprietary clinical models represent the lifelong data recorded in health records;

- Improved data liquidity—shared models that can support native exchange, integration, aggregation, and reuse of health information. As shared models increase in prevalence, requirements for patching between systems with messages will become less necessary; clinicians and domain experts ensure that health and related data models are fit for health purposes.
- Establishment of clinical knowledge governance processes and asset management for foundational clinical content, including versioning and publication lifecycle governance. The same methods can be downscaled and applied to manage adaptations and maintenance of local jurisdictional or any specialist clinical knowledge domain requirements.

The current stranglehold of proprietary clinical systems, especially the mega-suite enterprise systems with associated vendor lock-in, will be disrupted. Additionally, it can set some revolutionary changes in play—breaking down silos of data and encouraging the development of an agile and innovative ecosystem of applications based on a foundation of shared clinical models. Further, the impact of smaller vendor applications could be amplified by sharing data instances stored within a common clinical data repository or platform.

An open, global health data ecosystem facilitates the efficient use of scarce and expensive resources as it:

- Avoids duplication of effort by designing and developing a clinical model once; reuse in multiple contexts;
- Leverages or adopts existing international clinical models to kick-start or enhance the national library of clinical models;
- Minimises expensive and time-consuming face-to-face meetings by leveraging online collaboration portal tooling;
- Enables rapid and responsive reviews of each clinical model, with publication achievable within weeks to months;
- Fast-tracks review of whole messages or documents, with publication achievable within weeks to months, rather than months to years, because of the previously validated and published clinical model components.

For professional clinical colleges, an ecosystem of clinical models will enable:

- each organisation to champion and drive transformational change in health data, through active engagement in clinical model development. College members can design, publish, and govern clinical models relevant to their unique, specialist discipline. These archetypes can be added to the growing model collection, and members can lead or participate in the joint publication and governance of clinical models shared between specialities or common to all clinical care.
- transforming their responsibility for high-quality paper practice records into digital health records.

For clinical system vendors and application developers, a clinical model ecosystem will support:

- Transformation of business models away from siloed application vaults to competing on innovation and clever use of technology—user interfaces, workflow processes, security, data capture/storage/querying, and clinical decision support—rather than local data design.

- A coordinated and cohesive data road map, ready for future application enhancements and extensions.
- Native interoperability between clinical systems using the same model ecosystem
- Shared data models will also boost code reuse and more efficient software development, especially for new integration or exchange projects.

In a secondary data use environment, the ecosystem will enable:

- A once-only mapping of existing data archetypes and subsequent transformation into a validated and consistent format; new data can be captured and aggregated according to the same archetype definitions.
- Secondary use data design to leverage and align with the shared clinical models, such that extraction of data or derivations and transformation can increasingly be automated.
- Data stored in a common, agreed representation can be more easily aggregated for analysis or integrated from multiple source systems.

Patients and consumers will potentially be able to become more active in the management of, and contribution to, their own health record:

- Data for life—for the first time, health data can be persisted and reused from 'cradle to grave' because it is untethered to a proprietary format or single application.
- Data created and stored in a shared, standardised, and nonproprietary representation support the potential for application-independent data records that can persist for the life of the patient.
- Improved data 'liquidity'—so that data can flow between healthcare providers and systems to where the patient needs it.

Conclusion

Our health data is a priceless asset, yet most is fragmented, incoherent, poorly defined, and of uncertain quality. Day by day it grows in volume, variance, and complexity, creating an urgent obligation on us to act. Ensuring our health data is fit for purpose is both our challenge and our responsibility. We need to face this challenge head-on; the status quo is no longer acceptable.

After more than 60 years of experience with electronic health records and glacial progress in digital health, it has become increasingly evident that one of the major inhibitors to progress is the lack of standardised data models that represent common clinical knowledge, health processes, and workflow requirements. Our goals must shift towards creating a coordinated health data ecosystem as the foundation for digital health, comprising a standardised health information model, shared clinical models and open platforms.

Over the past decade the openEHR clinical modelling community has established a methodology for atomic health data standardisation that underpins the development of a digital health lingua franca and a coherent health data ecosystem. Further research is required to determine how far this approach can be expanded and progressed.

References

[1] T. Beale, Why Using HIT Standards Fails to Achieve Interoperability, *Woland's cat*, 2019. [cited: 2021-06-17]. Available from: https://wolandscat.net/2019/09/06/why-using-hit-standards-fails-to-achieve-interoperability/.

[2] HL7 International, HL7 FHIR Foundation Home [Internet], HL7 International, Ann Arbor, 2021. [cited: 2021-06-17]. Available from https://fhir.org/.

[3] openEHR Foundation, openEHR Foundation Home [Internet], openEHR Foundation, London, 2021. [cited: 2021-06-17]. Available from: http://www.openehr.org/.

[4] openEHR Foundation, Clinical Models Program [Internet], openEHR Foundation, London, 2021. [cited: 2021-06-17]. Available from: http://www.openehr.org/programs/clinicalmodels/.

[5] openEHR Foundation, openEHR Clinical Knowledge Manager [Internet], openEHR Foundation, London, 2021. [cited: 2021-06-17]. Available from: http://www.openehr.org/ckm/.

[6] S. Garde, P. Knaup, E. Hovenga, S. Heard, Towards semantic interoperability for electronic health records, Methods Inf. Med. 46 (3) (2007) 332–343, https://doi.org/10.1160/ME5001. 17492120.

[7] T. Beale, S. Heard, An ontology-based model of clinical information, Stud. Health Technol. Inform. 129 (Pt 1) (2007) 760–764. 17911819.

[8] openEHR EHR Information Model [Internet], openEHR Foundation, 2020. The Standard Instruction State Machine (ISM); [cited: 2021-06-17]. Available from: https://specifications.openehr.org/releases/RM/latest/ehr.html#_the standard_instruction_state_machine_ism.

[9] openEHR Foundation, Medication Order, Published Archetype [Internet], openEHR Foundation, 2018. openEHR Clinical Knowledge Manager [cited: 2021-06-17]. Available from: https://ckm.openehr.org/ckm/archetypes/1013.1.3124.

[10] openEHR Foundation, Medication Management, Published Archetype [Internet], openEHR Foundation, 2017. openEHR Clinical Knowledge Manager [cited: 2021-06-17]. Available from: https://ckm.openehr.org/ckm/archetypes/1013.1.123.

[11] Medication statement, Draft archetype [Internet]. openEHR Foundation, openEHR Clinical Knowledge Manager [cited: 2021-06-17]. Available from: https://ckm.openehr.org/ckm/archetypes/1013.1.4949.

[12] HL7 International, HL7 FHIR Resource—Medication Statement R4.0.1 [Internet], HL7 International, Ann Arbor, 2019. [cited: 2021-06-17]. Available from: https://www.hl7.org/fhir/medicationstatement.html.

[13] Medication screening questionnaire, Draft archetype [Internet]. openEHR Foundation, openEHR Clinical Knowledge Manager [cited: 2021-06-17]. Available from: https://ckm.openehr.org/ckm/archetypes/1013.1.4677.

[14] Medication summary, Draft archetype [Internet]. openEHR Foundation, openEHR Clinical Knowledge Manager [cited: 2021-06-17]. Available from: https://ckm.openehr.org/ckm/archetypes/1013.1.2825.

[15] H. Leslie, openEHR archetype use and reuse within multilingual clinical data sets: case study, J. Med. Internet Res. 22 (11) (2020), e23361. Available from: https://www.jmir.org/2020/11/e23361.

[16] Screening questionnaires project [Internet]. openEHR Foundation, openEHR Clinical Knowledge Manager [cited: 2021-06-17]. Available from: https://ckm.openehr.org/ckm/projects/1013.30.90.

[17] D. Kalra, A. Tapuria, T. Austin, G. De Moor, Quality requirements for EHR archetypes, Stud. Health Technol. Inform. 180 (2012) 48–52. 22874150.

[18] ISO/TS 13972:2015 Health informatics—Detailed Clinical Models, Characteristics and Processes, International Organization for Standardization, Geneva, Switzerland, 2015.

[19] ISO/TS 18864:2017 Health Informatics—Quality Metrics for Detailed Clinical Models, International Organization for Standardization, Geneva, Switzerland, 2017.

[20] FHIR Profile and IG Balloting [Internet]. Ann Arbor: HL7 International; [cited: 2021-06-17]. Available from: https://wiki.HL7.org/FHIR_Profile_and_IG_Balloting.

[21] H. Leslie, S. Ljosland Bakke, Peer review of clinical information models: a web 2.0 crowdsourced approach, Stud. Health Technol. Inform. 245 (2017) 905–909. 29295231.

[22] openEHR Foundation, Blood Pressure, Published Archetype [Internet], openEHR Foundation, 2018. openEHR Clinical Knowledge Manager [cited: 2021-06-17]. Available from: https://ckm.openehr.org/ckm/archetypes/1013.1.3574.

[23] L. Dahlander, H. Piezunka, Why crowdsourcing fails, J. Organ. Des. 9 (2020) 24, https://doi.org/10.1186/s41469-020-00088-7.

[24] openEHR Archetype Identification [Internet], openEHR Foundation, 2019. Lifecycle Model; [cited: 2021-06-17]. Available from: https://specifications.openehr.org/releases/AM/latest/Identification.html#_lifecycle_model.

[25] openEHR Archetype Identification [Internet], openEHR Foundation, 2019. Versioning; [cited: 2021-06-17]. Available from: https://specifications.openehr.org/releases/AM/latest/Identification.html#_versioning.

[26] Preston-Werner T. Semantic Versioning 2.0.0 [Internet]; [cited: 2021-06-17]. Available from: https://semver.org/.

CHAPTER

12

Identity—What is in a name?

Heather Grain[a,b,c]

[a]eHealth Education Pty Ltd, Melbourne, VIC, Australia [b]Global eHealth Collaborative (GeHCo), Melbourne, VIC, Australia [c]LG Informatics Pty Ltd, Melbourne, VIC, Australia

Names are not enough

Quality healthcare is reliant upon information that relates to:

- People, who might be patients, family, carers, or clinicians
- Organisations, providing services or guidelines, requesting information, paying for healthcare services, educating clinicians, researching care practices or services.
- Things used in healthcare including:
  - Pharmaceuticals
  - Devices
    - Implanted into bodies
    - Tool for treatment or care (e.g. endoscope)
    - Tool used for communication and data entry (e.g. smart phone)
  - Surgical supplies (e.g. drains, catheters, IV lines)
  - Linen (e.g. bed sheets)
  - Equipment (e.g. thermometers, wheelchairs)
- Authority to receive care or deliver care

Identity systems in healthcare are primarily built to meet standard business requirements; they tend not to be designed for, nor fit for, use in healthcare environments.

In business, identifying information about organisations and other things are commonly dealt with by the use of bar codes, Q-codes, and radio-frequency identification definition (RFID), such as codes used on packets at the supermarket or items in stores. These same processes are used in healthcare to identify packets of drugs, implant devices, and equipment. However, identification of individuals and organisations in healthcare has some extra complexities, risks, and impacts.

Identification is not just an administrative process in healthcare. It is an essential requirement for patient safety, efficient and effective patient care, and service provision. Consider the impact of errors in identification of an individual, an organisation, a drug, or a tool.

Roadmap to Successful Digital Health Ecosystems
https://doi.org/10.1016/B978-0-12-823413-6.00004-5

Impact of wrong identification

The impact of either identification of one person as another or the failure to identify that a record already exists for a person can impact patient safety, efficiency, and resource use. Healthcare must be able to identify people when they are not conscious or able to give information a normal business would expect them to be able to do when they interact with that business.

Special cases that healthcare must manage are the following:

- Identification of a person unable to provide details.
- Identification of a person where only those with them can provide details, and those people may not know the individual's formal name or identity information. Coworkers would not be likely to know all the information necessary to accurately identify someone with whom they work.
- Identification over time—where identifying information changes. Healthcare is improved if a person's health record details are available for his or her lifetime so that clinicians can assess historical information about situations which might now be affecting the care of the individual. Creation of a new record when the old record can't be found is a potential point of care failure that is not as important in other business settings.
- Identification of the individual across different healthcare organisations or services for information sharing.
- Identification of the individual prior to undertaking a surgical procedure or administering medication.

The impact of incorrect or inadequate identification of individuals in healthcare can result in the following:

- Information about previous conditions and treatment not being available to the current healthcare provider,
- Provider decisions may be based upon inaccurate information,
- Wrong, delayed, and, in some cases, dangerous care, e.g. operating on the wrong leg
- Administering a medication to the wrong person,
- Failure of financial claims for services,
- Incorrect data sharing or availability leading to services being duplicated which has a safety as well as a financial impact,
- Incorrect information being put into another person's record.

Wrong identification of a healthcare provider or organisation

If a healthcare provider or organisation is incorrectly identified in care documents or communication, this can result in the following:

- Failure of care transfer or referral which can delay a person's care,
- The inability to find information about previous care of an individual,
- Failure to meet the requirements for the rights to undertake specific activities, such as prescribing, undertaking, or approving services, leading to prescriptions not being filled or care not being given.
- Financial implications, where the service costs may not be accepted by insurers.

Wrong identification of items

Wrong identification of a drug can result in incorrect drugs being provided to patients; this can be the actual wrong substance or the wrong strength or out-of-date products being provided. There are safety implications to such errors.

Tool or equipment identification may impact the provision of care, such as when a specific tool is needed for a procedure and it is not booked to be present in the surgery, the procedure may be delayed or cancelled, which has safety implications for the patient, but significant resource waste for organisations.

What is identity?

In its simplest form, identity is the establishment of a unique, single identifier for a thing based upon defining attributes of that thing. For a person, the attributes used to establish identity vary based upon the practices of the country/business environment. These attributes are what distinguish one thing or person from another.

For example, in Australia, the government has established an identity establishment process which requires a minimum of three government issued documents (such as passport, driver's license, healthcare card) or more documents such as electricity bills which prove that the person named lives at the specified address.

These methods are not discussed in detail here as they are all based upon a business identification model where people are motivated and have the time and resources to obtain these documents and provide them to confirm who they are. In healthcare, such an identification method is often impractical, particularly in emergencies. There are also examples where a person in one family shares his or her identity documents with others in that family to obtain healthcare services, especially for financial reasons.

Identifiers are codes used to represent an individual. In the past, these identifiers were often associated with physical records, and each record had its own identifier which was used not only for identity but also as a filing sequence mechanism. Identifiers are created by organisations to manage their clients, patients, providers, and items. As such any person, organisation, or item may have multiple identifiers each used by a different organisation or purpose. In the past, there was a strong desire to hold all information together for patients wherever possible; this concept was called a unit record or UR. This concept is still desirable. With electronic systems, the data no longer need to be in one place or physically in a folder if data are associated with the one person and the computer can access the pieces of information relevant to the specific care requirements.

Computers make this process much simpler than in the past. A computer system can hold details of a person with all of their different identifiers, by working as a register of records held about the individual. In this way, there can be identifiers for different purposes and organisations without needing to renumber or change the 'old' number in the system.

The following is a fictitious example of how identifiers should work. Mary Smith has the following identifiers; all are associated with her healthcare identity in the national identification system. Table 1 shows a selection of possible identifiers for this individual, though there may be many more, and there is usually more metadata collected about identifiers (discussed later in this chapter).

TABLE 1 Identifiers for an Individual.

Name of identifier	Identifier	Issuing organisation	Use case
National individual identifier	12345678ABC	National government	Often used for aggregation of data at the national level and for information exchange between different systems
National insurance identifier	3014 5551 3322 2	National Department of Health	Number used for financial claims made upon the national healthcare system or to establish eligibility to use a national service
State healthcare identifier	V3545681	Government of Victoria	Issued to support data sharing within the state and for data aggregation within the state healthcare system.
Driver's license number	032588991	Government of Victoria	Government approved identifier for verification of identity and to authorise the eligibility of this individual to drive a vehicle
Local doctor's UR number	66343411	Ashwood Medical Centre	The identifier used by Mary's local healthcare provider to find the record of her care in their local healthcare system.
Big hospital UR number	5533112	Big Hospital—Victoria	The identifier used by 'Big Hospital' to identify Mary's local record in their healthcare system.

The availability of identifier registers reduces the effort to communicate between services without the need to modify the local system/s. This is the most flexible and sustainable way to manage identifiers for people, but also for organisations, and things.

Identifiers are often structured with special components, such as check digits. Check digits are the last character in an identifier. The last digit is calculated based upon the other items in the identifier and is used to reduce errors in transcription of identifiers. A very simple check digit could be calculated as the last digit of the sum of all preceding numbers:

123456 would have a check digit of (sum = 21, Therefore, the check digit is 1)

The identifier is 1234561

There are many standards and more complex check digit systems used.

Digital identities such as fingerprints and facial recognition are also being used increasingly in healthcare to reduce the impost of identity management. In India, the national identity system includes facial recognition which has been used very effectively in some healthcare systems. Such systems do not require the individual to be conscious, or to carry special identity documentation. Though they are still imperfect, they seem to be offering a way forward.

In some cases, identifiers are built from meaningful information. This is not a recommended process as it would mean the need to change the identifier if a component of the information changes. For example, If a national identifier included the state in which the person resides, a new number would have to be issued every time the individual moves across a state border.

Identity infrastructure

Identity management and system design need to be governed with a view to the needs of healthcare. An understanding of the different types and uses of identifiers throughout healthcare is essential for any system.

Where governments create identity management systems, they need to consider not only the design elements discussed later in this chapter but also the practical implications for systems including:

- encouraging collection of all names used by the entity (organisation, individual, etc.)
- Identifying preferred names so that the system can work in a practical manner without imposing unrealistic and often dangerous processes which work in a business environment but do not suit healthcare well.
- Representation of preferred names in the system—respecting the individual's requirements.
- Representation of names considering cultural imperatives.
- Establishing the requirements for systems to communicate with each other and share identifiers so that each system can share and establish identity safely and efficiently.
- The needs for identification over time, identity is not static but can and will change as people move to new locations, change their names, etc.
- The motivators for keeping information current. Identity management systems which use address as a key identifier worked well when the individuals were dependent upon mail, such as refund cheques, reaching them at their address. Since financial payments have been made directly to bank accounts, the bank details have become reliable, while addresses are less reliable, as there is no motivator for individuals to inform the identity management system of changes.
- Relationships between individuals and organisations to support healthcare.

Though many systems may consider one identifier as the 'essential or most important' identifier, that concept is a delusion. The most important identifier at any point in time is the one needed to achieve a specific task. All identifiers are of value and should be able to be managed in our systems.

Requirements for personal identity

It is a primary responsibility of caregivers to check the identity of patients and match them with the correct care prior to delivering such care. Patient misidentification is a primary cause of error and adverse events. There are specific components to the identity of individuals as detailed in the ISO standard on *Identification of subjects of healthcare* [1] including the following:

- Names
- Demographic information
  - Date of birth
  - Date of death
  - Place of birth

- Sex at birth
- Gender
- Marital status
- Living arrangement
- Family linkages

- Address—physical
- Address—electronic
- Phone numbers
- Identifiers, codes associated with the passport, driver's license, etc.
- DNA
- Biometrics

Some of these components are essential for identity management; others are stored with identity for convenience as they relate to the individual over potentially many visits to carers (they aren't specific to the episode of care). Other elements of these data were once collected for communication only but have now proven to be reliable long-term identifiers (e.g. mobile phone numbers).

Each of these components will be investigated to determine their relevance to identity management. Some of these data are reasonably static (persistent), and this makes them highly appropriate for identity management. This includes items such as sex at birth and place of birth (unlikely to change), while marital status or physical address are more likely to change.

GS1 has developed a brochure explaining its system of standards enabling precise automated patient identification. These are particularly useful for the accurate administration of medications [2].

Names

Different cultures structure and use names in different ways. The international standard on identification of subjects of care recognises the need for a range of component parts to names in addition to the representation of names using different character sets (language) and is the basis for much of the content in this chapter. Such requirements may seem irrelevant to a local system in your area, unless your area services people from multiple cultures. If you exchange information with other organisations in your country or in other countries, they may be more relevant than you imagined. There are also different cultural and pronunciation issues or linguistic presentation issues required to be managed where the information pertains to people from multiple cultures or are shared between different countries.

Table 2 shows the concepts used to represent names. Recurring indicates whether a name may have multiple occurrences of the item listed. For example, the first occurrence of given name = first given name, but there may be any number of additional given names.

There is more information needed to manage and use names in healthcare, including knowing the name/s applicable at a point in time, the individual's preferred name and the name associated with specific use cases or identifiers.

Name usage information

It is important to know when a name was/is in use. This information is useful for identification but also for communicating with the patient. The date a new name starts to be used (such as a name after marriage) or ends.

TABLE 2 Constituents of name.

Data element name	Recurring	Description	Relevance to identity in healthcare
Title	Yes	Cultural or honorary recognition of an individual	Rarely used to identify an individual but often used by clinical staff to address the patient with respect. Cultures vary in the importance or sensitivity of individuals to the use of their title. The relationship of the title to gender and marital status should also be considered as these can cause offence if used in a manner inconsistent with the patient's expectations
Given name	Yes	The name allocated to the individual (usually at birth) to differentiate them from other members of their immediate family	Significant identifier, though often shortened or alternative version or preferred name replaces this in everyday use—see preferred name functionality requirement Some cultures have given registered names and unregistered 'calling' 'pet' or 'nick' names always used to refer to the person.
Family name prefix	No	Often meaning of or part of the family or that the family has some aristocratic background	Cultural prefix to a name which indicates honorifics used historically and often considered today to be part of the basic family name. Examples include de, von, o'
Family name	Yes	The name/s shared by the individual with others in their family	Some cultures have more than one of these names (e.g. mother and father's family names—such as the Spanish or Portuguese), others hyphenate their names. In India, initials are frequently used in place of long and multiple family names Family names are strongly recommended to be displayed in upper case to clearly distinguish them from given name, irrespective of the cultural background of the individual or the use case
Name suffix	Yes	Indicators of qualifications, family differentiators, such as Junior, or honours represented after the individual's name	These suffixes are often represented as abbreviations

Usage information can also include information which can be used to identify known spelling errors for this name, temporary names used when identity can't be established (such as Broken LEGG). Best practice is to use a condition as the temporary name for an individual rather than Jane or John DOE—as there could be people with the name Jane or John DOE where that name is valid. The use of the condition is particularly valuable in clinical care where there are multiple injury events.

Preferred name and other usage types

A person's preferred name is an essential component for healthcare identity management particularly when the patient is not conscious to confirm formal name information. Those who may accompany a person to healthcare may only know his or her preferred name, and healthcare systems need to be able to cope with this situation.

Preferred names may arise for many reasons and an assumption that they are used to defraud is not appropriate. These names may be used for the following:

- To provide anonymity
- To differentiate between another family member with the same or similar name.
- To shorten the given name such as Kathy instead of Kathleen
- To use a preferred name which could be a middle or alternative given name or nickname.
- To create a name which is culturally 'easier' than their given name. Migrants will often use an alternative given name to fit in to their new environment. Some cultures adopt an English name for international communication.

Such names are essential to healthcare systems. When family members, friends, and workmates come to visit a patient in the hospital, they are likely to use the preferred name and, in some cases, might not even be aware of the formal name of the individual—yes many of our systems do not support preferred names. When a patient is coming out of anaesthesia, the clinical staff will often talk to them to confirm their health state. Using the name that the person knows best is an essential clinical requirement.

There are other usages of names, individuals may have professional names or names used for specific purposes. Some cultures have requirements that if their name was used by someone in their family who has died with whom they shared a name (usually a given name), they will change their name.

Names and identifiers

Formal identifiers such as insurance policy number usually require the use of the name associated with this identifier. The computer system needs to be able to collect and correctly associate the right name with the identifier to which it applies.

Demographics

These details are used in identity management and indicate key attributes of the individual for statistical analysis of services provided. Demographic details are also used to assist clinical care. Information, such as the biological sex of the individual, is important for clinical care while information about the individual's gender (social representation of sexual identity) is used for addressing the patient and respecting their wishes. Both concepts are essential to healthcare.

The date of birth is one of the most used identifiers of individuals. In some countries and situations, this data is exceptionally reliable, in others less so. Some developing countries do not routinely register every birth.

Example 1Person born in 1918 had birth registered by the father some days later. The mother insisted they were born on 2nd November, but the formal registration completed by the father was 3rd November. The individual discovered this when he or she applied for a passport.

Example 2Highly mobile populations not connected to the Internet or not located in other modern environments may know the season in which a person was born but not the specific date or month.

Identity management systems need to be able to indicate clearly and accurately which components of the date are known with certainty. If this approach is taken, computer search facilities can be built to take this uncertainty into account.

Death details may not seem immediately relevant for identity management; however, if several people's details stored in the system being used, share similar identifying information, it is important to know if any of them are deceased. If they are deceased, they are not likely to be the individual standing at the desk seeking care. However, there have been cases where people's families have reported the patient as deceased when they wish to avoid payment for services. Like any other data, it is possible to incorrectly record the occurrence of death. The source of death notification can be significant in confirming the accuracy of the data. It is common practice in many healthcare systems to confirm death details with national or jurisdictional death registers.

Concepts such as marital or living situations are rarely used for identity, but linkages to family individuals can be used in that way. Common family linkages maintained in healthcare identity management systems include the mother, father, siblings, children, aunt, uncle, cousin, or grandparent.

Such systems also need to be able to indicate if that relationship is biological, legal, social, or all of the above. Individuals who know they were adopted may indicate their relationship to a father, but that relationship is social and legal, not biological. Relationships like this can be used to indicate defacto partners or close friendships. Inaccurate or assumed information of this type can be clinically important, particularly as clinical care moves more and more into genetic associations with disease. This approach also needs to be able to support representation of artificial insemination, surrogates and similar yet more complex biological and legal relationships.

These relationships can also be used to manage security protection. For example, where information about the individual is only to be made available to those with a defined, accepted relationship or never to be given to specifically named person or external agency.

Address

Increasingly, address information is being divided into atomic components, the building, the floor, the apartment, the street number, the street, and so on. This approach also supports the use and storage of Global Positioning System (GPS) location addresses. Healthcare services such as ambulance and district nursing, care service planning, and public health monitoring benefit from using this information.

The current address of the individual is a major identifier, but not reliable. For care service providers and insurers' communication, the capture of current address details is commonly done through electronic means and may not be well maintained. Systems need to make address currency easy to maintain and manage for patient evaluation and update.

Electronic address

Electronic addresses include email addresses, phones, and mobile phones, skype names and similar identifiers. These addresses are often more reliable than a physical address as numbers such as mobile phone numbers are often more stable than the physical address and are being used as identifiers in many organisations. This is not appropriate for those who make use of pay as you go SIM cards.

Identifiers

Unique identifiers are used throughout healthcare. Every organisation has their customer and supplier identifiers. Jurisdictions, nations, and insurers each make use of these in different ways. It is important to consider these as simple keys to the patient information. A person can have many identifiers. Healthcare systems need to be able to monitor the following for each of those identifiers for different types of things identified.

It is common practice for healthcare to seek to have a single identifier. This was excellent practice when the record was a hard copy so that there was a single place to find all information. However, a single record is no longer relevant. In a digital age, data can be obtained from different sources—the key is to be able to amalgamate data with different identifiers when they relate to the one person. Many countries have developed national identifiers to serve this purpose. Organisations do not need to renumber their existing identifiers, but they do need to be able to link the disparate records electronically. An identifier index is the most efficient mechanism to achieve this with minimal modification at the local system level. Identifier data are listed, described in Table 3.

The right to privacy

Once identity is established, systems need to be able to protect that identity. There are several common scenarios where this requirement emerges.

Respectful use of the information in healthcare by those working in the system is required. Awareness of this responsibility has improved over the last 20 years. This problem is not well handled by those who work in healthcare. Unless our systems can provide a mechanism to use temporary names to support private care, it is often difficult for those who work in healthcare to obtain private care. The bureaucratic approach which often insists upon the individual's 'formal' name being used throughout his or her care does not recognise this right to privacy. This needs to be better addressed in healthcare systems around the world.

The international standard on identity management indicates how a name can be associated with the patient's formal identity numbers and names while still using a different name. The use of appropriate identification management techniques can resolve this issue, but the illusion that identity is simple or just a business activity has meant that most of the healthcare software products around the world are not built to deal with this issue appropriately.

The other area to be considered is when information is gathered for reporting, research, and care management purposes. Information technology has increased the ability to gather and analyse data. When this is done, the identity of individuals must be protected. International standards are available to manage de-identification requirements in healthcare systems. ISO 25237:2017 Health Informatics—Pseudonymization [3] is a valuable resource when contracting services for data mining or when governments request data from healthcare providers as well as to support research activities. This standard includes guidance on de-identification of data for data aggregation but also on the requirements for establishing and maintaining privacy of individuals from a technical perspective.

TABLE 3 Identifier data.

Data component	Description	Comment
Identifier	The unique key linked to the identity of a person, organisation, or thing	
Identifier issuing organisation	Identification of the organisation which issued this identifier	This can be used to establish the use cases relevant to the identifier
Identifier currency	Though most organisations seek a single identifier for an individual, errors do occur and temporary numbers are used. When this happens, a number may be retired (called merge) and linked to the retained number This information can be used to indicate if the physical record still exists or has been transferred to another media Date of destruction can also be recorded	Because healthcare systems must handle emergency identification well, merging capabilities in healthcare systems are essential The best systems also offer methods to unmerge content where such merges are made in error
Status	Indication of whether this identifier is currently active (start/end dates are usually applied as well as the specific status of the identifier). Status may include whether active, archived	
Format	The media of the record may be held in the identity information	This can indicate if the record is paper based, digitised, or electronic and whether it is a copy or the original source information
Link to who/what is being identified	Whether a person, group of individuals, or thing	
Identifier certainty	The healthcare system must know the certainty that this identifier is	If this is a temporary identifier, then the healthcare system and its users cannot be assured that all relevant information is available

Clinicians as individuals

Healthcare providers must also be able to identify those who are permitted to undertake specific tasks. Such identification includes all the requirements of the individual already discussed plus the identification of the individual's work locations and organisation's relationships at any point in time. Communication methods to support transfer of care to specific providers are also required to support continuity of care. Another need is the ability to manage authority to practice making use of for example:

- Specific qualifications of the individual
- Whether that qualification is still current and recognised
- Scope of practice associated with the set of qualifications
- Level of authority, e.g. case manager, allocated responsibilities

- Whether there are any restrictions on practice, such as those imposed due to malpractice in the past
- Whether the individual is appropriately immunised for the activity being undertaken

Clinicians and other staff from a healthcare organisation require all the same identity management data as patients. This information is used to record who provides or plans treatment or accesses patient information in healthcare systems. Recognition of formal names as well as preferred names is relevant to clinicians who might use preferred names as business names to differentiate from others in their family.

It is essential that the following information be maintained for clinicians:

- Qualifications
- Professional organisation recognition, e.g. fellowships
- Immunizations
- Links to organisational places of operation
- Registration
- Continuing professional development
- Languages spoken

Details of qualifications usually include the actual qualification and issuing organisation as well as the date of qualification. Individuals may have many qualifications in different specialty areas.

Information to be maintained for clinicians includes details of recognition such as fellowship, dates of approval to practice and specific areas of practice or any restrictions on practice. This information may be managed by government agencies or the relevant professional body—the approach differs in different countries.

Immunisation details include the specifics of the vaccine and period of coverage required for those working with patients or in healthcare settings. The requirements will differ depending upon the professional group. Though often stored by the organisation, this information could be held in the individuals' healthcare record or with their professional registration.

Links to organisations where the individual has worked may be held by the organisation or the individual and should include the date started and ended for each location and role undertaken by the individual. These types of data are usually stored in human resources (HR) information systems.

Searching to identify individuals

In the past, registers of individuals in healthcare were kept in books or files of cards. This limited the options for finding those individuals. Books were kept in registers for admissions to hospital, appointment books, and lists of specimens received in laboratories. Cards were usually used for simple patient records, such as those in older primary care services, or for registers of patient identity where the cards were filed alphabetically by the family name, given name/s. Such systems focused on either date/service or family name as the primary search criteria.

With the introduction of computer systems, search options have improved but are not yet as flexible as suggested in the international standard. Some of the improvements include the use of biometrics, including use of sound and partial matching and common name matching approaches to find typographical variations.

Searches can use name, date, sex, gender, address information, and identifiers and need to consider information about the certainty of each item. Improved searching options include recognition of current and past addresses, including electronic address search options.

Online search facilities to assist healthcare providers and subjects of care to find relevant care providers are a technique being used more commonly. Some national provider registers are now published online.

Organisations

Healthcare organisations need to be able to manage the identity of individuals seeking care or for whom care has been provided as well as details of the care providers in their employ. Organisations also need to clearly define their own identifying information from a business perspective, with the ability to clearly identify to their community:

- the services they provide,
- the location of services provided, potentially with map-related discovery options, such as google map display of services, easy to follow navigation systems,
- Hours of operation of services,
- Contact information for services.

Organisations may be hospitals, governments, care providers, not for profit, and public or private business entities. These entities have identity just as individuals do, including the details provided in Table 4.

Identification of things

GS1 is a not-for-profit standards development organisation, as described in Chapter 4, that develops and maintains global standards for business communication, including bar codes used extensively for the identification of things to support the global supply chain. There are many things that need to be identified in healthcare, each with its own unique characteristics that determine the most useful identification method to be adopted. These items include the following:

- Ambulances and other mobile services (such as district nurses)
- Pharmaceuticals
- Devices, and
- Equipment/tools
- Surgical supplies
- Implants

TABLE 4 Data for organisation identification.

Component of data	Description	Comment
Name, identification	This includes the formal and any informal or abbreviated names of the organisation Identifiers may include organisation numbers for taxation or healthcare communication or legal registration	It is also common to include source information related to the organisation's establishment as a company
Address	The address for the business headquarters but also addresses for other site/s where the organisation operates	Addresses may be linked to specific services
Electronic addresses	Website, phone, email, etc.	Including contact information
Departments or services	The specific services available, their addresses and contact details, the times and dates of operation	This information is not only relevant for general service management but also useful to external carers to support referral to the organisation's services. Increasingly, this information is being made openly available to assist patient access
Employees	Identifiers of employees, their employment dates, and roles undertaken Systems should link to professional currency, immunizations and work ready information, such as inhouse training records	Clinicians usually include medical, nursing, diagnostic, pharmaceutical, and allied health professionals, but in some organisations may include administrative and other clinical staff particularly where local systems must maintain details of who has accessed healthcare information as well as those who have contributed directly to the patient's care and record

Mobile services

Identification of mobile services involves the use of GPS either in the transport vehicle (ambulance, district care transport vehicle) or can be carried on the individual. In some situations, phones are being used to identify the location of staff operating outside a hub service.

Pharmaceuticals

Pharmaceuticals are usually identified nationally as each country manages the approval of the pharmaceuticals legally available for use in their country. There are different forms of identity for pharmaceuticals, what the product is and how it is sold. The international standard on Identification of Medicinal Products [3] should be the source of information on identity of pharmaceuticals and is described here as a very simple summary. Pharmaceuticals are identified using a large set of data including, but not limited to, the data in Table 5.

Identifiers for medicinal products are included by law on packaging in most countries either by bar codes or Q-codes to support scanning for inventory management and correct recording of drug administration.

TABLE 5 Data to represent pharmaceutical identity.

Data component	Description	Comment
Substance	The chemical compositional elements of the pharmaceutical	Usually identified using a recognised terminology identifier such as a ConceptID in SNOMED CT which has the advantage of being internationally consistent
Method of delivery	Tablet, capsule, injection, etc.	This can also include binding agents and emulsifiers used in the manufacturing process of substances
Single dose size	For example: 500 mg	These measures are usually identified using numbers and a unit of measure
Pack size	Either the number of single dose units (e.g. 20 tablets) or the volume of the pack size such as 250 g	
Manufacturer	Organisation name and id of the organisation responsible for manufacture and distribution of the product	This can include both a source provider and a local provider
Batch identifier	A code indicating the production line group to which this product belongs	
Product name	Unique and generic name for the pharmaceutical product	Names may be quite different to the name of the actual substance included in the drug and is usually proprietary In some countries, it is required that the substances must be clearly listed while the product name is optional

Devices and equipment

Devices may be a tool used directly in care or to support care. Many devices are uniquely identified (each individual device has its own id). Individual identifiers are used for devices such as diagnostic machines. Other devices may be produced in batches similar to pharmaceuticals and require a batch identifier to be included in the identifier of the item.

Device identifiers are usually identifiable electronically either through bar codes, Q-codes, or radio-frequency identification (RFID). These devices use electromagnetic fields to identify and track devices or equipment. In some surgeries, RFID software is used to monitor and record the equipment entering and leaving a surgical area to ensure all items are cleared and cleaned from the area.

Other identification

Healthcare uses a wide range of identifiers; these include codes to represent administrative, financial, clinical care, planning, and all other activities. For more information on the identifiers used in code systems, see Chapter 9 on data quality.

Requirements for successful identity systems

Identity, like much of healthcare, requires respect for the individual and the ability to represent that individual appropriately to his or her social, legal, and cultural status and needs.

Improvement is needed in existing systems of identity management as most are built upon standard commercial business identity management requirements with little ability to manage the complexity or longevity requirements of healthcare practice in addition to administrative processes.

Just as in all other areas of healthcare, identity management needs to be designed and maintained with a thought of the whole patient journey and the whole organisational and data supply chain. Identity management is a complex task requiring special skills beyond that which might be imagined.

Your digital health roadmap needs to evaluate the ability of your identity management system/s to:

- Represent all the attributes of identity appropriately and consistently,
- Manage identity data over time,
- Motivate for up to date identifying information and make collection of current data easy and efficient,
- Present data in a culturally and respectful manner for all individuals and organisations,
- Use tools to make identity management easier and safer. This might include facial recognition, RFID of devices, and bar coding of drugs.

The key is to remember that there is no one identifier more important than another, and that accuracy, security, and respect are essential in the management of identity.

References

[1] ISO/TS 22220, Health Informatics—Identification of Subjects of Healthcare, ISO, Geneva, 2011.

[2] GS1, Good Morning, Mr Jones, Here is Your Medication—Standards for Precise Patient Identification, 2018, Available from: https://www.gs1.org/sites/default/files/docs/healthcare/broch_a4_patient_caregiverid.pdf.

[3] ISO/11238, Health Informatics Identification of Medical Products (IDMP)—Data Elements and Strauctures for the Unique Identification and Exchange of Regulated Information on Substances, ISO, Geneva, 2018.

CHAPTER

13

Electronic health records and essential technology paradigms

Evelyn Hovenga[a,b,c], *Thomas Beale*[d], *and Heather Grain*[b,c,e]

[a]Digital Health, Australian Catholic University, Melbourne, VIC, Australia [b]eHealth Education Pty Ltd, Melbourne, VIC, Australia [c]Global eHealth Collaborative (GeHCo), Melbourne, VIC, Australia [d]ARS Semantica, London, United Kingdom [e]LG Informatics Pty Ltd, Melbourne, VIC, Australia

Many shades of electronic health records (EHRs)

Our health industry is knowledge-intensive and constantly changing as a result of ongoing healthcare research, technology developments, and service delivery innovations. Its core business relies heavily on clinical data, collected from patients and multiple healthcare providers, stored in a variety of formats, systems, and locations. Such information needs to be electronically retrievable from consolidated patient-centric electronic health or medical records [1]. This requires clinical data/information to have a defined digital representation, otherwise known as an information model.

Record keeping is a fundamental healthcare-related activity reliant on the collection of data and information. Patient safety will be compromised by software that processes clinical data inaccurately. Electronic health records (EHRs) collect and store lots of detailed data, also referred to as 'little' data, which feed 'big data'. We need a common understanding of the scope of all that little data (various levels of data granularity) within the context of a world view to enable data to be entered once, and used many times at any level, for multiple purposes within a digital health ecosystem. We need to ensure that data integrity is maintained at all times. This starts at every point of data entry, is continuous throughout each data processing cycle, and is used during the entire data supply chain, including data transfers and linkages.

For example, a common essential group of data every healthcare provider needs to know about are those about allergies and alert about potential allergies or contraindications. When collecting and designing these data, it is important to recognise that there are two different groups of data used to represent allergy information: first, reporting an event (which may or may not have considerable detail) and, second, the assessment of the risk of future adverse events. There is also a need to be able to differentiate between allergies and intolerance levels,

https://doi.org/10.1016/B978-0-12-823413-6.00022-7

which is especially applicable relative to food. Those working in general practice often require no more than a simple Boolean representation (e.g. is allergic to penicillin: True/False) of this type of clinical information while others, such as immunologists, require a more detailed expression of the same concept, such as a physiological response relative to a given substance.

The clinical knowledge domain, represented by healthcare data, rules, and relationships, that needs to be able to be represented in an EHR is huge. An estimate is based on the openEHR international archetypes (around 10,000 data points in 650 archetypes).

Using a consistent scalable representation of this domain content is central to achieving EHR benefits and safety. The availability of a representational formalism for expressing and making them openly available is a key to achieving this. Such standards include an agreed standard reference information model (ideally national or international), an agreed set of data types, and standard terminology binding to represent the concepts contained in EHRs. This gives software developers consistency to support their software development into the future and allow them to focus on user experiences, functionality, and interfaces. All EHR/EMR products should represent their data based on a single set of domain content definitions, ideally managed by a dedicated organisation(s), independent of vendors.

Types of health records

Health information about any individual needs to be patient-centred, recorded, and managed in a manner that enables ease of documentation (data/information collection), storage, access, retrieval, and timely and multipurpose use. Our primary focus is therefore the patient information record—the container for healthcare data that has traditionally been paper-based but now needs to be consistently structured to meet the demands of digital healthcare. There are a number of different types of records in use, and although people tend to refer to all as if they are equal in terms of functionality and/or ownership, this is not the case.

The acronym 'EHR' is a general term in common use that should only be used to describe a 'computable' health record. A digital health record can be a scanned-paper record consisting of images without any computable functionality. A patient or personal health record refers to a record maintained and controlled by a person (patient), while a *medical record* is owned and maintained by a healthcare provider, and these can be paper-based or electronic medical record (EMR).

Information and knowledge about our own health and the healthcare services received during our lifetime are usually documented in records kept by ourselves and the many different healthcare service providers we see, and where available, by us. Now that we have entered the electronic era, and we expect these records to be maintained in an electronic, and preferably computable, form. Achieving this in a manner that truly is worth the investment and provides real benefit to us requires a paradigm shift in thinking [2].

This chapter's primary focus is on patient and health system centred functional operational data needs. It explores challenges associated with the need to harmonise health data, its collection, interchange, linkages, use, and storage with the best available technologies to optimise the value of these investments within any digital health ecosystem.

A national example of EHR adoption

In President Bush's State of the Union address in January 2004, he outlined a plan to ensure that most Americans have electronic health records within the next 10 years. The Health

Information Technology for Economic and Clinical Health (HITECH) Act of 2009 was designed to encourage widespread use of EHRs. By 2015, EHR systems had been installed in 96% of hospitals [3]. Such records are usually not kept forever. Yet, ideally our health records contain all the information about our past, present, and future planned care. By 2020, we ask the following: Are these records still fit for purpose following their implementation and several years of use? What has changed over this time and were these records able to adjust to these changes? Was information collected in 2004 available and retrievable in 2020?

It is very difficult to predict what information is of use for future clinical decision-making. For example, a minor flu-like illness may indicate the onset of a chronic viral condition sometime later; a tick bite gives the clue of Lyme disease or a gastrectomy many years previously indicates that vitamin B12 deficiency might be the cause of neurological symptoms. Such past information needs to be easily retrievable when required.

Healthcare information is ideally kept up to date and complete to ensure continuity of care. Data availability needs to consider possible future relationships as demonstrated by these previous examples, to avoid unnecessary investigations and timely intervention, and most importantly to ensure the safety of the care provided. In the absence of such access, people have been expected to simply remember all significant health events such as allergies, vaccinations, surgical interventions, not to mention their current medications, whereas computers are wonderful tools for getting the information that we need at a particular moment overcoming this reliance on memory.

Future proofing our health records—A need for transitional arrangements

Data pertaining to any individual's health status represent the most valuable assets within the health industry. Such primary data are collected by, and on behalf of, individuals via pen and paper, or any type of device or application. Such data are stored in repositories by healthcare providers in a variety of formats determined by product vendors. Every type of repository has its own structure irrespective of format.

In recognition of current diversity, the ISOTC215 committee has reached an agreement regarding definitions [4] enabling a consistent interpretation of the type of record (repository) to be considered when developing methodologies for data sharing between providers and data linkages for secondary data use. This diversity is a major current limitation to optimal data use for any purpose. When reference is made to 'health records', it is important to recognise that this is an overarching term likely to be interpreted differently by individuals to mean such things as a healthcare record, a medical record, or a personal record. The term is also frequently used to refer to a system that makes use of health record content.

Why wait for Governments to recommend the standards as part of their national digital health strategies? Why can't the software industry work collaboratively to adopt the same foundational standards so that each software developer has the opportunity to develop their own unique applications that are completely interoperable with all EHRs, yet have a competitive advantage? Large EMR vendors have little motivation to solve EHR-related issues.

EHR structures need to be managed separately from EMR applications that make use of EHR content. That way we can make far better use of the 'little' data, enabling improved decision-making at all levels of the health industry, including the automation of reporting, monitoring public health, and improving the ease of clinical/best practice research, to achieve a sustainable cost-effective, high-quality national health system. This requires the adoption of an agreed standard platform infrastructure.

Health record attributes and definitions

First, we need to differentiate between health records, data obtained by individuals using various apps, and healthcare records. The latter refers to documentation provided by healthcare providers which records observations, test results, judgements, and clinical decisions made on behalf of the patient. This documentation also serves as a medico-legal record of care, whereas the former are the result of individuals observing and recording their own health-related physical, physiological, mental, and functional attributes. These are commonly referred to as Personal Health Records (PHRs), defined as a health record controlled by the person to whom it pertains. These are increasingly being integrated into patient-centric health records, able to accept patient-entered data.

PHRs are usually established independently by people who simply wish to maintain their own health records, such as immunisation records, although this type of persistent information is best stored with a person's identity records. There are many PHR applications which individuals can use. Alternatively, a person can simply use their own computer, phone, or smart watch, to maintain such a record and make it available to any provider as required via a Bluetooth–WiFi connection, memory stick, or printed copy. In the simplest form, it could consist of a printout listing current medications or allergies carried in one's pocket, briefcase, or handbag, or documented in WHO's international certificate of vaccination booklets carried with passports. PHRs may be electronic (computable or stored electronically) or handheld (paper-based). A PHR stored in any format may have contributions from the individual, from the providers of healthcare, or from their carer/s. The legal rules for a health record still apply to this record.

Records held by healthcare providers

Our focus is on healthcare records maintained by healthcare providers. There are situations when it is desirable to be able to link PHR content with that person's healthcare record. Patients may be given written access to their EHR in a controlled way so as not to confuse or compromise healthcare-provider-created data. A healthcare record is 'produced for and used within a healthcare organisation' [4]. This refers to traditional medical records which are now available in a number of different formats, paper-based, scanned, and stored digitally, or in a variety of different computable formats or in transitional (hybrid) formats. All electronic versions are commonly referred to as electronic health records (EHRs) or electronic medical records (EMRs), generally understood as storing and using institutional episodic data; yet, each type of record has its own functionality. Widespread adoption of computable records is the ultimate goal, as this is a fundamental requirement for a well-connected, cost-efficient, and cost-effective digital health ecosystem.

Both primary healthcare providers supporting ambulatory care, hospital (residential), and other healthcare providers need to be able to amalgamate data obtained from any number of other internal and external service providers into one patient identifiable record. Required record structures, processing functionality and connections, differ by type of service provider. For example, those providing a laboratory or imaging service usually maintain a record for each patient that may cover numerous events/encounters over time, which are not necessarily related to any one condition being treated. Image service providers used to share their reports with X-ray films or via CDs transported by the patient despite their own record storage computable format. This needs to be accessible online by other healthcare providers.

Shareable and integrated health records

Shareable or federated health records are 'able to be shared between healthcare organisations without risk or change to meaning' [4]. To achieve this, the healthcare record (EMR) requires a standardised format of both structure and content. Such records support federated, integrated care information systems. Shareable healthcare records are usually electronic in format. Even though some may be digitised, their records do not necessarily enable the transfer of data with the correct meaning of the recorded concepts.

Hospital-based healthcare records may be referred to as a shareable record, given that multiple internal service providers need to make use of and contribute to its content. In many instances, this need to share records also applies to external hospital service providers. There is a need to differentiate between the records (repositories) themselves and the systems that make use of their content and enable record sharing.

The 2005 ISO technical report that defines the EHR scope and content (TR20514) [5] noted that:

> *the difference between a nonshareable EHR and a shareable EHR is analogous to the difference between a stand-alone desktop PC and a networked PC where the latter adds enormous benefits in terms of locating, retrieving and exchanging information using the internet, an intranet, email, workgroup collaboration tools* etc.

This technical report defined a shareable EHR into three different levels as follows:

- between different clinical disciplines or other users,
- between different applications at a single EHR node, i.e. at a particular location where the EHR is stored and maintained, and
- across different EHR nodes, i.e. across different EHR locations and/or different EHR systems.

An integrated care EHR was defined as follows:

> *a repository of information regarding the health status of a subject of care in computer processable form, stored and transmitted securely, and accessible by multiple authorised users. It has a standardised or commonly agreed logical information model which is independent of EHR systems. Its primary purpose is the support of continuing, efficient and quality integrated health care and it contains information which is retrospective, concurrent and prospective.*

The requirements for global use of EHR content should ensure that its data can be used, shared safely and efficiently, and exchanged between clinicians of all disciplines, across all sectors of health, different countries, and different models of healthcare delivery, to support multiple uses such as research, epidemiology, population health, health administration, financing, and health service planning. This requires local, national, and global digital health ecosystem architectural structures.

EHR definitions reflecting functional perspectives

Computer-based patient records were first defined in the United States by Institute of Medicine (IOM) in 1997 [6]. Definitions reflecting multiple functional perspectives have continued to evolve. The Health Information Management Systems Society (HIMSS) definition of the EHR does not distinguish between the record (repository) and the system that contributes to and makes use of the data contained within the record itself, as follows:

The Electronic Health Record (EHR) is a secure, real-time, point-of-care, patient centric information resource for clinicians. The EHR aids clinicians' decision making by providing access to patient health record information when they need it and incorporating evidence-based decision support. The EHR automates and streamlines the clinician's workflow, ensuring all clinical information is communicated and ameliorates delays in response that result in delays or gaps in care. The EHR also supports the collection of data for uses other than clinical care, such as billing, quality management, outcomes reporting, and public health disease surveillance and reporting [7].

This definition reflects a common interpretation that has resulted in many poorly designed clinical systems. This limits data-sharing possibilities due to the structural impossibility of achieving seamless data exchange between EHR systems. This view is an instrumental precursor of the continuing myth that the adoption of messaging standards is all we need to achieve semantic interoperability to generate fully integrated health records able to support every patient's own life-long health journey.

HIMSS lists EHR attributes, essential requirements, and evidence that again demonstrates this mixed view as some refer to the record itself and others refer to system-processing requirements as shown in Box 1.

None of these definitions have made it clear that individual patient-centred EHRs represent a complex data repository system that needs to be accessible, retain context, and operate within a complex healthcare delivery ecosystem, using numerous additional complex software applications, which needs to be optimally interconnected.

This complex health knowledge domain is made up of numerous embedded knowledge domains, each with its own conceptual relationships, that from a software engineering perspective needs to be digitally managed. Software architectures provide the necessary foundations for any successful software system. There is an urgent need for the healthcare industry

BOX 1

HIMMS EHR attributes and essential requirements and evidence for assessing the extent to which an organisation is using an EHR.

1. Provides secure, reliable, real-time access to patient health record information, where and when it is needed to support care.
2. Captures and manages episodic and longitudinal electronic health record information.
3. Functions as clinicians' primary information resource during the provision of patient care.
4. Assists with the work of planning and delivering evidence-based care to individual and groups of patients.
5. Supports continuous quality improvement, utilisation review, risk management, and performance monitoring.
6. Captures the patient health-related information needed for reimbursement.
7. Provides longitudinal, appropriately masked information to support clinical research, public health reporting, and population health initiatives.
8. Supports clinical trials.

to consider how to make our EHR repositories fully interoperable within a digital health ecosystem from the start.

Health data standardisation and interoperability leadership

A previous chapter covered the result of a 2012 WHO Forum on this topic. This forum was followed in 2014 with a Joint Inter-Ministerial Policy Dialogue on eHealth Standardisation and a second forum on this topic [8]. The latter forum's aim was to facilitate further dialogue on the need for policy and governance mechanisms for the adoption of health data standards and to draft a policy and governance framework for full adoption by its member states. Its concluding outcomes of relevance to the content of this chapter are represented in their summary of recommendations relating to the use of health data standards and health data exchange between systems, as presented later.

Key points identified were that eHealth policy for standardisation and interoperability in a national context should:

- Adopt appropriate electronic health information exchange (HIE) technology.
- Set health data and health IT standards to ensure interoperability at data, device, and system levels, in a framework containing a fixed core set of maintained standards allowing for a degree of innovation outside that core set.
- Use existing international standards where possible and adapt specific standards to suit national contexts. This has resulted in numerous countries with clinical systems contributing to the number of data silos, as their data cannot be queried.
- Provide unique identifiers for patients, healthcare workers, and healthcare facilities, with verification and authentication procedures.
- Ensure the safety of interoperable medical devices, and ensure security, defining privacy, and security policies addressing technology use in healthcare delivery.
- Build capacity from country and ministry levels down to that of frontline health workers. This includes financial and academic capacity as well as technical and human resource capacity.
- Ensure good governance, balancing top-down and bottom-up approaches, and encompassing equity and accessibility, legality, user rights in line with human rights, privacy, responsibility, and accountability to citizens and to the state. Compatibility of technologies, efficiency, open dialogue and a shared vision on use of data are necessary for implementation. In monitoring compliance, clear goals and key indicators for monitoring and evaluation are needed, with mechanisms for social participation.

Unfortunately, these recommendations were based on the outmoded paradigm that the adoption of standards would magically glue noninteroperable systems together. We continue to observe the implementation of this never-ending costly approach relying primarily on messaging standards.

Electronic health information exchange (HIE) technologies represent key components of the interoperability concept. These technologies have evolved during the last 60 years as shown in Fig. 1.

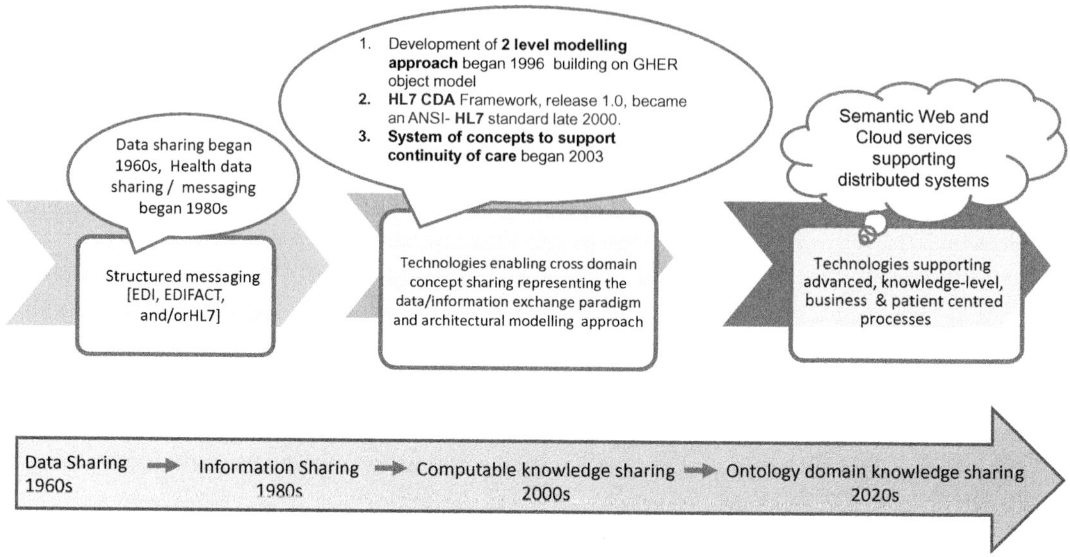

FIG. 1 Health data exchange historical technical developments.

This began with structured messaging making use of Electronic Data Interchange (EDI), Electronic Data Interchange For Administration, Commerce and Transport (EDIFACT), and later with Health Level Seven (HL7) messaging standards. This was followed by technologies that enabled knowledge sharing with technologies that represent concepts of care delivery, i.e. models defining data groups, as specified in the openEHR archetypes and ISO 13940, a standard that describes a taxonomy of concepts of care delivery to support continuity of care, to cooperation at the application level including Web services.

All of these technology solutions focus on information and communication technologies (ICTs) plus system interoperability schema using terminologies and ontologies for representing data, information, concepts, and knowledge. This demonstrates the evolution of three interoperability levels: a) foundational, b) structural, and c) semantic interoperability.

This digital era requires ICT systems to be more closely integrated in the real world of operational processes. This requires technology support for advanced, knowledge-level, and business process-focused interoperability between all principals acting in digital health ecosystems, including persons, organisations, devices, applications, components, or objects necessary to achieve common health service delivery objectives. These represent a number of different interoperability levels as described in Table 1.

Technologies referred to here are covered in further detail in other chapters. It is important to note that many EHR/EMR systems in current use (legacy systems) were originally designed during anyone time within this evolving historical technical era. These legacy foundational structures are ill-equipped to optimally accommodate later developments requiring these systems to transition to new technical platforms better able to meet the needs of, and process new medical/health, business, and patient-centred knowledge.

There is a need to achieve the highest possible level of interoperability to safely, effectively, and efficiently support patient-centred care. We now need to technically progress from

TABLE 1 Interoperability levels relative to instances [9] (wording modified from ISO 23903:2020).

Information perspective		Organisation perspective
Interoperability level	**Instances**	**Interoperability level**
Technical	Technical plug and play, signal and protocol compatibility	Lightweight interactions
Structural	Simple transactions, envelopes (EDI & HL7)	Data sharing
Syntactic	Messages and documents with the inclusion of agreed terms, vocabulary	Information sharing
Semantic	Advanced messages with common standard information models and terminologies	Knowledge sharing at IT concept level able to be processed by receiving computer
Organisation/service	Common operational process	Knowledge sharing at operational concept level. Agreed service function level cooperation
Knowledge-based	Multidomain processes, ontology focused	Knowledge sharing at ontology domain level. Cross-domain cooperation
Skills-based	Individual engagement in multiple knowledge domains	Knowledge sharing in individual (patient) context. Moderated end-user collaboration

data model, information model, and ICT domain knowledge perspectives to the knowledge perspective. Knowledge needs to represent the operational processing domain. This domain makes use of health data supply chains to facilitate automated secondary data use.

Health data standards play a very significant role as explored in other chapters. It needs to be recognised that to represent a highly complex, highly dynamic, multidisciplinary/multiknowledge domain healthcare system, by making use of one terminology/ontology, is not realistic.

One architectural solution

The ISO 23903:2021 standard [9] enables the advancement of interoperability from the data/information exchange paradigm to knowledge sharing at decreasing levels of abstraction. It defines a model and framework for a harmonised representation of existing or intended systems with a specific focus on ICT-supported business systems. The Interoperability and Integration Reference Architecture supports ontology harmonisation or knowledge harmonisation to enable interoperability between systems. This architecture also supports the integration of systems, standards, and solutions, at any level of complexity without the demand for continuously adapting/revising those specifications.

It argues that the adoption of a Meta Reference Architecture is suitable for a digital health ecosystem. It is considered to be the only solution to realising advanced interoperability, for flexible, scalable, business-controlled, adaptive, knowledge-based, intelligent health and social ecosystems. This Meta Reference Architecture follows a systems-oriented, architecture-centric, ontology-based, and policy-driven approach [9]. This fits with the adoption of a knowledge-based platform which needs to be a key foundation for any digital health ecosystem's architecture.

In addition to this, solution is a need for workforce capacity building to enable this Meta Reference Architecture to be adopted in a successful manner. Another chapter presents further details about workforce capacity building needs.

Learning health systems

From a national perspective, it is highly desirable to make use of a standard EHR reference information model as the EHR concept needs to be the centre piece for the design of a national System of Systems Architecture Framework (SoSAF) for its health system. Such a framework defines how the many systems are best able to work together to meet ecosystem objectives, whilst considering the details of the individual systems and their impact on ecosystem performance or functionality.

Such systems may be defined or referred to as *'learning health systems'*. [10]. The 2007 IOM workshop participants noted the need for 'Universal electronic health records' able to support *'comprehensive deployment and effective application of the full capabilities available in EHRs as an essential prerequisite for the evolution of the learning healthcare system'*.

The adoption of a suitably designed health industry SoSAF, consisting of one open-source agreed national or global technical knowledge-based platform, devices, and software applications, is expected to be able to support decision-making at all levels, advanced data analytics, including artificial intelligence, precision medicine, and automation.

Documentation supporting patient journeys

People now accustomed to managing their lives in an electronic environment have increased the demand for access to their own health records. Having control of one's own destiny contributes to a person's own sense of well-being. Healthcare concerns all of us. It is about maintaining our physical, functional, emotional, mental, and spiritual well-being within our environment, within its own physical, social, and family contexts. We need to know about preventing ill health or injury and how to best manage any genetic, acquired, or environmental limitations or disabilities in a manner that optimises our sense of well-being.

There is more information about our body and how it functions, more research and more effort optimising its function than any other aspect of life. When a person engages with anyone in the health industry, they communicate by providing information that needs to be interpreted by the receiver. Shannon (1948) referred to such communication as a 'mathematical information theory of communications' [11].

The transmission of information about ourselves is fundamental to our autonomy, including when to seek a healthcare service. The ability to exercise such control is known to contribute to better overall health outcomes. Information systems need to be able to support such patient empowerment. Information is the result of an interpretation and organisation of data based on our ability to understand the message. This has to do with not only the language used but also the way data (the message) is presented verbally or electronically.

Individuals, their parent or guardian or carer/bystander/rescuer, decide when to make contact with the health system. Patients may have a say in where and by whom they wish to be treated, but when severe ill-health episodes or injuries occur in any location, they may lose that control. First responders need to be able to access data as clues to the likely clinical need or potential risk. This may be provided by the patient needing attention or by anyone else in the vicinity who may be able to contribute relevant information, such as a family member or friend. In its absence, the first responder is unaware of existing medical conditions, medications taken, or circumstantial evidence or allergies that represent critical historical information. Good use of available technologies can contribute to the ability of first responders to provide optimum health services.

Healthcare professional and national perspectives

Healthcare professionals rely on all types of information presented by the patient (or other) and use all their senses and prior knowledge to interpret such information and to document the relevant points into the record. All informative content is what needs to be stored in an organised manner within individual health records. Healthcare records need to contain, and be able to share, all content required to support decision-making regarding services to be provided in support of any individual's specific health journey(s). Such journeys range in time from a single visit to a multitude of visits over any specified time period. Each visit may include encounters with any number of healthcare service providers. All relevant information needs to be shared between these healthcare providers as and when appropriate.

This is particularly important for those requiring health and care support services associated with the management of longstanding (6 months or longer) health problems, where continuity of care is of critical importance, and for the use of precision medicine. Longstanding health problems or diseases include individuals experiencing any disability, difficulty with basic living activities, any number of chronic morbidities. Self-perceived health and healthy life years are related concepts. Chronic, noncommunicable diseases represent the leading causes of ill-health, disability, and death globally of persons aged 15 years and above. Those of pensionable age are more likely to have at least two comorbidities. More than 40% of the population is estimated to require this type of continuing health management support.

With changing lifestyles and ageing populations, this group is increasingly the cause of most of any nation's burden of ill-health representing the biggest health challenge. This growing health burden is in addition to acute episodes, overall public health, and the management of communicable diseases, including pandemics. Longstanding health problems are increasingly responded to, within the ambulatory/primary care sector, supported by short, often day only, inpatient care.

National health system characteristics, including workforce scope of practice and financing arrangements, plus environmental, social, and lifestyle influences, all impact on possible healthcare delivery design methods for this population. All associated information and data collections, including all individual patient health and healthcare records, form the foundational asset for any national digital health ecosystem.

Information is a key to successful problem-solving and health service delivery. Problem-solving includes the application of prior knowledge or evidence-based clinical guidelines.

It is about deciding on an investigative strategy, making a correct diagnosis, assessing care needs, and choosing the most appropriate treatment option.

Ideally information is verifiable, free from bias, clearly communicated, appropriate, and accurate. Some information may be quantifiable, comprehensive, or timely. All necessary information and supporting knowledge needs to be accessible at points of care. These characteristics are the precursors or requirements of electronic information systems [2]. One issue yet to be resolved is a person's health information governance, ownership, control, access, and use.

Digital health ecosystems need to be structured in a manner that not only deals with urgent acute situations but also needs to be able to:

- optimise our collective ability to manage individual patient health journeys as well as our growing global health burden,
- facilitate complex responses in support of individual patient health journeys, frequently over a prolonged period of time,
- support the coordination of inputs from a wide range of health professionals, support services, and external healthcare provider and local support services,
- enable best practice evaluation and research, and
- promote patient/client empowerment and participation.

Lessons learned

Few if any health records' structures, and information systems making use of their content, were designed to address these challenges. So what needs to change?

Optimising the use of information and communication technologies to support the management of health and the delivery of healthcare services requires the engagement of visionaries, computing experts, innovators, risk takers, movers and shakers, and a change of thinking by managers, purchasing decision-makers, legislators, policy developers, and implementers.

Health data, stored within healthcare records, require an ability to clearly represent all health concepts that need to be electronically managed in a standard manner to minimise loss of meaning. Our wish to maximise the use of these data contained in these records requires us to analyse the many different characteristics or attributes associated with health records, the adoption of a standard record structure, and the reference information model, and to store and represent all clinical- and health-related content as governed standard data.

Standardised health record content

Back in the 1960s, Professor Larry Weed designed the problem-oriented medical record (POMR) that was adopted by many; in fact, it became ubiquitous in clinical documentation. It provided a structured framework for the recording of patient information in paper-based records that enabled better medical decisions to be made.

We now need an international standard for storing clinical data electronically, to enable reliable and accurate electronic communication and sharing of all relevant clinical data, between records held by every healthcare provider contributing to a patient's health journey. There are several fundamental standard requirements whose international, or minimally national, adoption will ensure that these benefits can be realised. They are presented in summary form in Table 2. These concepts are discussed in greater detail in other chapters, including the standards required for a knowledge-based platform, clinical data types, terminology, clinical domain models, unique identifiers, and security.

A single logical information model needs to represent a generic health record content structure that defines data types. Health/medical records collect data that can be organised according to type, such as entries that refer to actions, evaluations, observations, instructions, or to administrative entries. These concepts need to be defined within data models (such as archetypes) consisting of re-usable data points to enable content to be defined.

Other contributions are compositions such as a care plan or an event summary or a medication list or demographic data. These compositions (may be referred to as templates) are where data (value) sets) are defined. These represent operational requirements able to be used to link data from legacy systems via an Application Programming Interface (API) or other computable formats.

This methodology enables health information systems to work together within and across organisational boundaries to support the necessary communication and documentation flows in support of any individual patient health journey. This in turn supports an integrated national health system where individual healthcare providers and support services work together, efficiently and effectively due to seamless electronic support services whilst providing patient-centred care.

Clinical systems contributing to and using record data

Clinical information systems are software applications using clinical data and information retrieved from health record repositories, to support all clinical activities such as care planning, prescribing, medication management, order entry for pathology or diagnostic imaging or other services, contributing to and making use of EHR content.

TABLE 2 Fundamental Standards Enabling the realisation of benefits via the use of EHRs [2].

Standard required	Explanation
EHR structure/architecture/information model	A structured record facilitates ease and accuracy of data extraction including data types suitable for holding clinical content
Set of clinical content models and terminology	To ensure that all clinical data stored within these records are structured in a standard manner and use standard terminology, thus enabling data/information, retrieval, comparison, and aggregation without loss of meaning
Unique patient identifier	To ensure that data from one provider record can be added to another provider's or patient's held record that belongs to the same patient
Unique individual and organisational provider identifier	To ensure that patient records maintained by an individual provider will be associated with that provider only and no other

Clinical software applications may provide decision support services, access to clinical practice guidelines, or an online data/information collection service. They may be designed to enable advanced analytics support functions, or to meet routine reporting requirements. Ideally such health record applications include telehealth and home monitoring data. Clinical software applications may be incorporated within or linked with one enterprise system, commonly referred to as EMR/EHR Systems.

Clinical applications

Clinical software applications have the potential to compromise patient safety [12]. Anyone able to program can develop a clinical application and make it available for use by clinicians. There are few if any controls or regulatory mechanisms in use to ensure data integrity and patient safety. Clinical outcomes relative to clinical application use are rarely measured, making it difficult to detect errors.

For example, the introduction of any new drug requires rigorous review and evaluation to prove a drug's safety and efficacy before it is authorised for use. We don't have similar processes in place for clinical software or decision support systems. Systems are generally adopted on the basis of their coverage of required functions, rather than on measurable benefits.

A key clinical requirement is for EHRs to be able to support the evaluation of clinical interventions relative to patient outcomes, to identify best practice, potential issues, or make practice suggestions. This requires the adoption of agreed metadata standards relative to clinical specialty knowledge domains. These metadata need to minimally represent participant/patient characteristics and health settings, technical specifications, interventions, and valid outcome measures.

To date, no widely agreed definition of a health information technology evaluation framework exists [13]. The United Kingdom's National Institute for Health and Care Excellence has developed an evidence standards framework for digital health technologies that could be more widely adopted [14]. Its intent is to ensure new technologies are clinically effective and that they offer economic value. Its design supports the relevant principles of its Department of Health and Social Care code of conduct for data-driven health and care technology.

EHR quality evaluation and certification

There are many human biases that can adversely affect the delivery and perception of a software development project. These include changing priorities or not having clear definitions of 'success'. The ISO's Systems and Software Quality Requirements and Evaluation (SQuaRE) standard [15] was developed to address this by defining:

1. A 'quality in use model' composed of five characteristics, effectiveness, efficiency, satisfaction, freedom from risk, and context coverage, that relate to the outcome of interaction when a product is used in a particular context of use, applicable to both computer systems in use and software products.

2. A 'product quality model' composed of eight characteristics, functional suitability, performance efficiency, compatibility, usability, reliability, security, maintainability, and portability, that relate to static properties of software and dynamic properties of the computer system.

These criteria form a useful foundation for EHR certification purposes which is likely to influence adoption. EuroRec has developed a quality assessment suite that enables the licensee to prepare and manage certification, documentation, and procurement of Health IT products in general, mostly focused on EHR systems [16]. Certified EHR systems can receive a EuroRec seal, used primarily to support the process of harmonisation between EHR systems in Europe and cross-border interoperability. The level-1 seal indicates trustworthiness of the clinical data, whereas the level-2 seal encompasses 50 functional quality criteria that address essential functions such as access and security management, basic functional requirements on medication, and clinical data management.

The US Certification Commission for Health Information Technology (CCHIT) was established in 2004 to accelerate the adoption of robust, interoperable health information technology [17]. Certifiable EHR products have a 'systems' focus. Its certification criteria were established using a consensus-based approach with its stakeholders. These certification criteria did not relate to the previously mentioned ISO quality standard. The CCHIT was a US Department of Health and Human Services-authorised testing and certification body, that originally focused on certifying compliance with the 'meaningful use' requirements and continued to regulate certifications to fit the needs of a growing marketplace until it ceased all certification operations in 2014.

US Certification is now only provided by the Office of the National Coordinator (ONC) for Health IT Certification Program launched in 2010 and managed by the ONC-Authorised Certification Body (ACB) [18]. Its aim is the promotion of interoperability programs. This certification program design made use of ISO/IEC conformity assessment, testing, and calibration laboratory standards and National Institute of Standards and Technology (NIST) standards.

These criteria do not specifically relate to EHRs. They reflect API documentation data, relative to products certified to the API criteria based on a manual review by ONC of a developer's API documentation. This can be problematic when applied to clinical documentation as APIs tend to be designed to suit specific connected networks that are not necessarily compliant with agreed national or global data standards. They may not be able to connect with other external networks. Some islands of excellence exist in the United States although this is not the norm.

This variety of certification criteria refers to systems and their functionality. These criteria are largely the result of EHR system definitions most readers are familiar with, as presented previously. Separating the records (patient-centred data repositories) from the software applications themselves does not appear to be regarded as a significant requirement from a system design perspective. Yet, this has major implications for a system's ability to share or link and process data obtained from multiple distributed systems incorporating a multitude of applications, as is required by EMR/EHR systems. This lack of consistency remains a big challenge, compromising the ability to share and aggregate health record data, both for individual patients and for patient cohorts across multiple healthcare enterprises, or to maximise automation and reduce costs.

Managing EHR data access

We need to decide on the type of controls required to manage access to our EHRs, the electronic data repository that is akin to the paper record. Choices to make include how and where EHRs are to be stored or accessed from, and by whom, for what purpose. Any amount of EHR data can be stored on any server with web access, any handheld device, including our mobile phone or smart watch with or without a local access barrier, or in distributed systems at the institutions where we seek healthcare. Online records in one centralised repository in the cloud may be safely accessed simply by patients themselves as well as by those providing care at any location.

How should such data be aggregated into one accessible complete, accurate, and up-to-date record, accessible by any type of healthcare provider to suit any specific purpose? Should individual records consist of a module or sets of patient-centred reports generated by a large integrated information system to match the retriever's access rights, based on role as defined within legislative requirements? Should patients and/or their providers be able to decide what data to share with whom or another system, including decision support tools?

There are numerous niche applications (Apps) in use that can provide source data for inclusion into a person's EHR, such as to support an individual's own diabetes management or to keep a record of any physiological or locational monitoring with built-in actionable intelligence. The variety of Apps now available is endless. Apps are developed to suit specific groups of 'use cases'.

Decisions need to be made regarding the use of best available clinical evidence, knowledge and algorithm governance, data sharing and linkage with other systems, API access management, data storage and retrieval facilities, as well as data access management, in compliance with applicable health data and exchange standards and jurisdictional requirements.

These choices impact on legislative requirements and methodologies to be adopted for the management of information security, confidentiality, and privacy. Differences between types of EHRs and associated data collection applications relate to the content stored and accountability for anyone who compiles, owns, manages, and has access to any component of the record. The reality inevitably represents many variations regarding all of the above.

These are complex issues that relate to data sharing, cyber security, privacy, and confidentiality management arrangements yet to be resolved in an agreed consistent manner. In any event, every individual needs to be informed about what data pertaining to them, collected by others, is accessed and used by whom and for what purpose. In some situations, it will also be necessary for individuals to provide written consent.

For example, every group of use cases generates their own specific data-sharing needs:

- A personal health record for a child with 'special healthcare needs' may be developed specifically for the child's parents/guardian and carers, enabling them to help their children receive the best possible care [19].
- Healthcare providers are legally bound to record and maintain information they have offered, and they require ongoing access to this for medico-legal purposes.

- Some information, such as severe allergy, or food intolerances are best made available to every service provider using standard meaningful presentation methods.
- A link between some EHR data sets for real-time public health monitoring or for contact tracking for the management of any communicable disease (pandemic) outbreak. Some of these data may need to be accessible by other public service sectors for action.
- Mandated reporting of specific types of health conditions and communicable diseases.
- Data sharing needs vary and need to be arranged to suit individual health conditions, and their treatment and care protocols to support timely and continuity of care throughout any individual's health journey.

Differing views on the appropriateness of centralisation of information and access control will force Governments to abandon single-source systems as a solution in this space. Sharing distributed EHRs with various healthcare providers requires the adoption of the best available health data exchange methodology to enable data sharing. Such data sharing is best facilitated via a centralised unique identification service able to manage the source of truth. However, this apparently straightforward step ideally requires that each system shares a single logical information model applicable to its local digital health ecosystem.

A precursor of achieving information sharing and access control is a standard structure for EHR and personal health data content. The value of adopting effective data sharing is to enable the delivery of patient-centred care that is respectful of, and responsive to, individual patient preferences, needs, and values [20].

Strategies for patient data mobility by design are now being developed to create new value [21–23]. This includes the development and adoption of governing frameworks, a suitable infrastructure, and a set of standards able to support desired products and services. Data mobility needs to be safe, easy, and valuable for all its users, within a global digital health ecosystem.

Ecosystem data supply chain

To adopt the strategy of entering data once and using it many times requires a careful assessment of the desired data supply chain to suit individual patient journeys, functional use to support operational management within healthcare provider organisations and local health ecosystems, as well as national and global data needs to suit multiple use cases. This includes data access, retrievability, and exchange able to facilitate knowledge discovery, processing and application, requiring extensive knowledge development, and management expertise.

Fig. 2 provides an overview of current key functional data use requirements. Once we have managed to make use of a well-connected digital ecosystem, this data supply chain is expected to change over time.

For example, separate registries will become redundant as it will be possible to extract such national activity as a specified report. Direct knowledge about any individual patient or health condition or treatment option outcomes, results of data analytics, patterns, trends, and uses, will become more readily available.

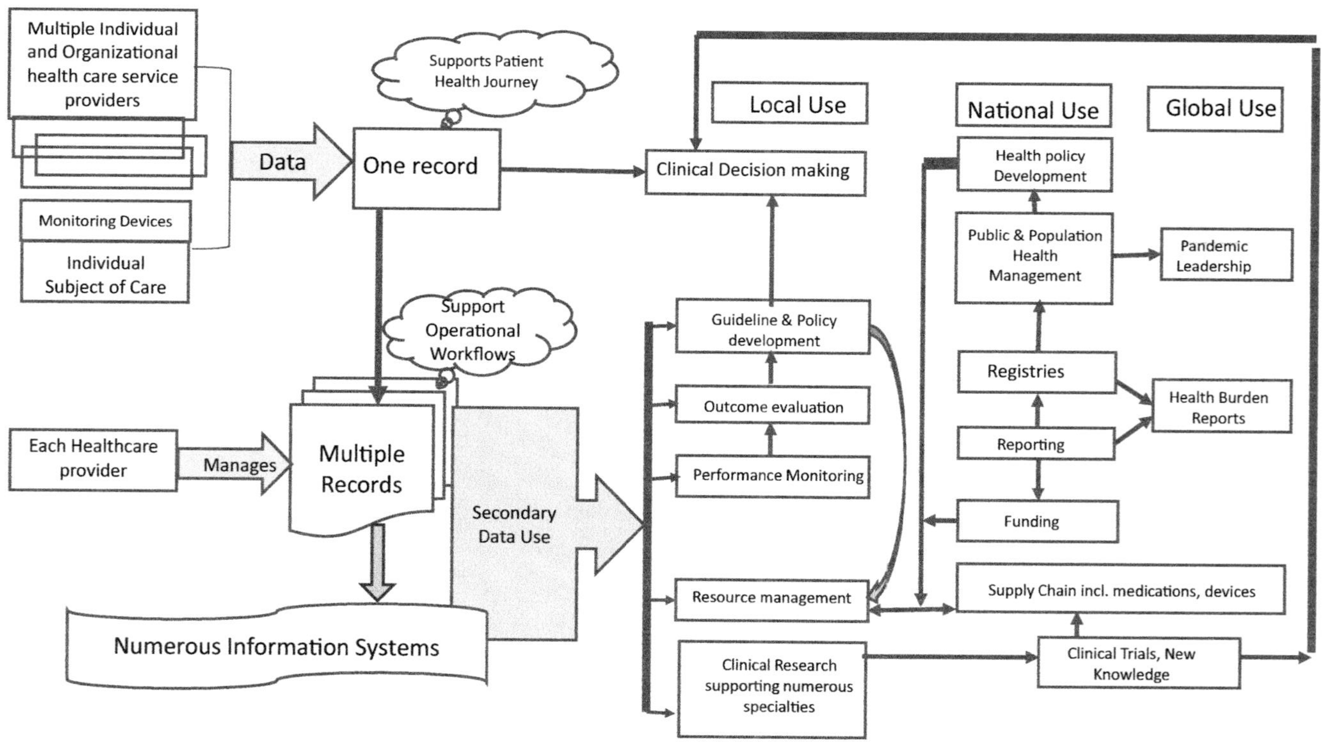

FIG. 2 Overview of a 2020 clinical data supply chain.

One person's health record data supply chain

A closer look at the data flow shown at the top left of the diagram requires us to consider all possible patient journeys. This represents the beginning of the data supply chain where the data are collected at each first or subsequent encounter with the health system. This is followed by possible use cases of these data throughout the data supply chain.

For example, data concerning the reason for the encounter such as a persistent cough, or anything else representing signs and symptoms, is recorded in the person's electronic health record, where the text can be mapped to any terminology including a SNOMED-CT code , used to support clinical decision-making and billing, or to register a notifiable health issue or inform the next service provider, etc. Once such data/information is available and represented as computable code, and its use can be automated to suit a specific use case. In addition, at any health encounter, the responder may also need to collect or access contextual data, such as the individual's date of birth, geographical location, mental health, health literacy, socio-economic status, health/lifestyle behaviours, and functional and physiological status.

Such data form part of the clinical assessment process that leads to a request for further diagnostic services or a treatment or care plan. The clinical assessment, planning, implementation, and evaluation process are repeated throughout any person's health journey until an outcome is recorded. Throughout each journey, there are numerous opportunities to generate new information that is used as actionable intelligence either by the person themselves or by any of their service providers. In every instance of data transfer actioned to support that person's continuity of care, the data need to retain a link to the person's identifier, as well as the individual and/or organisational health service provider for medico-legal purposes.

Clinically, these individual 'patient journey use cases' can be identified relative to a 'disease burden' classification such as communicable, noncommunicable disease or injury. Each of these can be further categorised into specific conditions requiring the knowledge and expertise of any of the many clinical specialties or subspecialties, including diagnostics. Each of these specialties has a definable body of knowledge. Each specialist is able to identify the set of patient types they see and the frequency of re-occurring health conditions they treat, for which they have developed any number of diagnostic, treatment/care options and protocols.

This knowledge forms the foundation for the development of clinical models as well as for published clinical guidelines which are the result of research and new knowledge discovery activities. These models and guidelines reference multiple additional service providers with expertise in any complimentary clinical discipline, contributing to individual patient journeys. In addition, the information to be made available at every step along the way determines resources and equipment requirements, and data/information use for operational and quality management purposes.

Data/information needs associated with every step along any person's health journey determine identifiable data transfer, sharing, linking and use requirements. Such data can also be linked to the human and physical resources required and their associated supply chains within each healthcare facility and beyond. The same data/information linkage possibilities apply to treatment and caring methods adopted. Each 'use case' determines if and when data needs to be identifiable in terms of treatment/care recipient, provider, geographical location, resource or equipment used.

Directories, terminologies, data standards and data value sets required to optimally support data use along these data supply chains need to be governed and used consistently within any digital health ecosystem.

By identifying each of these specific data points relative to different types of person/patient health journeys (use cases), one is able to see the emergence of data and communication flow patterns, as well as terminology requirements for each type of instance in terms of input, processing, and outcome data at each data point. Output data from one activity becomes input data for the next. The ability to provide digital support for these individual health journeys is particularly important to facilitate the provision of patient-centred care and precision medicine. The latter is especially critical for cancer care.

These patterns represent generic data flows enabling multiple use of any number of universal data models [24], and inform the development of a metadata map that informs the reference information model which is a component of the overall architectural (system design) framework. This serves as an information roadmap to identify and synchronise data from multiple systems as is required to populate individual person/patient electronic health records.

Multiple persons' health records data supply chain

Every healthcare service provider contributing to individual health journeys contributes to that person's record. In doing so, they are also able to make use of aggregated information from multiple de-identified records for secondary use, such as departmental resource and operational performance management, including practice evaluation and research, and to meet internal and external reporting requirements where relevant, as shown at the bottom of data supply chain diagram.

Again, each type of provider is able to identify their own use cases and identify common patterns of data transfer, linkage, aggregation, processing, and use needs, and identify commonly occurring patterns among provider types. Within a digital health ecosystem, some data (value) set selections need to be transferred and linked with the same data sets from other providers to meet any number of clinical specialty, provider type, local, national, or global ecosystem information needs for any number of use cases. Such needs are met by numerous information systems that make use of both health record and provider-specific data that result in big data repositories. Context is provided by identifiers linked to every unique entity. One critical use case is the provision of timely real-time data for public health purposes, such as the aggregation of signs and symptoms data at every first encounter with the health system, and to, for example, identify infection or disease clusters. This is especially useful for pandemic identification and management.

These data supply chains start with highly granular data, at various levels of detail, used for clinical decision-making, care planning, and to serve as a medico-legal record of service in the first instance. Such data are grouped into types of data, such as by disease or procedures, aggregated and linked via any number of unique relationships to represent new higher level concepts.

Data supply chains are representative of a data/information hierarchy and parent–child relationships that collectively make up the health industry's knowledge domain. There is a need to identify which set of standard terminologies are best able to represent each domain.

Consistent adoption and use of data value sets enables successful streamlining and automation of all data and information flows to meet any number of use cases.

Secondary data uses—Data supply chains

Data linkage and data integration processes require the (mostly) de-identified data from multiple health records over different time periods to be brought together to see the effect of events that have occurred across any defined time span. Linked and integrated data represent information from two or more data repositories. The purpose of combining data is to create new, more detailed data sets able to tell a much more meaningful story regarding any specific set of variables, which would be possible from a single data source. Shared data combinations enable secondary data use which has the potential to make healthcare research more cost- and time-effective despite any required ethical and policy constraints. Greater automation enables faster research result distribution and use.

An examination of the flow of clinical data from collection through to the many uses by numerous stakeholders, which may be referred to as the 'data supply chain', results in the identification of numerous secondary data uses and users. Many of these uses were traditionally made possible by manually extracting data from paper-based or digitised systems or from specific data collections such as surveys and from any number of reports from various databases, followed by data re-entry, data cleansing, and processing.

With better designed systems, it is possible to automate these data flows, not only improving data accuracy and timeliness, but also saving an enormous amount of human processing time. This requires an examination of the end point of each data flow from which one is best able to discern where the source data needs to be collected, by whom and where.

Registries and their relevance in the digital era

One traditional secondary data use example is the use of national and international registries. In our current digital era, registries represent historical artefacts consisting of distributed data silos in need of a major overhaul to enable the provision of data extraction and query services. Current clinical registries are databases that systematically collect health-related information, within an overall governance and management structure, about individuals who are:

- treated with a particular surgical procedure, device or drug, e.g. joint replacement,
- diagnosed with a particular illness, e.g. stroke, or
- managed via a specific healthcare resource, e.g. treated in an intensive care unit.

Many types of registries are in use. Some registries are used to seek participants who may be willing or may benefit from being a participant in a clinical trial. Others are established to compile nationally mandated data collections. Every registry has a specific focus and is created to serve a specific purpose. They may offer a historical view of real-world healthcare or other data sources. Registry data are used for decision-making, including policy development relative to three factors, such as stakeholders, the primary scientific or policy question, and the context [25].

Registries are used for public health surveillance purposes such as biosurveillance security, chronic diseases, and for diseases with rapid treatment advancement such as the blood cancers. Registries provide valuable epidemiological, demographic, and social indicators about the health of a population. Ideally, such registries contain information needed for health policy development or analysis to enable the implementation of appropriately planned responses to meet health service demands.

National repositories are dependent upon the adoption of a standard metadata schema that enables data linkage from multiple data sources. These repositories and registries all require data normally stored in patient health records. These data are most commonly, manually extracted from those records; yet, with the right technical infrastructure, this can be fully automated.

The World Health Organisation (WHO) collects national statistics for 50 core indicators organised into six major areas: mortality and burden of disease, health service coverage, risk factors, health system inputs, differentials in health outcome and coverage, as well as basic socio-demographic statistics from 193 member nations to monitor trends of global public health.

We no longer have a need to collect data for storage in numerous repositories. We now have the opportunity to have access to one distributed yet connected (global or national) big current data repository able to meet the needs of each and every old-fashioned registry. This enables greater application of evidence-based practice principles.

The availability of these data, provided by individual nations, enables the WHO to provide leadership in all matters that are critical to the health of these nations and the world generally. Many nations collect data additional to that required by the WHO for their own public health purposes. For example, Australia has a minimum data set collection for admitted patient episodes and numerous additional registries, many of which also share their data with international public health organisations such as for example the Centre for Disease Control (CDC) or the International Association of Cancer Registries (IACR). Other registries are clinical specialty-specific, established to enable best practice research.

The ability to link existing data repositories enhances their value, but this is dependent upon standardisation and the adoption of a generalised registry infrastructure for each domain. Data to populate these registries come from clinical systems such as pathology or imaging or from paper or e-health records using a variety of data collection methods. To clearly understand the types of knowledge represented in these data collections, one needs to analyse the relationships between the data entities. Only then, is it possible to develop the system requirements suited to maximising the usability of the data collected, taking advantage of today's technology [26–28]. Standardisation of clinical content is also relevant in this context.

The adoption of agreed standardised representation of clinical content in eHealth records not only allows more accurate retrieval of data for reporting purposes, it also provides the means of establishing and automating the reporting process. For instance, a national collection which is deemed critical may require that a person's weight is recorded in order that other values can be interpreted safely. Having agreed on the representation of body weight in the health record and the other values, it is possible to request that these values are always present in a health record of the target population. Thus, standardisation of clinical content should take the needs of this 'secondary' data collection into account. This is explained further in another chapter that focuses on the development of clinical knowledge models.

Documentation supporting billing and funding methods

Billing and funding methods are based on documentation contained in every electronic health or medical record generated by any health provider. One common billing or funding method is the use of the International Classification of Diseases (ICDs) and nationally specific procedure or benefits schedule codes. In addition, there are many and varied lists of service interventions or benefit schedules identified as product indicators and used for billing or funding purposes.

Coding system designs, together with the data standards adopted by terminology systems in use, and their link to operational data collected as part of healthcare service provision, determine the ability to automate coding and reporting. The design structures of some coding rules and/or associated classification systems in use for this purpose have significant limitations that compromise consistency. Similarly, the diverse range of terminologies and data standards in use limits the national reporting automation. Adoption of a standard approach globally would result in significant cost savings for software suppliers.

Data Lakes, data hubs, data marts, and data warehouses

Health Data Science is an emergent discipline. It is about the science and art of generating data-driven solutions through a comprehension of complex real-world health problems, employing critical thinking and analytics to derive knowledge from big data. This science intersects with the (bio)statistics, the computer science, and the health science disciplines.

Challenges to be considered when choosing data storage formats include the data catalogue, metadata, and components such as lineage, quality, ingestion, analytics, and data security. Large data storage facilities need to be capable of significant scaling, there should be no capacity limits (big data), and they need to have high durability, excellent data robustness, and the ability to store all types of data in one repository. Its core storage layer should not dictate a fixed schema.

Cloud computing is a term frequently used to describe the now popular use of data centres, data lakes, data hubs, and data marts, available to many users over the Internet, replacing the use of traditional data warehouses as a storage facility. Technology advances have slowly increased the uptake of networks, the Internet, distributed computing, grid computing, parallel computing, and cloud computing, all of which may be in use at anytime anywhere connecting any number of data providers and users within the digital health ecosystem. Each of these computing methods may be adopted to not only store data but also facilitate retrieval and/or data processing.

Widespread adoption of technology and functional differences between technologies are:

- distributed computing—software system components are shared among multiple computers to improve efficiency and performance,
- grid computing—computer processes from various domains are combined within a processor architecture to reach a main objective,
- parallel or super-computing—several processors (central processing units (CPUs)) execute and process an application, or they compute simultaneously within a computing architecture, to divide the workload to perform large computations, and

- cloud-based storage—Information technology services are delivered on demand using resources retrieved from the Internet through web-based tools and applications, without direct active management by the user, to make it possible to save and store data to remote databases managed and owned by an internal (enterprise specific) or an external (public/commercial) provider. Large clouds (data storage facilities) may have functions distributed over multiple locations from central servers. It relies on resource sharing and the minimisation of upfront IT infrastructure costs.

With increasing Internet access, cloud computing has become very popular as an evolving paradigm. Many healthcare facilities are establishing their own clouds or working with a number of vendor suppliers within any digital health ecosystem to make this available for EHR data-sharing purposes between healthcare providers as the means to improve the delivery of patient-centred care throughout anyone's individual health journey.

In the absence of a standard Government –/Jurisdictional-led health data cloud, we are now witnessing the implementation of multiple clouds, each making use of their own cloud storage architectures. This has the potential for all the same benefits and weaknesses of traditional data silos and data fragmentation. Clouds can be deployed as a private, community, public, or hybrid cloud, and they are popular due to a number of beneficial features including [29]:

- users' flexibility,
- conversion of capital expenditures to operational expenditures,
- device and location independence,
- ease of maintenance,
- resource sharing,
- performance monitoring by IT experts from the cloud service provider,
- support business continuity and disaster recovery, and
- scalable on-demand use.

Issues of possible concern are security complexity, and user access to security audit logs may be difficult or impossible. Such concerns have motivated some health facilities, Governments/Jurisdictions to establish a private health cloud to enable the retention of control over the infrastructure, information, and data security. The US National Institute of Standards and Technology (NIST) [30] has identified and described the following essential characteristics for cloud computing:

- On-demand self-service
- Broad network access
- Resource pooling
- Rapid elasticity
- Measured service

Cloud clients make use of a web browser, and any type of mobile application, device or terminal emulator connected to an enterprise system. NIST [30] described the cloud computing service models, which are arranged as layers in a stack. These are as follows:

- Software as a Service (SaaS), customer relationship management (CRM), email, virtual desktop, communication, games, etc.,
- Platform as a Service (PaaS), execution runtime, database, web server, development tools, etc., and

- Infrastructure as a Service (IaaS), Virtual machines, devices, servers, storage, load balancers, network, etc.

Successful planning to move to the cloud is primarily about data-retrieval methods, which requires cloud neutrality, multicloud architectures, storage format specifications that expect data and associated metadata, entities and privileges to be application-independent, independent (nonvendor) control of your data, management of your own data governance/quality, and the nonuse of code in the cloud to meet these requirements.

This requires the ability to make use of independent third-party tools and service-level agreements. It is critical to avoid cloud vendor lock-in, which includes the use of proprietary Application Program Interfaces (APIs). It is important for the global health industry not to perpetuate the existence of health data silos. The use of big data, where the data collected in health records form the foundation, needs to be enabled.

It is of major concern that securing the data in the Distributed File Systems (DFSs)–cloud ecosystem continues to be a challenge due to the interconnection of so many diverse hardware as well as software components [31]. The synchronisation of data stored in DFS is now a well-known issue. As the degree of data access concurrency is increasing, synchronisation could certainly be a performance bottleneck. It now appears to be standard practice to establish API gateways linking to a cloud as the solution to solving the interoperability problem within any digital health ecosystem.

Clouds managed by Amazon (AWS), Google, and Microsoft (Azure) make use of object-based data stores considered the best choice for storage. Additional layers usually have some added structure to assist effective use for reporting or analytical purposes. Architectures adopted by cloud vendors determine an array of elements such as file sizes, types of storage (row vs columnar), degree of compression, indexing, schemas, and block sizes. Data infrastructures determine the view of its data assets and the ability to ingest, consolidate, and process raw, unmodelled data in native format or lightly processed data. Technology neutral data storage requires its data structure to be defined when the data is needed, to enable innovation. This opens the door to exploratory analysis and data mining.

The Apache Hadoop consists of a collection of open-source software utilities that make use of a network of many computers to solve problems involving massive amounts of data and computation [32]. Its framework composition includes the Hadoop Distributed File System (HDFS), designed to reliably store very large data sets and to stream those data sets at high bandwidth to user applications. Many Hadoop tools [33] are now available for crunching big data. Data hubs may be designed to act as a point mediation and governance with a focal point for data sharing and governance of all critical data. The ability to read, decompress, and process only the values that are required for a current query is made possible by columnar file formatting (ORC format).

Discussion and conclusion

This chapter has focused on an analysis of the various definitions in use to describe electronic health records. These definitions reflect world views of EHRs, their purpose, and use. As such, they influence system design which in turn determines their capacity to accommodate both technical and medical advances over time. These data/information repositories (records) contain much of the source data used throughout, by what is now known as the 'health data supply chain' supporting numerous primary and secondary use cases.

The absence of an internationally agreed standard detailing a high-level information model, such as a 'chart of accounts' used for accounting, to guide a content structure for EHRs, has resulted in serious fragmentation of health data. An examination of individual patient records' scope, in terms of content and functionality, again revealed numerous differences. This continues to limit our ability to make optimal beneficial use of all the source data stored in such records. Another limitation is an inability to develop meaningful guidelines that represent a common view of the quality of any EHR. It is important to separate the storage structures used for the records themselves from the many complex systems that make use of the content.

Health records' repositories represent a most valuable asset for anyone in need of healthcare, as well as for their individual and organisational service providers. Health data stored in these repositories also represent a significant data source for the health data supply chain. An examination of data and communication flows associated with individual patient health journeys revealed the many primary and secondary actual and potential uses along the way. This demonstrates that knowing subsequent uses of the primary data collected at the first point of care determines what data needs to be collected, stored, processed, transferred, linked, deleted, or archived at what point throughout the data supply chain, to provide the necessary information at specific points of decision-making and use.

These requirements determine what technical solutions need to be adopted to optimise value from the use of these assets. A sound understanding of one's health data supply chain informs the need for supportive global, national and jurisdictional clinical, technical, legal, and ethical governance infrastructures. This is a primary foundational requirement for a successful digital health ecosystem, without which *ad hoc*, noninteroperable developments will continue to proliferate, severely limiting the benefits to be realised from digital health. Similar strategies have been successfully adopted by the airline, banking, and telecommunication industries. All represent well-connected global operations, which is also a health industry requirement.

One major dinosaur to be considered for reformation within this big picture is the continuing use of registries and national statistical reporting, requiring the manual extraction of data. This activity is a major cost factor for the health industry, not only in terms of resource usage but also in terms of their ability to provide timely and accurate information.

The establishment of registries became the norm historically. They now represent the continuation of numerous data silos and data fragmentation. Registry content is difficult and time-consuming to access, or to make use of to facilitate data linkage, data aggregation, or data sharing. Registry repository structures are not well suited to this digital era. There are a number of other historical developments due for major reformation to better suit the digital era, including for example some terminology/classification system structures such as the widely used International Classification of Diseases (ICD) as explored in other chapters.

Successful digital health ecosystems need to be structured to maximise automation, reduce data processing time and cost, improve knowledge discovery, and enable the realisation of beneficial health service outcome reporting within sustainable national health systems. The adoption of a suitable supportive foundational infrastructure, technology neutral data storage capacity, metadata standards, standard data models, and templates containing standard data value sets are some of the key requirements to enable this to be realised. Data collections undertaken to support billing and funding methodologies will also benefit from these adoptions. Meeting the necessary digital ecosystem structural design needs will require massive changes in mindsets, widespread adoption of digital thinking, and workforce capacity building, followed by well-managed local, national, and global transformations.

References

[1] E. Hovenga, S. Garde, Electronic health records, semantic interoperability and politics, Electron. J. Health Inform. 5 (1) (2010) e2. https://web.archive.org/web/20180410184331/http://www.ejhi.net/ojs/index.php/ejhi/article/view/116.

[2] E.J. Hovenga, S. Heard, E-health records and future healthcare, Stud. Health Technol. Inform. 151 (2010) 30–42.

[3] J. Henry, Y. Pylypchuk, T. Searcy, V. Patel, Adoption of Electronic Health Record Systems among U.S. Non-Federal Acute Care Hospitals: 2008–2015, The Office of the National Co-ordinator for health Information technology: Health IT Dashboard, 2016. [cited 2020 4 October]. Available from: https://dashboard.healthit.gov/evaluations/data-briefs/non-federal-acute-care-hospital-ehr-adoption-2008-2015.php.

[4] JIGSH. Joint Initiative for Global Standards Harmonization (JIGSH), Health Informatics Document Registry and Glossary, Standards Knowledge Management Tool (SKMT) [web page], Available from http://www.skmtglossary.org/.

[5] ISO/TR20514, Health Informatics—Electronic Health Record—Definition, Scope and Context, 2005, Available from https://www.iso.org/standard/39525.html.

[6] IOM, The computer-based patient record: revised edition: an essential technology for health care, in: R.S. Dick, E.B. Steen, D.E. Detmer (Eds.), The Computer-Based Patient Record: Revised Edition: An Essential Technology for Health Care, Institute-of-Medicine-Committee-on-Improving-the-Patient-Record, National Academies Press (US), Washington (DC), 1997.

[7] T. Handler, R. Holtmeier, J. Metzger, M. Overhage, S. Taylor, C. Underwood, HIMSS Electronic Health Record Definitional Model Version 1.0, 2003, Available from http://www.providersedge.com/ehdocs/ehr_articles/HIMSS_EMR_Definition_Model_v1-0.pdf.

[8] WHO, Joint Inter-Ministerial Policy Dialogue on eHealth Standardization and Second WHO Forum on eHealth Standardization and Interoperability, World Health Organization, Geneva, 2014. Available from https://www.who.int/ehealth/events/standardization_forum_2014/en/.

[9] ISO-23903:2021 Health Informatics—Interoperability and Integration Reference Architecture-Model and Framework, International Organization for Standards (ISO), Geneva, 2021. Available from https://www.iso.org/standard/77337.html.

[10] IOM, The Learning Healthcare System, National Academies Press, 2007. Available from https://www.nap.edu/catalog/11903/the-learning-healthcare-system-workshop-summary.

[11] R. Nelson, Major theories supporting healthcare informatics, in: S.P. Englebardt, R. Nelson (Eds.), Health Care Informatics: An Interdisciplinary Approach, Mosby, St Louis, 2002.

[12] E. Coiera, J. Ash, M. Berg, The unintended consequences of health information technology revisited, Yearb. Med. Inform. 25 (1) (2016) 163–169, https://doi.org/10.15265/IY-2016-014.

[13] M.T. Neame, G. Sefton, M. Roberts, D. Harkness, I.P. Sinha, D.B. Hawcutt, Evaluating health information technologies: a systematic review of framework recommendations, Int. J. Med. Inform. 142 (2020) 104247.

[14] NICE, Evidence Standards Framework for Digital Health Technologies, National Institute for Health and care Excellence, 2019. Available from https://www.nice.org.uk/about/what-we-do/our-programmes/evidence-standards-framework-for-digital-health-technologies.

[15] ISO/IEC25010. Systems and software engineering—Systems and software Quality Requirements and Evaluation (SQuaRE)—System and software quality models, International Organization for Standards, 2011. Available from https://pdfs.semanticscholar.org/57a5/b99eceff9da205e244337c9f4678b5b23d25.pdf. https://www.iso.org/standard/35733.html.

[16] EuroRec, Promoting the Use of high Quality Electronic Health Record Systems, Available from https://www.eurorec.org/.

[17] CCHIT, Certification Commission for Health Information Technology, Available from https://www.cchit.org/.

[18] ONC, Certification of Health IT, Available from https://www.healthit.gov/topic/certification-ehrs/about-onc-health-it-certification-program.

[19] R.A. Rocha, A.N. Romeo, C. Norlin, Core features of a parent-controlled pediatric medical home record, Stud. Health Technol. Inform. 129 (2007) 997–1001.

[20] IOM, Crossing the Global Quality Chasm: Improving Health Care Worldwide, Institute of Medicine, The National Academies press (US), Washington DC, 2018.

[21] CtrlShift, Data Mobility: The Personal Data Portability Growth Opportunity for the UK Economy, UK Department for Digital, Culture, Media & Sport, 2018. Available from https://www.ctrl-shift.co.uk/wp-content/uploads/2019/07/DCMS_Ctrl-Shift_Data_mobility_report_full.pdf.

[22] CtrlShift, Data Mobility Infrastructure Sandbox Safety: System Design & Engineering Workstream Report, CtrlShift, 2019. Available from https://www.ctrl-shift.co.uk/wp-content/uploads/2019/06/1905_DMIS_System-Design-Engineering-Final1.pdf.

[23] Nictiz, Decentralized Initiatives for the Exchange of Patient Data, MedMij, VIPP, PatientConnect & Registratie aan de Bron Netherlands, 2021. Available from https://www.nictiz.nl/english/exchange-of-electronic-patient-data-in-the-netherlands/decentralized-initiatives-for-the-exchange-of-patient-data/.

[24] L. Silverston, P. Agnew, The Data Model Resource Book Volume 3 Universal Patterns for Data Modeling, Wiley Publishing, Indiana, 2009.

[25] R.E. Gliklich, N.A. Dreyer, M.B. Leavy, Registries for Evaluating Patient Outcomes: A User's Guide, Internet, third ed., Agency for Healthcare Research and Quality (US), Rockville (MD), 2014. Available from https://www.ncbi.nlm.nih.gov/books/NBK208643/.

[26] A. Geissbuhler, C. Safran, I. Buchan, R. Bellazzi, S. Labkoff, K. Eilenberg, A. Leese, C. Richardson, J. Mantas, P. Murray, G. De Moor, Trustworthy reuse of health data: a transnational perspective, Int. J. Med. Inform. 82 (2013) 1–9.

[27] G. Hripcsak, M. Bloomrosen, P. Flatelybrennan, C. Chute, J. Cimino, D. Detmer, M. Edmunds, P. Embi, M. Goldstein, W. Hammond, G. Keenan, S. Labkoff, S. Murphy, C. Safran, S. Speedie, H. Strasberg, F. Temple, A. Wilcox, Health data use, stewardship, and governance: ongoing gaps and challenges: a report from AMIA's 2012 Health policy meeting, J. Am. Med. Inform. Assoc. 21 (2013) 204–211.

[28] C. Safran, M. Bloomrosen, W.E. Hammond, S. Labkoff, S. Markel-Fox, P.C. Tang, D.E. Detmer, P. Expert, Toward a national framework for the secondary use of health data: an American Medical Informatics Association White Paper, J. Am. Med. Inform. Assoc. 14 (2007) 1–9.

[29] Wikipedia, Cloud Computing, Available from https://en.wikipedia.org/wiki/Cloud_computing.

[30] NIST, P. Mell, T. Grance, The NIST Definition of Cloud Computing: Recommendations of the National Institute of Standards and Technology, US National Institute of Standards and Technology, 2011. Available from https://nvlpubs.nist.gov/nistpubs/Legacy/SP/nistspecialpublication800-145.pdf.

[31] S. Mazumdar, D. Seybold, K. Kritikos, Y. Verginadis, A survey on data storage and placement methodologies for cloud-big data ecosystem, J. Big Data 6 (2019) 15.

[32] Apache Hadoop. Wikipedia, Available from https://en.wikipedia.org/wiki/Apache_Hadoop.

[33] ScienceDirect, Hadoop Ecosystem: Introducing Big Data Technologies, 2018, Available from https://www.sciencedirect.com/topics/computer-science/hadoop-ecosystem.

CHAPTER

14

Health data security and privacy: Challenges and solutions for the future

Kassaye Yitbarek Yigzaw[a], Sílvia Delgado Olabarriaga[b], Antonis Michalas[c], Luis Marco-Ruiz[a], Christiaan Hillen[d], Yiannis Verginadis[e,f], Marcela Tuler de Oliveira[b], Dagmar Krefting[g,h], Thomas Penzel[i], James Bowden[g], Johan Gustav Bellika[a,j], and Taridzo Chomutare[a]

[a]Norwegian Centre for E-health Research, Tromsø, Norway [b]Amsterdam University Medical Centers, University of Amsterdam, Amsterdam, The Netherlands [c]Tampere University, Tampere, Finland [d]Secura, Eindhoven, The Netherlands [e]Institute of Communication and Computer Systems, Athens, Greece [f]Athens University of Economics and Business, Athens, Greece [g]HTW Berlin—University of Applied Sciences, Berlin, Germany [h]University Medical Center of Göttingen, Göttingen, Germany [i]Charité—Universitätsmedizin Berlin, Berlin, Germany [j]UiT The Arctic University of Norway, Tromsø, Norway

Introduction

The widespread adoption of technologies such as electronic health records (EHRs) and new sources of health information, such as wearables and health apps, has led to the collection of large amounts of health-related data [1,2]. These data have a huge importance in the direct care of patients and in the advancement of the science of medicine and healthcare [2,3]. However, continuous use of identifiable data to follow patient journeys for patient-centred care and the secondary use of health data raise concerns about security and privacy risks. Studies show that security incidents are increasing significantly, with data breaches caused by hacking increasing 15-fold between 2010 and 2016 [4].

Health institutions, as part of a digital health ecosystem, should therefore implement risk management strategies to properly manage their risk. Risk management is an ongoing process of identifying, assessing, and responding to risk. The National Institute of Standards and Technology (NIST) at the U.S. Department of Commerce has two key elements in its

https://doi.org/10.1016/B978-0-12-823413-6.00014-8

definition of risk, namely, the impact of an adverse event and the likelihood of the adverse event occurring. Institutions may mitigate, transfer, avoid, or accept the risk, depending on its potential impact, available resources, and business and legal requirements. The overall goal should be to reduce risks to a level acceptable to the institution while maintaining the benefits of data or systems.

Once an institution understands its risk, it determines the acceptable level of risk for achieving its business objectives and legal requirements, called risk tolerance. Risk tolerance is the basis for prioritising its cybersecurity activities, for example the reconfiguration of a vulnerable system, and implementing important security controls (e.g. encryption of data at rest). There are different international standards for risk management processes, for example International Organisation for Standardization (ISO) 31,000:2018 [5], ISO/International Electrotechnical Commission (IEC) 27,005:2018 [6], and NIST Special Publication (SP) 800-39 [7].

The NIST cybersecurity framework [8], one of the more influential, is a risk-based approach to managing cybersecurity risk using the principles and best practices of risk management processes. The framework is voluntary in the United States, but it is a useful tool for communicating risk throughout an organisation and across sectors, from agriculture to commerce. In this chapter, we position this framework in the context of healthcare to frame our discussion on cybersecurity.

The framework core contains five functions—the primary pillars for a successful and holistic cybersecurity program:

(i) Identify
(ii) Protect
(iii) Detect
(iv) Respond
(v) Recover

Identify, the first core function, supports institutional understanding of the business context, available resources (e.g. system, data, and people) that support critical functions, and the related cybersecurity risks. Identification helps focus and prioritise institutional efforts. For a better understanding of the context, the first part of the chapter explores cybersecurity in the healthcare environment and identifies common risks, health data use challenges, and legal constraints, that should normally be considered.

The second core function, Protect, supports the ability to limit or contain the impact of a potential cybersecurity event using relevant security controls. The latter part of the chapter reviews some of the state-of-the-art methods in access control, cryptography, de-identification, and privacy-preserving distributed data mining. These technical solutions, combined with user awareness and training, can be part of an effective protective strategy.

The Detect function supports timely discovery of cybersecurity events, including continuous monitoring and detection processes, while the Respond function relates to taking appropriate action after a cybersecurity incident. The Recover function, on the other hand, ensures timely recovery to normal operations to reduce the impact of a cybersecurity incident.

The remaining sections in this chapter will focus on the first two core functions (Identify and Protect), as they are vital for developing the technical building blocks necessary for cybersecurity. The rest of the core functions—Detect, Respond, and Recover—fall outside the scope of this chapter. The chapter concludes by presenting emerging security and privacy solutions and their implementations for solving challenges in three, real healthcare use cases:

(i) Case study 1: Audit and feedback on antibiotics prescribing.
(ii) Case study 2: Data sharing for stroke acute care.
(iii) Case study 3: Inpatient and outpatient sleep medicine.

Cybersecurity in healthcare

The primary concerns of cybersecurity are the confidentiality, integrity, and availability of data, although nonrepudiation, authenticity, and authorisation should also be considered (see Table 1). Garg and Kohnfelder [9] described the kind of threats that violate each of the six desired properties of a secure system, as illustrated in Table 1. Shostack [10] later popularised these threats in his threat modelling approach called STRIDE, which is a process to build a more secure system through analysing potential threats and vulnerabilities to a system and mitigate them.

TABLE 1 STRIDE threat modelling and the corresponding desired properties of cybersecurity, threat examples, and mitigations.

STRIDE threats	Description	Desired property	Examples	Mitigations
Spoofing	Posing as someone else or claiming a false identity	Authenticity	Leave credentials on a public space or sharing; reveal credentials to someone (e.g. social engineering and phishing attack)	Strong authentication, encryption, and cryptographic protocols, such as TLS/SSL
Tampering	Malicious modification of data or processes	Integrity	System admins or healthcare professionals accidentally modify data because of inappropriate access control	Strong authorisation, data hashing and signing, and secure communication channels
Repudiation	The ability to deny that an action or an event has occurred	Nonrepudiation	Patient or healthcare professionals deny not receiving, writing, or editing data	Strong authentication and secure audit trails
Information disclosure	Data leaks or data breaches	Confidentiality	Eavesdropping; stolen or lost devices; unauthorised access because of weak access control	Encryption, secure communication channels, and strong authorisation
Denial of service	Causing a service or a network resource to be unavailable to its intended users	Availability	Denial of service attack; ransomware attack; full disk space or memory or CPU time	Regular backups, load balancing, and firewalls
Elevation of privilege	Gaining access that one should not have	Authorisation	Unauthorised access using stolen or shared credentials; system admins or healthcare professionals gain unauthorised access to patient data	Proper authorisation, principles of least privilege

Depending on the purpose of the system and the data that it will process, some of the properties will be more important than others. Health data storage and sharing, or 'processing', as defined by the general data protection regulation (GDPR) [11], come with significant risks to the confidentiality of the information because access needs to be governed. But not only confidentiality is at risk here, any unregistered change to patient records, such as removing a severe allergy or a do-not-resuscitate (DNR) statement, can have dire consequences.

Patient data need to be available to authorised people on time. Security breaches may lead to the unavailability of patient data and could decrease the quality of care or result in inadequate treatment. The impact of security breaches includes social and economic loss for involved institutions and patients. For example, in 2020, a ransomware attack on the University Hospital of Düsseldorf in Germany led to data being unavailable, and, as a result, the hospital could not accept emergency patients for some time. Major news outlets like the Associated Press carried unconfirmed reports of the death of at least one patient whose ambulance was diverted to another hospital, delaying treatment by an additional hour.

Cybersecurity incident types

Healthcare institutions of all types and sizes continue to experience significant security incidents, both malicious and benign, as illustrated in Table 2. Long supply chains often expose institutions to compromised hardware and software from low-tier vendors, and this can be exploited by hackers to breach an organisation's infrastructure. Malicious attacks are intentional actions such as phishing, ransomware or other malware, and social engineering attacks. Phishing is the primary means for compromising systems and networks because recipients of phishing messages unknowingly open a malicious attachment, click on a malicious link, or respond to an email that has elicited sensitive information. The attackers then exploit the trusted access of their victims. The Office of the Auditor General (Riksrevisjonen) reported a simulated cyberattack on healthcare institutions in Norway and revealed that around 39% of the recipients of a phishing email clicked on a link they received in the email [12].

Ransomware and other types of malware remain significant challenges for healthcare institutions [13]. Most ransomware attacks consist of sending a piece of malware into the wild with the hope that some of the victims pay the ransom. However, paying the ransom is not necessarily a guarantee for the safe return of the data or that the data will not be used for

TABLE 2 Cybersecurity incident types.

Malicious actions	Other nonmalicious actions
Malware (e.g. virus, ransomware)	Natural phenomena (e.g. flood, earthquake)
Hijack (e.g. network/session)	Supply chain failure (e.g. network, power failure)
Social engineering (e.g. phishing, baiting)	System failure (e.g. overload, inadequate maintenance)
Theft (e.g. device or data)	Human errors (e.g. noncompliance)
Medical device tampering	
Skimming	
Denial of service	

Adapted from ENISA Cyber security and resilience for Smart Hospitals. https://www.enisa.europa.eu/publications/cyber-security-and-resilience-for-smart-hospitals (accessed 16 Mar 2021).

other illegal purposes or be sold on the black market. Counter measures such as offline and regular backups help minimise the potential impacts of such incidents.

The accidental loss of information stored on mobile devices is another common issue with healthcare security [4]. It could be the loss of a USB stick or CD-ROM or a laptop left on public transport. Therefore, using proper encryption on all devices should be a part of the security policy of any institution involved with sensitive data.

The primary goal of healthcare practitioners working in the medical field is helping people, and it is this mindset that can be used against them in social engineering attacks, for instance. For practitioners, anything (e.g. security protection) that comes between them and caring for their patients can cause irritations. It is not uncommon to find passwords on Postits glued to terminals, access cards being shared, or systems simply left unattended without being locked, just because that way of working causes less friction. An institution is as secure as its weakest link. Therefore, regular security awareness training of personnel is important to prevent some attacks.

In spite of all the increasing security incidents, cybersecurity does not get enough priority within healthcare institutions, despite its important role in patient safety. A recent study in the United States reported that a typical cybersecurity budget is 6% or less of the information technology (IT) budget of healthcare institutions. This is considered a relatively small amount and constrains the cybersecurity posture of institutions, according to a 2020 cybersecurity survey by HIMSS [13].

Regulatory framework and compliance

There are different organisational, national and international policies for cybersecurity. However, policies, such as laws alone, do not stop criminal activities, even though there may be clear penalties. It is still possible to breach policy, even with deterrents in place. In contrast, a system that prevents processing stops the user prior to performing an unauthorised action. Clearly, protection on a system level is superior to the protection offered by policies, but not everything can be protected in this way and policies will always be required as well.

The strongest form of policies are laws, such as the General Data Protection Regulation (GDPR). However, laws can be very ambiguous. The GDPR requires, for instance, the usage of state-of-the-art security protocols for processing data (GDPR Art. 32.1), but the term "state-of-the-art" is open to interpretation [14]. The Health Insurance Portability and Accountability Act (HIPAA) [15] was enacted in 1996 in the United States and was primarily designed to protect personal health information. Both GDPR and HIPAA require data controllers (e.g. healthcare institutions) to conduct risk assessments of data processing undertaken in their institutions. This helps reveal areas where the institution's IT could be at risk. Data controllers determine the purpose and means of personal data processing, they should periodically take measures to address the risks and comply with the applicable laws.

Fig. 1 shows the main thematic areas that must be considered when reviewing an organisation's compliance with the relevant laws. This figure also shows an inexhaustive list of actionable points for regularising the organisation's compliance. For most institutions, this starts with a risk assessment and privacy impact review, where data processing activities are systematically described and justified, and the potential risks to user privacy are evaluated.

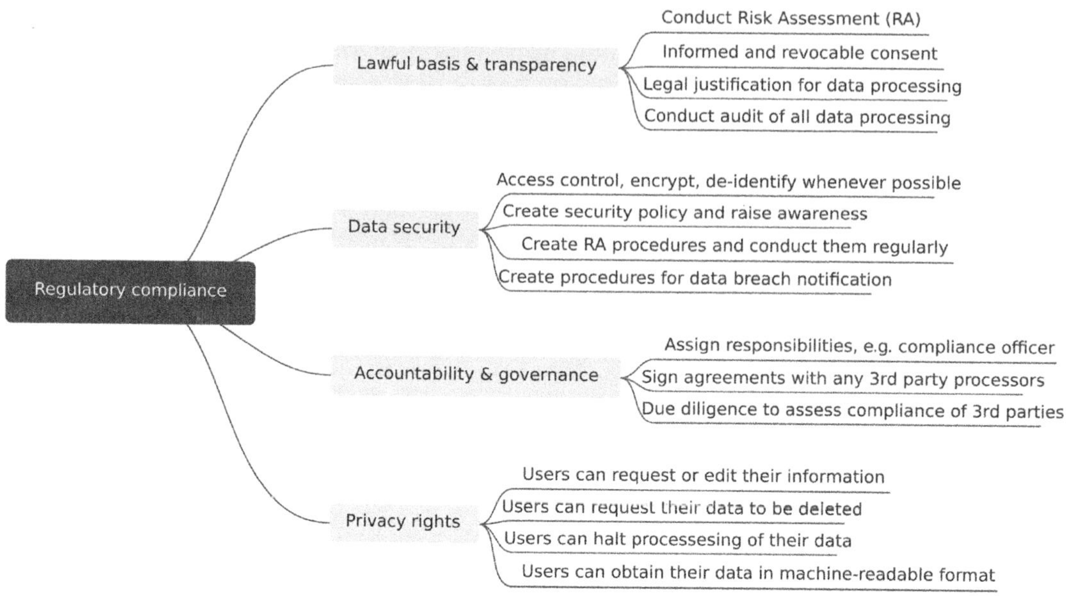

FIG. 1 Main thematic areas and a partial list of respective action points for organisational regulatory compliance reviews. *Adapted from GDPR Compliance Checklist. GDPR.eu. https://gdpr.eu/checklist/ (accessed 19 Mar 2021).*

Privacy rights, now a part of modern legislation, seek to empower users as owners of their own data. Previously, institutions virtually owned user data, but today, users have the right to request all information about them, halt processing of that information, or even ask for the information to be deleted. For some institutions, this new dynamic introduces a compliance gap, and it may be necessary to upgrade existing systems to fill such gaps.

Continual review of compliance gaps is vital because the legal environment is dynamic. For instance, the Court of Justice of the European Union passed judgment in 2020 on a popular "Schrems II" case [16] that invalidated the adequacy of the EU-US data protection shield (Privacy Shield), a mechanism to comply with data protection requirements when transferring personal data. Thus, data transfers from the EU to the United States require other security guarantees since, according to the ruling, the United States did not offer an equivalent level of protection as the GDPR.

Health data use challenges

The widespread adoption of technologies such as EHRs and new sources of health information, such as wearable displays and health apps, has led to the collection of large amounts of health-related data [1,2]. These data need to be shared between healthcare actors involved in the direct care of a patient to ensure the best care delivery. There is also huge importance and interest for the reuse of these data for the advancement in the science of medicine and to support the processes of health care [2,3]. The following sections analyse security and privacy challenges related to these two important uses of health data, namely, for primary (individual care) and secondary (reuse) purposes.

Data use for individual care

Health data collected within healthcare institutions have primary use for individual care. Studies have demonstrated that providing timely access to these data by healthcare actors involved in patient care improves healthcare efficiency [17,18], reduces medical errors [19], decreases cost [20,21], and increases patient satisfaction [22]. Exchanging health data across institutional boundaries and across healthcare levels is vital for collaborative care, but full interoperability is still difficult to achieve. For instance, accurately matching the health records of patients across multiple healthcare providers, to achieve data integrity and quality, is still a major challenge, since there are no universal personal identifiers. Systems that exchange patient data need to share semantic meaning; otherwise, the integrity of the data at the point of care will be compromised.

Many healthcare institutions today exchange patient summary [23,24], also known as a continuity of care document, which contains just the basic information related to allergies, medications, family history of diseases, previous diagnoses, etc. This setup is far from satisfactory because it affects the availability of patient data, which can raise patient safety and quality of care concerns. We need to be able to maximise the sharing of detailed identifiable data to follow patient journeys and to support continuity of care.

In addition, patients often do not have any way of granting access or consenting to this information exchange. Ideally, a patient should be able to give access to one part of the EHR record but not the other. For example, a patient may decide to conceal psychiatric records from a nephrologist, and the EHR system should be able to maintain this confidentiality. The complexity of designing fine granular data consent policies for users remains a research topic. Significant developments are required in EHR software architectures.

Another major challenge is the use of e-health tools (e.g. wearables, sensors, mobile apps, and health websites) outside traditional healthcare settings, which is increasing across the globe. *E*-health technologies have transformed traditional sources of health information for individual care, presenting new challenges to both patients [25] and the healthcare system [26]. These tools are sometimes not secure, often lacking full compliance with the latest regulations [27], and it is common for users to not know what data they have shared, where the data are stored, and who the legal data controller of their data is. However, e-health tools, such as symptom checkers [28], are important for both self-management of disease and collaborative care, since they allow patients to communicate health data to care providers and also allow care providers to provide recommendations to their patients.

Data reuse

Reuse of health data has huge potential in the context of research, but this data also has a broad use in clinical (e.g. quality and safety), governmental (e.g. biosurveillance), and business settings (e.g. actuarial science for health insurance) [29]. Clinical data repositories have been used in direct patient care to support the findings of how similar patients were diagnosed and what treatments were effective. Efficient reuse of historical data repositories allows us to ask questions, such as 'how was a particular group of patients treated in the past?', 'what was the outcome?', and 'should we consider a different approach for these cases?', thus helping to improve patient care by learning from previous experiences [29].

To use health data that contains personal data, data controllers need to comply with legal, ethical, behavioural, and organisational constraints [3]. Reuse of these data always requires, with few lawful exceptions, that data is pseudonymised or anonymised as prescribed by laws, such as HIPAA and GDPR [30]. Difficult decisions must be made regarding how the data is shared, as well as regarding an acceptable risk threshold associated with disclosing the data (see the section on De-identification).

After data is sufficiently anonymised, further steps may be required to make data sharing efficient, for example, transformation into normalised models that ensure interoperability and control by institutional review boards (IRB) when the use of the data is deemed clinical research rather than quality improvement (QI). The differences between clinical research and QI can be unclear because of overlapping definitions of research and QI [31]. In countries, such as Norway, Regional Committees for Medical and Health Research Ethics, REK (equivalent to IRB) reviews ethics and the informed consent processes for research on human subjects, but ethics reviews do not currently apply to the majority of QI projects.

The major challenge for large data reuse infrastructures is the need to securely aggregate data from multiple health institutions. A single institution usually does not have all the data necessary to understand a disease or prevalence; therefore, a more complete picture can only be achieved by aggregating data from many different clinical settings [29]. The challenge is to develop infrastructures that can securely aggregate distributed data while preserving the privacy of patients and individual institutions [32].

The challenges of obtaining consent from patients in big data projects are well known since getting consent from data subjects is one of the lawful reasons for processing personal data. Four data reuse initiatives that integrate data from several university hospitals across the German territory began a dialogue with the German authorities regarding patient consent. It was deemed sufficient to collect a general-purpose consent, where patients consented to the use of their data in future research. This was a major turning point because today it is no longer necessary to specify research questions or specific purposes at the time of giving consent; this move has wide support from the population [33].

The privacy of institutions can also be a major concern, in addition to that of patients and healthcare workers. Health institutions may have their own competitive advantage or intellectual property they wish to protect or they may fear the scrutiny or possible litigation [34].

Secure and privacy-preserving solutions

This section discusses the techniques used to improve the cybersecurity posture of healthcare institutions. From the many possible cybersecurity solutions [35], we single out two primary methods for protecting health data used in individual care:

- Access control
- Cryptography

and two auxiliary methods for protecting health data shared for reuse purposes:

- De-identification
- Privacy-preserving distributed data mining (PPDDM)

Access control

Access control is a mechanism by which a system grants or revokes the right to access information and tools. It is critical in digital healthcare ecosystems, where timely access to necessary health information by legitimate healthcare professionals can be a matter of life or death, while at the same time unauthorised access could also have severe negative consequences [32,36]. Data protection regulations such as the GDPR and the HIPAA require that data controllers (e.g. healthcare institutions) obtain the consent of individuals to collect, use, or disclose personal health information. In addition, access privileges provided to healthcare professionals should vary with their role (e.g. doctors, nurses, and laboratory technicians), context (e.g. scheduled appointment and emergency situation), and purpose of use (e.g. patient treatment, booking an appointment, and laboratory testing). These requirements depend on dynamic access control mechanisms that efficiently and effectively capture and enforce access policies at a discrete level [37].

Of all basic access control models used in authorisation techniques, role-based access control (RBAC) and attribute-based access control (ABAC) are considered the most flexible enough to support elevated security challenges in the healthcare sector [38,39]. However, the RBAC [40] model does not allow as much contextual information (e.g. place and time, as well as a requestor's role) as needed in access control decisions of sensitive healthcare data. The ABAC model [38,41] is a more appropriate model to address the need for efficient and flexible authorisation, as it allows the application of sufficient contextual parameters.

The eXtensible Access Control Markup Language (XACML) [42] is one of the well-known reference implementations of ABAC. XACML access control requests involve a set of attributes to describe a situation under which an incoming request is perceived. These attributes are fine-grained in extensive vocabularies and ontologies in order to better map to contextual circumstances relevant to e-health authorisation processes [36,43]. User-friendly XACML editors allow patients and data controllers to define unambiguous access policies [36]. Consent-centric access control approaches [37] are considered practical for involving patients in access control decisions because they protect against errors while providing the desired access auditability.

The main challenge to achieving adequate access control in healthcare is the need for semantic interpretation of user attributes, such as identity and context information, across multiple healthcare actors. The federation of access control across multiple actors is highly dependent on the actors using the same attribute vocabulary, which is difficult to achieve (e.g. location expressed as numerical values of latitude and longitude vs. location expressed as the text of the physical address). Interoperability thus often depends on ad hoc mechanisms for mapping between attribute vocabularies, which is resource-consuming and error-prone in practice.

One of the most promising approaches for an effective access control mechanism is a clear separation of concerns between policy definition and enforcement; that is, changes to access control policies should not require any kind of change in the system being protected [37]. This approach alleviates the security concerns associated with e-health systems regarding authorisation frameworks. Studies have shown that combining the ABAC authorisation model with modern encryption techniques such as attribute-based encryption (ABE) (see Cryptography section) enables an additional layer of protection to data [36].

Cryptography

Encryption schemes transform plaintext data into ciphertext in such a way that the ciphertext reveals little or no information about the original plaintext. It is commonly used to protect data in transit and data at rest. Each time a user uses online banking or buys something online, encryption is used to protect the information being transmitted where the recipient decrypts the received ciphertext, transforming it to plaintext, before processing the actual data. This section discusses emerging cryptographic schemes (see Table 3) such as symmetric searchable encryption, attribute-based encryption, functional encryption, and homomorphic encryption that allow authorised users to perform certain computations directly on ciphertexts without decrypting them. This property makes these encryption schemes useful for processing sensitive data in an untrusted third-party environment (e.g. a public cloud) while protecting the security and privacy of the underlying data.

Symmetric searchable encryption

Symmetric searchable encryption (SSE) [44] is an attractive encryption technique that allows outsourcing the storage of sensitive data to an untrusted third party (e.g. a public cloud) in an encrypted format and, subsequently, searching over ciphertexts using keywords without decryption. SSE encryption protects data as well as search keywords. SSE schemes can also be relevant even when data are stored in-house for protecting sensitive data from insider attacks due to malicious system administrators or negligence. SSE schemes make a trade-off between efficiency, functionality, and security. The functionality of an SSE scheme is measured by the expressiveness of search queries and whether the scheme allows single or multiple users, which refers to the number of users who can submit queries to search on ciphertexts. The majority of current secure SSE schemes allows users to search for records containing specific keywords but does not yet support operations such as multiple keywords, sorting, and range queries. SSE schemes that support multiclient models do not provide the same level of security and privacy guarantees as schemes that support the single-client model.

Studies show that many SSE schemes may leak sensitive information, although an ideal SSE scheme leaks no information during any operation [45]. Even small leakages can lead to

TABLE 3 The benefits and drawbacks of emerging cryptography technologies.

Cryptographic technique	Benefits	Drawbacks
Symmetric searchable encryption (SSE)	• Can outsource the storage of sensitive data securely • Can search over encrypted data	• Limited functionality and inefficiency • Some schemes can leak information
Attribute-based encryption (ABE)	• Supports flexible, granular access control management	• Inefficient for large data
Functional encryption (FE)	• Supports computation on encrypted data • Output can be in unencrypted form	• Still new and largely unexplored area • Supports only simple calculations
Homomorphic encryption (HE)	• Supports computation on encrypted data • Output can be in encrypted form	• Still new and largely unexplored area • Some schemes can be inefficient for large data

several attacks that can either directly or indirectly disclose sensitive information. For example, a corrupted user may learn sensitive information based on the results of multiple legitimate queries. Recent SSE schemes are:

- forward private, meaning that new record entries cannot be related to past search results, and
- backward private, meaning that no information about deleted records is revealed based on the results of past searches [46].

However, only a limited number of SSE schemes are both forward and backward private. Practical uses of SSE schemes have been limited despite their potential, mainly because most of the current implementations are either inefficient or provide very limited functionality. In the last years, we have seen good progress towards the design of SSE schemes that are more efficient and provide better functionalities, as well as their adoption in healthcare settings [47] (see Secure and privacy-preserving data use examples section).

Attribute-based encryption

Attribute-based encryption (ABE) [48] is a type of public key encryption that allows a data controller to encrypt data using its private key, where the resulting ciphertext contains policies for who can decrypt it. The data controller provides an ABE key, containing a set of attributes for each user that is intended to read the data. A secret key can only decrypt a ciphertext if the attributes included in the key (e.g. name, date of birth, employer, and role) match the policies defined for the ciphertext. In traditional public key cryptography, a message is encrypted with a public key of the intended recipient, and the decryption of a ciphertext requires a private key that matches the public key. In contrast, ABE allows the decryption of a ciphertext using any ABE key containing a set of attributes that matches the policies defined in the ciphertext.

While ABE schemes offer great flexibility in terms of access management, since they are asymmetric encryption schemes, they are inefficient for encrypting large volumes of data. The size of the produced ciphertexts and the time required to both encrypt and decrypt data grows with the complexity of the underlying policy [49]. However, ABE is practical for protecting the security of symmetric encryption keys, such as an SSE key, which is comparatively smaller in size [50] (see Secure and privacy-preserving data use examples section).

Functional encryption

Functional encryption (FE) is an emerging cryptographic technique that allows a secret key holder to compute a function on encrypted data without learning any information about the data, apart from the results of the function [51]. Despite its potential to provide reliable solutions to many security and privacy-related problems, it is still an unexplored area and, as a result, faces certain limitations. There has been good progress in the development of FE schemes [52,53]. However, the more efficient FE schemes involve executions of distributed protocols among participating parties that require the parties to be online for a certain amount of time [54]. In addition, FE schemes have been used only for simple calculations, and current schemes do not support computing advanced functions [55].

Homomorphic encryption

Homomorphic encryption (HE) is a public-key encryption scheme that allows users to perform certain computations on encrypted data using a public key. HE schemes are grouped into partial HE (PHE) [56], somewhat HE (SHE) [57], and fully HE (FHE) [58]. PHE schemes support an unlimited number of operations, either addition or multiplication on ciphertexts, whereas SHE schemes support a limited number of operations. FHE supports any computation on encrypted data. PHE and SHE schemes are more efficient than FHE schemes. The first FHE scheme was a trillion times slower than performing the same computations on plaintext data. Although there has been good progress in the design of more efficient FHE schemes [59], they are still not ready for practical use.

Although both FE and FHE allow the computation of arbitrary functions over encrypted data, they have an important difference. The output of running a function using FHE is still encrypted and thus requires access to the private key for decryption. The results, therefore, need to be sent to the data controller for the decryption, whereas the output of a computation using FE is directly accessible as a plaintext. This difference makes the two schemes suitable for different kinds of applications and services.

De-identification

De-identification involves removing identifying information from data. Synonymous with de-identification is anonymisation, which can be thought of as the ideal goal of de-identification, where it is impossible to re-identify data subjects. Another closely related term is pseudonymisation, which refers to replacing personal identifiers with pseudonyms in a dataset.

De-identification methods can be categorised broadly as perturbative (i.e. data swapping, adding noise, and generating synthetic data) and nonperturbative (i.e. generalisation and suppression). Perturbative methods distort the truthfulness of records so that only selected statistical properties are preserved, whereas nonperturbative methods make data values less precise.

A typical de-identification pipeline is illustrated in Fig. 2, where the process starts with pseudonymisation, that is, removing direct identifiers (e.g. using masking). Direct identifiers have a 1:1 relationship with the subject, for example, email or national ID number. Quasi-identifiers, on the other hand, have a 1:many relationship, for example date of birth or postal code, and may indirectly identify a subject. After removing direct identifiers, data transformation focusses on quasi-identifiers until data is ready to be disclosed.

HIPAA Privacy Rule's Safe Harbor method requires the removal of 18 direct and quasi-identifiers, designated as protected health information (PHI). An alternative to the Safe Harbor method is expert determination, where a duly qualified expert applies statistical or scientific principles on the data until an acceptable disclosure risk is achieved. Both HIPAA and GDPR allow de-identified health data to be disclosed without the consent of living or deceased patients. It is recommended, however, that such de-identification be undertaken by someone within the care team [60]. Legal obligations such as consent or data breach notifications are not required for anonymised data [61].

Although de-identification is certainly a valuable tool for protecting sensitive data, in many instances the risk of re-identification is nonzero [62]. One way to conceptualise risk minimisation strategies is through risk-based anonymisation, as described by El-Emam et al. [61]

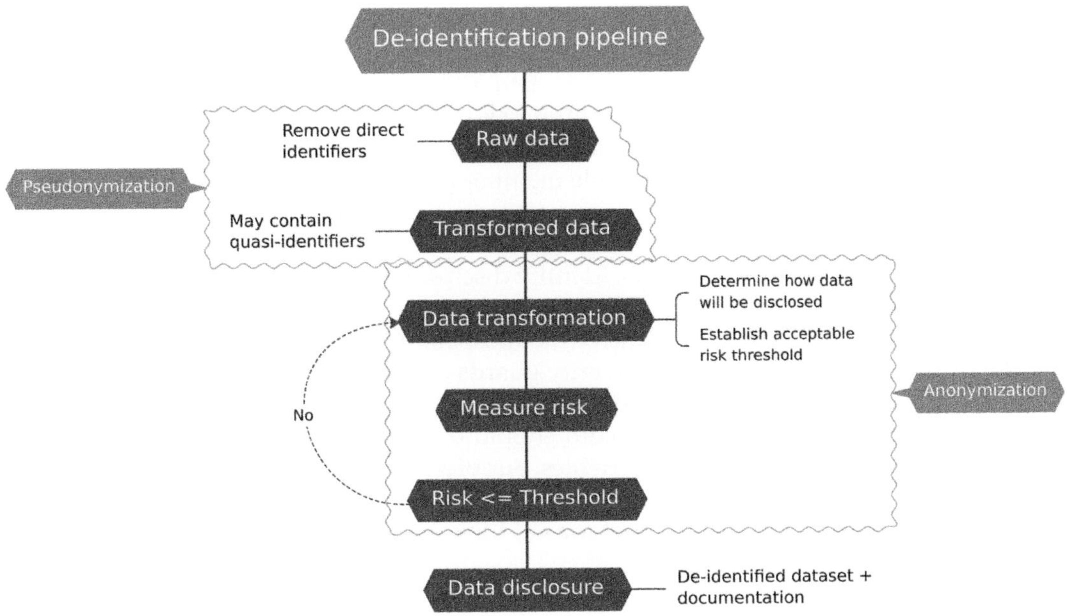

FIG. 2 A typical de-identification pipeline.

The key principle of this approach is to view re-identification risk as a function of the de-identification transformation of the data and also as a function of the context within which the data is disclosed or shared. To help shed light on this conceptual framework, the rest of this subsection will consider two aspects that characterise the framework, namely, data transformation and data disclosure.

Data transformation

The impact of re-identification risks can be viewed in terms of the sensitivity of the data in question. The consequences of re-identified health data in some cases can be more severe than financial data. An increasingly important consideration when studying sensitive health data is the disproportionately large amount of unstructured health data, such as free-text clinical notes (e.g. discharge letters) or medical images. In many instances, re-identification risk remains comparatively higher in de-identified clinical texts [63]. Similarly, because of genetic data, it may be difficult to completely de-identify data in cases where biospecimens are collected. The current legal framework and blanket guidelines provide an easy guide for data controllers but these clearly also have their limitations. For instance, a study found that the re-identification risks associated with data de-identified using HIPAA's Privacy Rule vary with the size of the US state [64]. Perhaps, an auxiliary solution is to quantify and evaluate acceptable risk in each local context.

For each situation, an acceptable risk threshold can be agreed upon, and the risk must be quantitatively measured using appropriate computational metrics [64] and then evaluated. Disclosure limitation techniques can then be applied to the data using either perturbative or nonperturbative techniques discussed earlier. This process continues in iterations until a satisfactory risk threshold is achieved. It is important to note that transformations on the data can reduce disclosure risk, but the transformation makes the data coarser, which could reduce utility. Duncan et al.'s [65] disclosure risk vs. data utility (R-U) confidentiality map is one of the many frameworks developed to study the risk-utility trade-off for disclosure limitation techniques. The main challenge is to produce a de-identified dataset that preserves confidentiality (by minimising disclosure risk) while maintaining sufficient data utility.

Identity disclosure is one of the most common disclosure risks associated with de-identified data sharing. We refer the reader to several references [66–68] for an extensive review of other types of disclosure risks. Identity disclosures occur when an adversary is able to assign the correct identity to a record in a de-identified dataset [69,70]. To limit identity disclosure risks, k-anonymity [71] is one of the most common privacy models implemented by de-identification algorithms. K-anonymity guards against singling out records that identify an individual, thereby reducing the ability to link records. Given a set of key variables, k-anonymity requires that a person's record should not be distinguishable from at least k-1 other records, with respect to the key variables. Suppression and generalisation are examples of methods that can be used to achieve k-anonymity to reduce the risk of re-identification. Ultimately, the risk should be viewed in terms of the context within which data are shared.

Data disclosure

So far, we have discussed operations on the data, that is, how sensitive the data is and how it can be transformed. We have to consider data sharing mechanisms and define the acceptable level of risk. The impact of cybersecurity incidents remains the same no matter how the data is shared because it is directly proportional to the sensitivity of the data. However, the likelihood of adverse events occurring varies with the context, and therefore, the method for data sharing needs to be known during de-identification, to produce a de-identified dataset appropriate for the sharing method. For instance, to release data in the context of public openness (e.g. the Internet), severe data transformation may be required, which reduces the utility of the data. A milder approach is to release the data to a limited set of specific individuals or institutions who are required to sign a data use agreement. Alternatively, high utility data, that is, data with minimal transformation, can be shared in the context of a data safe haven (DSH), which offers access control and other confidentiality safeguards, as illustrated in Table 4.

DSH, analogous to virtual data laboratory and data enclave, is a data-sharing mechanism that is receiving increasing interest. This is where a dataset is stored in a secure computing environment, and the data users get read-only access via secure access sites and/or remote

TABLE 4 Data sharing methods, implications, and potential impacts.

	How data will be shared		
	Open internet	**Data use agreement**	**Data safe haven**
Likelihood of an incident occurring	High	Medium	Low
Data transformation rigour required to reduce disclosure risk	High	Medium	Low
Utility of data after transformation	Low	Medium	High

connections. A data user may view microdata and conduct data manipulations and statistical analysis, but the dataset cannot be taken out of the secure environment. Results of the analyses are evaluated for disclosure risk before leaving the secure environment. The approval of data custodians and ethical board(s) is normally required to gain access to a dataset. It is usual to have a data use agreement that restricts any attempts to re-identify or transfer the data to a third party. We refer the reader to these references [72, 73] for further discussions on the pros and cons of DSH.

Privacy-preserving distributed data mining

Privacy-preserving distributed data mining (PPDDM), also known as privacy-preserving distributed statistical computation (PPDSC), is an emerging approach for processing data distributed across multiple data sources while protecting privacy [66,74]. PPDSC considers the problem of running statistical algorithms on confidential data divided across two or more different data sources, without allowing any party to view the private data of another data source. This method reveals statistics generated from the combined data for a group of data sources and does not reveal sensitive information about the input. PPDSC is a special case of a long-studied problem in cryptography called secure multiparty computation (SMC). The field of SMC deals with the question of how to securely compute a function on the combined data for a group of data sources without revealing anything apart from the output. However, it does not consider the question of how much information about the input is revealed in the output [75].

Goldreich et al. [76] showed that any function can be securely computed between multiple parties. However, the theoretical result was too inefficient for practical use. Since then, SMC has been widely studied to improve the building blocks and algorithms. As a result, specialised protocols for statistical problems have been designed to achieve better efficiency by utilising the specific properties of the statistical problem. Many statistical problems can be decomposed into similar subfunctions, such as summation, multiplication, scalar product, and comparison, at various stages [77,78]. Therefore, secure protocols are constructed for these basic functions. Computing a function by composing secure subprotocols is secure according to the composition theorem [79].

Secure summation, one of the most frequently studied problems, computes the summation of secret values of three or more data sources without revealing the private values, apart from the aggregated result. Secured intermediate information exchange (SINE) is one of the existing secure summation protocols, which also provides protection against the risk of collusion attack [80].

The existing secure protocols for computing statistical problems provide different security guarantees, efficiency, and scalability [66,74]. Secure protocols often provide security guarantees against honest-but-curious (semi-honest) adversaries. This adversarial model allows the designing of secure protocols that are more efficient and scalable, compared to protocols that are secure against malicious adversaries, for example [75]. The honest-but-curious adversarial model assumes that, even if corrupted parties follow a computation protocol specification, the adversary may try to use the internal state of the corrupted parties, including the messages exchanged during the protocol execution, to learn private information of other uncompromised parties. The honest-but-curious adversarial model is suitable for settings in which parties are trusted to follow a computation protocol specification but must run a secure protocol because of legal restrictions for data sharing. This model is also useful to prevent accidental leakage. An adversary that gained access to a party's database after the execution of a secure protocol is prevented from learning private information about legitimate parties [75]. Therefore, the adversarial model provides sufficient privacy guarantees in many contexts while enabling efficient and scalable computation.

SMC protocols do not alter the original data, such that the data utility is not affected, in contrast to de-identification that needs to make a balance between re-identification risks and data utility. The legal and ethical requirements for data processing using PPDDM may vary slightly, depending on the building blocks used. In general, the requirements of the approach are not different from the requirements of de-identified data processing. El Emam et al. [81] argued that the use of their security protocol in public health research does not require consent or ethical review, as long as the healthcare institutions have a data use agreement or internal policy that prohibits the exchange of secret keys used in the protocol.

Promising progress has been made in protocol design and the implementation of cryptographic primitives, fuelled by the constant increase in computing power. But even then, efficiency deteriorates as the number of data sources and records increases. However, protocols specialised in specific statistical problems will be more efficient and scalable than protocols for generic solutions. More and more practical uses of PPDDM in healthcare settings have been reported [82–84].

Secure and privacy-preserving data use examples

This section presents the security and privacy challenges of three healthcare use cases and solutions they implemented to address the challenges. The first use case reuses EHR data for generating feedback to clinicians and the other two use cases are related to health data sharing among healthcare actors in multiple institutions for individual health care. The three use cases are part of an EU-funded project, ASCLEPIOS [85], which aims to develop a secure platform for health data processing in the cloud.

Case study 1: Audit and feedback on antibiotics prescribing

The increasing emergence of antibiotic-resistant bacteria observed worldwide is lowering the success rates of infection treatment using antibiotics. The World Health Organisation (WHO) estimates 700,000 die annually because of antimicrobial resistance (AMR) [86]. Misuse and overuse of antibiotics are one of the main drivers of AMR, and for Norway, general practitioners (GPs) are responsible for approximately 80% of all antibiotic prescriptions [87]. Learning from data generated in routine clinical care at GP practices could be a major driver of quality improvement. Feedback on the clinical performance of a clinician, viewed in comparison with peers, has been shown to be an effective instigator of a change in prescribing behaviour [88,89].

Security and privacy challenges

The traditional way of aggregating data from multiple health institutions is through centralised or federated databases. In a recent review, Al-Issa et al. [90] discuss two major problems that could arise in this situation. First, a centralised database creates a 'one-stop honey pot' for attacks, and second, owners of the data may lose control over their sensitive data.

The first challenge could be to avoid the collection of sensitive data in any central location. However, execution of statistical operations on confidential data is still required, even when data is distributed across multiple institutions. Statistics aggregated from the combined data of all GP offices should not reveal sensitive information and should be accessible to all GPs. However, personal data on antibiotics prescribing of any one clinician should only be available to that clinician. The second challenge is to ensure that individual healthcare institutions do not lose control of their data while they contribute to aggregate statistics. Losing control increases exposure to several risks such as legal liability, malpractice lawsuits, intrusive marketing and inference or disclosure of income, performance, or competitive data [91], which consequently affect their willingness to participate.

Secure system for audit and feedback

Next, we describe an innovative system that solves these problems using a state-of-the-art technique: privacy-preserving distributed data mining (PPDDM)—see preceding section on Privacy-preserving distributed data mining. Fig. 3 shows the architectural design of the system. The system contains software components deployed at each healthcare institution, including a third party denoted as a coordinator. The coordinator can be any institution

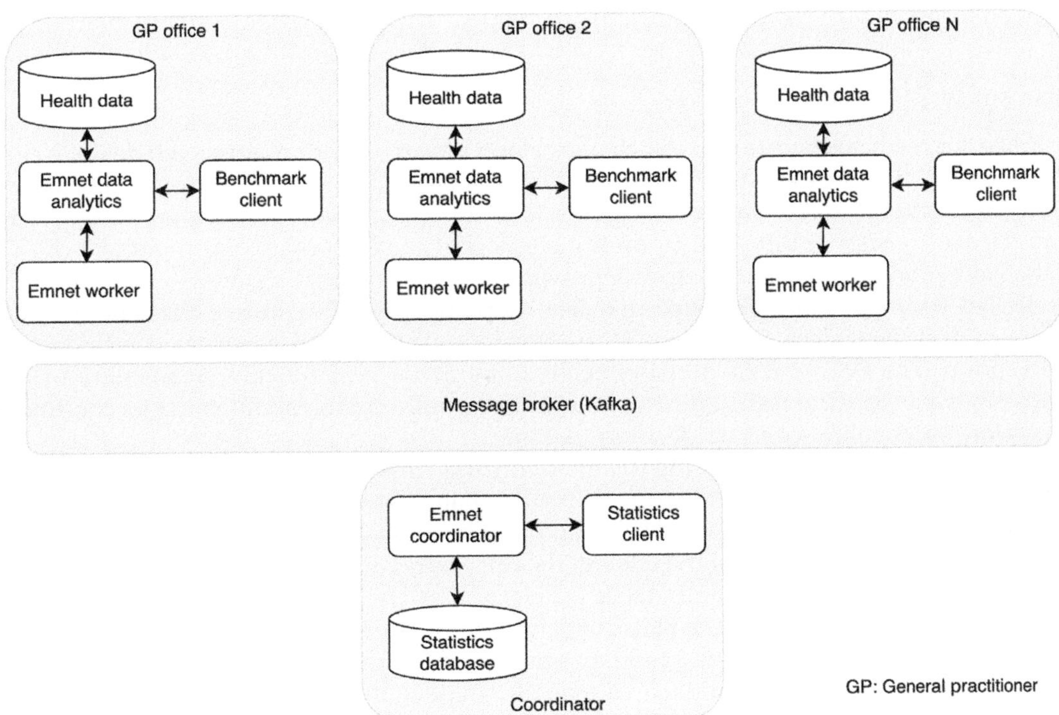

FIG. 3 Architectural design of the audit and feedback system.

trusted to follow protocol specifications, which are a standard security model known as the honest-but-curious adversarial model [92]. It does so without learning private information.

Secure communication between Emnet workers and the Emnet coordinator is supported through a message-oriented middleware (message broker) that uses SSL encryption. A combination of asymmetric (RSA) and symmetric (AES) encryption schemes is used for end-to-end encryption of messages between a sender and a receiver.

Administrative users schedule the computation of individual and group-level statistics using the Emnet client, a web application that interacts with the restful web service interfaces provided by the Emnet coordinator. The coordinator broadcasts the query to workers running at each institution. Locally running Emnet workers execute the query on their data. To compute group-level statistics, Emnet workers jointly execute a secure summation protocol; a building block to compute different statistical functions. The secure summation protocol, first described in Yigzaw et al. [93], is a distributed secure multiparty computation (DSMC), which involves dividing Emnet workers into neighbour peers. Each neighbourhood runs a secure protocol and aggregates the local results. The protocol keeps most of the message exchange among neighbour peers, effectively reducing complexity and bolstering scalability. A working protype was tested in 2019 with 20 GPs in Norway [32].

To sum up, the system solves key security and privacy challenges in at least two important ways. First, by storing data locally, the health institutions retain control of their data, as no sensitive data is disclosed outside the healthcare institution [32]. Second, data aggregation based on formally proven security attributes provides satisfactory confidentiality guarantees.

Case study 2: Stroke acute care

Stroke is a medical condition in which poor blood flow to the brain causes cell death. Acute care for stroke involves a complex collaboration of various parties, including the emergency call centre, ambulance services, and hospitals. Currently, treatment in the acute phase of ischaemic stroke consists of intravenous thrombolysis (IVT) and endovascular treatment (EVT). In the Netherlands, almost all hospitals provide IVT treatments. EVT, however, is a highly specialised treatment only provided in a few hospitals. It is difficult to determine the best treatment and then place the patient at the treatment centre with adequate conditions for intervention. The IVT vs EVT decision needs to be made very quickly. Availability of patient's electronic medical record (EMR), in addition to information about patient's condition, is therefore essential to speed up decision-making.

Various teams enter and leave the treatment during stroke acute care, which starts with a call to the emergency call centre. During the call, a trained healthcare worker follows a triage protocol to identify the suspicion of stroke and then requests an ambulance. Once the ambulance arrives and takes charge of the patient, the call centre professional hangs up the call and leaves the treatment. When the ambulance team examines the patient and confirms the stroke suspicion, a request is sent to the closest primary stroke care centre to receive the patient. The ambulance team contacts the hospital to inform the estimated arrival time and relevant patient conditions. When the patient arrives at the hospital, the assigned team takes charge of the treatment and the ambulance team leaves. If the patient is eligible for EVT, transportation to a comprehensive stroke centre by ambulance might be necessary, involving more teams in the treatment. Finally, when no longer at risk, the patient is discharged and the hospital team leaves the treatment.

Security and privacy challenges

From the first call to the emergency call centre to hospital discharge, the teams need to exchange information while treating the patient. Currently, this information is exchanged orally or by phone, as there is no unified EMR that can be shared by all professionals during treatment. Our proposal is to store the patient's EMR in cloud storage to enable ubiquitous access by the teams during treatment. However, cloud storage may raise questions about patient privacy due to unauthorised access by the cloud provider or healthcare professionals. Moreover, the GDPR requires that all processing of a patient's health information is proportionate and limited to what is necessary for care. Therefore, professionals involved in acute care must lose access to the patient's EMR immediately after completing their tasks.

Cloud-based secure solution

The EMRs are stored as ciphertext in the cloud, and the encryption keys are shared with the involved teams only during their involvement in the emergency treatment. Fig. 4 illustrates the components of the secure EMR system. The system protects patients' privacy by only storing health data in the cloud in encrypted form and by making encryption keys accessible only to healthcare professionals during their involvement in emergency treatment. Patients and healthcare professionals organised in acute care teams are the system users, who interact with the EMR system through a web application. The registration authority assigns attributes to users, and these are used to authorise (or not) access to patient data. A user receives an attribute-based encryption (ABE) key based on his/her own attributes from the master authority.

The patient generates a symmetric searchable encryption (SSE) key and encrypts the key using the ABE scheme containing access control policies that grant access to specific healthcare

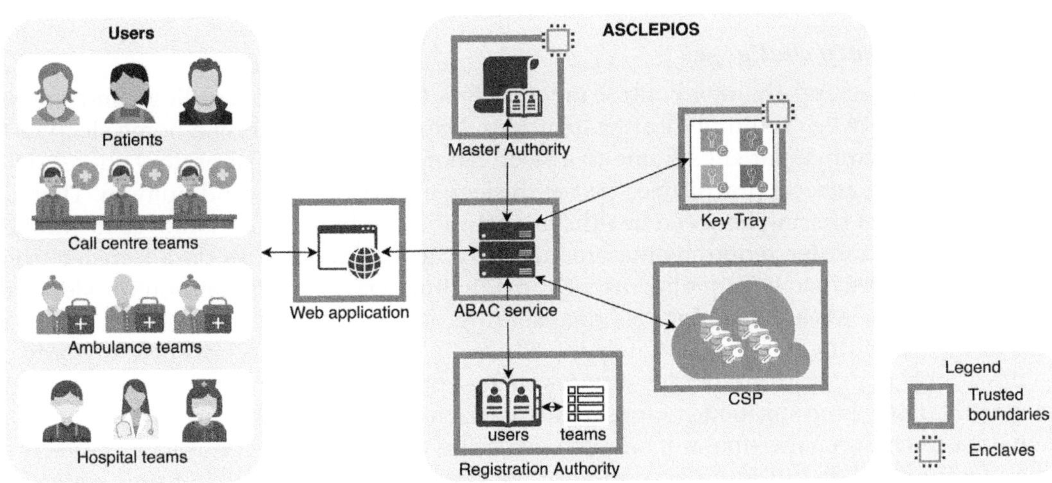

FIG. 4 Secure EMR system for acute stroke care data sharing.

professionals. Then, the patient sends the ciphertext of the SSE key to the key tray for storage. Only a user in possession of an ABE key that contains matching attributes can recover the SSE key and decrypt and encrypt data locally. The ABAC service provides an additional layer of access control to the EMR system on top of ABE, ensuring that only a health professional (e.g. ambulance team member) that is currently treating a patient gets access to patient data.

The proposed EMR system enables granting and revoking access on demand. This solution is expected to alleviate security and privacy concerns associated with cross-institutional data sharing that has been a barrier for collaborative care in the Netherlands and many other countries.

Case study 3: Inpatient and outpatient sleep medicine

Sleep medicine is a growing field in healthcare due to the high prevalence of sleep disorders. The International Classification of Sleep Disorders 3rd edition (AASM 2014) [94] contains 67 distinct sleep disorders. Besides these primary diagnoses, there are also secondary sleep disorders which may be related to medical or mental problems, including substance or drug use or misuse. The disorders are diagnosed via medical interviews and overnight recordings of multidimensional biosignals, either at a sleep laboratory using polysomnography (PSG) or at home using polygraphy (PG). PSG monitors many body functions including brain waves, eye movements, muscle activity, heart activity (ECG—electrocardiography, blood pressure, and pulse wave), respiration (airflow and respiratory movement), blood gases (oxygen saturation and carbon dioxide tension), snoring, and movement of the sleeping person. A PG uses a reduced set of sensors focussed on respiratory and cardiovascular events.

Once a sleep recording has been performed, data are transmitted to a specially trained sleep expert who manually scores the recordings on a computer screen to determine the sleep profile and relevant patterns and events [95]. A report compiled, based on the interpretations of all signals, is the basis for the treatment of the patient. Usually, a patient returns to the sleep physician at annual intervals because most sleep disorders are chronic conditions.

Security and privacy challenges

Sleep diagnostics and therapy control involve data (i.e. biosignal recordings and the reports) transfers between multiple healthcare actors. Data collected during a home sleep testing is transferred from a patient's home to a sleep laboratory. A sleep laboratory shares data with the attending general practitioner and with sleep experts for a second opinion. There is also a need for data sharing between healthcare institutions when a patient moves to another place. New data transfer requirements are currently upcoming such as data transfer from wearables to analysis facilities [96]. Currently, printed material is still common for sleep reports, while CD-ROMs are used for PSG data sharing.

Cloud-based solutions are being slowly introduced in sleep medicine, but currently only in a unidirectional manner, from therapy devices such as breathing support to the device manufacturer [97]. There is no solution yet for secure multiactor access and analysis of sleep-related health data. Such a scenario does not only require a fine-grained access control and protection of the identifying data but also encryption of the medical data itself, as not only video recordings but also long-term multidimensional biosignal recordings and reports are typically implicitly re-identifiable [98]. This, in particular, is the case for rare sleep disorders such as narcolepsy—a disorder where patients suddenly fall asleep.

Secure system for sleep medicine

The proposed system as shown in Fig. 5 builds upon an existing central data management system capable of storing sleep data [99]. The system processes metadata, which are structured data, including textual descriptions and sleep parameters within reports. Metadata are stored as key value pairs and encoded with HL7 FHIR, a common medical messaging standard. The system also processes unstructured bulk data such as biosignal recordings. Security controls have been applied to enable secure cloud-based sleep medicine while supporting the aforementioned data sharing requirements.

Metadata is encrypted on the client side using symmetric searchable encryption (SSE) before storage on the SSE-enabled cloud storage within the data management. SSE keys used for the encryption are also themselves locally encrypted using attribute-based encryption (ABE) before storage in the key storage. In addition, access to data and services is protected by attributed-based access control (ABAC). Biosignal recordings are encrypted before upload to a bucket storage in the cloud. Search for unstructured data is supported based on structured metadata of the file headers.

The current security system supports the main sleep medicine diagnosis process, from biosignal recording by the patient to access by the sleep lab professionals who score the recordings and create a report that can be shared with the general practitioner through the platform. The different actors, data, and security components are depicted in Fig. 5. The integration of privacy-enhancing services and encrypted storage increases the complexity of the system, which can be seen as a downside. On the other hand, the service-oriented architecture and standardised interfaces allow for the system to be extended. This opens up the possibility of integrating security-enhanced capabilities as a further development step.

Discussion

Cybersecurity threats to healthcare institutions and patient safety are real. Security incidents have several adverse consequences to patients and institutions, as well as patient safety. Therefore, cybersecurity must be a priority and institutions must make the investments needed to protect their patients. Institutions need to continuously identify their security risks and risk tolerance and, consequently, prioritise their cybersecurity activities to maximise the impact of each dollar spent. We describe next a stepwise cybersecurity roadmap, based on NIST's cybersecurity framework, an essential tool for healthcare institutions that seek to manage cybersecurity risk.

A 5-step cybersecurity roadmap

(i) The first step in developing a cybersecurity roadmap is understanding the current security posture of the institution, by indicating which security outcomes are currently being achieved. We highlighted some of the most common risks and challenges in healthcare, but it is up to each institution to assess whether they are

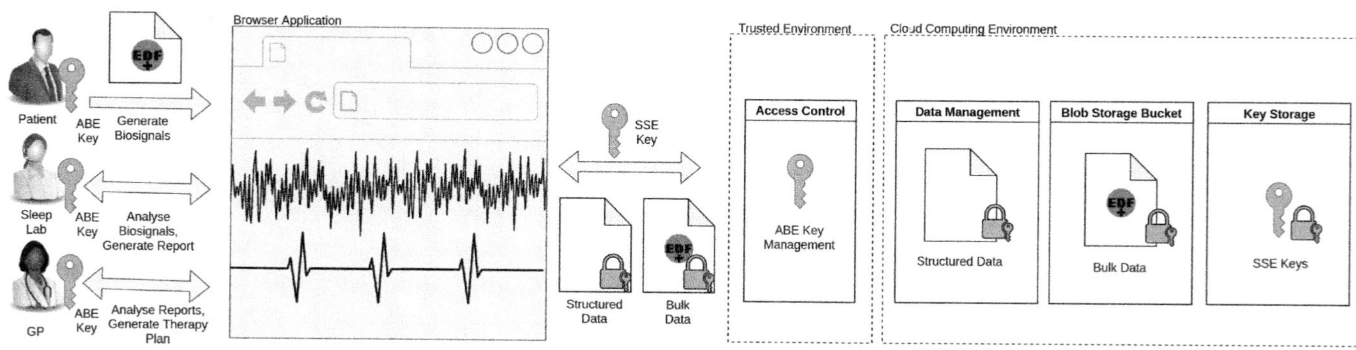

FIG. 5 Secure system for data sharing in sleep medicine.

protected against these adverse events. Attention must be paid to the institution's standing vis-à-vis important legislation such as GDPR and HIPAA.

(ii) The second step is conducting a risk assessment to determine the likelihood of a cybersecurity event and the impact that the event could have on the institution. An important consideration in risk management strategy is making a proper balance between handling risk and retaining necessary benefits of data or systems. For example, data protection regulations allow secondary use of de-identified health data although it is known that the risk of re-identification is nonzero. The re-identification risk of de-identified data varies with the context for data sharing, such as data available to the public on the Internet or to specific individuals or institutions who signed data use agreements. The former data sharing context entails more threat actors than the latter and, consequently, has a higher likelihood of re-identification attempts.

(iii) The third step is creating a target profile that describes the institution's desired cybersecurity outcomes. In addition to user awareness and training, institutions can be equipped with recent advances in access control, cryptography, de-identification, and privacy-preserving distributed data mining, to develop a profile appropriate for their risk appetite. For instance, institutions may choose to minimise the risk associated with accidental loss of data stored on mobile devices by providing access to the data in a secure data processing environment that contains proper access control.

(iv) In the fourth step, healthcare institutions analyse the gaps between the current status and target cybersecurity profiles, which helps create a prioritised cybersecurity action plan to address gaps based on its business and legal requirements, risk tolerances, and available resources.

(v) The final step is implementing the action plan. While identifying risk and deploying protections, security controls may be deployed with the intent of deterring or delaying an attack. It is critical to adequately monitor for anomalous behaviour or activities that are potential attempts of a security breach. Once an attack is detected, it is critical to implement appropriate actions regarding the cybersecurity incident and recover to normal operations in a timely manner to reduce the impact of a cybersecurity incident. The steps are repeated as needed to continuously assess and improve the institution's cybersecurity.

In this chapter, we presented cybersecurity in the context of healthcare, and we adopted a technical perspective that can contribute to the implementation of a cybersecurity strategy in health care systems. We identified common health data use challenges and presented state-of-the-art cybersecurity solutions. Some of these solutions were exemplified using real-life use cases, and we concluded with an actionable, stepwise cybersecurity roadmap.

Acknowledgement

This work was supported by Research Council of Norway Grant No. 248150/O70 and the European Commission's H2020 research and innovation programme Project No. 826093. The funding body did not have any role in the writing of the manuscript.

References

[1] C. Safran, M. Bloomrosen, W.E. Hammond, et al., Toward a national framework for the secondary use of health data: an American medical informatics association white paper, J. Am. Med. Inform. Assoc. 14 (2007) 1–9, https://doi.org/10.1197/jamia.M2273.

[2] S.M. Meystre, C. Lovis, T. Bürkle, et al., Clinical data reuse or secondary use: current status and potential future progress, Yearb. Med. Inform. 26 (2017) 38–52, https://doi.org/10.15265/IY-2017-007.

[3] L. Marco-Ruiz, T. Beale, J.J. Lull, et al., Towards open process models in healthcare: open standards and legal considerations, in: C. Fernandez-Llatas (Ed.), Interactive Process Mining in Healthcare, Springer International Publishing, Cham, 2021, pp. 81–99, https://doi.org/10.1007/978-3-030-53993-1_6.

[4] J.G. Bellika, A. Makhlysheva, P.A. Bakkevoll, A significant increase in the risk for exposure of health information in the United States: result from analysing the US data breach registry, in: Proceedings from The 15th Scandinavian Conference on Health Informatics 2017 Kristiansand, Norway, August 29–30, 2017, Linköping University Electronic Press, 2018, pp. 55–59.

[5] ISO 31000:2018. ISO. https://www.iso.org/cms/render/live/en/sites/isoorg/contents/data/standard/06/56/65694.html. (accessed 8 Mar 2021).

[6] ISO/IEC 27005:2018(en), Information technology—Security techniques—Information Security Risk Management. https://www.iso.org/obp/ui/#iso:std:iso-iec:27005:ed-3:v1:en. (accessed 8 Mar 2021).

[7] R.S. Ross, Managing Information Security Risk: Organization, Mission, and Information System View, Spec Publ NIST SP—800-39, 2011.

[8] NIST, Cybersecurity Framework, NIST, 2013. https://www.nist.gov/cyberframework. (accessed 8 Mar 2021).

[9] L. Kohnfelder, P. Garg, The Threats to Our Products. Threats Our Prod, 1999. https://adam.shostack.org/microsoft/The-Threats-To-Our-Products.docx. (accessed 8 Mar 2021).

[10] A. Shostack, Threat Modeling: Designing for Security, John Wiley & Sons, 2014.

[11] EUR-Lex—32016R0679—EN—EUR-Lex. https://eur-lex.europa.eu/eli/reg/2016/679/oj. (accessed 8 Mar 2021).

[12] Riksrevisjonen, Riksrevisjonens undersøkelse av helseforetakenes forebygging av angrep mot sine IKT-systemer. https://www.riksrevisjonen.no/globalassets/rapporter/no-2020-2021/undersokelse-av-helseforetakenes-forebygging-av-angrep-mot-sine-ikt-systemer.pdf. (accessed 26 Mar 2021).

[13] HIMSS, Cybersecurity Survey, 2020. https://www.himss.org/sites/hde/files/media/file/2020/11/16/2020_himss_cybersecurity_survey_final.pdf. (accessed 8 Mar 2021).

[14] B. McCall, What does the GDPR mean for the medical community? Lancet Lond Engl 391 (2018) 1249–1250, https://doi.org/10.1016/S0140-6736(18)30739-6.

[15] Health Information Privacy, HHS.gov, 2015. https://www.hhs.gov/hipaa/index.html. (accessed 8 Mar 2021).

[16] CURIA, Documents. http://curia.europa.eu/juris/document/document.jsf?text=&docid=228677&pageIndex=0&doclang=en&mode=lst&dir=&occ=first&part=1&cid=12312155. (accessed 8 Mar 2021).

[17] O. Ben-Assuli, I. Shabtai, M. Leshno, The impact of EHR and HIE on reducing avoidable admissions: controlling main differential diagnoses, BMC Med. Inform. Decis. Mak. 13 (2013) 49, https://doi.org/10.1186/1472-6947-13-49.

[18] P. Fontaine, S.E. Ross, T. Zink, et al., Systematic review of health information exchange in primary care practices, J. Am. Board Fam. Med. 23 (2010) 655–670, https://doi.org/10.3122/jabfm.2010.05.090192.

[19] N. Menachemi, T.H. Collum, Benefits and drawbacks of electronic health record systems, Risk Manag. Healthc. Policy 4 (2011) 47–55, https://doi.org/10.2147/RMHP.S12985.

[20] M.E. Frisse, K.B. Johnson, H. Nian, et al., The financial impact of health information exchange on emergency department care, J. Am. Med. Inform. Assoc. 19 (2012) 328–333, https://doi.org/10.1136/amiajnl-2011-000394.

[21] J. Valle, C. Gomes, T. Godby, et al., The feasibility of the nationwide health information network, Health Care Manag. 35 (2016) 103–112, https://doi.org/10.1097/HCM.0000000000000101.

[22] J.R. Vest, T.R. Miller, The association between health information exchange and measures of patient satisfaction, Appl. Clin. Inform. 2 (2011) 447–459, https://doi.org/10.4338/ACI-2011-06-RA-0040.

[23] L. Marco-Ruiz, R. Pedersen, Challenges in archetypes terminology binding using SNOMED-CT compositional grammar: the Norwegian patient summary case, Stud. Health Technol. Inform. 245 (2017) 1332.

[24] L. Marco-Ruiz, R. Pedersen, The patient summary case: challenges in archetypes terminology binding using SNOMED-CT compositional grammar, in: eTELEMED 2019: The Eleventh International Conference on eHealth, Telemedicine, and Social Medicine. Athens (Greece), 2019, pp. 49–55.

[25] A. Budrionis, R. Wynn, L. Marco-Ruiz, et al., Impact of the use of electronic health tools on the psychological and emotional well-being of electronic health service users (the seventh Tromsø study—part 3): population-based questionnaire study, J. Med. Internet Res. 22 (2020), https://doi.org/10.2196/13118, e13118.

[26] K.Y. Yigzaw, R. Wynn, L. Marco-Ruiz, et al., The association between health information seeking on the internet and physician visits (the seventh Tromsø study—part 4): population-based questionnaire study, J. Med. Internet Res. 22 (2020), https://doi.org/10.2196/13120, e13120.

[27] I.C. Señor, J.L. Fernández-Alemán, A. Toval, Are personal health records safe? A review of free web-accessible personal health record privacy policies, J. Med. Internet Res. 14 (2012), https://doi.org/10.2196/jmir.1904, e114.

[28] L. Marco-Ruiz, E. Bønes, E. de la Asunción, et al., Combining multivariate statistics and the think-aloud protocol to assess human-computer interaction barriers in symptom checkers, J. Biomed. Inform. 74 (2017) 104–122, https://doi.org/10.1016/j.jbi.2017.09.002.

[29] C. Safran, Update on data reuse in health care, Yearb. Med. Inform. 26 (2017) 24–27, https://doi.org/10.15265/IY-2017-013.

[30] J.L. Fernández-Alemán, I.C. Señor, P.Á.O. Lozoya, et al., Security and privacy in electronic health records: a systematic literature review, J. Biomed. Inform. 46 (2013) 541–562, https://doi.org/10.1016/j.jbi.2012.12.003.

[31] E. Bellin, N.N. Dubler, The quality improvement-research divide and the need for external oversight, Am. J. Public Health 91 (2001) 1512–1517, https://doi.org/10.2105/ajph.91.9.1512.

[32] K.Y. Yigzaw, A. Budrionis, L. Marco-Ruiz, et al., Privacy-preserving architecture for providing feedback to clinicians on their clinical performance, BMC Med. Inform. Decis. Mak. 20 (2020) 116, https://doi.org/10.1186/s12911-020-01147-5.

[33] G. Richter, C. Borzikowsky, W. Lesch, et al., Secondary research use of personal medical data: attitudes from patient and population surveys in The Netherlands and Germany, Eur. J. Hum. Genet. (2020) 1–8, https://doi.org/10.1038/s41431-020-00735-3.

[34] K. El Emam, J. Mercer, K. Moreau, et al., Physician privacy concerns when disclosing patient data for public health purposes during a pandemic influenza outbreak, BMC Public Health 11 (2011) 454, https://doi.org/10.1186/1471-2458-11-454.

[35] F. Rezaeibagha, K.T. Win, W. Susilo, A systematic literature review on security and privacy of electronic health record systems: technical perspectives, Health Inf. Manag. 44 (2015) 23–38, https://doi.org/10.1177/183335831504400304.

[36] E. Psarra, I. Patiniotakis, Y. Verginadis, et al., Securing access to healthcare data with context-aware policies, in: 2020 11th International Conference on Information, Intelligence, Systems and Applications (IISA), IEEE, 2020, pp. 1–6.

[37] H. Mousaid, Toward an Interoperable and Centralized Consent Centric Access Control Model for Healthcare Resources: Model and Implementation, Univ Waterloo, 2020.

[38] Y. Verginadis, A. Michalas, P. Gouvas, et al., PaaSword: a holistic data privacy and security by design framework for cloud services, J. Grid Comput. 15 (2017) 219–234, https://doi.org/10.1007/s10723-017-9394-2.

[39] A. Taherkordi, F. Zahid, Y. Verginadis, et al., Future cloud systems design: challenges and research directions, IEEE Access 6 (2018) 74120–74150, https://doi.org/10.1109/ACCESS.2018.2883149.

[40] S.M. Chandran, J.B. Joshi, LoT-RBAC: A location and time-based RBAC model, in: International Conference on Web Information Systems Engineering, Springer, 2005, pp. 361–375.

[41] Y. Verginadis, I. Patiniotakis, P. Gouvas, et al., Context-aware policy enforcement for PaaS-enabled access control, IEEE Trans. Cloud Comput. (2019) 1, https://doi.org/10.1109/TCC.2019.2927341.

[42] eXtensible Access Control Markup Language (XACML) Version 3.0. http://docs.oasis-open.org/xacml/3.0/xacml-3.0-core-spec-os-en.html. (accessed 8 Mar 2021).

[43] S. Veloudis, I. Paraskakis, Y. Verginadis, et al., Ontological templates for regulating access to sensitive medical data in the cloud, in: 2017 IEEE 30th International Symposium on Computer-Based Medical Systems (CBMS), IEEE, 2017, pp. 805–810.

[44] S. Kamara, C. Papamanthou, Parallel and dynamic searchable symmetric encryption, in: International Conference on Financial Cryptography and Data Security, Springer, 2013, pp. 258–274.

[45] D. Cash, P. Grubbs, J. Perry, et al., Leakage-abuse attacks against searchable encryption, in: Proceedings of the 22nd ACM SIGSAC Conference on Computer and Communications Security, 2015, pp. 668–679.

[46] R. Bost, B. Minaud, O. Ohrimenko, Forward and backward private searchable encryption from constrained cryptographic primitives, in: Proceedings of the 2017 ACM SIGSAC Conference on Computer and Communications Security, 2017, pp. 1465–1482.

[47] R. Zhang, R. Xue, L. Liu, Searchable encryption for healthcare clouds: a survey, IEEE Trans. Serv. Comput. 11 (2018) 978–996.

[48] A. Sahai, B. Waters, Fuzzy identity-based encryption, in: Proc. of EUROCRYPT'05. Aarhus, Denmark, 2005.

[49] M. Green, S. Hohenberger, B. Waters, Outsourcing the decryption of abe ciphertexts, in: Proceedings of the 20th USENIX Conference on Security, SEC'11, USENIX Association, Berkeley, CA, USA, 2011, p. 34.

[50] A. Michalas, A. Bakas, H.-V. Dang, et al., Access control in searchable encryption with the use of attribute-based encryption and sgx, in: Proceedings of the 2019 ACM SIGSAC Conference on Cloud Computing Security Workshop, 2019, p. 183.

[51] D. Boneh, A. Sahai, B. Waters, Functional encryption: definitions and challenges, in: In Proceedings of the 2011 Theory of Cryptography Conference (TCC) Conference, Springer, 2011.

[52] S. Goldwasser, S.D. Gordon, V. Goyal, et al., Multi-input functional encryption, in: Annual International Conference on the Theory and Applications of Cryptographic Techniques, Springer, 2014, pp. 578–602.

[53] A. Bakas, A. Michalas, A. Ullah, Functional sifting: a privacy preserving reputation system through multi-input functional encryption, in: Proceedings of the 25th Nordic Conference on Secure IT Systems (NordSec'20), 2020. Online.

[54] T. Dimitriou, A. Michalas, Multi-party trust computation in decentralized environments in the presence of malicious adversaries, Ad Hoc Netw. 15 (2014) 53–66.

[55] A. Bakas, A. Michalas, Multi-input functional encryption: efficient applications from symmetric primitives, in: Proceedings of the 19th IEEE International Conference on Trust, Security and Privacy in Computing and Communications, TrustCom, Guangzhou, China, 2020.

[56] T. ElGamal, A public key cryptosystem and a signature scheme based on discrete logarithms, in: G.R. Blakley, D. Chaum (Eds.), Advances in Cryptology, Springer, Berlin, Heidelberg, 1985, pp. 10–18, https://doi.org/10.1007/3–540-39568-7_2.

[57] J. Fan, F. Vercauteren, Arch. Somewhat Practical Fully Homomorphic Encryption, IACR Cryptol, 2012.

[58] C. Gentry, Fully homomorphic encryption using ideal lattices, in: Proceedings of the forty-First Annual ACM Symposium on Theory of Computing, 2009, pp. 169–178.

[59] N. Dowlin, R. Gilad-Bachrach, K. Laine, et al., CryptoNets: Applying Neural Networks to Encrypted Data with High Throughput and Accuracy, Microsoft Research, 2016.

[60] K. El Emam, M. Hintze, R. Boardman, Does de-identification require consent under the GDPR and English common law? J. Data Prot. Priv. 3 (2020) 291–298.

[61] M. Hintze, K. El Emam, Comparing the benefits of pseudonymisation and anonymisation under the GDPR, J. Data Prot. Priv. 2 (2018) 145–158.

[62] K.E. Emam, E. Jonker, L. Arbuckle, et al., A systematic review of re-identification attacks on health data, PLoS One 6 (2011), https://doi.org/10.1371/journal.pone.0028071, e28071.

[63] US Patent for System and method to reduce a risk of re-identification of text de-identification tools Patent, (Patent # 10,395,059 issued August 27, 2019) - Justia Patents Search https://patents.justia.com/patent/10395059. (accessed 12 Jan 2021).

[64] K. Benitez, B. Malin, Evaluating re-identification risks with respect to the HIPAA privacy rule, J. Am. Med. Inform. Assoc. 17 (2010) 169–177, https://doi.org/10.1136/jamia.2009.000026.

[65] G. Duncan, S.A. Keller-McNulty, S.L. Stokes, Disclosure Risk vs. Data Utility: The R-U Confidentiality Map, Published Online First, 1 February 2003, https://doi.org/10.1184/R1/6471215.v1.

[66] C.C. Aggarwal, P.S. Yu, A general survey of privacy-preserving data mining models and algorithms, in: C.C. Aggarwal, P.S. Yu (Eds.), Privacy-Preserving Data Mining: Models and Algorithms, Springer US, Boston, MA, 2008, pp. 11–52, https://doi.org/10.1007/978-0-387-70992-5_2.

[67] A. Gkoulalas-Divanis, G. Loukides, J. Sun, Publishing data from electronic health records while preserving privacy: a survey of algorithms, J. Biomed. Inform. 50 (2014) 4–19.

[68] B.C. Fung, K. Wang, R. Chen, et al., Privacy-preserving data publishing: a survey of recent developments, ACM Comput. Surv. 42 (2010) 1–53.

[69] L. Sweeney, k-anonymity: a model for protecting privacy, Int. J. Uncertainty Fuzziness Knowledge Based Syst. 10 (2002) 557–570.

[70] K. El Emam, E. Jonker, L. Arbuckle, et al., A systematic review of re-identification attacks on health data, PLoS One 6 (2011), e28071.

[71] K. El Emam, F.K. Dankar, R. Issa, et al., A globally optimal k-anonymity method for the de-identification of health data, J. Am. Med. Inform. Assoc. 16 (2009) 670–682, https://doi.org/10.1197/jamia.M3144.

[72] K. El Emam, Guide to the De-Identification of Personal Health Information, CRC Press, New York, NY, USA, 2013.

[73] L. Ohno-Machado, To share or not to share: that is not the question, Sci. Transl. Med. 4 (2012) 165cm15.

[74] Y.A.A.S. Aldeen, M. Salleh, M.A. Razzaque, A comprehensive review on privacy preserving data mining, Springerplus 4 (2015), https://doi.org/10.1186/s40064-015-1481-x.

[75] Y. Lindell, B. Pinkas, Secure multiparty computation for privacy-preserving data mining, J. Priv. Confid. 1 (2009) 5.

[76] O. Goldreich, S. Micali, A. Wigderson, How to play ANY mental game, in: Proceedings of the Nineteenth Annual ACM Symposium on Theory of Computing, ACM, New York, NY, USA, 1987, pp. 218–229.

[77] C. Clifton, M. Kantarcioglu, J. Vaidya, et al., Tools for privacy preserving distributed data mining, ACM SIGKDD Explor. Newsl. 4 (2002) 28–34.

[78] M. Kantarcioglu, A survey of privacy-preserving methods across horizontally partitioned data, in: C.C. Aggarwal, P.S. Yu (Eds.), Privacy-Preserving Data Mining, Springer, New York, NY, USA, 2008, pp. 313–335.

[79] R. Canetti, M. Fischlin, Universally composable commitments, in: Annual International Cryptology Conference, Springer, 2001, pp. 19–40.

[80] S. Wang, X. Jiang, Y. Wu, et al., EXpectation propagation LOgistic REgRession (EXPLORER): distributed privacy-preserving online model learning, J. Biomed. Inform. 46 (2013) 480–496, https://doi.org/10.1016/j.jbi.2013.03.008.

[81] K. El Emam, S. Samet, J. Hu, et al., A protocol for the secure linking of registries for HPV surveillance, PLoS One 7 (2012), e39915.

[82] K. El Emam, L. Arbuckle, A. Essex, et al., Secure surveillance of antimicrobial resistant organism colonization or infection in Ontario long term care homes, PLoS One 9 (2014), e93285.

[83] K.Y. Yigzaw, A. Michalas, J.G. Bellika, Secure and scalable deduplication of horizontally partitioned health data for privacy-preserving distributed statistical computation, BMC Med. Inform. Decis. Mak. 17 (2016), https://doi.org/10.1186/s12911-016-0389-x.

[84] J. Yuan, B. Malin, F. Modave, et al., Towards a privacy preserving cohort discovery framework for clinical research networks, J. Biomed. Inform. 66 (2017) 42–51, https://doi.org/10.1016/j.jbi.2016.12.008.

[85] ASCLEPIOS—H2020. https://www.asclepios-project.eu/. (accessed 8 Mar 2021).

[86] WHO, No Time to Wait: Securing the Future from Drug-Resistant Infections, 2019. https://www.who.int/antimicrobial-resistance/interagency-coordination-group/IACG_final_report_EN.pdf?ua=1.

[87] European Centre for Disease Prevention and Control, ECDC Country Visit to Norway to Discuss Antimicrobial Issues: 12–16 March 2018, Publications Office, LU, 2018. https://data.europa.eu/doi/10.2900/342180. (accessed 18 Jan 2021).

[88] C.L. Gong, J.W. Hay, D. Meeker, et al., Prescriber preferences for behavioural economics interventions to improve treatment of acute respiratory infections: a discrete choice experiment, BMJ Open 6 (2016), https://doi.org/10.1136/bmjopen-2016-012739, e012739.

[89] M. Hallsworth, T. Chadborn, A. Sallis, et al., Provision of social norm feedback to high prescribers of antibiotics in general practice: a pragmatic national randomised controlled trial, Lancet 387 (2016) 1743–1752, https://doi.org/10.1016/S0140-6736(16)00215-4.

[90] Y. Al-Issa, M.A. Ottom, A. Tamrawi, eHealth cloud security challenges: a survey, J. Healthc. Eng. 2019 (2019), https://doi.org/10.1155/2019/7516035, e7516035.

[91] K. El Emam, J. Hu, J. Mercer, et al., A secure protocol for protecting the identity of providers when disclosing data for disease surveillance, J. Am. Med. Inform. Assoc. 18 (2011) 212–217, https://doi.org/10.1136/amiajnl-2011-000100.

[92] Y. Lindell, B. Pinkas, Secure multiparty computation for privacy-preserving data mining, J. Priv. Confid. 1 (2009), https://doi.org/10.29012/jpc.v1i1.566.

[93] K.Y. Yigzaw, J.G. Bellika, A. Andersen, et al., Towards privacy-preserving computing on distributed electronic health record data, in: Proceedings of the 2013 Middleware Doctoral Symposium, Association for Computing Machinery, New York, NY, USA, 2013, pp. 1–6, https://doi.org/10.1145/2541534.2541593.

[94] M.J. Sateia, International classification of sleep disorders-third edition, Chest 146 (2014) 1387–1394, https://doi.org/10.1378/chest.14-0970.

[95] S. Rosenberg Richard, V.H. Steven, The American academy of sleep medicine inter-scorer reliability program: sleep stage scoring, J. Clin. Sleep Med. 09 (2013) 81–87, https://doi.org/10.5664/jcsm.2350.

[96] M. de Zambotti, N. Cellini, A. Goldstone, et al., Wearable sleep technology in clinical and research settings, Med. Sci. Sports Exerc. 51 (2019) 1538–1557, https://doi.org/10.1249/MSS.0000000000001947.
[97] P.A. Cistulli, J.P. Armitstead, D. Liu, et al., Real world PAP adherence: results from a big data approach in more than two million patients, in: B109. Treatment Options in Sleep Disordered Breathing: Adherence and Health Outcomes, American Thoracic Society, 2018, p. A4391, https://doi.org/10.1164/ajrccm-conference.2018.197.1_MeetingAbstracts.A4391.
[98] M. Pelc, Y. Khoma, V. Khoma, ECG signal as robust and reliable biometric marker: datasets and algorithms comparison, Sensors 19 (2019) 2350, https://doi.org/10.3390/s19102350.
[99] M. Beier, C. Jansen, G. Mayer, et al., Multicenter data sharing for collaboration in sleep medicine, Futur. Gener. Comput. Syst. 67 (2017) 466–480, https://doi.org/10.1016/j.future.2016.03.025.

CHAPTER

15

The digital needs of genomics resulting from pandemics

Kathleen A. McCormick

SciMind, LLC, North Potomac, MD, United States

Introduction

This has been one of the most trying times of the last century with a pandemic that touched every continent. When the last 1920s came in, the world called it the Roaring Twenties; now, the 2020s have become the new Roaring Twenties with attention to the COVID-19 pandemic. The COVID-19 pandemic may be referred to as the catalyst for healthcare information technology (HIT), or the accelerator for advances in healthcare information technology (HIT). This chapter will describe how a virus, once sequenced, required the same biotechnology and IT support as the Precision Health for cancer and other diseases. Those research centres and hospitals with sufficient biotechnology capabilities sprung into action detecting the virus and beginning the quest for treatments matched to the virus genomic structure, and vaccine development began through 'Operation Warp Speed'. Those research centres, and hospitals with informatics integration platforms and standards, were prepared for COVID-19 within 3 weeks. They will be referenced and described later in this chapter. When the genetics of the COVID-19 virus is described in this paper, the term SARS-CoV-2 will be used.

This chapter highlights the unique precision informatics necessary for an individual response and a population response. A model is presented that expands upon the previously published Precision Health model in two books in 2020 as demonstrated by McCormick [1], and McCormick and Calzone [2] to include the expanded needs in integration through the continuum of care, the data standards unique to COVID-19, the expanded surveillance of populations and individuals through mobile devices, and the voluminous distribution requests for telehealth in the ICU, primary care clinics, homes, and nursing homes. The bioinformatics components of this model are adapted from Regan et al. [3].

Months into the pandemic with millions of persons affected, the databases and repositories that were developed and expanded in integrated networks were mined with artificial intelligence (AI) and machine learning (ML) to determine who were the at-risk populations and

Roadmap to Successful Digital Health Ecosystems
https://doi.org/10.1016/B978-0-12-823413-6.00018-5

what were the most beneficial treatments affecting outcomes in early, middle, and advanced disease resulting from COVID-19.

Now as we approach the 9th month of experience, the vaccines that began development in March 2020 are now ready to follow vaccine distribution, guidelines are written for successful treatments by the National Institutes of Health (NIH) [4], and recommendations are written for reopening schools, businesses, and new measurement technologies are in place to evaluate the effectiveness.

The final component of this chapter is a look back at the lessons learned and a look to the future utilising the innovations of the COVID-catalysed HIT, integrated education, and research on the genomics of the survivors and non-survivors, and hospitalisation outcomes and chronic conditions resulting. The next wave of innovations in education, businesses, and competencies needs to care for persons of future pandemics and the future needs of a 4th Informatics Revolution in big data, AI, ML, robotics, and telehealth.

A model for person-centric and population-centred precision health

The current model builds upon the models developed by McCormick in 2020 [1], and McCormick and Calzone [2] also in 2020 for Precision Health that begins to identify a persons' genetic profile or their tumour profile. The omics components identified using bioinformatics of this model are expanded and adapted from Regan et al (2018) [3]. This model is updated with the large blocks to include Population Precision Health with expanded surveillance and contact tracing; integration, IHE, and standards; telehealth including tele-ICU and telehealth between health professionals in offices, home, and nursing homes; big data, artificial intelligence, machine learning, and robotics; sequence databases especially for virus genomics; and tools for analysis (Fig. 1).

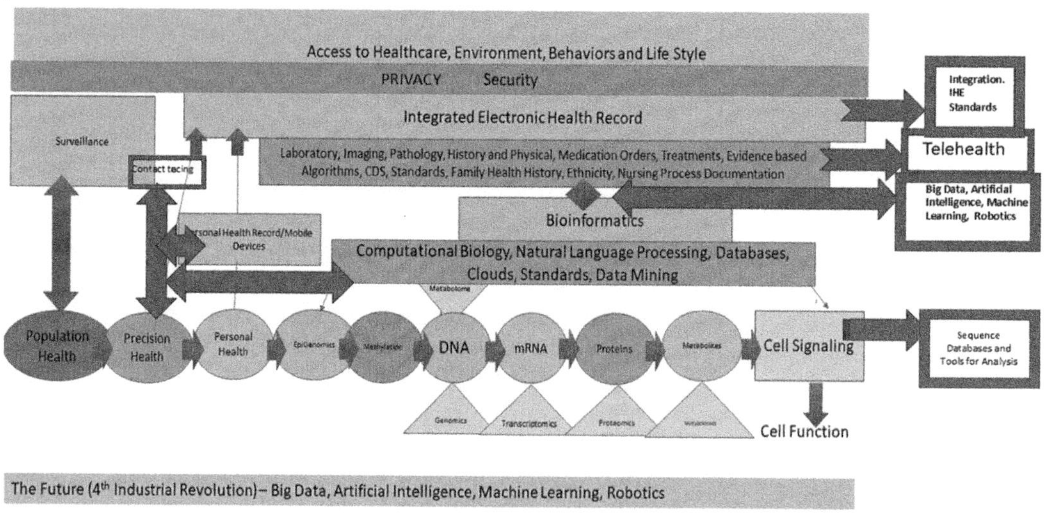

FIG. 1 Relationship between omics science and informatics.

As soon as the virus genetic structure was identified, the diagnosis, treatments, and vaccine developments began with Operation Warp Speed. The basic technologies utilised result from the bioinformatics and computational biology components of biotechnology that are required to produce the results. Large international databases were expanded to identify strains of the virus, mutations, and unique characteristics. In the United States, President Trump signed the Coronavirus Aid, Relief, and Economic Security (CARES) Act into law, which provided $2 trillion to stimulate the fight for testing, drug development, and vaccine development in the COVID-19 pandemic (Public Law 111-148 111th Congress Act) [5].

Biotechnology and computational biology

Bioinformatics and computational biology

The model in Fig. 1 model describes the components of bioinformatics and the computational biology used to analyse the results. This science has been developed and perfected over the past 20 years for the detection of genes identifying cancer and other diseases in Precision Health. Once the gene for the virus was identified in COVID-19, the same informatics supports were utilised.

The virus: Diagnosis

As early as January 2020, when the RNA of the virus was identified as the aetiologic agent of the disease soon to be named COVID-19, international teams of scientists immediately sequenced its genome (Andersen, Rambaut, Lipkin, Holmes, and Garry) [6]. The virus had a strong identity to SARS-CoV and a sequence identity with MERS-CoV. This coincided with severe clinical symptoms being identified as acute respiratory distress syndrome (ARDS). Consortium around the globe was established in early 2020, and the need for guidelines for testing and the development of public repositories during a pandemic for determining genomic sequencing were created.

To determine that the virus was present, there were three types of tests that evolved: (1) rapid antigen testing; (2) molecular tests such as genetic, RNA, and PCR; and (3) antibody tests. The rapid antigen tests are immunoassays from nasopharyngeal samples to detect the presence of COVID-19 protein antigens. When the protein antigens are present, it means the person has active COVID-19. This test is rapid, and results are seen in 15 min, but they can produce false positives and false negatives and are less sensitive than the molecular tests. The molecular tests are also from nasal samples but are tests for the COVID-19 RNA, converted to the DNA and Polymerase Chain Reaction (PCR) amplification, and determine if the person has active COVID-19. Since the immune system produces proteins, antibodies, in response to active COVID-19, blood samples can determine if antibodies are present in response to an infection. This test is not sensitive specifically to COVID-19, so does not assure that the virus is present. However, if the person has previously had or been exposed to COVID-19, the antibody test will determine if the person has developed immunity to COVID-19.

The virus treatments

Fig. 2 summarises the state of the science of treatments from the NIH treatment guidelines as of December 3, 2020 [4]. They were developed by an expert panel who continuously review the clinical trial literature. The guidelines in the electronic form can be updated quickly as new evidence is available.

Treatments can be classified into those for persons with mild to moderate symptoms, and for more severe and acute persons. For those with mild symptoms, something for virus treatment like the flu is not yet available for oral or nasal inhalation. However, several are under development.

For those patients with mild symptoms, not requiring hospitalisation, SARS-CoV-2–specific approaches, such as antiviral antibodies, are also being developed. Convalescent plasma, hyperimmune γ-globulin, and polyclonal antibody products are being tested in a wide range of studies including in participants with mild to moderate disease. Additionally, monoclonal antibodies are in development by several companies and academic investigators. Early-phase clinical trials sponsored by Eli Lilly (Ly-CoV555) and Regeneron (REGN-COV2) have yielded promising results suggesting that monoclonal antibodies may be effective in decreasing viraemia and improving clinical outcomes in patients with early COVID-19.

While treatment options for patients with severe disease requiring hospitalisation are now available, with corticosteroids emerging as the treatment of choice for critically ill patients, interventions that can be administered early during infection to prevent disease progression and longer-term complications are urgently needed.

Both remdesivir and dexamethasone have now been endorsed globally by multiple COVID-19 treatment guideline committees and have led to improvements in patient outcomes among those patients receiving them. They have been used with patients requiring high-flow oxygen or mechanical ventilation. Several antivirals approved or in development for other viral infections, such as HIV, hepatitis C virus, and ebolaviruses, are under investigation for early treatment of COVID-19 requiring hospitalisation.

Vaccine development

Immune-modulating drugs are being extensively examined for treatment of moderate to severe COVID-19. Even though these agents are less likely to be as beneficial as antivirals during early infection, this approach is also being explored for early, mild disease. Accentuating the body's immune system to fight the COVID-19 virus resulting illness requires the immune system to develop immunity to the virus. The goal is to get the T-lymphocytes and the B-lymphocytes to recognise the virus and fight it in the future. According to the Centers for Disease Control in the United States (CDC, 2020), there are currently three types of vaccines that have been developed to prepare persons for immunity: (1) mRNA vaccines give the person the ability to recognise the protein in the virus that is harmful and build T-lymphocytes and B-lymphocytes to fight the COVID-19 virus; (2) protein subunit vaccines include harmless pieces of the virus that cause the COVID-19 and the body produces T-lymphocytes and antibodies to fight the virus; and (3) vector vaccines contain a weakened version of a live virus that includes genetic material that can be inserted into the COVID-19 [7]. Once the viral vector is inside the persons' cells, the genetic material gives the cells instructions to make the protein that is unique to the COVID-19 causing virus. The persons' cells make the protein that prompts the body to build T-lymphocytes and B-lymphocytes that remember how to fight the COVID-19 virus.

DISEASE SEVERITY	PANEL'S RECOMMENDATIONS
Not Hospitalized, Mild to Moderate COVID-19	There are insufficient data to recommend either for or against any specific antiviral or antibody therapy. SARS-CoV-2 neutralizing antibodies (**bamlanivimab** or **casirivimab plus imdevimab**) are available through EUAs for outpatients who are at high risk of disease progression.[a] These EUAs do not authorize use in hospitalized patients. **Dexamethasone** should not be used **(AIII)**.
Hospitalized[a] But Does Not Require Supplemental Oxygen	**Dexamethasone** should not be used **(AIIa)**. There are insufficient data to recommend either for or against the routine use of **remdesivir**. For patients at high risk of disease progression, the use of remdesivir may be appropriate.
Hospitalized[a] and Requires Supplemental Oxygen (But Does Not Require Oxygen Delivery Through a High-Flow Device, Noninvasive Ventilation, Invasive Mechanical Ventilation, or ECMO)	Use one of the following options: • **Remdesivir**[b,c] (e.g., for patients who require minimal supplemental oxygen) **(BIIa)** • **Dexamethasone**[d] **plus remdesivir**[b,c] (e.g., for patients who require increasing amounts of supplemental oxygen) **(BIII)**[e,f] • **Dexamethasone**[d] (e.g., when combination therapy with remdesivir cannot be used or is not available) **(BI)**
Hospitalized[a] and Requires Oxygen Delivery Through a High-Flow Device or Noninvasive Ventilation	Use one of the following options: • **Dexamethasone**[d,f] **(AI)** • **Dexamethasone**[d] **plus remdesivir**[b,c] **(BIII)**[e,f]
Hospitalized[a] and Requires Invasive Mechanical Ventilation or ECMO	**Dexamethasone**[d] **(AI)**[g]

Rating of Recommendations: A = Strong; B = Moderate; C = Optional
Rating of Evidence: I = One or more randomized trials without major limitations; IIa = Other randomized trials or subgroup analyses of randomized trials; IIb = Nonrandomized trials or observational cohort studies; III = Expert opinion

[a] See the Panel's statements on the FDA EUAs for bamlanivimab and casirivimab plus imdevimab. These EUAs do not authorize use in hospitalized patients.

[b] The remdesivir dose is 200 mg IV for one dose, followed by 100 mg IV once daily for 4 days or until hospital discharge (unless the patient is in a health care setting that can provide acute care that is similar to inpatient hospital care). Treatment duration may be extended to up to 10 days if there is no substantial clinical improvement by Day 5.

[c] For patients who are receiving remdesivir but progress to requiring oxygen through a high-flow device, noninvasive ventilation, invasive mechanical ventilation, or ECMO, remdesivir should be continued until the treatment course is completed.

[d] The dexamethasone dose is 6 mg IV or PO once daily for 10 days or until hospital discharge. If dexamethasone is not available, equivalent doses of other corticosteroids, such as prednisone, methylprednisolone, or hydrocortisone, may be used. See the Corticosteroids section for more information.

[e] The combination of dexamethasone and remdesivir has not been studied in clinical trials.

[f] In the rare circumstances where corticosteroids cannot be used, baricitinib plus remdesivir can be used **(BIIa)**. The FDA has issued an EUA for baricitinib use in combination with remdesivir. The dose for baricitinib is 4 mg PO once daily for 14 days or until hospital discharge.

FIG. 2 Pharmacologic management of patients with COVID-19 based on disease severity. Doses and durations are listed in the footnote.

Genomic epidemiology

Several large private-public partnerships have developed international repositories of sequence data for SARS-CoV-2. One of the largest has almost 200,000 full genome sequences of the genome of COVID-19. This is hosted by GISAID (Global Initiative on Sharing All Influenza Data) (https://www.gisaid.org/) [8]. Data analytic tools such as NextStrain and NextClade provide the users to create visualisations.

In the United States, the National Center for Biotechnology Information (NCBI) continues to lead efforts making SARS-CoV-2 sequence data sharable and accessible [9]. This resource also contains analytic tools for pharmacologic and scientific research. The sequences are submitted to GenBank or a Sequence Read Archive (SRA). Also, in the United States, but open internationally, the CDC has a surveillance program for sequence of SARS-CoV-2 called Sequencing for Public Health Emergency Response, Epidemiology and Surveillance (SPHERES). This platform provides public health agencies and other stakeholders the ability to describe methods and procedures and challenges related to studying the epidemiology of COVID-19. Genomic epidemiology accelerated post-COVID-19 to understand this infectious disease and its transmission to persons and populations. Many US states have their own networks of public health laboratories that share sequence data and track the mutations and strains of SARS-CoV-2 spreading in their state.

Surveillance

The need for surveillance beyond the public health departments' capacity to surveille infections and viruses expanded rapidly during the pandemic to include testing results, hospitalisations, bed capacity, staffing capabilities, equipment availability, deaths, recoveries by age and sex, and contact tracing. Different models of surveillance proliferated—regional, public health, personal, and contact tracing. As early as April 2020, an expert panel developed *A National COVID-19 Surveillance Systems: Achieving Containment* report [10].

Regional surveillance

Every region of the US country was tasked with (1) testing and tracing the virus for widespread diagnosis; (2) syndromic surveillance building upon existing public health infrastructures to detect spikes and falls in the virus with timely and transparent reporting out outbreaks and surge capacity at the local level; (3) serologic testing capacity to identify the virus and markers of immunity; and (4) rapid response for capacity to isolate, treat, contact tracing, and quarantine. The rapid response also included the capacity to treat persons at home or in the hospital for COVID-19.

Public health syndromic surveillance

Over the past decades, a national syndromic surveillance system has been developed that includes most hospitals, data mappings, analytic methods for determining outbreaks, state and local connections, access, user controls, audit, security, and privacy protections.

This system was expanded to include early detection of community outbreaks, as well as emergency and urgent care visit, admissions to hospitals, and discharges and transfers (ADT data). Timely feedback to local and state governments was accelerated.

Personal

Large budget shifts were immediately required for not only the rapid surveillance, but also the payment of testing and personnel, and the electronic data sharing to support test results and contact tracing. The CARES ACT [5] included money for the Centers for Medicaid and Medicare (CMS), the Centers for Disease Control (CDC), and the Office of the National Coordinator for Health Information Technology (ONC) to provide standard content to enable the online reporting of laboratory test results using existing automated electronic reporting infrastructures. It was recommended that the data include patient demographics, geographic locations, and test results.

Contact tracing

The ability to trace positively tested persons provides the ability to quarantine individuals and timely identification of other persons exposed. Isolation or hospitalisation of sicker individuals provides the ability to treat individuals. For example, in one state alone 1000 persons were trained to support contact tracing. In many countries, individuals' watches, smartphones, and geographic locators were used to trace individuals with exposure. The contact tracing required expansion of directories, multichannel messaging applications, real-time translation services, symptom reporting, and isolation monitoring.

The integrated healthcare documentation system and standards

For decades, in the United States, the Integrated Healthcare Documentation Systems have been developed locally and regionally. These systems rely on HL7, LOINC, and ICD-9CM Standards. Recently, FHIR and SMART standards have expanded the ability of integrating electronic health records from different vendors.

As an example, two Integrated Health Exchanges (IHEs) accelerated their systems to become COVID-19 content ready within three weeks. In Indiana, their IHE had all but 4 hospitals out of 104 connected to thousands of doctors' offices, and public health departments into a central database [11]. The network was used for statewide situational awareness of COVID-19 using LOINC terms of test results from hospitals into state health departments. They used predictive modelling of the data in the database to track the progression of the virus in one state, and an interactive dashboard with statistics related to the virus for all the participants in the network [12]. The group stresses that such a network cannot exist without a culture of trust and collaboration of those participating in the network.

In another part of the country (California), Atul Butte converted the Health Act Network funded by the NIH National Center for Advancing Translational Sciences (NCATS) to collect uniform data for COVID-19 persons [13]. The network took de-identified data from the electronic health record from a medical network of leading academic health centres. The focus of this network is the 480 million data point accessible to researchers in the entire University of California (UC) System. The UC researchers created a COVID Research Data Set (CORDS) to be Health Insurance Portability Accountability Act (HIPAA) compliant and exclude direct patient identifiers. The data represent the states' diverse population. A significant benefit of the CORDS data is that it is real-time and represents a broader population of patients than any one hospital network. The review of the data allowed the UC system to develop a risk assessment of patients with COVID-19. Using the risk assessment data, the researchers identified and published a paper identifying angiotensin-converting enzyme 2 (ACE2) expression modulators and cardiomyopathy as a risk factor for mortality in 28,6650 COVID-19 patients. They developed GENEVA (GENe Expression Variance Analysis) to explore massive amounts of RNA-sequence datasets using a semi-automated framework [14].

Data mining

Thus far, a consortium of genetic scientists and researchers have identified five genes associated with the most severe forms of COVID-19 disease which provide the opportunity to conduct new clinical trials on treatments matched to the underlying genetic structure [15]. The GenOMICC Consortium identified the following: TYK2, DPP9, DPP4, IFNAR2, CCR2, and OAS1,2,3 from 2244 intensive care COVID-19 patients that caused lung inflammation. Finally, at the New York Genome Center, a team of scientists have used CRISPR (gene splicing) technology to identify genes that can protect human cells susceptible to COVID-19 which has led to the development of models that help researchers to understand the immune response to the disease [16]. They identified ACE2, CTSL, RAB7A, CCDC22, PIK3C#, ATP6AP1, and VPS3 where genes could undergo CRISPR knockout and survival could increase.

Artificial intelligence and machine learning

Artificial intelligence has been one of the accelerated digital health technologies during the pandemic. The above use of bioinformatics relies on artificial intelligence for the massive databases internationally that hold the virus information, mutation information, population surveillance, and patient contacts. In a new chapter in Essentials of Nursing Informatics, vol. 7, Koski and Murphy describe the multiple definitions and uses of artificial intelligence [17]. In their chapter, they classify the uses of artificial intelligence in information synthesis, in augmenting human performance, and in surveillance. They highlight that during the pandemic, the need for AI was accelerated because of the need for syndromic surveillance, drug repurposing in drug treatment and vaccine development, and lung image analysis of lung damage from the adult respiratory distress syndrome in the lungs of critically ill COVID-19 patients. AI was also used to analyse databases of patients to determine risk assessments which were identified early in the May-June timeframe as heart disease, hypertension, obesity, and age. Adaptive computerised decision support (CDS) tools were used to extract data from the electronic health records to: (1) manage surge capacity, (2) incorporate best practice treatments and guidelines for mechanical ventilation, (3) integrate cardiopulmonary monitoring of patient outcomes, (4) predict patients' risks, and (5) collect the genetic data of those patients who died. Consumer-focused AI was used in the development of chatbots to provide consumers with reliable, scientific, trustworthy information related to the pandemic and in assessing anxiety and depression [18].

The combination of AI and ML has been used to study the outcomes of patients from radiographic and laboratory data, to train algorithms to predict those patients who would be at the highest risk of intubation and death. This was helpful for patients who tested positive but had no symptoms of acute lung disease. Machine learning was also used in the development of treatments and vaccine since discovery involved the analyses of multiple databases.

Telehealth and tele-ICU

The surge capacity, surge capability, staffing, and severity of illness of patients requiring intensive care required the use of more training of healthcare staff to take care of critically ill patients in settings outside the intensive care units. In a recent chapter in the 7th edition of Essentials of Nursing Informatics, Rincon and Sugrue, experts in telehealth, describe the use of telehealth and tele-ICU with concurrent rapid implementation and training of hospital staff during the COVID-19 pandemic [19]. Concurrently, government policies were changing to support broader use of telehealth, and financing of expanded telehealth applications was generated. They describe the use of tele-ICU in the areas of expert professional oversight and surveillance of physiologic profiles to alert staff to patient's physiologic deterioration, dissemination of evidence-based expert guidelines, provide expert advice and guidance, and collect, analyse, and feedback for quality reporting. The authors also provide the competencies required for nurses to provide leadership in telehealth in this chapter. Written during the pandemic, these authors described the need for optimisation of telehealth, implementation logistics, and training required to cross-train nurses that did not usually work in the intensive care unit.

Telehealth also surged during the COVID-19 pandemic between primary healthcare professionals and consumers as healthcare providers worked from home, and patients/consumers stayed at home. Doctors and nurses also utilised telehealth to care for patients in nursing homes, senior centres, and skilled nursing care facilities. Routine visits and prescription monitoring took place using telephone calls and different web-enabled video conferences. This format also required several policy and payment reforms. Rincon and Sugrue provide examples of best practice using smartphones, iPads, and video with consumers who were not critical.

Wrapped in an envelope of privacy and security

As the personal information of the individual includes his or her genomics linked to his or her phenotype (how a person's genes, biology, and environment come together to make up a person), the security of these data became a pinnacle test of a person's willingness to participate in the testing, diagnosis, treatments, and vaccines for COVID-19. New HIPAA legislation is proposed. The issues regarding the security surrounding the pandemics could take up the entire chapter on digital health from the pandemic. This author recommends the chapter by Dr. Dixie Baker 2020 in Essentials of Nursing Informatics, 7th edition [20]. The chapter describes a seven-layer trust framework that includes protection essential for establishing and maintaining trust in the healthcare information systems. The safeguards include Health Insurance Portability Accountability Act (HIPAA) standards. Physical, operational, and architectural safeguards are described to mitigate risk to protect patient's safety and individual privacy, to ultimately assure quality of care and financial stability of the healthcare enterprise.

Challenges and future directions

As we move forward with having developed new drugs and vaccines at Warp Speed, as we witnessed greater than 90% of millions of patients surviving COVID-19, and as we mourned the loss of a million persons globally to the pandemic, there are enormous lessons learned in digital health moving forward. The excellence of biotechnology, interoperability and standards, telehealth, and surveillance in centres that readied for COVID-19 within the first three months demonstrated that real-time analytics of electronic health record data is possible and practical in assessing outcomes and quality. This highlighted that data had to be decentralised and available in databases locally for such a rapid response. The availability of guidelines within 9 months of the surge in cases was outstanding.

The challenges facing us remain in the distrust of research, healthcare, and science that occurred because of false information relayed by the media and social networks. That there was not a corporate ethics department or international ethics monitoring groups to advise the public was a lost opportunity that may have to be rectified. The ethics of personal protection equipment (PPE) distribution, treatment use and distribution, and vaccine distributions by lottery became apparent.

The COVID-19 data glaringly highlighted the disparities in healthcare internationally in the private payer and national payer systems for ethnic groups by race and age and health conditions. The differences in approach to diagnosis, testing, and vaccine distribution magnify the cultural diversity of states, provinces, countries, and continents.

We became aware that we could work from home (with robust Internet support and bandwidth) and communicate with patients (with robust internet support and bandwidth). We could hold webinars and conferences on web platforms (with robust Internet support and bandwidth) and communicate with millions of persons who were computer literate and digital health literate. However, the training, education, and research needs of those who were computer illiterate became glaring. While we had online American Nurses Credentialing Center (ANCC) certification courses in nursing informatics, nationally and internationally, when it came to cross training in real hospitals during the pandemic, there was no time to take a course in telehealth, AI, informatics, genetics and bioinformatics, surveillance, and using electronic media at home and at work. More training is needed to provide enough healthcare professionals with information and digital health knowledge moving forward. More nurses certified in informatics will be required in the future.

Research pointed out the issues regarding privacy occurring during the pandemic and highlighted the data use, data reuse, and data sharing policies needed to be developed, implemented, and enforced for public health and scientific research data. On October 29, 2020, the National Institutes of Health (NIH) released the Statement on Final NIH Policy for Data Management and Sharing [21]. For protecting the scientific data of public health and population health, the Centers for Disease Control and Prevention (CDC) and the Agency for Toxic and Substances and Disease Registry (ATSDR), the US principal disease prevention and health promotion agencies, previously released a policy on Releasing and Sharing Data on September 7, 2005 [22]. More recently, the CDC released a data sharing statement specifically for the sharing of data related to Vaccine Administration. In that statement, they developed a COVID-19 Vaccine Reporting Specification (CVRS) document [23].

In summary, at the time of writing this chapter, almost 200,000 scientific articles have been written on COVID-19 since February 2020. It is expected that analyses and new innovations will continue for years to come. What will be needed are new partnerships between geneticists, genetic counsellors, healthcare professionals, statisticians, and AI, ML, CDS, health service researchers, and evidence-based researchers to continue to analyse the data on survivors versus non-survivors, chronic disease resulting from COVID-19, and the effects of vaccines' just to name a few. New innovations that have been developed will continue to progress in healthcare. Those healthcare chains that were not a part of integrated networks are quickly seeking entrance into the IHEs. Those companies that have skyrocketed in users during COVID-19 are conducting mergers and acquisitions to broaden corporate capabilities for the next pandemic(s). Only through healthcare information technology and genomics have we been able to accelerate the identification, diagnosis, treatment, and vaccinations of this pandemic.

References

[1] K.A. McCormick, Precision health and genomics, in: W. Carroll (Ed.), Emerging Technologies for Nurses: Implications for Practice, Springer, New York, NY, 2020, pp. 155–184.

[2] K.A. McCormick, K.A. Calzone, Nursing's role in genomics and information technology for precision health, in: V.K. Saba, K.A. McCormick (Eds.), Essentials of Nursing Informatics, seventh ed., McGraw-Hill, New York, NY, 2020, pp. 635–652.

[3] M. Regan, M.B. Engler, B. Coleman, S. Daack-Hirsch, K.A. Calzone, Establishing the genomic knowledge matrix for nursing science, J. Nurs. Scholarsh. 51 (2018) 50–57, https://doi.org/10.1111/jnu,12427.

[4] NIH COVID-19 Treatment Guidelines, December 3, 2020. https://www.covid19treatmentguidelines.nih.gov/.

[5] Public Law 111–148 111th Congress an Act—Congress.gov (CARES ACT). https://www.congress.gov/111/plaws/publ148/PLAW-111publ148.pdf.

[6] K.G. Andersen, A. Rambaut, W.I. Lipkin, E.C. Holmes, R.F. Garry, The proximal origin of SARS-CoV-2, Correspondence to the Editor, Nat. Med. 25 (2020) 450–455.

[7] Center for Disease Control and Prevention (CDC). 2020. Different Types of Vaccines. https://www.cdc.gov/coronavirus/2019-ncov/vaccines/different-vaccines/how-they-work.html?CDC_AA_refVal=https%3A%2F%2Fwww.cdc.gov%2Fcoronavirus%2F2019-ncov%2Fvaccines%2Fabout-vaccines%2Fhow-they-work.html.

[8] Global Initiative on Sharing All Influenza Data (GISAID). December 28, 2020. https://www.gisaid.org/.

[9] COVID-19 Sequence Database. National Center for Biotechnology Information. December 28, 2020. https://www.ncbi.nlm.nih.gov/genomes/VirusVariation/Database/nph-select.cgi?taxid=1335626.

[10] M. McClellan, S. Gottlieb, F. Mostashari, C. Rivers, L. Silvis, A National COVID-19 Surveillance System: Achieving Containment, Duke Margolis Center for Health Policy, 2020. https://www.centerforhealthsecurity.org/our-work/publications/a-national-covid-19-surveillance-system-achieving-containment.

[11] S. Grannis, HIE Strengthen COVID-10 Response, 2020. https://www.regenstrief.org/article/grannis-hie-strengthens-covid-19-response/.

[12] Indiana Ongoing COVID Dashboard. 2020. https://www.regenstrief.org/covid-dashboard/.

[13] A. Butte, University of California System Creates Centralized DTA Set to Accelerate COVID-19 Research, 2020. https://www.universityofcalifornia.edu/press-room/university-california-health-creates-centralized-data-set-accelerate-covid-19-research.

[14] N. Kaur, B. Oskotsky, A.J. Butte, Z. Hu, Mining Transcriptomics and Clinical Data Reveals ACE2 Expression Modulators and Identified Cardiomyopathy as a Risk Factor for Mortality in COVID-19 Patients, 2020.

[15] E. Pairo-Castineira, S. Clohisey, L. Klaric, A.D. Bretherick, K. Rawlik, et al., Genetic mechanisms of critical illness in COVID-19, Nat. Acceler. Rev. (2020) 2020, https://doi.org/10.1038/s41586-020-03065-y.

[16] Z. Daniloski, T.X. Jordan, H.H. Wessels, et al., Identification of required host factors for SARS-CoV-2 infection in human cells, Cell (2020), https://doi.org/10.1016/j.cell/2020.10.030.

[17] E. Koski, J. Murphy, Artificial intelligence in healthcare, in: V.K. Saba, K.A. McCormick (Eds.), Essentials of Nursing Informatics, seventh ed., McGraw-Hill, New York, NY, 2020, pp. 601–610.

[18] A.S. Miner, L. Laranjo, A.B. Kocaballi, Chatbots in the fight against the COVID-19 pandemic, NPJ Digit. Med. 3 (2020) 65, https://doi.org/10.1038/s41746-020-0280-0. PMID: 32377576; PMCID: PMC7198587.

[19] T.A. Rincon, M.D. Sugrue, Telehealth: healthcare evolution in the technology age, in: V.K. Saba, K.A. McCormick (Eds.), Essentials of Nursing Informatics, vol. 7, McGraw-Hill, New York, 2020, pp. 611–628.

[20] D.B. Baker, Trustworthy systems for safe and private healthcare, in: V.K. Saba, K.A. McCormick (Eds.), Essentials of Nursing Informatics, vol. 7, McGraw-Hill, New York, 2020, pp. 163–180.

[21] National Institutes of Health 2020. Statement on Final NIH Policy for Data Management and Sharing. https://www.nih.gov/about-nih/who-we-are/nih-director/statements/statement-final-nih-policy-data-management-sharing.

[22] Center for Disease Control and Prevention (CDC), 2005. CDC/ATSDR Policy on Releasing and Sharing Data. https://www.cdc.gov/maso/policy/releasingdata.pdf.

[23] Center for Disease Control and Prevention (CDC). 2020. https://www.cdc.gov/vaccines/covid-19/reporting/requirements/cvrs-whats-new-2.5.html.

CHAPTER

16

Health informatics in the Middle East and North Africa

Najeeb Al-Shorbaji[a] *and Dari Alhuwail*[b,c]

[a]eHealth Development Association, Amman, Jordan [b]Information Science Department, College of Life Sciences, Kuwait University, Kuwait City, Kuwait [c]Health Informatics Unit, Dasman Diabetes Institute, Dasman, Kuwait

Diverse nations

According to the World Bank and other United Nations agencies, the Middle East and North Africa region (MENA) is made of Algeria, Bahrain, Egypt, Djibouti, Iran, Iraq, Jordan, Kuwait, Lebanon, Libya, Morocco, Oman, Qatar, Saudi Arabia, Sudan, Syria, Tunisia, United Arab Emirates, West Bank and Gaza (State of Palestine in UNICEF), and Yemen. The following countries are not included in the above categorisation: Afghanistan, Mauritania, Pakistan, Turkey, and Somalia. It is worth noting that Afghanistan, Pakistan, and Somalia are part of the World Health Organisation (WHO) Eastern Mediterranean Region (EMRO), while Mauritania is in the WHO African region and Turkey is in the WHO European region. For the sake of this analysis, the geographic scope defined by the World Bank will be used.

The MENA region is a diversified geographic entity that has several features that make it unique and most diversified. According to the World Bank, the MENA region 'is in turmoil as Syria, Iraq, Libya, and Yemen are in a civil war, causing untold damage to human lives and physical infrastructure. Fifteen million people have fled their homes, many to fragile or economically strapped countries such as Jordan, Lebanon, Djibouti, and Tunisia, giving rise to the biggest refugee crisis since World War II'. Conflicts in the Region contributed to a doubling of extreme poverty, resulting in uneven educational attainment and access to basic infrastructure.

The MENA Region has multiple political systems that include emirates and kingdoms, elected governments, and military rulers. The Region has a very diversified economic income as expressed by the gross domestic product (GDP) [1] as well as human capital in terms of education, professions, and employment. One can define three groups of countries in the Region according to the level and sources of income: The Gulf countries, North African countries,

Roadmap to Successful Digital Health Ecosystems
https://doi.org/10.1016/B978-0-12-823413-6.00029-X

and then the other countries. The World Bank report on the MENA region published in 2013 [2] identified trends in the risk factors that account for the most death and disability in the region. Poor diet, high blood pressure, and high body mass index were the major risk factors contributing to heart disease, stroke, and diabetes in the region. Each of these health conditions increased by more than 50% between 1990 and 2010. The report added that 'Among countries in the region, the leading causes of disease burden were as diverse as preterm birth complications in Algeria and Palestine, depression in Jordan, and ischemic heart disease, or coronary artery disease, in Egypt, Iran, and Lebanon.

In five countries (Yemen, Pakistan, Sudan, Palestine, and Egypt), more than 25% of the population are living below the international poverty line. [3] The Universal Health Coverage (UHC) in the Region also has its challenges. The WHO/EMRO report on 'Health Financing Atlas' 2018 in the Eastern Mediterranean Region, including Afghanistan, Pakistan, and Somalia, said that 'Low public spending on health, despite high fiscal space. Public spending on health in the Region is low despite the general availability of public resources, as demonstrated by the level of fiscal space in the Region' [4]. Life expectancy at birth in different MENA countries varies from 80.1 years in Qatar to as low as 66.1 years in Yemen; the average is 74 years. This demonstrates not only the lifestyle, the social determinants of health but also the shift from communicable diseases to a pattern of noncommunicable diseases, which resulted in having a common set of health challenges, including a rise in the burden of noncommunicable diseases as we see in the Gulf Cooperation Council countries [5]. The number of DALYs (Disability-Adjusted Life Years) due to diabetes rose by more than 1000% in the United Arab Emirates, for example [6].

Digital Taxonomy—Before embarking on an analysis of and presentation of policies and practices regarding the issue of digital health in MENA countries, it's worth describing the topic. In 2005, the WHO defined eHealth as 'the cost-effective and secure use of information and communication technologies in support of health and *health-related fields*, including healthcare services, health surveillance, health literature, and health education' [7]. In 2018, the WHO decided to use the term 'Digital Health' to refer to the concept of using information and communication technology for health [8]. In 2005, a comprehensive listing of definitions of eHealth was provided in a systematic review that covered 51 definitions of the term. In 50 definitions, these health and information and communication technologies (ICT) were mentioned; however, in 11 definitions, eHealth was referred to in terms of commerce, suggesting that eHealth is 'health care's component of business over the Internet', the 'application of e-commerce to health care and pharmaceuticals', or as 'new business models using technology' [9].

The diversity in the Region meant diversity in the way ICT for health has been regulated, planned, used, monitored, and evaluated. In the absence of a national digital health strategy, several issues emerge as lacking or not properly handled, including infrastructure, human resources, interoperability, system integration, data security, privacy, and confidentiality. This absence has resulted in a batch work in countries by which the digital health infrastructure is taken hostage or at least highly influenced by the national telecommunication infrastructure, without proper consideration of the health sector's input. This diversity of applications of eHealth in countries of the Region, based on needs, availability of resources, maturity level, and motivation and skills of individuals, resulted in a diversity of applications and experience in and within countries. This diversity in laws, regulations, health systems, health challenges, and ICT regulations has resulted in diversity in the application of digital health in the region.

Health informatics in the Region is highly influenced by the availability and cost of ICT infrastructure, especially the Internet and mobile phone penetration. In 2015, 43% of the mobile phone population accessed the internet from their mobile device. This figure in MENA was expected to grow to 51% in 2019 [10]. Recent (2018) statistics show that the average Internet penetration stands at 68% being as low as 27% in Yemen, while it has reached 100% in Kuwait and Qatar [11]. As for smartphones which are considered the best tools to connect to the Internet on the move and for mobile applications use, it was said that 'Today (July 2020), 60% of Internet access in the Middle East and North Africa is done through the use of smartphones, adding that this is the smallest share of access worldwide in comparison to other regions [12]. Overall, the number of Internet users in the MENA region reached approximately 285 million. The internet penetration rate in the Gulf Cooperation Council is the highest among all subregions of the Middle East and North Africa'. The forecast of the market expects that by 2025, 69% of the population will be mobile subscribers, only slightly behind the global average of 71% [13].

Health information technology utilisation in the region

As has been indicated earlier, the geography of the MENA and EMR adoption as per the World Bank and WHO is different, in addition to the extreme diversity of income between countries of the Region. These two factors need to be considered in this analysis. The intention here is to give examples from different countries as to how ICT in health has been planned and utilised. For an indication of the development of a national eHealth strategy and national telemedicine strategy in the countries of the Region, we will use the data from the World Health Organisation Global Observatory for eHealth (WHO GOe) [14] and as they appear in Annex I.

Algeria—has neither developed a national eHealth policy or strategy nor a national telehealth strategy or policy [15]. This has resulted in an unplanned and uncoordinated effort to introduce ICT applications for health. Aouachria, Aouache, and Oudjoudi [16] indicated that 'Lay down a national strategy (in Algeria), which will allow the skills development of well-trained personnel in the e-health domain, particularly in interoperability standards'. An experiment of telemedicine was initiated in June 1996, where a hospital in the north was connected to another in the south. Medical images were transmitted and received via telephone lines between the two hospitals. The experiment was evaluated as successful but not sustained [17]. Anane concluded that '*medical resources are concentrated in the large cities and patients of rural areas and the south must traverse long distances to benefit of these resources. A national telemedical system permits to distribute medical resources equally over the country. Medical informatics technology holds great promise for the Algerian primary health care sector and the general idea has been initiated by the Center for the Development of Advanced Technologies (CDTA) in June 1996*'.

In 2006–08, Algeria received a grant from the International Development Research Centre (IDRC, Canada) [18] to develop a telemedicine project that consisted of a pilot network connecting two hospitals in Algiers with one hospital in a rural area and another in the extreme south (Sahara Desert). The network began by providing consultation, ongoing medical education, and diagnostic support in the area of paediatrics. In 2015, several issues were discussed in the context of tele-homecare for diabetic patients, which to a great extent apply to digital health in general in Algeria [19]. Tele-homecare (THC) in Algeria is still facing many challenges especially in rural areas such as:

1. Unavailability of the required ICT infrastructure for THC (e.g., internet connection)
2. High cost of THC solutions.
3. Lack of patients' awareness and acceptance of receiving healthcare services via THC networks and applications.
4. Lack of uniform standards for data-sharing: Unified format and content of data storage are the basis for communication between systems. Information classification and data standards are not strictly uniform.
5. Security, safety, and privacy of data exchange.

Bahrain—The Kingdom of Bahrain has developed a national eHealth policy and a national telehealth strategy [20]. Since the late 1980s, the Kingdom has embraced the process of automation of hospitals to gain efficiency. A committee at the Ministry of Health was formed in 1989 to adopt and implement a health information system (HIS) [21]. The HIS was developed initially from hospital information systems and included modules for radiology, pathology, pharmacy, physician orders, and communications. In 2017, Bahrain published its first National Health Data Dictionary (NHDD), which provides national data standards for the health sector. The NHDD was later revised, and version 2 was published in 2019 with a focus on the national digitalisation transformation experienced by the Kingdom. *'The NHDD continues to be the basis for the National Electronic Medical Repository (NEMR) as well as the National Health Insurance Information System (NHIIS), two of the systems that comprise the Sahhati (my health) Universal health coverage transformation program'*. [22] In response to their own question *'Can Bahraini patients accept e-health systems?'*, the authors found that *'The most factors affecting e-health adoption are trust, health literacy, and attitude. Additionally, people in the private and government sectors understand e-health benefits'*. [23].

Egypt—has developed a national eHealth policy [24]. The strategy relates to the telecommunication and information technology sector in Egypt and not specifically dedicated to the health sector. Health is mentioned along with education as areas that have the potential for computerisation, including eHealth, eLearning, and the use of the Internet for health [25]. It is interesting to note that Egypt does not have a strategy for telemedicine despite all the different telemedicine activities in the country. Hussein and Khalifa provided A SWOT (2012) analysis of telemedicine in Egypt and identified four future trends in telemedicine by 2020 from governmental, financial, technological, and medical perspectives, which were aligned with the global trends in telemedicine. [26] Several attempts were made to implement an electronic health record (EHR) system in Egypt. These attempts failed for different reasons, as per Abdel Gaber and others [27]. A number of barriers were identified:

1. Management-related barriers such as cost, awareness, and commitment;
2. Technical barriers related to data migration, standardisation, training, the usability of systems, and security;
3. Organisational barriers, including workflow issues, policies, legislation, planning, leadership;
4. Psychological barriers, including culture, trust, user satisfaction, and doctor-patient relations.

They suggested a roadmap to develop EHR in Egypt using the following steps:

1. Establish an independent and specialised organisation to manage the implementation of EHR nationwide.
2. Setup EHR Standards and Accreditation conditions; the start point of the organisation mentioned in step (1) is the accreditation authority.
3. Construct EHR training authority which should be responsible for putting and planning the training programmes for all healthcare actors.
4. Finding sources of funding.
5. Develop policies that lead to the implementation of EHR.
6. Set up the security criteria to protect the confidentiality and privacy of EHR data.
7. Set up clear plan to migrate the health old data from paper-based to electronic systems.
8. Set up accreditation procedures for vendors that implement EHR software.
9. Announce the EHR benefits through the national media inside the society.

Farahat, Hegazy, and Mowafy concluded their study by indicating that ICTs resources are underutilised by health information professionals in Egypt and that the three most common barriers to investing in eHealth were also identified, namely lack of funds, risk of instability of the electric supply, and lack of incentives for ICTs [28].

Telemedicine in Egypt has been implemented in a most unplanned and uncoordinated manner. Several projects have emerged with no connection, and more importantly, these projects are not part of the healthcare system in Egypt. Telemedicine projects in Egypt have been in place for a long time, but in all cases lack sustainability, institutionalisation, standardisation, and infrastructure. One major effort has been the establishment of Ain Shams University Virtual Hospital (ASUVH) as the national hub of the Egyptian-African Telemedicine Network [29]. It has adopted a model by which health medical services are offered by the ASUVH. These started by offering services to individuals and were expanded to offer service packages to universities, hospitals, the ministry of health, and medical societies among others.

Djibouti—does not have a national eHealth strategy and has been facing multiple problems in terms of implementing eHealth applications. Among the cited perceived barriers to implementing eHealth applications are leadership, governance and policy, legal frameworks, standards, and financial barriers [30]. Even though different agencies have been providing financial and technical support for the transformation into a digital economy, it is still facing major challenges. The World Bank announced in 2018 new support for Djibouti's ongoing efforts to leverage digital technology to bring the government closer to citizens and improve the impact, transparency, and efficiency of its public administration.

The project was meant to support the roll-out of digital systems to make it easier for citizens to access services, and for more efficient tax, customs, and administration to boost government revenues. eHealth was not part of the package as it stands lower in the priorities of the government [31]. Djibouti has a large refugee population that requires special attention for their health support. While the government's strategy to improve the quality of provision has taken the form of fairly straightforward ICT integration, equipment updates, and foreign partnerships in the medical field, its moves to increase accessibility have been more complex [32].

Iran—The Islamic Republic of Iran has developed its national eHealth strategy but has not developed a national telemedicine strategy [33] even though it was not possible to access the strategy itself. In their review of six hospitals' eHealth work in Iran, Jahanbakhsh, Sharifi, and Ayat concluded that '*The findings of this study emphasise that almost all of these mentioned hospital case studies do not have any e-Health development strategies at all*'. [34]. A national association of eHealth has been established in the country that aims to bring all efforts together. The country has taken steps towards health/medical informatics education as programmes exist at both undergraduate and postgraduate levels. Attempts have been made to create a national electronic health records system and to introduce telemedicine services.

Medical informatics education is provided by several universities in Iran. A typical description of how they approach medical informatics as a multidisciplinary topic is relatively new in the country. It embraces a variety of trends in computer science, including computer science, computer engineering (software, hardware, artificial intelligence, IT, mathematics in computer applications, and even related fields in Electrical Engineering in the country). The main mission of a department of medical informatics is to train qualified and committed faculty members and scholars to service this discipline and the country.

Darvish and Far concluded that '*The major obstacles to establish e-health in Iran are lack of a specific strategy, ambiguous and complicated information technology infrastructure, two cultural and educational problems concerning training and practice in information technology and communication skills, rapid changes of managers especially in Ministry of Health and Welfare, inability to attract skilled information technology technicians in e-health field, an unclear mechanism to fund e-health system, lack of codified technical standards and weak health service implementation methods. The infrastructure of information and communication technology in Iran is complex and ambiguous and e-health strategies and priorities have not been determined yet*'. [35].

Iraq—is a country that has been either under embargo or in a war for the last 40 years. The country has never recovered from these man-made disasters and consequently the health sector in general and eHealth in particular. In their comparison between Iraq and other countries, Abdel Ghani and Jaber indicated that Iraq has poor ICT infrastructure, the majority are not aware of telemedicine services, they have resistance to change towards telemedicine, but have enough funding to support telemedicine programmes [36]. In a survey that aimed to give an overview of the use of the Internet and telemedicine in Iraqi hospitals, while assessing the favourability of circumstances for wider introduction of telemedicine in Iraq, the survey found that the majority of hospitals were provided with internet access, which is used mainly for administrative purposes [37].

The most commonly reported obstacles were lack of telemedicine knowledge and expertise, scarce funding (contrary to the previous report), lack of national telemedicine laws and guidelines, and the busy schedule of physicians that make them reluctant to try something new to them. Access to HINARI, a WHO programme set up with major publishers to enable low- and middle-income countries to gain access to one of the largest collections of biomedical and health literature, was granted, free of charge, to Iraqi (Group B) institutions because it is considered a low-income country and it is not possible to transfer funds from there [38]. Neamah concluded that in Iraq, like in most other developing countries, finding reliable patient data is challenging. This is due to a lack of functional infrastructure and human resources [39].

Jordan—The strategy of Electronic Health Solutions (EHS) has been considered as the national strategy for eHealth in Jordan [40]. EHS was founded in early 2009 to advance the healthcare sector in Jordan as an innovative technology-driven, private, nonprofit company that effectively provides automated solutions to enhance the quality and efficiency of Jordanian public healthcare services [41]. EHS renamed the solution to 'Hakeem' (Doctor), which was described as a world-class implementation of EHRs across the Kingdom and has been deployed in all public sector hospitals and primary healthcare centres.

Hakeem as an EHR system collects health data and shares data at the national level to create and sustain a patient-centric health record, which is used to provide clinical services to patients nation-wide, focusing on preventive healthcare for the patient, patient safety, and improving patient experience as a whole. Matar and Al-Nabhan concluded that Jordan is facing barriers to implementing eHealth system models due to limited resources and poor ICT infrastructure in some areas [42]. Additional constraints were also identified, including a lack of hospital administration awareness and expertise towards e-health, limited financial resources, and restrictive hospital policies. In an editorial in *the Annals of Medicine and Surgery* [43], it was concluded that Jordan is ready to establish a nation-wide telemedicine project after a few small telemedicine projects have been established in the Kingdom.

The national health strategy of Jordan indicated that Hakeem will contribute to improving health services in Jordan '*through the creation of an electronic health file for each citizen and facilitate the access of users of the system to any medical facility using the national number and linking the Department of Civil Statistics database with the system database so that the file contains procedural and surgical comprehensive reports, current medications and the notes taken in each visit to the hospital or clinic*'. [44] The strategy recognised the need for evidence, as presented in research, the need for better health information systems, and it proposed some specific actions be taken into consideration of the following:

1. Weakness in the adoption of evidence-based policies and decisions;
2. The lack of allocations for studies and research and publishing of research in scientific journals;
3. The absence of a national strategy for health information and research;
4. Weakness in the computerisation of the health system;
5. Weakness in the electronic Modern Health Systems Applications (*E*-Health);
6. Weakness in access to data and information of the private sector;
7. The absence of a national reference entity for research and health studies.

The EHS, through its Electronic Medical Library (ELM), has provided access to a number of databases and electronic medical/health journals in response to the need for such access by researchers and policy-makers in the country [45]. The vision of ELM is to '*become the central online hub for health information across Jordan*'.

Private hospitals and clinics use their own fragmented, unconnected applications and in most cases locally developed solutions. These solutions focus mainly on administrative aspects, including financial management, which has created different types of issues, including data exchange, standardisation, and interoperability.

Jordan established its national eHealth Development Association (eHDA) in 2018 aiming at bringing all efforts of eHealth professionals in the country to collectively work together to ensure that health informatics applications fit the needs and are ethically implemented [46]. Additionally, the use of mobile phones for health (mHealth) is accelerating in Jordan. Mobile applications for appointments, education, and well-being are flourishing. In response to the coronavirus pandemic, the Government of Jordan released an application called 'Aman', which is a privacy-conscious, exposure detection application developed for MoH to keep the community and Jordan safe from coronavirus [47].

Kuwait—The State of Kuwait has embarked on its journey towards adopting and leveraging health information technology since the 1980s. This adoption was accelerated by the necessity of the 1990 political situation. The published evidence suggests that 'the US military tested a store-and-forward ophthalmic telemedicine system in 1998, successfully transmitting digital images from Kuwait City to 140 international e-mail addresses via T1 commercial satellite link'. [48] Additionally, Kuwait has focused many of its health information technology solutions to combat the high wave of noncommunicable diseases such as diabetes. In 2010, the Kuwait–Scotland eHealth Innovation Network was established by the Dasman Diabetes Institute, the Ministry of Health along with the University of Dundee, to address alarmingly high rates of diabetes by exchanging health information, online clinical education, and research [49].

In 2013, Kuwait issued its health IT strategy [50] focusing on the development of modules to manage data and information at the Ministry and its healthcare facilities. The Kuwait Health Informatics Association (KHIA) was established in 2011, and since then, it had gone through a major transformation and under its new leadership elected in 2019 has set a clear direction in taking a lead in enhancing the field of health informatics in the State of Kuwait. *'With the support of MENAHIA, KHIA's was able to become a recognised society member in the International Medical Informatics Association (IMIA)'*. [22] The KHIA continues to play a pivotal role in supporting the Ministry of Health with its digital transformation across the nation through the participation of its members in national-level projects.

Lebanon—established, within the Ministry of Public Health, its National eHealth Programme without having a nationally adopted strategy for eHealth or telemedicine, despite the fact that one of its responsibilities is to introduce telemedicine services in the country [51]. The Programme aims to develop the health sector and enhance the quality of health care through the use of ICT as per the following specific tasks:

1. Improve the health information systems and databases, and establish a data centre for health information;
2. Supervise the management and implementation of the *E*-government (for the health sector);
3. Ensure interoperability;
4. Adopt M-health technology;
5. Develop E-prescription system;
6. Introduce Telemedicine;
7. Provide E-learning;
8. Abide by the ethics of E-health;
9. Cooperate with the national and international stakeholders.

An attempt to reach consensus in Lebanon regarding the different elements contributing to building an electronic health record system [52] found that to be able to reach that consensus several prerequisites have to be met, including regulation and coordination, legislation, standards for interoperability, database management, capacity building, and other resources. In an eHealth readiness study, Saleh and others (2016) [53] confirmed the need for the readiness of healthcare providers to implement eHealth as the majority of healthcare providers at primary healthcare centres across Lebanon are ready for eHealth implementation. They added that efforts should be directed towards capacity building for healthcare providers.

In the absence of a formal health/medical informatics education in Lebanon, the American University of Beirut provides a nursing informatics certificate. The certificate is designed to provide students with the knowledge and skills necessary to take a leadership role in the selection and implementation of healthcare information management systems and in applying the knowledge gained from the information generated from these systems. [54] In response to the coronavirus pandemic, the Ministry of Public Health launched a mobile application called 'Ma3an' (*Together*) for Contact Tracing [55] that aims to facilitate the cooperation with other ministries, the private sector, and the civil society in Lebanon. The application was developed in collaboration with one of the universities in the country, which demonstrates the link between the public and the private sectors in digital health applications.

Libya—The Libyan health authorities attempted to adopt an EHR system over two decades ago. These attempts were faced with several challenges that made it impossible to introduce such systems. Khalil and Al-Bousify [56] believed, back in 2008, that 'While the Libyan health system is quite far from this level (introducing an EHR system) and still needs a lot of basic e's before e-health, can be easily accessed, such as: early intervening, effective, efficient, emergency oriented, equity-providing, ethically practiced, evidence-based and educationally competent! And that the limited resources provided to the health sector would be more useful if implemented into the medical process directly. An electronic system now is an unnecessary leap that will overload the health system causing more harm than benefit'. In 2017, the MoH, the European Union, and WHO organised a workshop on developing [57] the national health information system (HIS) in Libya, which discussed the weaknesses, including:

1. About one-third of the attributes of a functional HIS are not present, nor distributed across all components;
2. Libya does not have a comprehensive costed Monitoring and Evaluation (M&E) plan;
3. There are no Standard Operating Procedures (SOPs) for data management, institutionalised data quality assessments, or even a functioning, integrated web-based HIS system;
4. Effective mechanisms for review and action such as independent reviews of data, linkages between health sector reviews, disease and programme-specific reviews, and active engagement of civil society in-country reviews, are also nonexistent; and
5. The HIC lacks an M&E unit and a unit responsible for collecting and processing emergency information.

The report, among a number of recommendations, made few that are closely related to the use of ICT for health, including:

1. Strengthen routine health information system (HIS) by training MoH cadres at all levels and purchasing ICT equipment;
2. Pilot DHIS-2 with potential for rolling out nationally depending on the outcome of the pilot;
3. Complete the human resources registry to improve decision making regarding human resources for health.
4. Develop a comprehensive hospital management system.

Morocco—has developed a national eHealth strategy, but not a national telemedicine strategy [58]. In Morocco, health informatics is part of a national effort as part of the Moroccan digital plan to integrate Information and Communication Technology (ICT) into several programmes and processes at the service of its citizens and businesses. The Health Portal, as per the government strategy of 2011, would allow citizens to find out about the healthcare services offered by region or by type of care [59]. It would also allow them to make appointments online at the level of a health facility. Online health files, including reimbursement, would allow people of Morocco to remotely consult an insurance agency to regain reimbursement status, thus reducing their trips and improving the quality of the services rendered to them. Pension management portals allow better dissemination of social information (certificates pensions, contribution situation, etc.) for the benefit of citizens and greater efficiency in the lower-cost delivery of services. The above confirms the 'business side' of healthcare delivery as per the national strategy.

Furthermore, the more recent Moroccan eHealth status could be illustrated by the 2012–16 strategy as Morocco instituted a new constitution ratifying health care as a fundamental right of the Moroccan people [60]. The Ministry of Health (MoH) introduced a 2012–16 health strategy. Improvement of the health system governance was the pillar axis of this strategy. It is focused on the usage of Information Technology and its contribution to improving health quality within healthcare organisations. While discussing the electronic health records implementation in Morocco, Parks et al. identified the five main categories of challenges due to silo strategies such as:

- **(i)** EHRs selection and weak bargaining power,
- **(ii)** identical errors repeated across silos,
- **(iii)** a lack of interoperability standards,
- **(iv)** insufficient human and financial, and
- **(v)** missed cooperation and collaboration opportunities [61].

Ghazal et al. identified the legal issues to deliver telemedicine services from the European Union countries to Morocco [62]. They concluded that for Morocco to allow cross-border telemedicine, the policymakers are called to update and extend the national law of telemedicine. Moroccan health informatics professionals formed a national association of health and medical informatics. The Association aims to disseminate and transfer knowledge regarding health information technology by organising scientific meetings and promoting the efficient use of medical information technology to improve patient care [63].

Palestine—made up of the West Bank and Gaza, has been either under occupation or under siege since 1967. Importation of equipment, services, software, and other digital health solutions is very limited. Several nongovernment organisations, funding agencies, and United Nations agencies have been providing support, as best as occupation permits. Abu Hamra assessed the primary healthcare centres in Gaza, which showed a level of weakness in the provision and use of ICT infrastructure [64]. The study uncovered the need to heavily invest to promote eHealth readiness in Gaza. The study concluded that efforts are required, especially in awareness creation, training of health workers, infrastructure development, and promotion of the computerisation of the medical health record system. In their concept note, El Jabari et al. proposed a federated electronic health record system (FEHR) based on the Clinical Document Architecture (CDA), an XML-based markup standard intended to specify the encoding, structure, and semantics of clinical documents for exchange, which is different from other electronic health record models (full document vs. a piece of data) [65].

West Bank and Gaza are two of the five operational fields (in addition to Jordan, Lebanon, and Syria) that UNRWA (United Nations Relief and Works Agency) health services cover, through a collective of 143 health centres, of which 22 are in Gaza and 42 in West Bank [66,67]. Jabari and Adwan reported that there were more than 400 government primary care centres (separated physically by occupation, the separation wall, and politics) connected to the District of Health Information Systems (DHIS) [68]. UNRWA Health Program has a vision to realise '*Upon its completion, e-Health will provide quality improvement in the healthcare provided to Palestine refugees and will reduce the cost of the service*'. 'Replacing the paper-based system, increasing efficiency, reducing cost, improving services, and mitigating all the risks have been the guiding principles in the in-house development of the system'. The coverage includes the deployment of the eHealth system that was developed to manage health services in these areas [69].

As part of its effort to serve the Palestine refugees using eHealth technology, UNRWA released its mobile health for noncommunicable diseases (NCD) in October 2020 [70]. The mobile health application supports NCD patients in their efforts to control their conditions, and for non-NCD users who can monitor their health status to prevent the development of an NCD in the future. Moreover, it is designed to be used by any user in the world who speaks Arabic.

The Ministry of Health in Gaza launched a mobile application for telemedicine programme called 'Sehhaty' to decrease the load on the hotlines used by people to inquire about health issues, COVID-19 symptoms, follow-up test results, etc. [71]. 'Sehhaty' (my Health in Arabic) has the same concept of the hotlines through providing teleconsultations for both patients with chronic conditions and other patients. It is also used to share the results of the coronavirus polymerase chain reaction (PCR) test securely and more efficiently than sending results as text messages.

Oman—The Sultanate of Oman has adopted health information technology early and relied on building its national capacity by building systems in-house and training locals to run and manage these systems. The Sultanate developed its national eHealth strategy [72]. Reducing the cost of healthcare services and bringing efficiency were among the major goals of the Ministry. As Oman boosts its Universal Health Coverage, the health facilities in the Sultanate, through digitisation of medical records, reached 86% of all government hospital and healthcare facilities having been linked electronically to a central database [73]. The Omani MoH and the Ministry of Technology & Communications leveraged artificial intelligence for breast cancer detection at the Royal Hospital. Following a successful pilot, MoH was encouraged to expand the system across 'five tertiary care hospitals around Oman to provide the necessary early healthcare to the diagnosed women' with high accuracy diagnostic rates (96%) [22].

Qatar—The State of Qatar has been making strides in developing its health informatics ecosystem through several initiatives mainly to address the country's national 2015 e-health vision 'to improve the overall health in Qatar and deliver healthcare of the highest standard by providing the public, patients, and clinicians with appropriate and timely information' [74,75]. The Health Informatics Association of Qatar (HIAQ) has been established to provide a platform for adequate representation of a professional body to allow for serving the needs of the community. Development of a strategic framework for the Association and the presentation of the preliminary findings of this effort have been among its major objectives [76]. Several joint research projects among university partners, healthcare authorities, and hospitals have taken place, intending to harness the power of health informatics and artificial intelligence in the diagnosis and treatment of patients relating to several national priority healthcare needs such as mental health and chronic diseases [22].

Saudi Arabia—The Kingdom of Saudi Arabia has been pioneering many health informatics initiatives in the Region. As one of the nations with a high concentration of health informatics higher education programmes [77,78], the Kingdom continues to lead many initiatives. The eHealth strategy developed by the Saudi Arabian MoH focuses on patient-centric care through dedicating major investments for the development and deployment of e-Health applications and services [79]. The Saudi MoH implemented many service programmes such as the health performance programme, the e-prescription system, and a central health appointment system (Mawid) to enable patients to book, cancel, or reschedule a medical appointment in healthcare facilities. '*The Saudi MoH also created the "Sehha" application, which provides solutions to enable individuals to receive audio-video medical consultations at their homes provided by MoH's specialists*' [22].

During the Riyadh Global Digital Health Summit on 11–12 August 2020, the Riyadh Declaration on Digital Health was formulated and issued. The Declaration '*articulated seven key priorities and nine recommendations for data and digital health that need to be adopted by the global health community to address the challenges of the COVID-19 pandemic and future pandemics*'. [80]. The Saudi Health Informatics Association (SAHI) was founded in 2006 with a vision '*to be a trustworthy enabler and leading organisation for health informatics discipline in Saudi Arabia*'. SAHI has organised, in partnership with other national organisations and the private sector, several national health informatics meetings, and exhibitions [81].

Sudan—has developed a national eHealth strategy [82]. The strategy was developed in 2005 aiming at increasing the uses of ICT to support the Sudan National Health Care Services, including its clinical, public health, and managerial aspects [83]. In July 2012, the WHO and the International Telecommunication Union (ITU) organised the National eHealth Strategy Development Meeting which revisited the 2005 strategy to validate and update [84]. It concluded that Sudan made several achievements and progress, including putting telemedicine solutions, teleradiology applications, and teleconsultation telepathology services. It also pointed to the development of a health information management system and Data warehouse and health system observatory.

A more recent report identified several challenges facing Sudan eHealth implementation. These challenges were classified into four components: political decision, economic status, socio-cultural behaviour, and technical barriers [85]. The report noted that several scattered 'eHealth' projects were found to be running in parallel with an evident overlap of efforts and lack of collaboration leading to a waste of resources, including a surveillance project at the Epidemic unit, Human Resources observatory database, and the 'e-Health' project (a hospital information system) under the e-government project. It also suggested that all applications are web-based, in their initial pilot stage and each with their own established networks and regrettably again working in isolation.

Sudan Doctors' Union developed a discussion paper that considers the potential benefits of implementing eHealth interventions during the COVID-19 [86]. The paper listed the challenges facing eHealth development in the country, including:

1. IT platforms access indicating the impact of the embargo on the country and the Access to reliable Internet coverage at acceptable speed is variable in the country;
2. Professional registration, which aims to protect patients and healthcare professionals providing advice and guidance, as healthcare professionals should be registered with an appropriate professional licensing body in Sudan;
3. Indemnity issue, which raises the concerns by healthcare professionals that the care and advice they provide will not be covered by their existing medical societies in their home countries;
4. Funding of the projects, as there is no clear understanding of how this project will be funded. Possible sources of funding could be from the government of Sudan, charitable donations from the Sudanese diaspora, and businesses within Sudan.

The report recommended several eHealth solutions, including telemedicine services during COVID-19 and beyond, use of web services for public health information, and other outpatient services based on ICT.

Syria—as a country, has been in a crisis due to civil war since 2011. It has neither developed a national eHealth policy/strategy nor a national telehealth strategy/policy [87]. With the current conflict in Syria, it is very unlikely to get information on the eHealth situation in that country. This is despite the fact that before the civil war, Syria had its national health information system, health decision support system, and a health informatics association [88]. Syrian refugees in Lebanon, Jordan, Germany, and Turkey have been the focus of eHealth applications development aiming at monitoring their health situation and providing efficient and cost-effective services under these conditions. This population has previously enjoyed excellent healthcare, sound nutrition, high living standards, and a wonderful education system [88].

The Syrian American Medical Society (SAMS) provided telemedicine services, including medical support and training to Syrian populations using telemedicine. In 2016, SAMS expanded their telemedicine programme, creating connections between doctors in the United States and Syria to treat critical medical cases as they provided 7930 medical services via Skype and WhatsApp [89].

The Editorial of Telemedicine and e-Health concluded that '*Telemedicine clearly can get there (Syria) and we suggest we should*' [88]. Talhouk et al. concluded that '*the differences in levels of resources available across PHCs to be a key factor that needs to be considered when integrating and scaling a health technology within the primary health care system. Additionally, the mobility of refugee communities was found to be a disincentive consideration for the integration of technologies specifically for refugee reproductive health. Lastly, participants recounted experiences of engaging with refugees in which they identified varying views regarding refugee technology and health literacies. Such views are key considerations that should be accounted for when designing technologies that specifically connect refugees to the health care system*' [90].

Tunisia—attempted to develop its national eHealth strategy in 2015 [91] through holding a workshop organised by the Ministry of Public Health and WHO but due to several factors, mainly instability, Tunisia has not developed a national eHealth policy/strategy nor a national telehealth strategy/policy. Sadok et al. (2016) [92] identified a considerable number of difficulties in Tunisia, such as lack of a political agenda, structural reforms, and a coherent investment programme, as well as market regulations that affect telemedicine and e-Health development.

The Tunisian Society of Telemedicine and eHealth [93] was established (SITTeS) in 2000 as a scientific society to advance the development of digital health, uniting the efforts by professionals in Tunisia and further collaboration with relevant bodies abroad. The STTeS has played a key role in the development of eHealth with the Ministry of Public Health, ensuring continuity albeit political evolutions and revolutions. The Society is also a strong advocate for positioning Tunisia as a platform for Teleradiology and eHealth services towards Sub-Saharan Africa. Tunisia also hosts the Arab Society for Telemedicine [94], which was established in 1999 aiming at promoting the development of telemedicine, telecare, and telehealth as part of e-health in the Arab world. Tunisia considers telemedicine and eHealth as an important tool for healthcare's main objectives specified in its national health strategy: social protection and health of poor people, the health of the elderly, public-private complementarity, emergency services, prevention, and reduction of regional disequilibrium. The 2016 Strategic Plan 'Digital Tunisia' included an eHealth Strategy Plan.

The National Digital Health Strategic Plan 2020 has several components in support of the ambitions of the national health system reform programme of Tunisia [95]. The Plan has the following components:

1. The acceleration of the deployment of the National Health Network—New Generation (RNS-NG) and the upgrading of local information technology infrastructures within the health facilities. The strategy indicated that this would be the base programme for all digital health projects;
2. Modernisation of the Health Information System in Tunisia;
3. Upgrading of the Hospital Management Information System by providing an electronic health record system;
4. The development of teleservices (teleappointments, online directories, mCessation, mHealth, etc.), mentioned later;
5. The development of a cross-border agreement for eHealth and telemedicine experiments;
6. Support of the Department of Pharmacy and Medicines and the National Laboratory for Medicines Control in implementing their digitisation strategies.

As part of the collaboration between WHO and the ITU in the project titled 'Be Healthy, Be Mobile' (BHBM) and in line with its national digital health strategy, Tunisia adopted the mTobaccoCessation, mDiabetes programmes [96]. Additionally, Tunisia has been among the first countries in the Region to launch its COVID-19 contact tracing mobile phone application in May 2020 [97]. The application called 'Ihmi' in Arabic (protect) identifies, alerts users who may have had contact with others infected with the virus.

United Arab Emirates—The United Arab Emirates (UAE) has been making great progress towards health information technology adoptions. It has to be acknowledged that the UAE is composed of seven emirates, not all have the same level of adoption of digital health solutions. The country has been at the forefront in the utilisation of information and communication technology in health. Salameh believed that *'In the last few decades, UAE is moving rapidly from traditional medicine to robot medicine. Employing the most modern and advanced tools, instruments and applications are well known in different health organisation and authorities'*. [98].

The Government has initiated the work on National Unified Medical Records 'Riayati' [99]. It is a digital health platform providing updated data for patient records. The platform is designed to promote and empower the healthcare sector in the country and develop health information systems in addition to the management of healthcare facilities. The platform will connect to and use data from other electronic health record systems used in the country (Nabidh 'Pulse' and Malaffi 'My file' in particular). Data exchange standards are being applied in the process. The national strategy for artificial intelligence (AI) covers different areas that have an impact on the well-being of people in the UAE [100]. Health has been no exception, as the strategy focuses on managing big data, including medical devices, predictive modelling, and block-chain technologies.

The Emirates Health Informatics Society (EHIS) was launched in 2014 as a specialised part of the Emirates Medical Association to support the UAE's health leadership initiatives in the area of eHealth [101]. The Society has been emphasising the development of a national health informatics workforce. After debating the advantages and problems of eHealth, mHealth, and electronic health records utilisation in the UAE, Salameh [98] provided several recommendations for consideration by the health authorities and policymakers in UAE to bring more advancement in the digital healthcare service, through:

1. Finding technological solutions to ensure that it will be real-time responses to solve issues and complaints of patients;
2. Employing artificial intelligence that can achieve a revolution in mobile application tools for detecting the symptoms with great accuracy and also suggesting medications;
3. Finding solutions that can increase healthcare data security.

Yemen—is another country in crisis as it has been at war since 2014. It has neither developed a national eHealth policy or strategy nor a national telehealth strategy or policy. Coupled with the current conflict, many of the health-related activities are being affected, except some emergency health services. Taking into consideration the attempt of Yemen to find solutions to improve the quality of healthcare delivery and services and to overcome the challenges that prevent it, Al-Fadhli and others (2015) [102] proposed telehealth technology as a solution to these problems. The proposed model recognises the challenges facing Yemen in implementing it, including the ICT and technology infrastructure, the absence or inadequate accesses to technological resources at many rural hospitals, lack of infrastructure, services, and expertise, limited resources, low literacy levels and health professionals threatens the country's aspiration for an improved living situation. Al-Fadhli and others [103] developed the Yemen Telehealth Readiness Assessment Framework (YTRAF) to assess the readiness for telehealth in Yemen. They concluded that the Ministry of Public Health requires considerable efforts to raise the level of its readiness in several areas to be ready for successful adoption of Telehealth. These areas include improving the processes at the MoH, building the infrastructure, creating awareness of the problem and the solution, working on the skills of people in the health sector, developing proper applications and awareness of the cultural, ethical, and behavioural issues.

Digital health innovations and future directions in MENA region

An ecosystem is a term that is being borrowed from the environment as there has been almost an agreement that it is a community of living and nonliving things that work together—it consists of abiotic (soil, water, air) and biotic parts (flora, fauna) [104]. On that basis, the ecosystem for digital health in MENA reflects the ecosystems in the countries of the Region.

Diversity is the key notion that governs the Region where countries differ in leadership style and support to digital health, info-structure, laws and regulations, standardisation and interoperability enablers, human resources, and on top of all that that level of literacy specifically in digital health applications and in general.

The above analysis has demonstrated that the Region has a long way to go to create an acceptable and conducive ecosystem for innovation and to set the direction for digital health that matches other countries' innovations. Despite all this, it is clear that some countries in the Region have made strides in that direction. To create a supporting and sustainable ecosystem for digital health innovation, a country needs to:

1. Commit itself to a national plan of digital health that takes into consideration the full integration of the different components for managing knowledge and innovation including people, technology, and procedures. Without this roadmap, it will be creating more chaos in application development, deployment, and use.
2. Enable access to and utilisation of digital health applications standards in support of interoperability. A major hurdle that impedes digital health implementation is the lack of both semantic and syntax interoperability between systems not only at the institutional (healthcare facility) level but also at national and global levels. The absence of applying the standards meant enforcing the mentality of silos, pilots, and standalone applications. Creating an ecosystem that enables and sometimes forces people to apply standards is the way forward for countries of the Region.
3. Introduce legal and ethical frameworks and principles that govern digital health implementation represent a major aid to substantially improving the uptake of digital health in a country. This simply means introducing the necessary legal frameworks to legally accept a digital version of documents, to make manufacturers of tools and medical devices accountable for errors resulting from these tools, ensuring that medical responsibility goes beyond local or even national boundaries when introducing telehealth services. Laws to protect privacy, confidentiality, and data ownership will support the introduction of digital health in a country. Ethics of provision of healthcare services using the technology that promotes equity and enabling managing health data and its ethical access and sharing, are integral to building that ecosystem. The ethical framework includes manufacturers of equipment and software developers; operators of these and the ministries of health that give licenses for these to operate in the country.
4. Develop and sustain human resources for digital health. Building human resources capacity for digital health is a major component in creating the ecosystem necessary for digital health to flourish. It will never be enough to have the most advanced technology, the best laws, and regulations, etc. without having the right human resources to manage. Human resources include three elements to form the ecosystem:
   - **(a)** The leadership and management that has to be both understandings of the role and value of digital health as an enabler to achieve better quality, more cost-effective, and equitable health services.
   - **(b)** Health services and technology operators who are on the ground doing the job. These include all types of the health workforce and all types of ICT operators. The combined effort of health workers and ICT experts forms a force that will allow the ecosystem to sustain and expand.

(c) The general public includes both healthy citizens and patients. This group of people needs to recognise the value and role of digital health in healthcare delivery, education, information support, and achieving equity.

5. Introduce digital health educational programmes to meet the needs of today and tomorrow. These programmes should be designed and implemented based on the needs and full engagement of academic institutions, professional health informatics societies, and patients. These programmes should allow for growth in a professional sense and cater to the needs of healthcare institutions.
6. Set priorities to contribute to a more acceptable digital health introduction. Evidence-based decisions and continuous monitoring and evaluation will ensure better acceptance of digital health adoption. Allocation of funds as part of the national budget, ensuring sustainability, avoiding pilots, and driving digital health investment based on needs will all contribute to acceptance and support by the different stakeholders and actors in the field.
7. Build partnerships with all potential stakeholders in the country and globally. Solo development of applications will create an environment of alienation and results in resentment of digital health. Support and full engagement of the people, healthcare and information technology professionals, government ministries and agencies, academic and research community, private sector, nongovernment organisations including the United Nations agencies and the media will ensure a favourable ecosystem for digital health.
8. Applying innovative ideas and technologies that attract citizens and patients especially those people who are familiar with the technology. Technology is not strange to these people (native digital) whether on the healthcare providers' side of the recipients of services. Having an environment where people are part of the innovation process will ensure acceptance and endorsement when it is mature enough to adopt. Using an electronic health record system, a mobile application for contact tracing or an artificial intelligence and machine learning tools will only be successful as long as target people are part of that, as they represent a major component in the digital health ecosystem.
9. Develop and adapt culturally acceptable solutions. Awareness and familiarity may not be enough to make the solution acceptable by people. Acceptability may be influenced by language, where multilingualism is important, by the way, equipment and software are operated and the reasons for that, the data that is being collected and its secondary use, and the consent that people have to sign before using the solution.

Conclusion

Six of the MENA countries are in crisis due to civil wars, embargo or occupation (Iran, Iraq, Libya, Palestine, Syria, and Yemen), which obliges these countries to pay less attention to health development and more attention to using telehealth and telemedicine services to provide health care to people. The ICT infrastructure varies in the MENA countries where Gulf countries enjoy the best penetration rate to the Internet in general and at hospitals in particular, mobile phone subscriptions are the highest in the Region as well. Among MENA countries, there have been 14 countries (Bahrain, Egypt, Iran, Iraq, Jordan, Kuwait, Lebanon,

Morocco, Oman, Qatar, Saudi Arabia, and Sudan) that developed national eHealth strategies or included eHealth as part of their national information and communication technology plans. Since these policies range from a fully fledged strategy to an announcement of using a digital health platform or a website meant that many of these strategies and policies lack presenting a future vision of how the country will deal with eHealth in the future and also failed to discuss issues of governance, interoperability, integration, collaboration, stakeholders' engagement, and evaluation. Algeria, for example, lacks a national strategy that is hoped to include a human resource developed plan to support eHealth. Lay down a national strategy, which will allow the education of well-trained personnel in the e-health domain, particularly in interoperability standards. Lay down a national strategy, particularly regarding the need to adopt interoperability standards.

Different laws and regulations need to be introduced to remedy the fragmentation and lack of a unified vision for eHealth in these countries. The WHO Global eHealth Survey [14] showed that six countries (Bahrain, Iraq, Jordan, Lebanon, and Morocco) indicated that they have a legal framework that 'Protects the privacy of personally identifiable data of individuals irrespective of whether it is in paper or digital format'. At the level of application development and installation in healthcare centres, most countries of the Region have done so in a fragmented, disconnected, and unplanned manner. This has been the case when looking for health data on patients during their visits to healthcare centres for diagnosis or treatment.

Hospitals in the private sector have either developed or adopted individual solutions since each hospital competes with others, aiming to make more profit. This does not mean these hospitals lack the interest and dedication to provide excellent healthcare services. This simply means the patient has to provide data every time they visit one of these hospitals due to having policies and systems that do not support data exchange and sharing, connectivity, integration, and interoperability. In many cases, privacy and confidentiality are used as an excuse for not sharing health data. In the absence of governance, it becomes the right of the administrator to decide on issues of public concern. In the public sector, where healthcare is provided by multiple providers, there is room for the implementation of multiple solutions and fragmentation. Ministries of health and military organisations are among the more prominent public sector healthcare providers. COVID-19 and the curfews have resulted in limited access to the literature on the topic (except the Internet) and individuals who are in charge of digital health projects.

References

[1] The World Bank, Middle East and North Africa, 2021. https://data.worldbank.org/country/ZQ. Accessed on 25 January.

[2] World Bank, Fairness and Accountability: Engaging in Health Systems in the Middle East and North Africa, © World Bank, Washington, DC, 2013. https://openknowledge.worldbank.org/handle/10986/16109. License: CC BY 3.0 IGO.

[3] WHO/EMRO, Health and Well-Being Profile of the Eastern Mediterranean Region: An Overview of the Health Situation in the Region and Its Countries in 2019, World Health Organization, Regional Office for the Eastern Mediterranean, Cairo, 2020. 143 p.

[4] WHO/EMRO, Strengthening Health Financing Systems in the Eastern Mediterranean Region Towards Universal Health Coverage: Health Financing Atlas, 2018. https://applications.emro.who.int/docs/EMROPUB_2019_EN_22347.pdf. Accessed on 25 January 2021.

[5] A. O'Neill, Life Expectancy at Birth in the Middle East and Africa Region in 2019, Statista, New York, 2019.

[6] Institute for Health Metrics and Evaluation, The Lancet: Latest Global Disease Estimates Reveal Perfect Storm of Rising Chronic Diseases and Public Health Failures Fuelling COVID-19 Pandemic, Institute for Health Metrics and Evaluation, 2021. healthdata.org. Accessed on 10 January.

[7] Resolution WHA58.28, eHealth, in: Fifty-eighth World Health Assembly, Geneva, May 16–25, 2005. Resolutions and Decisions, Annex, World Health Organization, Geneva, 2005, pp. 108–110. WHA58/2005/REC/1 https://www.who.int/healthacademy/media/WHA58-28-en.pdf?ua=1. Accessed 20 October 2020.

[8] Resolution WHA71/A71, Digital health, in: Seventy-First World Health Assembly, Geneva, 2018. http://apps.who.int/gb/ebwha/pdf_files/WHA71/A71_R7-en.pdf. May 21–26, Accessed 20 October 2020.

[9] O.H. Hans, et al., What is eHealth (3): a systematic review of published definitions, J. Med. Internet Res. 7 (1) (2005). https://www.jmir.org/2005/1/e1/. Accessed 20 October 2020.

[10] Statista, MENA: Mobile Phone Internet User Penetration 2014–2019, Statista Research Department, 2020. https://bit.ly/3kzvs6q. Accessed on 10 September.

[11] The World Bank, Individuals Using the Internet (% of Population)—Middle East & North Africa, 2020. https://bit.ly/2TGML9X. Accessed on 10 September.

[12] Statista, Smartphone Market in MENA—Statistics & Facts, Statista Research Department, 2020. https://www.statista.com/topics/5338/smartphone-market-in-mena/. Accessed on 10 September 2020.

[13] Digital Marketing Community, The Mobile Economy: Middle East and North Africa, GSMA Intelligence, 2018. https://bit.ly/2J2H1oC. Accessed on 10 September 2020.

[14] World Health Organization, Atlas of eHealth Country Profiles 2015: The Use of eHealth in Support of Universal Health Coverage: Based on the Findings of the 2015 Global Survey on eHealth, WHO, Geneva, 2016. 392 p https://apps.who.int/iris/bitstream/handle/10665/204523/9789241565219_eng.pdf?sequence=1.

[15] World Health Organization, Atlas of eHealth Country Profiles 2015: The Use of eHealth in Support of Universal Health Coverage: Based on the Findings of the 2015 Global Survey on eHealth, WHO, Geneva, 2016, p. 18. https://apps.who.int/iris/bitstream/handle/10665/204523/9789241565219_eng.pdf?sequence=1.

[16] M. Aouachria, M. Aouache, I. Oudjoudi, The base frame of the Algerian E-health system: methodology to select the IHE integration profiles and the standards to control interoperability, Stud. Health Technol. Inform. 23 (271) (2020) 153–160.

[17] N. Anane, Telemedicine in Algeria: Current Status and Future Prospects, Center of Development of Advanced Technologies, Algiers, 2000. http://citeseerx.ist.psu.edu/viewdoc/download?doi=10.1.1.496.7203&rep=rep1&type=pdf. Access on 20 October 2020.

[18] International Development Research Centre, Telemedicine in Remote Areas of Algeria, 2008. https://www.idrc.ca/en/project/telemedicine-remote-areas-algeria. Accessed on 20 October 2020.

[19] Zarour, K. (August 2017). Towards a telehomecare in Algeria: case of diabetes measurement and remote monitoring. Int. J. E-Health Med. Commun.8(4):61–80. Accessed on 20 October 2020.

[20] World Health Organization, Atlas of eHealth Country Profiles 2015: The Use of eHealth in Support of Universal Health Coverage: Based on the Findings of the 2015 Global Survey on eHealth, WHO, Geneva, 2016, p. 36. https://apps.who.int/iris/bitstream/handle/10665/204523/9789241565219_eng.pdf?sequence=1.

[21] H. Moghaddasi, Hospital information systems: the status and approaches in selected countries of the Middle East, Electron. Physician 10 (5) (2018) 6829–6835. https://europepmc.org/article/PMC/6033126#id226946. Accessed on 23 October 2020.

[22] D. Alhuwail, et al., Middle East and North African Health Informatics Association (MENAHIA): strengthening health informatics ecosystems in the region and beyond, IMIA Yearbook of medical informatics, 2020. CC BY-NC-ND 4.0, Yearb. Med. Inform. 29 (01) (2020) 290–294, https://doi.org/10.1055/s-0040-1701968. Accessed on 30 November 2020.

[23] A. Zolait, et al., Can Bahraini patients accept e-health systems? Int. J. Health Care Qual. Assur. 32 (4) (2019) 720–730. https://www.emerald.com/insight/content/. https://doi.org/10.1108/IJHCQA-05-2018-0106/full/html. Accessed on 10 January 2021.

[24] World Health Organization, eHealth Egypt, WHO, Geneva, 2015. https://www.who.int/goe/publications/atlas/egy.pdf?ua=1. Accessed 30 October 2020.

[25] Egypt. Ministry of Telecommunications and Information Technology, Telecommunications and Information Technology National Strategy 2013–2017: Egyptian Digital Society in the Digital Economy, Ministry of Telecommunications and Information Technology, Cairo, 2012. 67 p.

[26] R. Hussein, A. Khalifa, Telemedicine in Egypt: SWOT analysis and future trends, GMS Med. Inform. Biomet. Epidemiol. 8 (1) (2012) 1–16, https://doi.org/10.3205/mibe000125. https://www.egms.de/static/en/journals/mibe/2012-8/mibe000125.shtml.

[27] S. Abd Elgaber, et al., A Roadmap to Implement EHR Nationwide in Egypt. Communications on Applied Electronics (CAE), vol. 7, Foundation of Computer Science, New York, 2017, pp. 18–28. No. 6, September.

[28] T.M. Farahat, N.N. Hegazy, M. Mowafy, Information and communication technologies in primary healthcare facilities in Egypt, Prim. Health Care Res. Dev. 19 (2018) 88–95.

[29] H. Wahba, T. Emara, A. Elbokl, The Egyptian-African telemedicine network: the treat and teach comprehensive model (Chapter 12), in: Telemedicine Technologies- Big Data, Deep Learning, Robotics, Mobile and Remote Applications for Global Healthcare, Elsevier Academic Press, Amsterdam, 2019, pp. 183–192.

[30] World Health Organization, Global Observatory on eHealth—Key Indicators, WHO, Djibouti. Geneva, 2016. https://apps.who.int/gho/data/node.goe.ki-DJI?lang=en. Accessed on 25 October 2020.

[31] World Bank, Djibouti Launches Digital Transformation to Improve Services to Citizens. https://www.worldbank.org/en/news/press-release/2018/04/25/djibouti-launches-digital-transformation-to-improve-services-to-citizens. Accessed on 25 October 2020.

[32] Oxford Business Group, Addressing Challenges and Facilitating Universal Health Care in Djibouti's Urban and Rural Areas. https://oxfordbusinessgroup.com/overview/renewed-focus-addressing-challenges-and-facilitating-universal-coverage. Accessed on 25 October 2020.

[33] World Health Organization, Atlas of eHealth Country Profiles 2015: The Use of eHealth in Support of Universal Health Coverage: Based on the Findings of the 2015 Global Survey on eHealth, WHO, Geneva, 2016, p. 162. https://apps.who.int/iris/bitstream/handle/10665/204523/9789241565219_eng.pdf?sequence=1.

[34] M. Jahanbakhsh, M. Sharifi, M. Ayat, The status of hospital information systems in Iranian hospitals, Acta Inform. Med. 22 (4) (2014) 268–275.

[35] A. Darvish, M.A. Far, A review on implementation of electronic health in Iran literature, MOJ Women's Health 5 (5) (2017) 294–298.

[36] M.K. Abdel Ghani, M.M. Jaber, A review on the theories of adoption telemedicine in Middle Ease: toward building Iraqi telemedicine, in: Proceedings of the 4th International Conference on Mathematical, Computational and Statistical Sciences (MCSS '16), WSEAS Press, Athens, 2016.

[37] K. Al-Moosawi, M. Nayyef, C. Waegemann, Survey study: reality check of internet and telemedicine use in Iraqi hospitals, J. Int. Soc. Telemed. eHealth 5 (2017), e7. 5,7-1-6 https://journals.ukzn.ac.za/index.php/JISfTeH/article/view/190/pdf. Accessed on 10 December 2020.

[38] World Health Organization, Hinari Access to Research for Health Programme, 2021. https://www.who.int/hinari/en/. Accessed 12 January 2021.

[39] A.F. Neamah, et al., E-Health state in Middle East countries: an overview, Turk. Online J. Des. Art Commun. (2018) 2974–2990. Special Edition.

[40] World Health Organization, Atlas of eHealth Country Profiles 2015: The Use of eHealth in Support of Universal Health Coverage: Based on the Findings of the 2015 Global Survey on eHealth, WHO, Geneva, 2016, p. 132. https://apps.who.int/iris/bitstream/handle/10665/204523/9789241565219_eng.pdf?sequence=1.

[41] Jordan, Electronic Health Solutions. https://www.ehs.jo. Accessed on 30 October 2020.

[42] N. Matar, M. AlNabhan, Evaluating E-health services and patients requirements in Jordanian hospitals, Int. Arab J. E-Technol. 3 (4) (2014) 250–257.

[43] A.Z. Obeidat, K. El-Salem, A national telemedicine program in the kingdom of Jordan: editorial, Ann. Med. Surg. 62 (2021) 145–149, https://doi.org/10.1016/j.amsu.2021.01.009. Accessed 15 January 2021.

[44] Jordan. High Health Council, The National Strategy for Health Sector in Jordan 2015–2019, The Council, Amman, 2015. 109 p.

[45] Jordan, Electronic Health Solutions, Electronic Library of Medicine, 2020. https://ehs.com.jo/electronic-library-medicine-jordan. Accessed on 30 October.

[46] Jordan, eHealth Development Association, 2020. www.ehda.org.jo. Accessed on 30 October.

[47] Aman, Aman.jo Jordan App COVID-19, 2020. https://play.google.com/store/apps/details?id=jo.gov.moh.aman&hl=en. Accessed on 30 October.

[48] A.S. Weber, et al., Systematic thematic review of e-health research in the Gulf cooperation council (Arabian Gulf): Bahrain, Kuwait, Oman, Qatar, Saudi Arabia and United Arab Emirates, J. Telemed. Telecare 23 (4) (2017) 2017, https://doi.org/10.1177/1357633X16647894. Accessed on 15 November 2020.

[49] D. Alhuwail, Understanding health information management practices in public hospitals in Kuwait, Health Inform. Manag. J. 49 (2–3) (2020), https://doi.org/10.1145/2107556.2107572. https://dl.acm.org/doi/abs/ Accessed on 15 November 2020.

[50] Kuwait, Ministry of Health. Department of Information Systems. IT Vision and Applications, The Ministry, Kuwait, 2013. 32 p.
[51] Lebanon, Ministry of Health, National eHealth Programme, 2020. According to decision No.1/227 of 04/03/2013 https://www.moph.gov.lb/en/Pages/6/2651/national-e-health-program. Accessed on 20 November.
[52] H. Hamadeh, eHealth Roadmap for Lebanon. Building Consensus on the Readiness for Electronic Health Record in Lebanon, AUB, MoH and WHO, Beirut, 2016.
[53] S. Saleh, Readiness of healthcare providers for eHealth: the case from primary healthcare centers in Lebanon, BMC Health Serv. Res. 16 (1) (2016) 644, https://doi.org/10.1186/s12913-016-1896-2. Nov 10. Accessed on 12 October 2020.
[54] American University of Beirut, Nursing Informatics Certificate, 2020. https://www.aub.edu.lb/rep/cec/Pages/nursing-informatics.aspx. Accessed on 20 November.
[55] Lebanon, Ma3an: Together Against Corona, Ministry of Public Health, 2020. https://moph.gov.lb/ar/ma3an. Accessed on 20 November.
[56] M.M. Khalil, A. Al-Bousify, Libyan health system and eHealth: a vital step or an unnecessary leap? Libyan J. Med. (2008) 080206. AOP.
[57] Libya. Ministry of Health, Libya Health Information System Assessment and Planning Workshop: Assessment and Roadmap of Priority Actions—Final Report 2017, Ministry of Health, Tripoli, 2017. 68 p.
[58] World Health Organization, Atlas of eHealth Country Profiles 2015: The Use of eHealth in Support of Universal Health Coverage: Based on the Findings of the 2015 Global Survey on eHealth, WHO, Geneva, 2016, p. 246. https://apps.who.int/iris/bitstream/handle/10665/204523/9789241565219_eng.pdf?sequence=1.
[59] Morocco, Programme eGouvernement, 2020. http://www.egov.ma/sites/default/files/Programme%20eGov%20Morocco.pdf. Accessed on 15 December.
[60] Morocco, Sectoral Strategy for Health for 2012–2016, Ministry of Health, 2012. http://www.sante.gov.ma/Docs/Documents/secteur%20sant%C3%A9.pdf. Accessed on 30 September 2020.
[61] R. Parks, et al., Electronic health records implementation in Morocco: challenges of silo efforts and recommendations for improvements, Int. J. Med. Inform. 129 (2019) 430–437. https://www.sciencedirect.com/science/article/abs/pii/S1386505619302096?via%3Dihub. Accessed on 20 September 2020.
[62] H. Ghazal, et al., Legal challenges of cross border telemedicine between Morocco and the European Union, in: 22nd Conference of the International Society for Telemedicine and eHealth and Moroccan Society for Telemedicine and eHealth 2017 (Casablanca 6-8 December 2017), 2017.
[63] Morocco, Moroccan Society of Medical Informatics and Health, 2020. http://smims.ma/index.php?item=mission. Accessed on 22 October.
[64] E. Abu Hamra, Assessment of Applicability of e-Health in Gaza Governorates: Barriers and Opportunities, Unpublished thesis, Al Quds University, Gaza, 2014.
[65] C. El Jabari, et al., Towards a new paradigm of federated electronic health records in Palestine, Informatics 7 (2020) 41. https://www.mdpi.com/2227-9709/7/4/41/pdf. Accessed on 25 November 2020.
[66] UNRWA, Health in the Gaza Strip, 2020. https://www.unrwa.org/activity/health-gaza-strip. Accessed on 7 November.
[67] UNRWA, Health in the West Bank, 2020. https://www.unrwa.org/activity/health-west-bank. Accessed on 7 November.
[68] C. Jabari, L. Adwan, Health informatics in the Arab world, in: I. Laher (Ed.), Handbook of Healthcare in the Arab World, Springer, Cham, Switzerland, 2019, pp. 1–12.
[69] Ballout, G., et al. (2017). UNRWA's innovative e-health for 5 million Palestine refugees in the near East. BMJ Innovations: Health IT, Systems and Process innovations, vol. 4: 3 (https://innovations.bmj.com/content/4/3/128.full). Accessed on 7 November 2020.
[70] UNRWA, 2020. UNRWA Launches *'YOUR HEALTH'* Mobile Application (https://bit.ly/3a5d6pM).
[71] Palestine. Ministry of Health, Sehhaty, 2020. http://apps.moh.gov.ps/sehatty/. Accessed on 10 December 2020.
[72] World Health Organization, Atlas of eHealth Country Profiles 2015: The Use of eHealth in Support of Universal Health Coverage: Based on the Findings of the 2015 Global Survey on eHealth, WHO, Geneva, 2016, p. 261. https://apps.who.int/iris/bitstream/handle/10665/204523/9789241565219_eng.pdf?sequence=1.
[73] Oxford Business Group, Oman's Health Care System Increasingly Ready to Meet the Population's Needs, 2021. https://oxfordbusinessgroup.com/overview/rise-growing-population-finds-health-care-system-ready-meet-its-needs. Accessed on 15 January.

[74] Qatar, Ministry of Public Health—Qatar National E-Health & Data Program, 2021. www.moph.gov.qa. Accessed on 10 January.
[75] World Health Organization, Atlas of eHealth Country Profiles 2015: The Use of eHealth in Support of Universal Health Coverage: Based on the Findings of the 2015 Global Survey on eHealth, WHO, Geneva, 2016, p. 285. https://apps.who.int/iris/bitstream/handle/10665/204523/9789241565219_eng.pdf?sequence=1.
[76] Yousif, F., et al., (2020). Health Informatics Association of Qatar (HIAQ): building a digital health ecosystem. Stud. Health Technol. Inform. Vol. 272. pp. 474–477. (https://doi.org/10.3233/SHTI200598). Accessed on 15 January 2021.
[77] H.A. Asiri, Challenges of the health informatics education in the Kingdom of Saudi Arabia: what stands in our way? J. Health Inform. Dev. Ctries. 8 (1) (2014) 26–35. https://www.jhidc.org/index.php/jhidc/article/download/114/158. Accessed on 28 November 2020.
[78] M. Altwaijiri, B. Aldosar, Health informatics master program at King Saud bin Abdulaziz University for Health Sciences, Riyadh, Saudi Arabia, IMIA Ybook. Med. Inform. (2008). https://pdfs.semanticscholar.org/7f36/93396fb523e0efebcc6be87be7edaa2aa77e.pdf. Accessed on 10 December 2020.
[79] Saudi Arabia, National E-Health Strategy—Ministry Vision "e-Health", Ministry of Health, 2020. https://www.moh.gov.sa/en/Ministry/nehs/Pages/default.aspx. Accessed on 9 December.
[80] B. Al Knawy, et al., The Riyadh declaration: the role of digital health in fighting pandemics, Lancet 396 (10262) (2020) 1537–1539, https://doi.org/10.1016/S0140-6736(20)31978-4. Accessed on 10 October 2020.
[81] Saudi Arabia, Health Informatics Association, 2020. www.sahi.org.sa. Accessed 25 November.
[82] World Health Organization, Atlas of eHealth Country Profiles 2015: The Use of eHealth in Support of Universal Health Coverage: Based on the Findings of the 2015 Global Survey on eHealth, WHO, Geneva, 2016, p. 330. https://apps.who.int/iris/bitstream/handle/10665/204523/9789241565219_eng.pdf?sequence=1.
[83] S. Mandil, Proposals for a Sudan eHealth Strategy, WHO, Geneva, 2005.
[84] WHO and ITU, National eHealth Strategy Development Meeting, ITU, Geneva, 2012. http://www.itu.int/ITU-D/cyb/events/2012/e-health/Nat_eH_Dev/Session%205/E-Health_Strategy_Exprience_Sudan.pdf. Accessed on 25 September 2020.
[85] D.S. El Zaki, E-Health in Sudan: Case Study, 2020, PowerPoint presentation, not dated https://slideplayer.com/slide/2333343/. Accessed on 25 September.
[86] Sudan Doctors' Union in the UK Branch, Sudan e-Health and Telemedicine: A No Regret Strategy During COVID-19 Pandemic: Scoping Document, The Union, London, 2020. 17 p https://cdn-cms.f-static.net/uploads/905961/normal_5f4a76d14d1cc.pdf. Accessed on 25 September 2020.
[87] World Health Organization, Atlas of eHealth Country Profiles 2015: The Use of eHealth in Support of Universal Health Coverage: Based on the Findings of the 2015 Global Survey on eHealth, WHO, Geneva, 2016, p. 339. https://apps.who.int/iris/bitstream/handle/10665/204523/9789241565219_eng.pdf?sequence=1.
[88] R. Merrell, C.R. Doar, Editorial: tales of telemedience in Syria, Telemed. E-Health 22 (12) (2016) 969–970, https://doi.org/10.1089/tmj.2016.29015.rcm. Accessed on 20 December 2020.
[89] Syrian American Medical Society, Annual Report, 2017. https://www.sams-usa.net/reports/annual-report-2016/. Accessed on 15 September 2020.
[90] R. Talhouk, et al., Integrating health technologies in health services for Syrian refugees in Lebanon: qualitative study, J. Med. Internet Res. 22 (2020) 7. https://www.jmir.org/2020/7/e14283. Accessed on 10 December 2020.
[91] N. Al-Shorbaji, Travel Report to Tunisia, (14–17 February 2015): Roadmap for the eHealth Strategy, 2015. Unpublished WHO Report.
[92] M. Sadok, R. Chatta, P. Bednar, ICT for development in Tunisia: going the last mile, Technol. Soc. 46 (2016) 63–69. https://www.sciencedirect.com/science/article/abs/pii/S0160791X16300756. Accessed on 18 October 2020.
[93] Société Tunisienne de Télémédecine & e-Santé, 2020. http://www.telemedecine-tunisie.tn/. Accessed on 20 November.
[94] A. El Matri, Role of Arab telemedicine society in support of e-health, in: 3rd International Conference on Information and Communication Technologies: From Theory to Applications, (Damascus, 2008), 2008, pp. 1–2, https://doi.org/10.1109/ICTTA.2008.4529926. Accessed on 20 November 2020.
[95] Tunisia, "Digital Health" Development Program in Tunisia, Ministry of Public Health, 2020. http://bit.ly/2M9P7Oi. Accessed on 10 December.
[96] WHO, Be Mobile—Country Programmes, Be He@lthy, 2020. https://www.who.int/ncds/prevention/be-healthy-be-mobile/countries/en/. Accessed on 15 December.

[97] Tunisia, Ministry of Public Health, IHMI, 2020. https://e7mi.tn/faq_ar.html. Accessed on 3 September.

[98] F.F. Salameh, Exploratory study of smart healthcare services (e-health & m-health applications) in the UAE: exploring related literature & websites contents, J. Content Community Commun. 10 (2019) 2395–7514. Amity School of Communication. Year 5.

[99] UAE, Ministry of Health and Prevention, 2020. https://www.mohap.gov.ae/en/MediaCenter/News/Pages/2725.aspx. Accessed on 15 December.

[100] UAE, The National AI Strategy 2031, 2020. http://www.uaeai.ae/en/. Accessed on 17 December.

[101] UAE, Emirates Health Informatics Society, About EHIS, 2020. http://www.ehis.ae/About-EHIS. Accessed on 30 November.

[102] A.A. Al-Fadhli, et al., Telehealth in Yemen: an overview and a proposed model, in: Proceedings of the 5th International Conference on Computing and Informatics, ICOCI 2015, Universiti Utara Malaysia, Istanbul, Turkey, 2015. 11–13 August, 2015 http://repo.uum.edu.my/15534/1/PID104.pdf. Accessed on 30 November 2020.

[103] A. Al-fadhli, et al., Developing a telehealth readiness assessment framework and tool for least developed countries: a case of Yemen, Indian J. Public Health Res. Dev. 9 (11) (2018) 293–302, https://doi.org/10.5958/0976-5506.2018.01660.1. Accessed on 30 November 2020.

[104] Ecosystems, 2020. https://sc-s.si/joomla/images/Ecosystems.pdf. Accessed on 5 December.

CHAPTER

17

A framework for regional health information systems interoperability: The Asia eHealth Information Network (AeHIN) experience

Alvin B. Marcelo[a], *Kristin Chloe Pascual*[b], *Jai Ganesh Udayasankaran*[c], *Fazilah Shaik Allaudin*[d], *and Boonchai Kijsanayotin*[e]

[a]University of the Philippines Manila, Manila, Philippines [b]Asia eHealth Information Network, Kowloon, Hong Kong [c]Sri Sathya Sai Central Trust, Puttaparthi, India [d]Ministry of Health, Kuala Lumpur, Malaysia [e]Faculty of Medicine, Ramathibodi Hospital, Mahidol University, Bangkok, Thailand

Background

In 2005, the World Health Organization defined eHealth as 'the use of information and communications technologies in health [1]'. Soon after, WHO began publishing the Global Observatory for eHealth or GOe [2], a report of eHealth initiatives and advances in different countries around the globe. The GOe contained country reports on the state of their eHealth initiatives.

About AeHIN

Asia eHealth Information Network (AeHIN) is a collaboration of advocates, created by the World Health Organization in 2011 to help Asian countries with digital health development. It has over 1377 professionals in eHealth, Health Information Systems (HIS), and Civil Registration and Vital Statistics (CRVS) from 68 countries and has helped build

Roadmap to Successful Digital Health Ecosystems
https://doi.org/10.1016/B978-0-12-823413-6.00005-7

capacity in the region by laying down national health information foundations: governance, architecture, programme management, standards, and interoperability (Mind the GAPS Framework). It envisions achieving interoperable digital health systems for universal health coverage and health systems in Asia. To achieve this vision, its mission is to promote learning, resource sharing, and knowledge exchange for strengthening digital health through the networking approach.

AeHIN has a pool of health and IT professionals from South and Southeast Asia committed to promoting better use of ICTs to achieve better health. AeHIN is a nonprofit incorporated in Hong Kong and maintains a secretariat team in Manila, Bangkok, and Kuala Lumpur.

Problem 1: Lack of standards

At the 2011 HIS interoperability conference in Vietnam, one of the key issues raised was the lack of standards. AeHIN then sought standards that countries can use and identified Health Level 7 (HL7) as one such standard. Consequently, AeHIN offered an HL7 training programme to its members, financially supported by WHO WPRO and Norwegian Agency for Development Cooperation (Norad). After the training, while participants showed appreciation for the course, they expressed challenges with not knowing how to use the HL7 standards without an accompanying blueprint. HL7 was analogous to standard building blocks, but how it is used was in the domain of enterprise architecture (EA).

Problem 2: Lack of enterprise architecture

Enterprise architecture is 'the practice of analysing, designing, planning, and implementing enterprise analysis to successfully execute on business strategies. EA helps businesses structure IT projects and policies to achieve desired business results and stay on top of industry trends and disruptions using architecture principles and practices, a process also known as enterprise architectural planning (EAP)' [3]. Realising this gap in knowledge, AeHIN sought out various enterprise architecture courses and selected The Open Group Architecture Framework or TOGAF. TOGAF is an International Standards Organization (ISO)-approved method for developing enterprise architectures. After training two cohorts in Mongolia and Malaysia, AeHIN asked for feedback from the participants. The participants agreed designing blueprints was important but argued it was time-consuming and complicated work that needed authorisation from the highest levels of the organization. Developing an EA without a clear governance mandate could result in a waste of effort.

Problem 3: Lack of governance

Accepting that leadership and governance have been the most common challenges across all the countries and that efforts at creating blueprints are dependent on the authority of an accountable body, AeHIN searched for training programmes that could help countries build this capability. While numerous governance frameworks were available, only the Control Objectives for Information and Related Technology framework (COBIT 5) had an accompanying training and certification programmed . Two cohorts were trained in Manila and Bangkok, producing certified professionals. These certified professionals went on to promote and champion COBIT 5 in their respective countries.

Problem 4: Lack of capacity to manage programmes and projects

Lastly, faced with the complexity of various simultaneous projects related to eHealth (fast communication networks by the Ministry of ICT, electronic medical records by the Ministry of Health, electronic claims by the national health insurance agency, digital civil registrations, and vital statistics, etc.), AeHIN members requested for certification training on project management. PRINCE2 (PRojects IN Controlled Environments) was selected because its training programmes were shorter but Project Management Professional (PMP) was also an option. A cohort was trained for PRINCE2 as part of its AeHIN's regional capacity building on digital health programme management in Malaysia. Thirty participants from 14 countries, namely Bangladesh, Bhutan, Cambodia, Indonesia, Lao PDR, Malaysia, Maldives, Myanmar, Nepal, Pakistan, Philippines, Sri Lanka, Thailand, and Vietnam attended the training. The participants have started applying their newfound knowledge managing projects in collaboration with other sectors.

Where are we now: Mind the GAPS

In February 2012, the WHO-ITU jointly published the National eHealth Strategy Toolkit [4]. The toolkit is a result of a collaboration between the World Health Organization (WHO) and the ITU (International Telecommunications Union). In 2013, the WHO-ITU National eHealth Strategy Toolkit was launched at the Second AeHIN General Meeting in Bangkok. The toolkit proposed a framework with seven components but fell short of prescribing a sequence by which these components should be developed. Guided by the toolkit, and after conducting a series of capacity building programmes from 2012 to 2019, AeHIN has summarised a systematic intervention that countries can adopt to strengthen their national eHealth development. This is AeHIN's Mind the GAPS approach (Fig. 1).

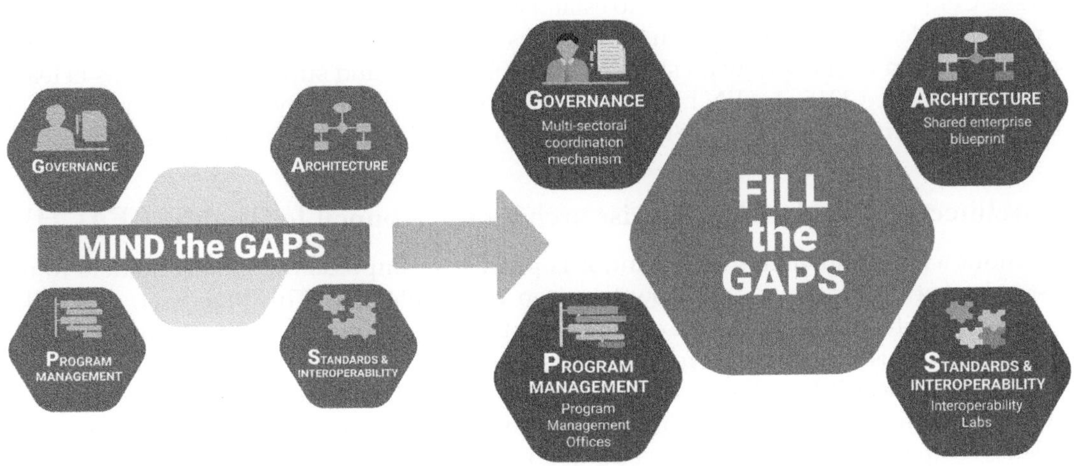

FIG. 1 Mind the GAPS, fill the GAPS framework.

The framework espouses that in order to develop their national eHealth programmes, countries need to systematically work on four important fundamental interrelated domains:

- Governance: Interoperability of healthcare applications will only be possible if they are governed by a multisectoral coordination mechanism. There should be an accountable group defining the benefits, risks, and resources of the national eHealth programme.
- Architecture: Healthcare applications become interoperable if they are able to reference a shared enterprise blueprint. In this manner, stakeholders in a country will understand the whole architecture and how they can contribute to it within their respective sectors.
- People and Programme Management: Interoperability requires close coordination of various projects in separate agencies. Programme management enables these stakeholders to synchronise their efforts with each other to achieve optimal impact. Moreover, strengthening health IT human capacity in country to develop, implement and effectively use eHealth/digital health is essential for the success of eHealth/digital health program.
- Standards and Interoperability: Interoperable healthcare applications become possible when stakeholders adopt the same standards.

Where do we want to go: Fill the GAPS

AeHIN's Mind the GAPS framework helped structure the network's capacity building initiatives. Over the years, AeHIN has crafted and co-designed activities (Fill the GAPS) that aim to strengthen each country's capacity to develop national eHealth programmes and address the persistent problems that have been plaguing interoperability.

On governance: Convergence workshops

With support from development partners, AeHIN helps Ministries of Health (MOH) gather ('converge') various stakeholders to support the national eHealth programme.

These Convergence Workshops aim to establish a multisectoral coordination mechanism for the national eHealth programme with the MOH as the lead agency. The term 'convergence' was specially selected to show how all stakeholders consolidate and support the MOH's eHealth strategy [5,6]. Since 2015, AeHIN has supported convergence workshops in Myanmar, Timor-Leste, Vietnam, Indonesia, Nepal, Bhutan, and Lao PDR (People's Democratic Republic).

On architecture: Regional enterprise architecture council for health (REACH)

Without a clear and shared blueprint, it is practically impossible for stakeholders to understand what is being built and how they can contribute to it. This blueprint, also called enterprise architecture, is a complex process that converts a vision (e.g. the national health strategy) into a visible graphical representation of the information technology components required to support it (e.g. the national eHealth blueprint). Because enterprise architecture is a crucial component for health information systems interoperability, the process is best guided by experts certified in the designing and building of these blueprints. From 2013 to 2015, AeHIN enabled the training of staff from various ministries of health and academe and formed them into a community of practice, the Regional Enterprise Architecture Council for Health or REACH. REACH is composed of AeHIN members who were trained and/or certified in the Open Group Architecture Framework (TOGAF) [7]. Together, they share knowledge and artefacts that enable reuse of proven architectures in healthcare.

On programme management: Network of national ehealth programme management offices

Because national eHealth systems require multisectoral coordination, AeHIN has proposed the creation of strong national eHealth programme management offices in each MOH to allow different country stakeholders to be informed of current eHealth policies and standards and conversely, update the government on the state of their applications. This coordination is crucial because achieving the expected benefits of eHealth is dependent on the successful implementation of projects of other sectors (e.g. broadband wireless access from the Ministry of Telecommunications and national identification systems from the civil registration authority). Furthermore, these national eHealth programme management offices need to be capable of overseeing all of these concurrent projects and advise the governance body of their status. To this end, AeHIN enabled the training on project management (Projects in Controlled Environments or PRINCE2) to equip them with the knowledge and skills to conduct multisectoral coordination work.

Standards and interoperability: Community of interoperability labs (COIL)

Governance, architecture, and programme management are important but to ensure interoperability, laboratories can be set up where stakeholders can obtain support and training on making their healthcare applications interoperable and compliant with the national blueprint. These laboratories can provide experts (teams), training, tools, and tests that will ensure compliance with the blueprint and interoperability of healthcare applications.

Country experiences

India

Based on the vision 'Health and wellbeing for all at all ages', of the National Health Policy (NHP) 2017, Ministry of Health and Family Welfare (MoHFW), Government of India (GoI) recognised the need for creating a framework for the evolution of a National Digital Health Ecosystem (NDHE), to support 'Continuum of care' for citizens. To create and enable digital health ecosystem and prioritise digital health in India, and to develop an implementation framework for the National Health Stack (NHS), MoHFW, GoI brought out the National Digital Health Blueprint (NDHB) [6] laying out the building blocks and an action plan to comprehensively and holistically implement digital health (Fig. 2).

The NDHB recognises the need to establish a specialised organization, called National Digital Health Mission (NDHM) [8] that can drive the implementation of the Blueprint, and promote and facilitate the evolution of NDHE. The Blueprint keeps the overall vision of NHP 2017 at its core and recommends commencing with a pragmatic agenda to start with, adopting the principle of 'Think Big, Start Small, Scale Fast'. To this end, it has been designed

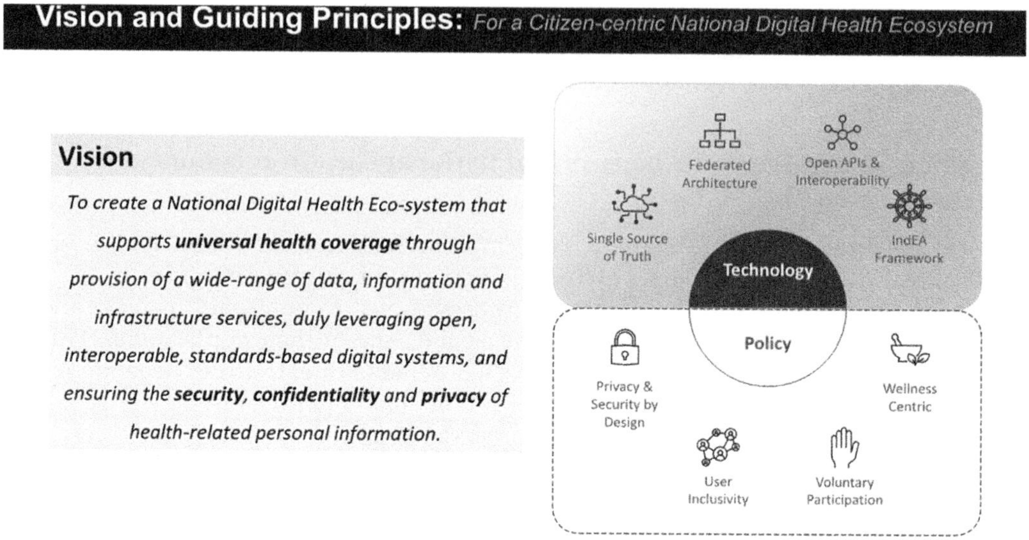

FIG. 2 Vision and Guiding Principles of NDHM (National Digital Health Mission). *Source: National Health Authority (NHA), Government of India (GoI).*

as a layered framework (Fig. 3), with the vision and a set of principles at the core, surrounded by the other layers relating to digital health infrastructure, digital health data hubs, building blocks, standards and regulations, and an institutional framework for its implementation.

For enhancing the security and privacy of personal and sensitive information of the patients while ensuring interoperability and technological flexibility and independence, the NDHB recommends a federated digital health architecture (Fig. 4) instead of large centralised systems for the management of health information. NDHB in addition to being an architectural vision also provides specific guidance on its implementation.

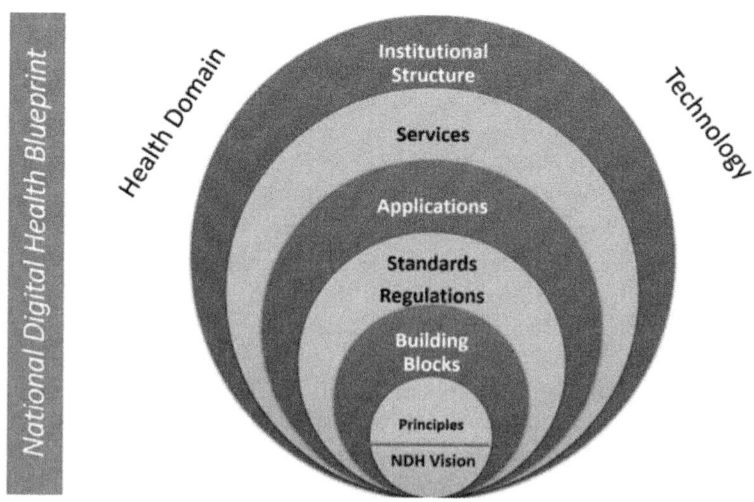

FIG. 3 Layered structures. *Source: National Digital Health Blueprint, MoHFW, GoI.*

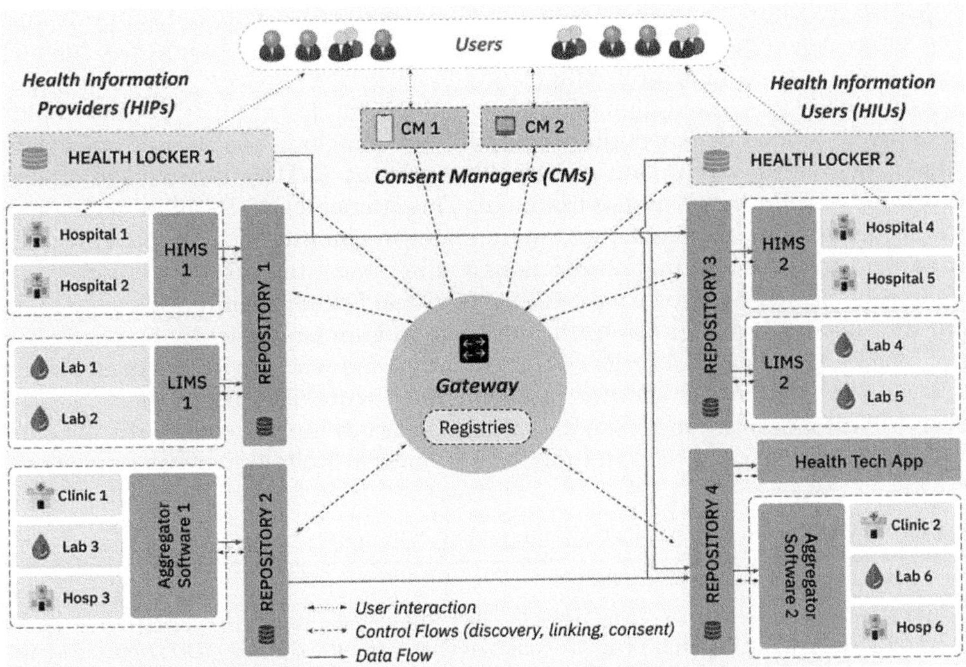

FIG. 4 Federated Digital Health Architecture. *Source: National Health Authority, GoI.*

The GoI has established NDHM, with MoHFW defining the policy, regulatory frameworks with the National Health Authority (NHA) being the implementation agency. Given the interlinkages and comprehensive scope, NDHM is a collaborative initiative between multiple ministries and departments within the ministries. The mission will have twin arms—one for regulation and other for implementation and operational management working under a defined governance framework. The governance structure proposed for implementation of NDHM [8] is depicted in Fig. 5.

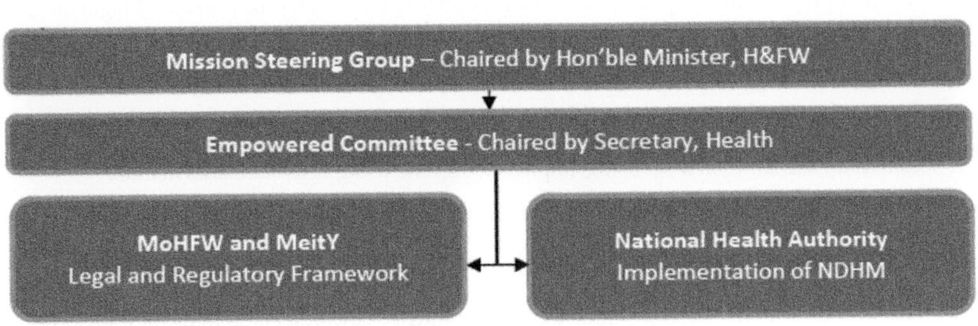

FIG. 5 Governance structure of NDHM. *Source: National Health Authority, GoI.*

The NDHM envisions to create a national digital health ecosystem that supports universal health coverage in an efficient, accessible, inclusive, affordable, timely, and safe manner, that provides a wide range of data, information and infrastructure services, duly leveraging open, interoperable, standards-based digital systems, and ensures the security, confidentiality, and privacy of health-related personal information. This will include adoption of open standards by all the actors in the NDHE. National Health Authority (NHA) has released an enabling framework for NDHM Sandbox Environment. This framework will allow technologies or products to be tested in the contained environment in compliance with NDHM standards and judge the consumer and market reactions to the same. This Sandbox Environment will help organisations intending to be a part of the National Digital Health Ecosystem to become a Health Information Provider or Health Information User or efficiently link with building blocks of NDHM [9]. The Sandbox Environment is live and ready to use by those who would like to create software services and products integrated with NDHM services.

The key milestones from various national digital health initiatives in India over the past two decades are summarised in a chronological manner in Table 1.

TABLE 1 Key milestones in national digital health initiatives in India.

Year	National digital health milestones
2003	Recommendation, Framework for development of Information Technology Infrastructure for Health (ITIH) In India by Department of Electronics and Information Technology (DIT) from the Ministry of Communication and IT (MCIT) Recommendations on Guidelines, Standards and Practices for Telemedicine in India by Department of Information Technology (DIT), Ministry of Communication and IT (MCIT)
2005	Ministry of Health and Family Welfare (MoHFW) sets up a task force for telemedicine and standards related to Electronic Medical Records (EMR)
2010	MoHFW sets up three subgroups (standards, data connectivity, and data ownership) to deliberate on standardisation of Electronic Medical Records (EMR)
2013	Recommendations on Electronic Medical Records Standards in India by EMR Standards Committee constituted by MoHFW and Federation of Indian Chambers of Commerce and Industry (FICCI) MoHFW notified the Electronic Health Record (EHR) Standards for India in September 2013
2014	India became a member of SNOMED International in March 2014 as part of approach to use terminology standards to achieve semantic interoperability in health systems
2015	MoHFW proposes to set up National eHealth Authority (NeHA) for the purpose of regulating, developing, and deploying digital health across the continuum of care
2016	MoHFW notified the revised Electronic Health Records (EHR) Standards for India [10]
2017	MoHFW established the National Resource Centre for EHR Standards (NRCeS), a Centre of Excellence to accelerate and promote adoption of EHR standards in India. NRCeS is a single point of contact for assistance in developing, implementing, and using EHR standards in India
2018	MoHFW sets up a fourth subgroup to take into consideration the legal and mandatory aspects of notified EHR Standards MoHFW drafts the Digital Information Security in Health care Act, 2018 (DISHA) specifically for the protection of Digital Health Data (DHD) National Institution for Transforming India (NITI Aayog), Government of India proposed 'National Health Stack (NHS)' with the aim to create digital health records for all the citizens of India by the year 2022

TABLE 1 Key milestones in national digital health initiatives in India—cont'd

Year	National digital health milestones
2019	NITI Aayog released a report on 'National Digital Health Blueprint (NDHB)' in July outlining the action plan to realise the development of National Health Stack (NHS). The report recommended the establishment of 'National Digital Health Mission (NDHM)' as an entity responsible for the successful implementation of NDHB The Personal Data Protection Bill 2019 (PDP Bill 2019) tabled in the Parliament by the Ministry of Electronics and Information Technology (MeitY) on December 11
2020	Board of Governors in supersession of the Medical Council of India (MCI) released 'Telemedicine Practice Guidelines: Enabling Registered Medical Practitioners to Provide Healthcare Using Telemedicine' [11] on March 25 Launch of pilot phase of National Digital Health Mission (NDHM) [8] announced during the 74th Independence day, on August 15 National Health Authority (NHA) released enabling framework for NDHM Sandbox Environment on August 18 [9] MoHFW approves the Health Data Management Policy of the NDHM on December 14
2021	National Health Authority (NHA) released the consultation papers on Healthcare Professionals Registry and Health Facility Registry on June 22 inviting comments from public NHA released the consultation paper on Unified Health Interface (UHI) on July 23 inviting comments from public Launch of nation-wide rollout of National Digital Health Mission now known as Ayushman Bharat Digital Mission (ABDM) announced on September 27 NHA released the consultation paper on proposed Health Data Retention Policy on November 23 inviting comments from public

Malaysia

Over the last two decades, the Ministry of Health Malaysia has been strengthening its digital health strategy and technology adoption, aligned to national health and economic transformation with a goal towards achieving Universal Health Coverage (UHC). The recent growth in digital technology driving the evolution of health care can play a significant role in the entire spectrum of care. It can improve population health and disease outcomes, enhance quality and safety, improve accessibility to care and skills, and improve access to information. In February 2021 , we launched the Malaysia Digital Economy Blueprint. This roadmap aims to transform the country into a digitally-driven, high income nation and a regional leader in the digital economy. At the same time, the blueprint aspires to enrich and improve the wellbeing of all Malaysians.

Over the years, the MOH has been actively pursuing various significant digital health initiatives which include:

Hospital information system, primary care clinical system, pharmacy information system, blood bank information system, surveillance and disease management, and case-mix system.
The Malaysia Health Information Exchange, a patient record interoperability platform as a precursor for public-private integration in health.
The Malaysia Health Data Warehouse as a key repository centre of health data for analytics, strategic planning, and evidence-based decision-making and research.
National level systems related to MOH regulatory role such as food safety and pharmaceutical.

Use of social media in health promotion, crisis and risk management, and dissemination information on health policy and related matters.
Online Health Services facilitated and enabled by technology platforms (Uberisation of Healthcare).
The wave of 4IR: Internet of things (IoT) to artificial intelligence, genomics, robotics, 3D printing, 5G use cases, virtual reality, and blockchain technology have also started to revolutionise the healthcare landscape in Malaysia. These were further accelerated in the pandemic.
Nationwide Electronic Medical and Health Records towards integrated and coordinated care have been initiated and will commence in 2021.
Digital health innovations and clinical trial/R&D.
New digital initiatives in COVID-19 to assist in data collection and reporting, analytics and modelling, contact tracing, telemedicine, and risk communications.

In the 11th Malaysia Plan (2015–20), we have focused on strengthening health system governance and organisational capacity; ICT readiness and integration; national standards and framework; and health ICT skills and expertise. The Information System Strategic Plan 2016–20 [12] is depicted in Fig. 6. The 12th Malaysia Plan (2021-2025) is expected to be launched in September 2021 with a focus on digital trajectory and innovation. The MOH Digitalisation Strategic Plan (2021–2025) will be launched in 4th quarter of 2021.

However, 4IR with telecommunication advances and the dawn of 5G technology are presenting multiple strategies and challenges ahead which include policy and regulation, clinical evidence, safety and quality, medical ethics, and clinical practice. Efforts are underway now to review and develop regulatory strategies and policies to foster digital health technologies, cloud infrastructure support, advocate translational research, spur innovations and boost the digital economy while at the same time, building an enabling ecosystem in Malaysia. The national digital infrastructure plan early this year outlines the needs of digital connectivity and to prepare the nation for a gradual transition to 5G technology. At the same time, the National

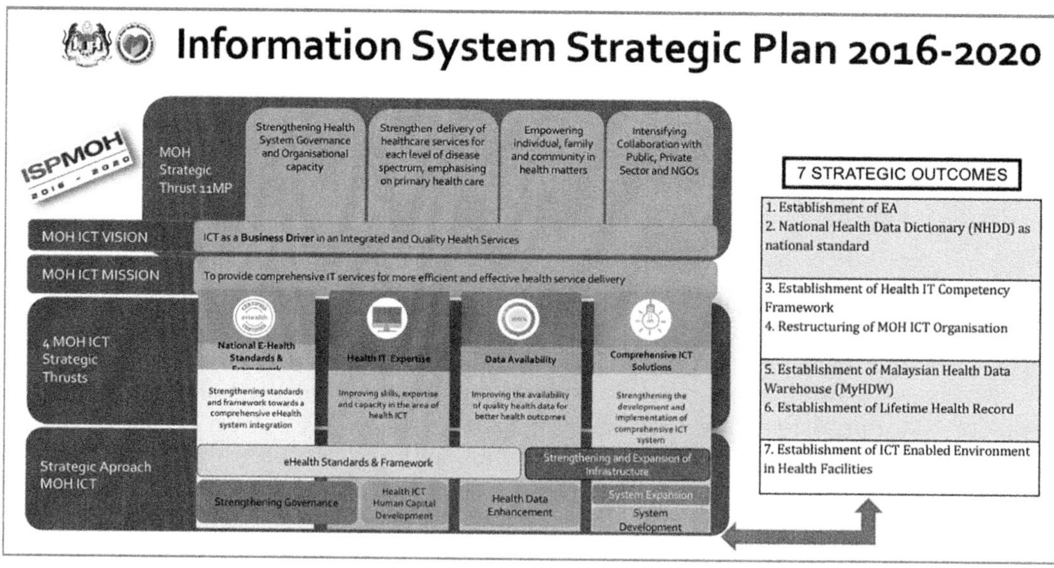

FIG. 6 MOH Malaysia information systems strategic plan 2016–20.

Technology and Innovation Sandbox launched recently will spur and catalyse innovation growth including medical and health technology.

Malaysia also recognises the success of digital health and requires partnerships and strategic collaborations at regional and global levels in building digital capabilities and competencies, policy/guideline development and research/innovation collaborations.

Moving forward, in the post-2020 agenda, the MOH is taking the lead in digital health leadership, digital health trajectory and innovations. The new MOH Digitalisation Strategic Plan (2021–2025) is in the final stage. Some of the focus areas of future healthcare include:

- Quality and safety of care
- Illness to wellness
- Healthcare back to communities, families, and individuals
- Inclusive innovations and smart solutions
- Collaborative partnerships
- Public-private integration

Philippines

The Philippines Department of Health (DOH) was the earliest adopter of IT governance at the national level. In 2013, the DOH Secretary with the Secretary of Science and Technology (precursor of the Department of Information and Communications Technology or DICT) formed the National eHealth Governance Steering Committee (NeHSG) and Technical Working Group (NeHTWG) to lead eHealth development in the country [13]. Co-chaired by the secretaries of DOH and DICT, the NeHSG approved the adoption of COBIT 5 as a governance framework and set strategic directions for the use of ICT in health. Taking these directions, the NeHTWG then defined programmes and projects in support of the national eHealth strategy. The NeHTWG formed the National eHealth Programme Management Office (PMO) to monitor the different eHealth-related projects and synchronised their activities to ensure benefits delivery (e.g. the Wi-fi investment of DICT will create greater value if DOH promotes telemedicine and electronic medical records in all health facilities) (Fig. 7).

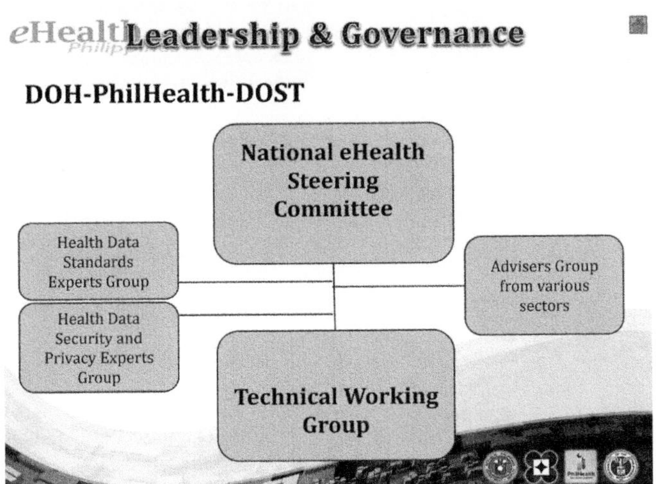

FIG. 7 Philippines National eHealth Governance Steering Committee and Technical Working Group.

A crucial development in the Philippine governance was the delegation of authority from the Steering Committee to the TWG to proceed with agreements if consensus had been reached. This delegated authority increased confidence among the members of the TWG and facilitated many decisions around the eHealth-related projects. Fig. 8 shows the escalation algorithm endorsed by the TWG and approved by the Steering Committee. The algorithm retains full accountability to the Steering Committee but empowers the TWG to proceed with decisions if there is unanimous agreement by all TWG members. For contentious matters, these should still be elevated for resolution by the Steering Committee.

Thailand

In 2010, the Ministry of Public Health (MoPH) of Thailand, with the support from WHO SEARO (South-East Asian Regional Organization), conducted research on the eHealth situation of Thailand [14]. The study found that Thailand lagged behind in laying down national eHealth foundations. The country's eHealth experts recommended that the country should put the development of eHealth foundations the priority. The recommendations are:

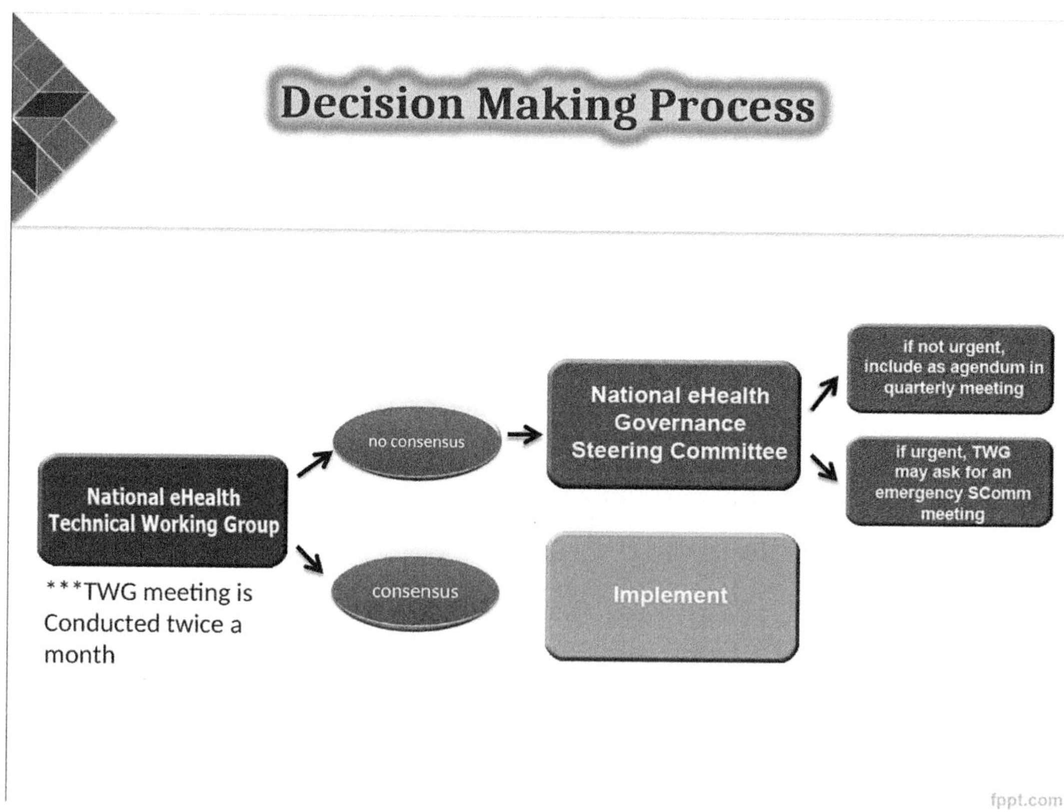

FIG. 8 Escalation algorithm for decision-making in the Philippines. This algorithm facilitated projects if consensus has been reached by all parties.

(1) Thailand should create a multistakeholder, national-level, eHealth governing authority to provide leadership and direction,

(2) Legislations related to health information security, privacy, and confidentiality should be acted to protect people

(3) National health information standards need to be developed to enable eHealth services interoperability and

(4) Systematic mechanism for capacity building of people who design, implement, operate, and use of the eHealth systems has to be planned and implemented.

Since that time, there have been several health IT-related stakeholders including the MOPH, the Ministry of Digital Economy and Society-MDES (formally ministry of information and communication technology), national health insurance agencies, and many health systems reform committees of the parliaments that have proposed and attempted to establish the national eHealth/digital health governing mechanism and/or agency. However, the attempts to create the multistakeholder, national-level, eHealth/digital health governing authority to provide leadership and direction have not been realised.

As to legislation related to health information security, privacy, and confidentiality, Thai government enacted the personal data protection act (PDPA) and the cybersecurity act in 2019. Both laws are general laws that cover all digital data from all sectors including the health sector.

The consensus-based eHealth/digital health national blueprint/architecture has also been an ongoing endeavour. In 2019, the MOPH's office of permanent secretary proposed the National Digital Health Platform (NDHP Fig. 9) to health systems-related stakeholders. The NDHP aims to integrate and standardised health data from all sources in health

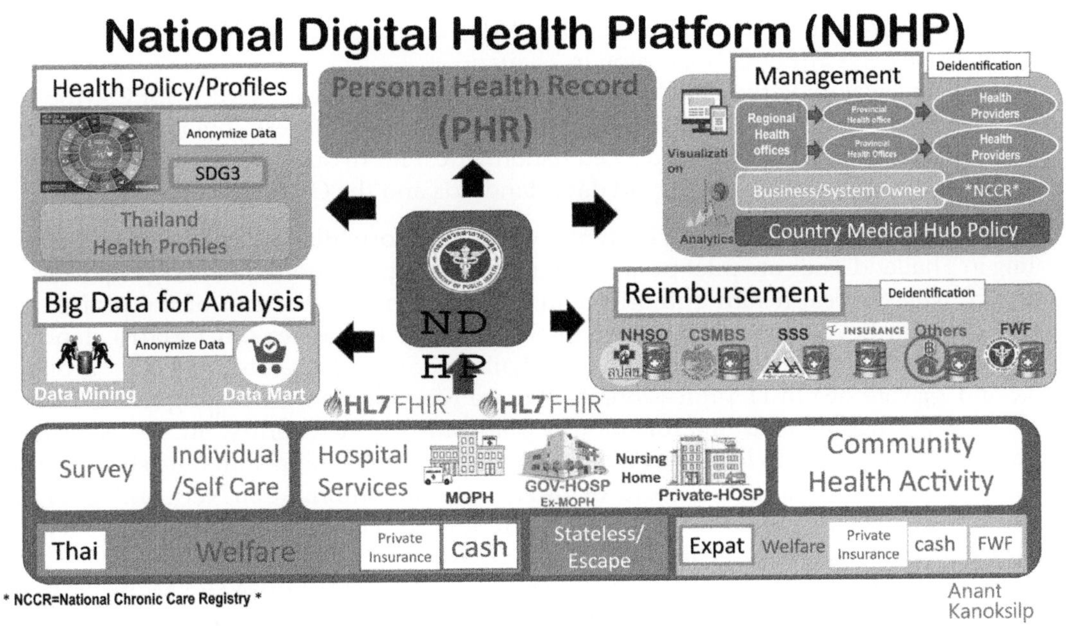

FIG. 9 Thailand National Digital Health Platform proposes by the Ministry of Public Health.

systems that will benefit every stakeholder including patients, providers, administrators, and public health. The NDHP follows the OpenHIE architecture specification [https://ohie.org/architecture-specification/] in its national digital health platform.

The Ministry of Public Health in Thailand has developed, adopted, and implemented several health data standards [15]. Most of these standards serve administrative purposes including:

- The integrated national citizen ID system and civil registration and vital statistics (CRVS) system were implemented in 1982. The national citizen ID has been used to identify an individual when the person transacts with both public and private organisations including healthcare organisations [16].
- We had adopted ICD-10 since its inception (1990). Thailand modified and extended the WHO ICD-10 international to ICD-10 TM (Thai Modification). The main use is for public health reports and reimbursements. The ICD-10-TM and ICD-9-CM (Clinical Modification) are used for coding diagnosis and health service intervention, respectively. Thai Diagnosis Related Group (TDRG) standard has been developed since 1992 as a financing tool for prospective payment systems. It progressively evolved and has been widely used for acute care in-patient reimbursement since 2002.
- The Ministry of Public Health also developed and maintained health facility registry and identification code (health facility IDs) [16].
- In 2013, Thai Health Information Standards development center (THIS) under MOPH's Health Systems Research Institute (HSRI) developed the national medicine terminology and coding system classed, Thai Medicines Terminology (TMT) in response to the request from the Comptroller General's Department (CGD), Ministry of Finance (MOF) to be used in the civil servant medical beneficiary system's (CSMBS) electronic claim information systems and later be implemented in MOF's electronic Governmental Procurement (eGP) information system. The TMT, a concept-based terminology model, is the pharmacy SNOMED-CT extension in Thailand.
- The THIS is developing the Thai Medical Laboratory Terminology (TMLT) for CSMBS claim information system. TMLT is Thailand's modified LOINC terminology system. Both TMT and TMLT now have been adopted nationwide. Fig. 10 depicts Thailand's approach to the development of national health data standards and the OpenHIE architecture.

In the past two years, there have been many Health Information Exchange (HIE) projects piloting in Thailand They are pilot projects from several organizations including departments in MOPH, MDES, local health authority and telecom companies. Most of the projects are adopting Health Level 7 Fast Healthcare Interoperability Resources (HL7 FHIR) standard. In 2019, Standards and Interoperability Lab Thailand (SIL TH) was formed by a group of public and private health IT professionals who are working and interesting in developing standard-based digital health interoperability platforms for the country. SIL TH is a digital health sandbox that aims to learn and develop interoperability and health data standards for Thailand's health systems. SIL TH is one of the Asia eHealth Information Network's (AeHIN) Community of Interoperability Labs (COIL). Currently, SIL TH is learning and implementing HL7 FHIR in Thailand's health care context.

The four eHealth/digital Health/HIS challenges, identified by 2010 WHO-MOPH eHealth report, Thailand must overcome to establish an integrated effective national eHealth and health information systems is still illusive. Although, in the past decade, many digital health/HIS programs and projects have been initiated and implemented to fill the four challenge gaps, Thailand health information systems are still fragmented and not be able to effectively

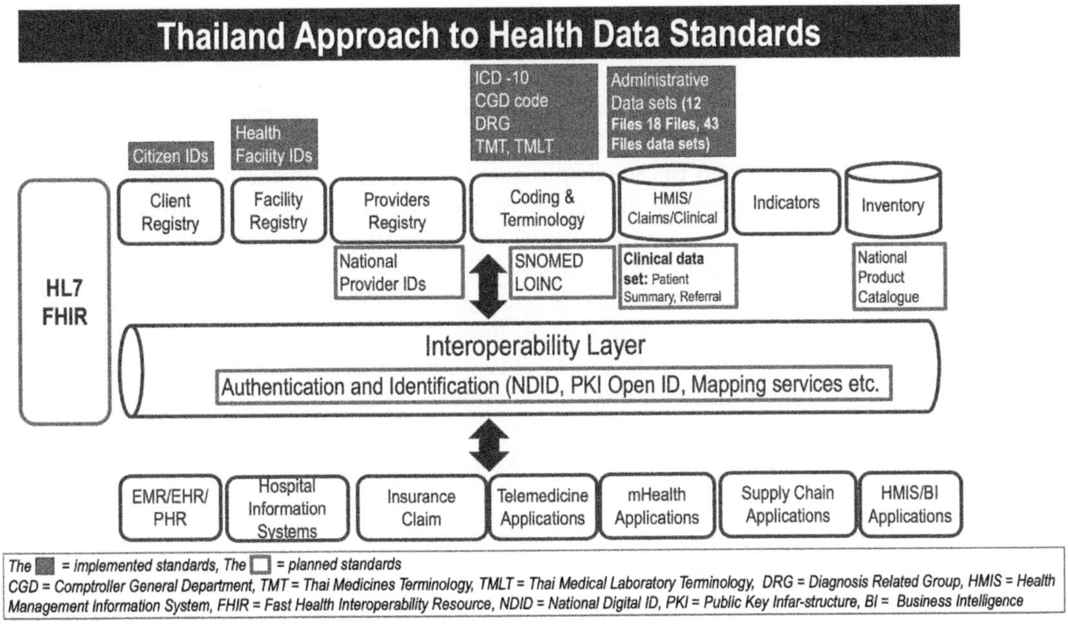

FIG. 10 Thailand Approach to Health Data Standards

support the country health care, public health, and health systems. The urgent need of the Thai health systems to create a collaborative digital health governance mechanism to lead the development of the country digital health and integrated health information systems is eminent. Development of national health information architecture and data standards are essential to integrate and interoperate the current fragmented health information systems. The third important area concerns PDPA and cyber security law on health data management and sharing. Compliance with the PDPA legislative requirements will require all health stakeholders to set-up and update systems for data security which will require a lot of technical supports and financial resources. Additionally, availability of data to conduct research, discover knowledge are important for informing healthcare service, public health, health policy and management. Thailand has committed to an open data policy. However, integrating, accessing, securing, and sharing health data while protecting people privacy remains an area that need to develop and support further in the country.

Conclusion

In summary, despite strong aspirations to use ICT in health, all countries experienced significant challenges with complexity, compliance, and costs. Their national health information system was fragmented and complex because it had many stakeholders with competing interests and varying digital capabilities. Privacy awareness was increasing among citizens resulting in stricter compliance regulations on cybersecurity. Poorly coordinated ICT investments resulted in failed or unsatisfactory projects that are unable to show significant benefits for the expenses incurred.

An important insight is that eHealth development is complex and no single sector can do it alone. It needs stakeholders from different sectors (health, ICT, finance, education, social protection, etc.) to collaborate and coordinate under the leadership of the Department of Health. Governments can better navigate this complexity through the adoption of IT governance frameworks. Without a guiding framework, it is easy to get lost in the complexity. With IT governance, stakeholders can refer to a shared reference as they progress with their respective programmes and projects.

Good governance, which can be as simple as maintaining regular meetings and jointly monitoring eHealth-related projects, is key to progress. Establishing a shared programme management office can help in communications between all stakeholders. Consensus building may take longer, but the results are more sustainable.

Taking the Mind the GAPS, Fill the GAPS approach, countries can systematically address their challenges with complexity, cost, and compliance. Governance provides a multisectoral coordination mechanism that ensures accountability for key decisions made in a timely manner. A shared enterprise architecture clearly shows stakeholders what are the important components that need to be built and how they fit into a comprehensive interoperable health information system. Programme management ensures all stakeholders are collaborating and coordinating their eHealth activities. Standards, when adopted by all stakeholders, enable interoperability.

References

[1] WHO, Global Health Observatory (GHO), 2019, Data [cited 2019 22 March]. Available from: https://www.who.int/gho/en/.

[2] WHO, Global Observatory for e-Health, World Health Organization, 2016. Available from: https://www.who.int/observatories/global-observatory-for-ehealth.

[3] S.K. White, What is Enterprise Architecture? A Framework for Transformation, CIO, Australia, 2018. Available from https://www.cio.com/article/3313657/what-is-enterprise-architecture-a-framework-for-transformation.html.

[4] WHO/ITU, National eHealth Strategy Toolkit, 2012, [cited 2019 17 April]. Available from: https://www.itu.int/pub/D-STR-E_HEALTH.05-2012.

[5] ADB, Digital Health Convergence Meeting Tool Kit, Asian Development Bank, 2018. Available from: http://dx.doi.org/10.22617/TIM189609-2.

[6] MOHFW, Final Report on National Digital Health Blueprint (NDHB), Ministry of Health and Family Welfare, India, 2019. Available from: https://main.mohfw.gov.in/sites/default/files/Final%20NDHB%20report_0.pdf.

[7] OpenGroup, Making Standards Work: The TOGAF Standard—Version 9.2, The Open Group, 2019. [cited 2019 6 February]. Available from: https://www.opengroup.org/.

[8] MOHFW, National Digital Health Mission (NDHM) Strategy Overview, Ministry of Health and Family Welfare, India, 2020. Available from: https://ndhm.gov.in//assets/uploads/NDHM_Strategy_Overview.pdf.

[9] NDHM, Sandbox Enabling Framework v1.0, National Digital Health Mission, National Health Authority, India, 2020. Available from: https://ndhm.gov.in//assets/uploads/Sandbox_Guidelines_v6.pdf.

[10] MOHFW, Electronic Health Record (EHR) Standards for India Standards Set Recommendations v2.0, Ministry of Health and Family Welfare, India, 2016. https://www.nhp.gov.in/NHPfiles/EHR-Standards-2016-MoHFW.pdf.

[11] MOHFW, Telemedicine Practice Guidelines, Ministry of Health and Family Welfare, India, 2020. Available from: https://www.mohfw.gov.in/pdf/Telemedicine.pdf.

[12] MoH-Malaysia, Information Systems Strategic Plan (ISPMOH) 2016–2020, Ministry of Health, 2016. Available from: https://www.moh.gov.my/moh/resources/Penerbitan/Pelan%20Strategik%20/2016-2020/buku_PSTM_KKM_2016_2020.pdf.

[13] GOVPH, National eHealth Governance, Phillipines Government, Phillipines, 2020. Available from: http://ehealth.doh.gov.ph/index.php/transparency/national-ehealth-governance.

[14] B. Kijsanayotin, N. Kasitipradith, S. Pannarunothai, eHealth in Thailand: the current status, Stud. Health Technol. Inform. 160 (2010) 376–380.

[15] B. Kijsanayotin, eHealth in Thailand: Interoperability and Health Information Standards, 2016, Available from: https://www.this.or.th/files/interopbook.pdf.

[16] S. Mills, J.K. Lee, B.M. Rassekh, M. Zorko Kodelja, G. Bae, M. Kang, S. Pannarunothai, B. Kijsanayotin, Unique health identifiers for universal health coverage, J. Health Popul. Nutr. 38 (2019) 22.

CHAPTER

18

The road to interoperability: openEHR modelling and implementation

Lu Xudong[a], *Nan Shan*[b], *Cai Hailing*[a], *Cheng Yexuan*[a], *Li Mengyang*[a], *Wang Li*[c], *and Min Lingtong*[d]

[a]College of Biomedical Engineering and Instrument Science, Zhejiang University, Hangzhou, Zhejiang, China [b]College of Biomedical Engineering, Hainan University, Haikou, Hainan, China [c]College of Computer Science and Software Engineering, Canghai Campus of Shenzhen University, Shenzhen, Guangdong Province, China [d]School of Electronics and Information, Northwestern Polytechnical University, Xian, Shanxi, China

Introduction

Can you imagine one electronic healthcare record (EHR) system that serves all humankind? One of the barriers obstructing this fantasy from coming true is 'interoperability', which is a forever-going issue in the HIT domain. The Institute of Electrical and Electronics Engineers defines the term as 'the ability of two or more systems or components to exchange information and to use the information that has been exchanged'. [1] That is to say, interoperability in the HIT domain is the ability for two or more information systems to exchange, interpret, and utilise data meaningfully.

As with all other recurrent themes in biomedical informatics, the exchange of information in healthcare is essential, especially to the quality of care that everyone receives and the advancement in medicine that everyone benefits from. Complete interoperability is fundamental and critical to the management of patient lifecycle medical information and also contributes to the reduction of medical errors and healthcare cost. Despite all benefits of seamless, effective interoperability in healthcare, it is still an undergoing endeavor due to the heterogeneity of data in terms of syntax and semantics.

Various solutions including use of standards, web services, healthcare service bus, interface engines, and ontologies have been proposed with a certain degree of success [2]. Among

Roadmap to Successful Digital Health Ecosystems
https://doi.org/10.1016/B978-0-12-823413-6.00027-6

them, use of standards to exchange data among systems seems to be the most effective one. However, the development of consensus-based standards is a much too complex and time-consuming process including developing, coordinating, promulgating, revising, amending, reissuing, and interpreting [3], and it will be even more complex considering the ever-changing knowledge in medicine and the implementation of standards for legacy systems.

openEHR, a technology consisting of open specifications, clinical models, and software, can be used to create standards and build information and interoperability solutions for healthcare [4] and has the potential to achieve a complete solution to the above problems. The approach descended from the European Union's GEHR-Project (The Good electronic Health Record Project) in the early 1990s. It proposed an effective, collaborative, and fast-responsive mechanism of building consensus on content of clinical information to realise continuous, consistent management of clinical data. Instead of having time-consuming processes on standardising limited entries based on the minimum requirements of clinical data exchange, the definition of openEHR clinical information models (CIMs), aka archetypes, is carried out as an open, collaborative process among multidisciplinary teams of clinicians and information architects [5], leading to a maximum data set that integrates consensus while reserves differences for each corresponding concept. Besides, with the multilevel modelling framework, openEHR also has the advantage of building systems with the excellent scalability to ever-changing medical knowledge.

The chapter begins with the introduction of the multilevel modelling paradigm of openEHR, followed by the approach of openEHR modelling and CDR implementation which led to the interoperability. Finally, a case study of CDS application based on the openEHR-based CDR is introduced to explain how the approach achieve the interoperability in actual situation.

Multilevel modelling paradigm of openEHR

Formal modelling of clinical content that can be made available internationally is one of the most promising pathways to semantic interoperability of health information [6], and openEHR does not only satisfy the requirement but rather succeed in a clever way by separating out each domain concept. One of the key paradigms which openEHR is based on is known as multilevel modelling, originally described in Beale [7] as 'two-level modelling'.

As a contrast, the biggest shortage of single-level modelling framework is that the domain concepts are hardcoded into the relational schema and tend to be ambiguous without proper engagement of clinicians and lack of international or regional expert consensus, making it difficult to realise semantic interoperability in wider area. Furthermore, the system based on a single-level model can barely handle with the changing data requirements without re-engineering; therefore, the system can only live a limited life span.

Under the multilevel approach, there are three levels of models required for a system:

Reference model (RM): A stable reference information model constitutes the first level of modelling;

- re-usable content element definitions: formal definitions of clinical content data points and groups, in the form of archetypes;

- context-specific data set definitions: formal definitions of use-case specific data sets used for forms, documents, messages, etc., created by combining required elements of relevant archetypes into openEHR templates [8].

The RM is a stable and formal information model implemented in software; it defines the basic structures and attributes including data types, data structures, and other components. The RM does not define any domain knowledge and thus separates data representation from concept expression and ensures synaptic interoperability. Archetypes are the foundation components to construct clinically meaningful entities and act as a gateway to semantic interoperability and data sharing between different systems and types of healthcare professionals. Templates aggregate and constrain the archetypes to create context-specific clinical content such as a screen form, document, report, or message. It is remarkable that text in archetypes can be constrained by a terminology like SNOMED CT, ICD, etc., to support specific clinical use cases (https://specifications.openehr.org/releases/BASE/latest/architecture_overview.html#_consequences_for_software_engineering).

While archetypes are generally broad content models and have very open compositional possibilities, templates are used to narrow the choices of archetypes for local or specific purposes [9]. By separating data representation from domain content, openEHR not only enables the mechanism to expedite proceedings of building extensively recognised CIM but also empowers the clinicians instead of IT personnel by featuring them as the primary drivers of clinical content models [10]. It also solves the common shortcoming of most single-modelling approaches where knowledge concepts were shoved into database models and interoperability was out of the question when this procedure was implemented by different irrelevant healthcare software providers. As for openEHR, the reference model ensures syntactic interoperability independent of any defined archetypes, and semantic interoperability is achieved by the definition and use of archetypes [11]. Fig. 1 shows the openEHR software engineering.

Archetypes are a way of adding domain semantics to information models, while avoiding endless growth and maintenance of the latter, as would occur if a textbook approach were used to modelling [12]. The types of archetypes are listed as follows:

- Demographic: defines generic concepts of demographic information; includes PARTY, ROLE, and relevant detailed classes.

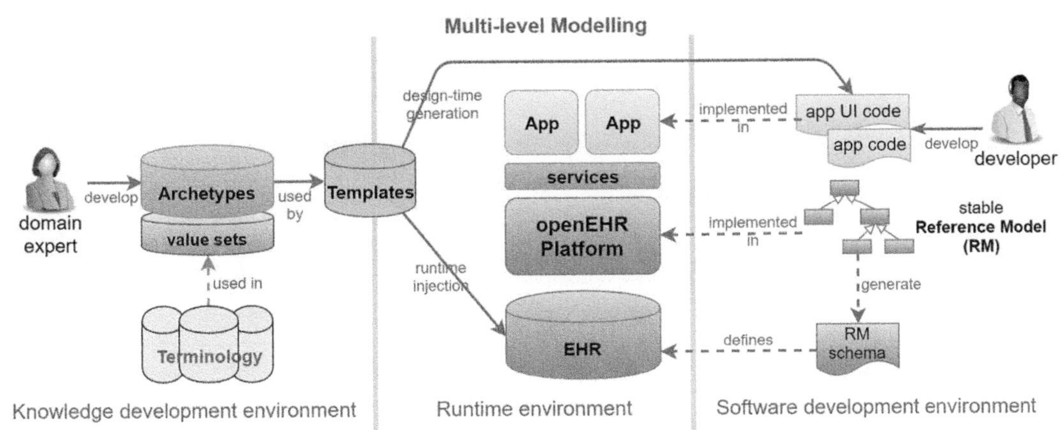

FIG. 1 The openEHR software engineering.

- Composition: the top-level structure and 'data container' containing section archetypes and entry archetypes, and it is considered equivalent as a clinical document.
- Section: a navigational structure that facilitates human access, which is similar to the table of contents of a document. A section archetype can contain section archetypes and entry archetypes.
- Entry: defines the generic structures for representing clinical statements, which has five descendants as follows: Observation—represents the observations that occurred to the patient in the past, including clinical observations, examinations, lab tests, and situations of the patient; Instruction—represents the interventions to be performed in the future, e.g., medication orders; Action—represents what has been executed, e.g., insertion of an intravenous cannula; Evaluation: represents opinions and assessments on the patient, such as diagnosis, risk assessment, goals, and recommendations; and Admin_Entry—used to capture administrative information, such as admission, appointments, discharge, billing, and insurance information.
- Cluster: represents reusable clinical content that can be embedded into entry archetypes or other cluster archetypes.
- Element: represents a single item to be reused in entry archetypes or cluster archetypes. An openEHR template assembles and constrains archetypes for context-specific purpose, which is closest to users and typically used to generate application programming interfaces (APIs), XML schema definitions (XSDs), user interface forms, storage schemes, etc. Design principles of openEHR are described in far more detail by the open specifications available from the openEHR homepage.

Then, an openEHR template assembles and constrains archetypes for context-specific purpose (as shown in Fig. 2), which is closest to users and typically used to generate application programming interfaces (APIs), XML schema definitions (XSDs), user interface forms, storage schemes, etc.

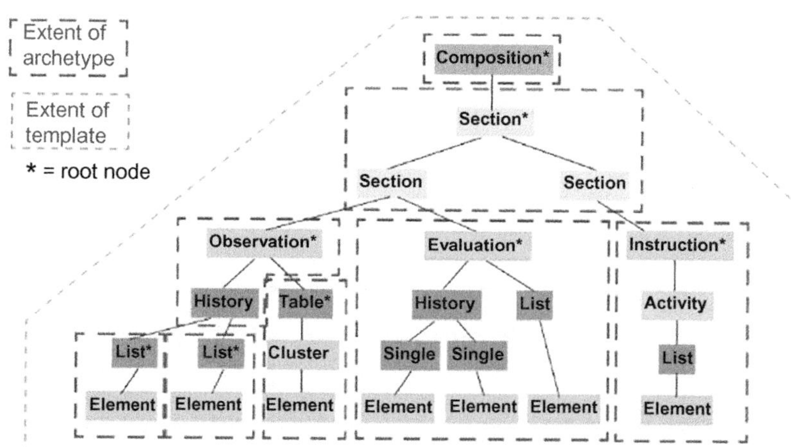

FIG. 2 Archetypes and templates.

The above-mentioned collaborative definition of archetypes is achieved in web platforms where experts review and publish the archetypes that will be used to define the EHR. That is where Clinical Knowledge Manager (CKM) and Healthcare Modelling Collaboration (HMC) comes in as formal governance of clinical models in order to actually achieve semantic interoperability.

openEHR modelling

As so important archetypes are to the interoperability, developing them in a rigorous and consistent way is of great significance. Through the practice-proven multistep approaches of archetype modelling described later in this section, it shows the possibility to keep the modelling process conforming with the openEHR specification and design intention, and so the possibility to achieve interoperability. Later on, with the developed openEHR archetypes and templates available in larger scale comes the necessity of model governance to justify them through audition and also a regional/global platform to help experts collaborate more efficiently to develop/update the clinical model.

Collaborative platform and clinical model governance

The core of the openEHR approach is multilevel modelling, whose specific implementation complies with the model-driven assumption to make the information system or application developed based on this more scalable, maintainable, adaptable to domain changes, and semantically interoperable. Building archetypes requires the knowledge of archetype, domain, and information at the same time. It is a process of expressing domain knowledge in a requirement-driven, multiparty collaborative and iterative manner. As a result, adequate organisation and utilisation of multiple resources are necessary and critical to the quality of archetypes. Consequently, an information modelling ecology where service providers, hospitals, clinicians, and domain experts can collaborate is really important.

As openEHR archetypes and templates have been developed, it has become apparent these clinical models require formal publishing and governance support to achieve semantic interoperability. The openEHR Foundation released an online archetype repository for community use in April 2009—known as the Clinical Knowledge Manager (CKM). By the same token, researchers and developers in China have designed and developed Healthcare Modelling Collaboration (HMC) since 2017. This platform is committed to supporting online authoring and collaborative management of openEHR archetypes and templates, facilitating multiparty cooperation and timely participation in healthcare information modelling.

HMC has utilised quite a few functions including user management (login, registration, user management, role management, permission management), archetype management (query, retrieval, creating, editing, upload, review process management, comparison and export), template management (query, retrieval, creating, editing, upload and export), statistics, and so on and so forth. The platform contains two kinds of repositories, namely public repository and subrepository. The public repository is a repository for models that can be either standardised or collected from subrepository. It was designed to be a stable, reliable, and comprehensive archetype repository to provide references and suggestions for various organisations and manufacturers. The subrepository provides each repository of localised or incubated clinical models for working groups and domain experts from different clinical fields. It also makes model development and management independent across clinical domains. However, the public repository administrators have the right to view any information in the subrepository and synchronise to the public repository.

As of September 2020, the HMC platform (as shown in Fig. 3) has attracted 135 online users to register, established 9 subrepositories vary from stomatology to HER, and contained 909 archetypes, among which 435 are translated from CKM for better promotion and use of archetypes in China. These translated archetypes are ready to be reviewed by domain experts.

In summary, the openEHR approach featured by multilevel modelling and expert collaboration can facilitate data sharing in healthcare via interoperability at both data and knowledge levels. CKM enables comprehensive semantic interoperability by reviewing, publishing, and maintaining archetype internationally, and HMC also helps to govern the openEHR models localised by Chinese professions. The feasibility of this approach was verified by the Chinese experience in openEHR modelling and experts collaborating as elaborated above. It is reasonably hopeful that the vision of a life-long interoperable EHRs is not fantastic nonsense.

Multistep approaches of developing archetypes

Developing archetypes is a rigorous and iterative process on account of its significant part in data sharing and semantic interoperability; informal modelling will lose the edge of openEHR and hinder the interoperability. The openEHR data modelling approach usually includes 6 steps: collecting data requirements, normalising data elements, organising domain concepts, searching corresponding archetypes, editing archetypes, and reviewing archetypes. Through a case study of modelling an EMR in China aiming to investigate the feasibility and challenges of archetyping a complete EMR data set with the openEHR approach, the mechanism of each step involved in the process will be illustrated and shown in Fig. 4.

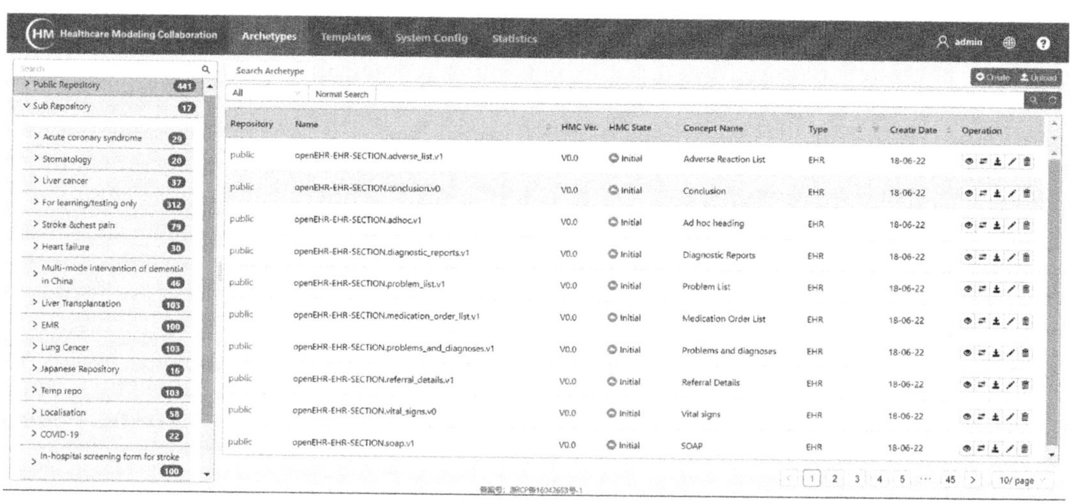

FIG. 3 Healthcare modelling collaboration.

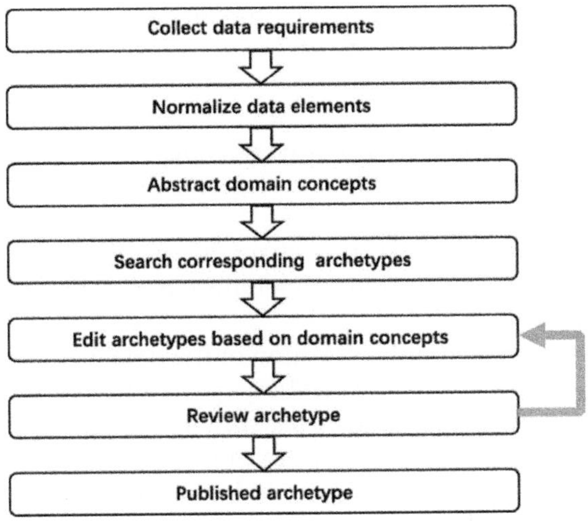

FIG. 4 The iterative process of archetype modelling.

Step 1: **Collecting data requirements**

Two representative EHR systems using relational database to store data were chosen as the sources to collect data requirements. To collect data requirements, the database schemes of the two implemented EHR systems have been interpreted to tables and fields. First, all the data fields in the relational tables were collected as the data elements required in an EHR. Then these data elements were grouped into the corresponding categories based on the tables they belong to and organised into domain concepts further.

Step 2: **Normalising data elements**

To acquire complete EHR data elements without semantic overlap, three EHR-related national standards in China were referenced. First of all, a complete set of data elements was acquired by complementing the data elements that were not included in the data requirement categories in the previous step but were defined within these standards. Then, the data elements of EHR requirements were normalised by referencing these three standards.

Step 3: **Organising domain concepts**

Based on the categories of EHR requirements and the normalised data elements described in previous steps, domain concepts were organised with three patterns: patient demographics pattern, clinical pattern, and nonclinical pattern. Each EHR requirement category corresponds to one of these three patterns.

Step 4: **Searching corresponding archetypes**

To reuse existing archetypes as much as possible, the searching step was executed to find out the corresponding archetypes for the domain concepts, which is of great significance for semantic interoperability. In addition, the searching step can also facilitate the improvement of domain concepts by referencing existing archetypes.

The corresponding existing archetypes were retrieved based on three kinds of relationships between existing archetypes and domain concepts. First, the concept and the existing archetype have the same semantics. Second, the semantics of the concept is one of the particular subsets of the existing archetype semantics, e.g., the concept 'operation request' and the existing archetype 'openEHR-EHR-INSTRUCTI-ON.request.v0'. Third, the semantics of the concept is more general than the existing archetype's, in this case, the domain concepts have to be refined for better retrieval.

Step 5: **Editing archetypes based on domain concepts**

If the existing archetype covered all the data elements and nothing needed to be changed, then the existing archetype was reused directly.

If the existing archetype covered all data elements but the metadata required to be refined, then a revision operation was executed, including translation, extending value sets and description.

If the existing archetype covered only a portion of the data elements, three potential modification choices were provided. The specialisation operation was executed when the clinical concept could be expressed by specialising the existing archetypes to make the semantics narrower, which required changing the identification information of the archetype. The extension operation was executed when some compatible modifications to the existing archetype were needed for expressing the clinical concept. A new version was created when some incompatible modifications to the existing archetype were needed, which changed the version information.

If no archetype could be found for certain concept, then a new archetype was designed. To be specific, a suitable archetype type and a proper archetype name should be chosen first. Then, the metadata of the archetype was edited, including concept description, keywords, purpose, use, and misuse. At last, the data element and relevant terminologies were edited.

Step 6: **Reviewing archetypes**

Reviewing archetypes is a pragmatic way to acquire consensuses and high-quality archetypes within the target domain, which is always executed by domain experts. In this method, two kinds of archetype statuses were designed, i.e., 'initial' and 'published'. The archetype with initial status is an initial or intermediate artefact, but the published archetype is the final product that can be implemented within EHRs. An archetype's status can only change to published when it passes the review step. Otherwise, the archetype will go back to the previous step, and the iterative process will be executed until domain experts successfully authorise the archetype.

Wrap up

The CKM and archetype modelling method enable the feasibility of the openEHR approach. On the one hand, CKM has accumulated enough archetypes to cover most of the EHR requirements, which support archetype sharing and reuse. On the other hand, the archetype modelling method enables the domain experts to design archetypes to represent their EHR requirements, especially, the new requirements that are not covered by existing archetypes. For meeting the EHR requirements in China, some existing archetypes were reused through different modification operations, and some new archetypes were designed from scratch. Based on the localised EHR requirements, the modifications consisted of language translation, value domain adjustment, data items supplementation, data items specialisation, and so on.

The openEHR-based CDR implementation

The openEHR approach can represent the information and knowledge semantics by multilevel modelling, and later by reusing existing archetypes with consistent semantics, it can improve the interoperability between different clinical information systems. The clinical data repository (CDR) is a core system that provides unified management and open access of clinical data to support clinical practice with information technology in healthcare environments [13]. These clinical data have been scattered in different clinical information systems, e.g., electronic medical record system, computerised physician order entry system, laboratory information system, etc. In this situation, the interoperability between these systems is inevitable but troublesome.

The characteristics of clinical data are too complicated, flexible, and changeable to capture a universal, comprehensive, and stable schema of information. Most of the CDR are designed and developed using a 'single level' approach, in which information model and domain knowledge are modelled directly in software and database models. Such CDR systems are usually large scale and difficult to maintain. With the medical knowledge and requirements changing, the model of the CDR has to be modified and re-designed. Frequent change will eventually cost a lot of manpower and resources. Semantic interoperability is a challenge when integrating the data from multiple information systems to CDR. The openEHR approach can represent the information and knowledge semantics by multilevel modelling, and it advocates the use of collaborative modelling to facilitate reusing existing archetypes with consistent semantics so as to be a potential solution to improve the semantic interoperability [14]. In order to protect CDR from changes in the healthcare domain, openEHR has published a series of specifications to guide the development of future-proof information systems. Solutions span from information modelling to system architecture, to meet the continually evolving needs [15].

This part presents an archetype relational mapping (ARM) persistence solution based on a relational database by providing a sophisticated mapping approach based on templates and archetypes in order to map archetypes to relational tables. First, the ARM approach is introduced in detail, including archetype modelling, template definition, and mapping rules. Then, a comparison of the generated ARM database, the conventional database deployed in the hospital, and the openEHR official Node + Path database is conducted. After analysis, issues and experiences encountered in the process will be discussed.

The ARM approach employs archetypes and templates to generate a persistence model and provide data access based on a relational database. The ARM is intended to fulfill several functionalities:

(1) Effective adaptation to changes in archetypes.
Archetypes reflect the changing realities of EHR, and existing archetypes are updated over time. Archetype changes result in several challenges to ARM, such as the application of new archetypes to the relational database and the possibility of incompatible versions among archetypes, among other potential challenges.

(2) Generation of customised persistence models for various requirements. ARM employs templates to customise the data persistence model. There are three steps in the implementation of the persistence layer: archetype modelling, template definition, and database mapping.

Archetype modelling

Archetype modelling selects and expands existing archetypes from the public archetype repository or defines new ones in order to meet all data requirements as discussed in the above parts.

Template definition

Archetypes define widely reusable components of information, and templates encapsulate the local usage of archetypes and relevant preferences. One corresponding template is designed for each archetype in order to add constraints, such as local optionality, archetype chaining, tightened constraints, default values, and ARM constrains. ARM constrains attempt to achieve better performance by aligning the concept-focussed archetype model and the data-focussed relational model. The ARM constraints are designed as follows:

(1) Assign identification data item. An identification data item is used to uniquely identify instances of each archetype. Only one identification data item can exist for each archetype.

(2) Assign query data item. Some data items may always be used as query conditions, particularly those with identical characteristics to the identification data item. These types of data items can be categorised as query data items, and an archetype may have multiple query data items to facilitate the data query. Collection data structures such as CLUSTER, ITEM_TREE, ITEM_LIST, and archetype slots cannot be used as identification data items or query data items. As basic units of internal structures of archetypes, collection data structures group-related data items and can thus be viewed as embedded archetypes with their own identification data items and query data items.

(3) Define mappings between generalised archetypes and specialised archetypes to facilitate data query. If a domain contains many concepts and data items with similar structures, a general concept with a data item can be used to store the name and all fields related to these concepts. Archetypes can clearly represent the general concept and specific concepts by specialisation. The name and fields of data items in specialised archetypes are mapped to a subset of fields of the corresponding data item in the generalised archetype. Then, the data stored as a generalised archetype instance can be queried with specialised archetypes, and vice versa, using the mappings.

Database mapping

Database mapping generates a relational database schema in order to automatically persist the data represented by archetype instances into relational databases using archetypes, templates, and ARM constraints. A set of mapping rules is designed as follows:

(1) Map each archetype to a table. According to the archetype semantic relationship specified in openEHR specifications [16], including revision, specification, and new version. The new and old version of the same archetype can be organised into a single data table.

(2) Map the basic data items represented by the archetype basic data type. If the upper bound of the data item occurrence is 1, this indicates a single occurrence data item which can be mapped as a common column. If the upper bound of the data item occurrence is *, this indicates a multiple occurrence data item which is mapped into a standalone table with two columns: one is a foreign key column referring to the identification data item of the current archetype, and the other is a common column mapped from the data item.

(3) Constrain the identification data item as the key column. Unique and clustered index constraints are mapped from the identification data item and added to the column. If there is no identification data item in an archetype, a data item named 'id' is generated and used as the identification data item to identify records in the database table; the generated 'id' is invisible and cannot be accessed using the archetype.

(4) Constrain the query data item as an indexed column. The nonclustered index constraints mapped from query data items are added to columns in order to accelerate data query.

(5) Map archetype slots. If the upper bound of the archetype slot occurrence is *, this indicates that the current archetype and the target archetype exhibit a one-to-many relationship, which is mapped as a foreign key column in the target archetype table, referring to the identification data item of the current archetype. If the upper bound of the archetype slot occurrence is 1, this indicates that the current archetype and the target archetype exhibit a one-to-one relationship, so the data items of the target archetype are embedded into the table of the current archetype, thus embedding the target archetype into the current archetype.

(6) Map collection data items according to the collection data structure. If the upper bound of the collection data item occurrence is 1, this indicates a single occurrence collection data item, so the data items contained in this collection data item can be mapped into the table of the current archetype and can be viewed as flattened. If the upper bound of the collection data item occurrence is *, this indicates a multiple occurrence collection data item that is mapped into a standalone table with one foreign key column, referring to the identification data item of the current archetype.

(7) Propagate query data items. An efficient method to reduce the recursive level deep in the archetype hierarchy tree when querying the leaf archetype is to store the most frequently queried data items of the ancestors in the descendants. For multiple-occurrence archetype slots and collection data items, the query data items in the current archetype can be mapped identically to the identification data item, into the target archetype data table of the archetype slot, or the standalone data table of the collection data items as foreign key columns.

(8) Naming. Naming rules vary slightly. The general principles are as follows:

- **(a)** The archetype name is used as the table name for tables mapped from archetypes.
- **(b)** The archetype name concatenated with the data item name is used as the table name for tables mapped from collection data items and multiple occurrence data items; the version portion of the archetype name should be removed, since all versions of an archetype are mapped into the same database table.
- **(c)** The path of the data item concatenated with the field name is used as the column name for a column mapped from a data item field, but the human readability of the path is poor. One alternative is to use the textual name provided within the archetype ontology section of the data item. However, the uniqueness of the textual name needs to be guaranteed.

(9) Add database constraints. Since archetypes and templates cannot precisely control specific database field properties of generated databases. One way is to provide a profile about these constraints, including whether to be a primary key, whether to add an index, whether to be unique, and whether to allow to be empty, the field type and length that you prefer, etc.

A mapping example of openEHR-DEMOGRAPHIC-PERSON.person-patient.v1 was shown in Fig. 5.

Database comparison

To determine whether the performance can meet the requirements of clinical practice, a performance comparison is conducted between the generated ARM database, conventional database for IV (integrated viewer), and the official Node + Path database. The conventional database has been deployed in a tertiary class A Chinese for three years. Several legacy systems have been integrated into the database. These systems represent typical centralised data-reporting applications in a hospital that presents patient information from all examination departments and encompasses most EHR data, thus serving as an ideal candidate against which to apply the ARM approach and conduct the comparison.

A data set is extracted directly from the conventional database, containing 103,320 imaging tests, 8,573,157 images, 654,213 laboratory tests, and 4,846,688 laboratory test result items for

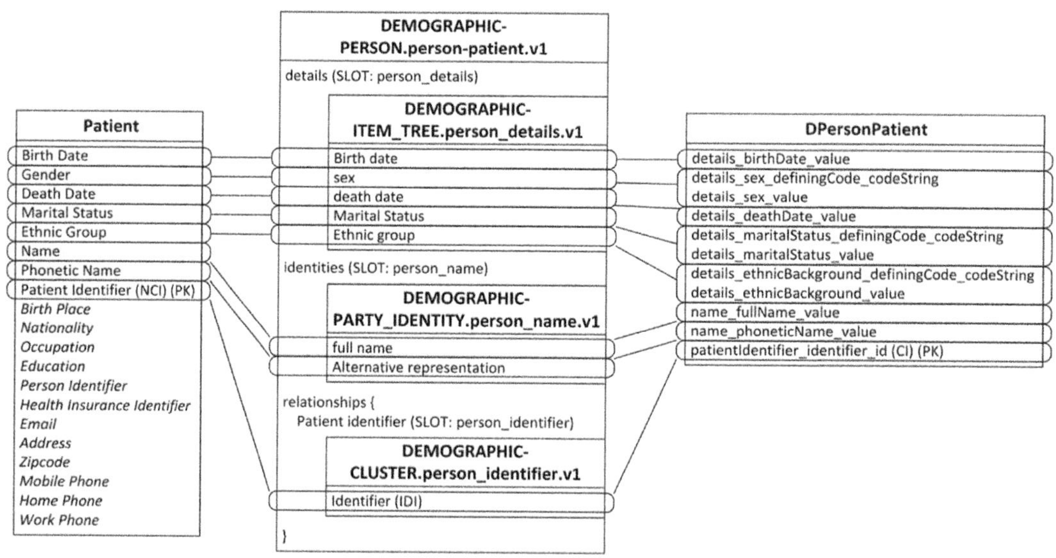

FIG. 5 Mapping of openEHR-DEMOGRAPHIC-PERSON.person-patient.v1. In all figures, PK stands for primary key. FK stands for foreign key and indicates the data item which current data item relates to. SLOT indicates target archetype which conform to current data item. Solid line indicates foreign key relationship between archetypes. Dash line indicates composition relationship through slot between archetypes. IDI stands for identification data item. QDI stands for query data item. CI stands for clustered indexed. NCI stands for nonclustered indexed. Data items in italic type are not covered by archetypes.

29,743 patients. All data has been de-identified. The data set is imported into three database systems. The Node + Path database is far greater than the space required by the other two databases. Although the Node + Path is much more efficient in storing sparse data, it also includes too many redundancies to store the path of each archetype data item.

Seven group data-retrieving tests including 20 queries are designed for the comparison, and the results are shown in Table 1.

The performances of the ARM database and IV data base were very similar in the execution of data-retrieving tests. ARM performed better in tests 1, 2, and 4, while IV performed better in tests 3 and 5. In conclusion, the ARM strikes a balance between automation and performance to generate data persistence with archetypes to facilitate the semantic interoperability.

There were several encountered challenges and issues in our implementation of the persistence layer.

TABLE 1 Query benchmark.

QUERY	IV(MS)	ARM(MS)	NODE+PATH(MS)
QUERY 1.1	80 (+ 74%)	46	5017
QUERY 1.2	91 (+ 54%)	59	5121
QUERY 1.3	196 (+ 15%)	170	5358
QUERY 2.1	221 (+ 16%)	191	24,866
QUERY 2.2	219 (+ 17%)	187	25,094
QUERY 2.3	474 (+ 129%)	207	26,158
QUERY 3.1	242	270 (+ 12%)	294,774
QUERY 3.2	224	299 (+ 33%)	297,388
QUERY 3.3	254	411 (+ 62%)	362,950
QUERY 4.1	198 (+ 13%)	176	127,547
QUERY 4.2	254 (+ 32%)	193	128,508
QUERY 4.3	1249 (+ 57%)	797	129,901
QUERY 5.1	113	186 (+ 65%)	328,181
QUERY 5.2	125	205 (+ 64%)	329,097
QUERY 5.3	139	239 (+ 72%)	388,727
QUERY 6.1	14,596 (+ 5150%)	278	5746
QUERY 6.2	16,340 (+ 5293%)	303	6029
QUERY 6.3	16,453 (+ 5140%)	314	6984
QUERY 7.1	14,582 (+ 1028%)	1293	41,217
QUERY 7.2	14,649 (+ 669%)	1904	53,352

The percentage values in IV and ARM columns are more time spent on each query in the slower database than the faster database. The Node+Path database is not included in the calculation.

ARM deployment

In ideal archetypes, the data type of data items should be as stable as the data items and remain unchanged. However, there are incompatible issues about the archetypes that are initially developed from local data requirements and later extended to a global scope. Two mapping rules are designed to automatically adapt to change. One rule maps each version of an archetype to a table; old versions of an archetype will gradually become outdated and obsolete and can then be safely moved to a backup database. The second rule maps all versions of an archetype to a new table and then imports all data from the old table into the new one; the old table is then removed.

De-normalisation

Several de-normalisations are introduced in ARM to achieve better performance. First, the de-normalisation of the granularity of archetypes is achieved by embedding archetypes by archetype slot mapping. Since the granularities of archetypes and data requirements are not always identical, archetypes composed by archetype slots with a single occurrence represent one concept and can be embedded together in data requirements. In this manner, query steps can be reduced and the joining of tables can be avoided.

Second, de-normalisation of the index redundancy of query data items propagation is achieved. Proper indices can greatly improve the data query performance of relational databases. However, it is inconvenient to add indices in archetypes due to their high normalisation. By introducing index propagation, indices can propagate through archetype slots in order to reduce the number of query steps.

Limitations of ARM

ARM approach may not meet the requirements of situations in which the databases must be highly tuned. When the hierarchy and structure of archetypes are similar to the data requirements, few reorganisations are necessary before application of the ARM approach. If the archetypes and data requirements exhibit structural differences, many reconfigurations are necessary to align the archetypes to the data requirements, possibly including extensions and modifications to the archetypes. This will require much more effort than the conventional database design approach.

Affections to archetypes and templates

In general, the ARM approach conforms to the design principles of archetypes and templates. Furthermore, the ARM places more emphasis on subtle details in order to achieve better design of archetypes and templates. ARM refines the design principles of archetypes and templates by considering their practical use and application in clinical practice.

Conclusion

In summary, ARM uses archetypes to generate relational databases and achieve similar performance as compared to conventional databases in data-retrieving queries. System components like ARM can facilitate the adoption of openEHR architecture in CDR systems.

CDS application based on openEHR CDR

The interoperability issue exists in various HIT application domains such as clinical decision support application, in which the standardisation of the input and output of computerised guidelines can be a tricky thing. As openEHR aims to facilitate the interoperability between information systems by providing archetypes that can be shared among organisations, it comes up with the Guideline Definition Language for better use of openEHR archetypes to author computer-interpretable guidelines (CIGs). This represents the capability of flexibility and extensibility of openEHR.

Using openEHR GDL for computerised guidelines

Clinical decision support systems (CDSSs) are computer-based programs that provide evidence-based prompts and reminders to help caregivers improve the quality of care [17]. In a CDSS, computerised guidelines (CGs) are developed to formalise the guideline knowledge in a machine-interpretable way [18]. Computerised guidelines are mostly represented as production rules in the form of 'IF conditions TEHN actions'. Many efforts, such as Ardern Syntax, EON, GLIF, Asbru, Gaston, and SAGE, have been proposed and validated in the past decades [19]. While these efforts have been proposed for decades, the interoperability issue still hinders the wide adoption of the CDSS. The interoperability issue in the CDSS domain is also known as the 'curly braces problem' [20]. The 'curly braces problem' refers to using nonstandard site-specific information models in CG. Nonstandard site-specific information models cannot be easily shared among organisation and therefore hampers the shareability of CGs.

It would be helpful if there is an approach that can standardise the input and output of CG and leverage existing resources. OpenEHR has such potential. From the technology point of view, openEHR aims to facilitate the interoperability between information systems. The openEHR archetype provides a standard information model that can be shared among organisations so that the 'curly braces problem' can be avoided. While from the domain knowledge perspective, it aims to bring informaticists and medical specialists together. Specifically, openEHR uses the archetype to capture the detailed and domain-specific clinical concepts that are modelled by clinical specialists.

The Guideline Definition Language (GDL) has been proposed by the openEHR community to facilitate using openEHR archetypes to author CIGs [21]. Recently, GDL has been upgraded to its second major version, known as GDL2. GDL improves the shareability of encoded CIGs to be shared among organisations cross-border [22].

This section briefly introduces the basic notion of GDL and the methodology of applying GDL for CG authoring. A case study of using GDL to capture the guideline knowledge of COVID-19 has been introduced subsequently.

The openEHR guideline definition language

GDL uses archetype models and reference models as data models to express clinical guidelines knowledge. It aims to provide a versatile guideline expression method to implement complex decision logic. GDL is based on the archetype models which is expected to achieve the bindings between the rule itself and the archetype data elements so that the constraint of the data elements can be better formulated.

A complete GDL rule mainly consists of three parts, namely the metadata part in the GDL header, the guide definition part, and the ontology definition part. The guide definition part includes a binding module with data elements in archetype models and a rule definition part used for rule editing, as shown in the Fig. 6. The four parts are described in details as follows:

- **Metadata:** The metadata part is based on the metadata description in the ADL, which includes mostly version information, language information, author information, keywords, rule usage, and using scopes.
- **Archetypes and Elements:** They are local variables that associate with the unique path of the archetypes or elements.
- **Rules:** The definition part of the rule is a part of the GUIDE_definition module. The keyword RULES guide the whole rule module, and the content part uses local variables to express the clinical concept. The keywords WHEN and THEN are used to guide the condition and action of each rule. Priority is used for setting execution order of the rules.
- **Ontology:** The concept represented by the gt code used in the rule definition is indexed with the clinical concept and can also support the bindings of external term sets like SNOMED CT and ICD 10.

```
(GUIDE) <
    gdl_version = <"...">
    id = <"...">
    concept = <"...">                                  1
    language = (LANGUAGE) <
        original_language = <[ISO_639-1::en]>
    >
    description = (RESOURCE_DESCRIPTION) <...>
    definition = (GUIDE_DEFINITION) <
        archetype_bindings = <
            ...                                        2
        >
        rules = <
            ...
        >                                              3
    >
    ontology = (GUIDE_ONTOLOGY) <
        ...                                            4
    >
>
```

FIG. 6 Basic structure of GDL.

The approach of applying GDL

While a comprehensive language and a user-friendly tool for GDL have already been published by the openEHR society, there is yet of a step-by-step approach of interpreting a narrative guideline to a GDL-based CG.

Based on the approach proposed by Zhou et al [23]., this section introduces a rapid CG develop and disseminate approach. The rapid CG development and disseminate approach has six steps (see Fig. 7), which are (1) create guideline knowledge specification, (2) extract guideline knowledge, (3) domain concept modelling, (4) clinical algorithm editing, (5) validate computerised guidelines, and (6) release and disseminate guidelines. The approach is explained in detail as follows.

(1) Create guideline knowledge specification. In this step, a joint CIG development team consisted of both medical specialists and informaticists has to be set up. Medical specialists provide clinical requirements for decision support. According to the requirements, the informaticists select related guidelines, read through each narrative guideline, segment interesting sections in the guideline, and finally confirm with medical specialists. Selected and segmented human-readable guideline files are organised as the output of this step.

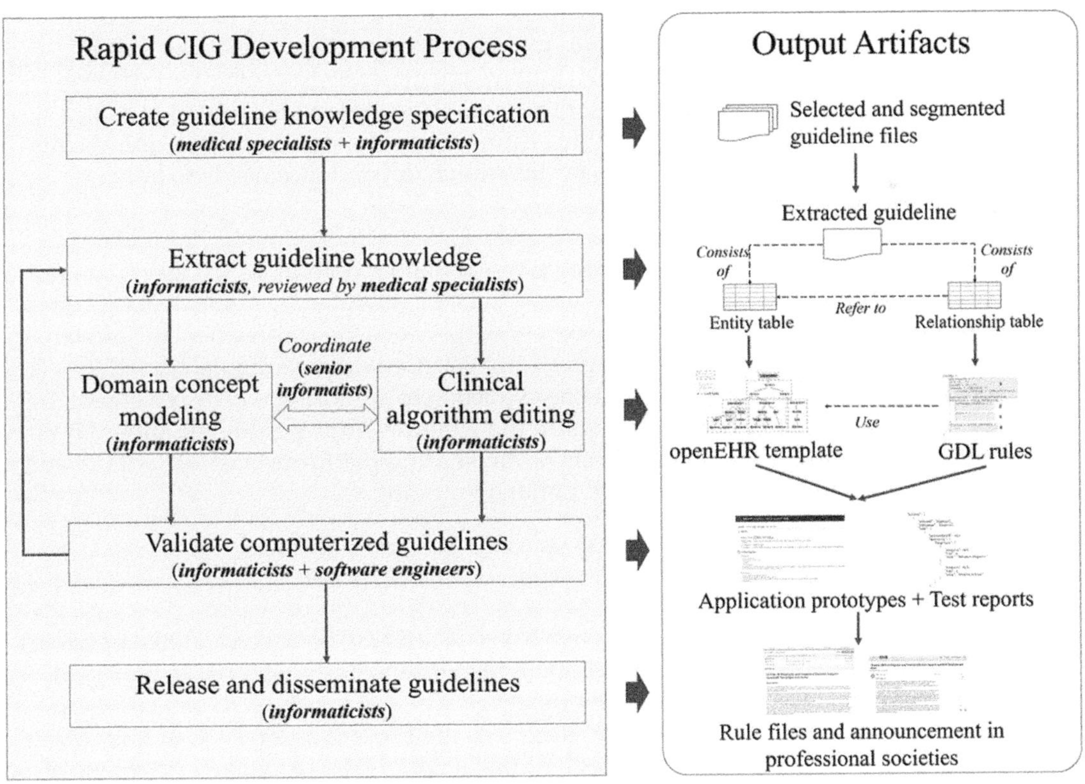

FIG. 7 The scheme of the rapid CIG development approach.

(2) Extract guideline knowledge. This step bridges the human-readable narrative guideline and computer-interpretable guideline by breaking narrative text to entities and relations. Informaticists read through the narrative guideline, break the guideline to small logic blocks, which can be represented by production rules. Each block has a left-hand side (LHS) representing the conditions and a right-hand side (RHS) representing the consequent actions. These relationships are collected as a relationship table and delivered as an artefact. Both LHS and RHS are further broken down from phrases to individual terms. Entities are marked up and extracted from those terms and form an entity table, which is also an artefact. The extraction results must be reviewed by medical specialists to ensure their correctness.

(3) Domain concept modelling and

(4) Clinical algorithm editing are carried out concurrently and collaborate by two groups of informaticists to accelerate the development pace. This process should be coordinated by senior informaticists to ensure consistency in both groups. Domain concept modelling refers to the openEHR template development process. Based on the aforementioned entity table, informaticists search through the openEHR archetype repository and select suitable archetypes to represent the entities best. Archetypes should be created if there is no appropriate archetype for specific entities. An openEHR template is developed to organise these archetypes. The detailed approach of openEHR template modelling is described elsewhere. In the clinical algorithm editing step, another group of informaticists translates the relationship table to GDL rules. The translation is straightforward since GDL follows the general structure of production rules. While editing clinical algorithms, the domain concept model is needed as the input. In the meantime, clinical algorithm editing raises domain concept requirements that need to be created or refined. This bi-directional dependence needs to be coordinated by senior informaticists.

(5) Validate computerised guidelines. In this step, the CIG is validated with clinical data. Guideline authoring tools usually contain a CIG validation module, which takes data from manual input by informaticists. The CIG can be further validated by implementing clinical decision support systems (CDSSs). This kind of work needs the involvement of software engineers. If the validation results have any inconsistency with the original narrative guideline, the whole process from guideline extraction to authoring will be reviewed.

(6) Release and disseminate guidelines. Finally, the well-developed guideline should be made publicly available by informaticists. Code-sharing platforms such as GitHub and forums of professional society are good options for doing so.

The COVID-19 CDS application

Across the year of 2020, coronavirus disease 2019 (COVID-19) has been a global pandemic, affecting over 200 countries and territories [24]. In response to this global crisis, the openEHR takes various actions, among which, developing and disseminating GDL-based COVID-19 CG is a remarkable one. This section briefly describes the process and results of this effort.

The Chinese COVID-19 guideline has been published by NHC of China with the purpose of sharing the latest evidence-based best practice regarding COVID-19 diagnosis and treatment and standardising caregivers' practice nationally [25]. In order to efficiently authoring the guideline, a guideline authoring team was organised. The team consisted of one medical expert, three informaticists, and two software engineers.

An informaticist reads the guideline and splits the sentences to a spreadsheet. Colour systems were used while splitting: the blue colour indicated the sentence was related to diagnosis or treatment; the black colour indicated the sentence was not suitable to be translated to a clinical rule; the red colour indicated the sentence contains vague content and needs to consult with medical specialists later on. Then, an informaticist annotated the entities in the text and extracted them in another spreadsheet. Repeated entities were merged together.

Based on the entity table developed in the previous step, two informaticists mapped the listed entities to the openEHR concepts. While mapping, the openEHR template was expanded accordingly. A bunch of archetypes in the format of ADL files were exported from the CKM.

Sentences in the narrative guideline were broken down into LHS and RHS by an informaticist in the guideline extraction step. ADL files describing the COVID-19 data requirements were imported to the GDL2 Editor. Following the structure of the extracted guideline file, an informaticist translates each LHS and RHS pair to a GDL rule in a WHEN…THEN… format.

The CIG developed in the previous step was validated both by GDL2 Editor and a prototype CDSS for COVID-19 (see Fig. 8). Archetypes in the ADL file format and GDL rules in the GDL2 format were exported from their editors and packaged together and committed to

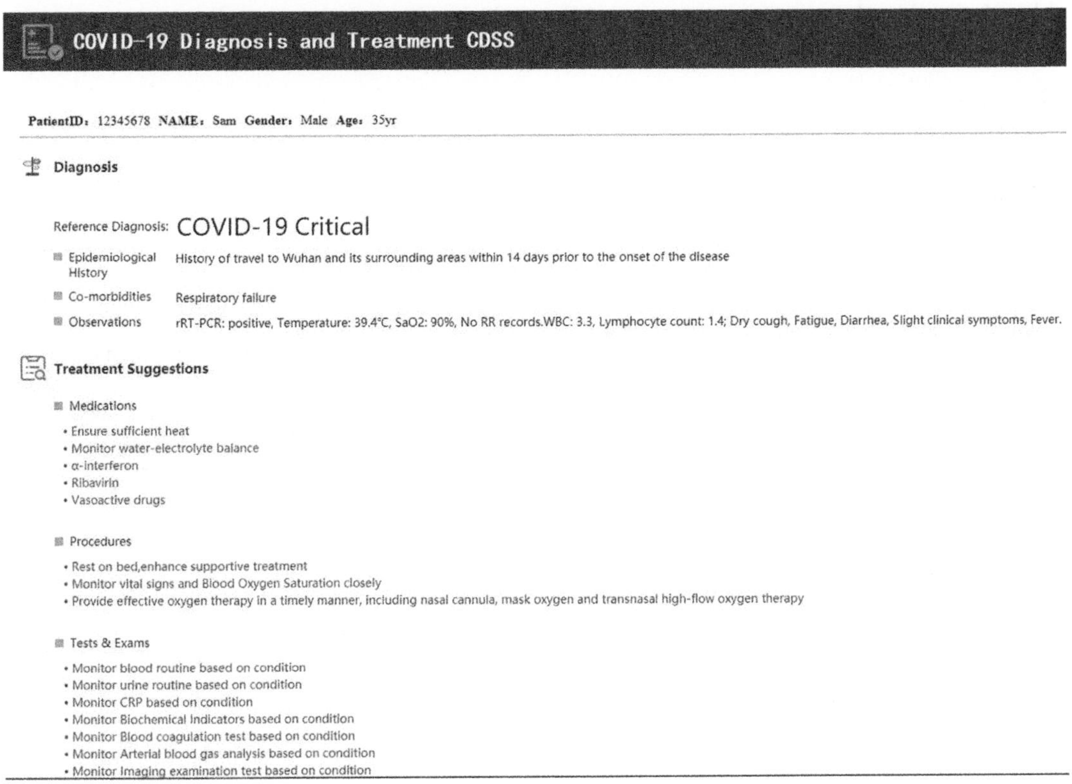

FIG. 8 Prototype application of the COVID-19 guideline.

GitHub. Java data models and Drools rules were also committed to GitHub to benefit non-openEHR users. Then, the dissemination was reported on the openEHR disclosure forum.

Wrap up

This section introduces the requirements and basic notion of the openEHR guideline definition language (GDL). Taking the advantage of openEHR techniques, GDL facilitates the interoperability of CDSS by eliminating the inconsistence of the information models of various information systems. Consequently, GDL has greater potential to be adopted in a broader range of organisations. A detailed guideline authoring approach using GDL has been introduced in detail. This approach provides a step-by-step guideline for those who aims to applying GDL for their studies and practices. Finally, a COVID-19 computerised guideline was developed as a demonstration of using GDL and the authoring approach.

Conclusion

This chapter introduced the approach of openEHR modelling and implementation to achieve the interoperability. The case study of the CDS application based on the openEHR CDR proved that openEHR approach has advantages in modelling a new medical application field and meeting the requirements of rapidly updating knowledge.

References

[1] Institute of Electrical and Electronics Engineers (IEEE), IEEE Standard Computer Dictionary: A Compilation of IEEE Standard Computer Glossaries, IEEE, New York, NY, 1990.

[2] O. Iroju, A. Soriyan, I. Gambo, J. Olaleke, Interoperability in healthcare: benefits, challenges and resolutions, Int. J. Innov. Appl. Stud. 3 (1) (2013) 262–270.

[3] Healthcare Information and Management Systems Society. Interoperability in Healthcare. HIMSS. https://www.himss.org/resources/interoperability-healthcare. 2021. Accessed March 5, 2021.

[4] Sam Heard T. What is openEHR?. Openehr.org. https://www.openehr.org/about/what_is_openehr. 2021. Accessed March 5, 2021.

[5] R. Pedersen, C. Granja, L. Ruiz, Implementation of OpenEHR in combination with clinical terminologies: experiences from Norway, Int. J. Adv. Life Sci. 9 (2017) 82–91.

[6] S. Garde, R. Chen, H. Leslie, T. Beale, I. McNicoll, S. Heard, Archetype-based knowledge management for semantic interoperability of electronic health records, Stud. Health Technol. Inform. 150 (2009) 1007.

[7] T. Beale, Archetypes: constraint-based domain models for future-proof information systems, in: OOPSLA 2002 Workshop on Behavioural Semantics, 2002.

[8] The openEHR Foundation. 2021 Architecture Overview. Specifications.openehr.org. https://specifications.openehr.org/releases/BASE/latest/architecture_overview.html#_multi_level_modelling_and_archetypes. Accessed March 5, 2021.

[9] Beale T, Heard S. Archetype Definitions and Principles. Specifications.openehr.org. https://specifications.openehr.org/releases/1.0.2/architecture/am/archetype_principles.pdf. 2021. Accessed March 5, 2021.

[10] H. Leslie, S. Heard, S. Garde, I. McNicoll, Engaging clinicians in clinical content: herding cats or piece of cake? Stud. Health Technol. Inform. 150 (2009) 125–129.

[11] S. Garde, P. Knaup, E. Hovenga, S. Heard, Towards semantic interoperability for electronic health records, Methods Inf. Med. 46 (3) (2007) 332–343, https://doi.org/10.1160/ME5001.

[12] The openEHR Foundation. 2021 Architecture Overview. Specifications.openehr.org. https://specifications.openehr.org/releases/AM/latest/Overview.html#_business_purpose_of_archetypes. Accessed March 5, 2021.

[13] M. Samore, D. Lichtenberg, L. Saubermann, C. Kawachi, Y. Carmeli, A clinical data repository enhances hospital infection control, Proc. AMIA Annu. Fall Symp. (1997) 56–60.

[14] K. Bernstein, M. Bruun-Rasmussen, S. Vingtoft, S.K. Andersen, C. Nøhr, Modelling and implementing electronic health records in Denmark, Int. J. Med. Inform. 74 (2–4) (2005) 213–220, https://doi.org/10.1016/j.ijmedinf.2004.07.007.

[15] The openEHR Foundation. Program o. openEHR—Specifications Start Page. Specifications.openehr.org. https://specifications.openehr.org/. 2021. Accessed March 5, 2021.

[16] Beale T. Archetype Semantics. 2021 Specifications.openehr.org. https://specifications.openehr.org/releases/1.0.2/architecture/am/archetype_semantics.pdf. Accessed March 5, 2021.

[17] I. Sim, P. Gorman, R.A. Greenes, R.B. Haynes, B. Kaplan, H. Lehmann, P.C. Tang, Clinical decision support systems for the practice of evidence-based medicine, J. Am. Med. Inform. Assoc. 8 (6) (2001) 527–534. BMJ Group BMA House, Tavistock Square, London, WC1H 9JR.

[18] M. Peleg, Computer-interpretable clinical guidelines: a methodological review, J. Biomed. Inform. 46 (4) (2013) 744–763, https://doi.org/10.1016/j.jbi.2013.06.009.

[19] P.A. de Clercq, Guideline-Based Decision Support in Medicine, informedici.nl, 2003.

[20] A. Wulff, B. Haarbrandt, E. Tute, M. Marschollek, P. Beerbaum, T. Jack, An interoperable clinical decision-support system for early detection of SIRS in pediatric intensive care using openEHR, Artif. Intell. Med. 89 (2018) 10–23, https://doi.org/10.1016/j.artmed.2018.04.012.

[21] Chen R. 2021 Guideline Definition Language v2 (GDL2). https://specifications.openehr.org/releases/CDS/latest/GDL2.html. Accessed March 5, 2021.

[22] N. Anani, R. Chen, T. Prazeres Moreira, S. Koch, Retrospective checking of compliance with practice guidelines for acute stroke care: a novel experiment using openEHR's guideline definition language, BMC Med. Inform. Decis. Mak. 14 (2014) 39, https://doi.org/10.1186/1472-6947-14-39.

[23] L. Zhou, N. Karipineni, J. Lewis, et al., A study of diverse clinical decision support rule authoring environments and requirements for integration, BMC Med. Inform. Decis. Mak. 12 (2012) 128, https://doi.org/10.1186/1472-6947-12-128.

[24] World Health Organization. 2021 Coronavirus Disease 2019 (COVID-19) Situation Report. https://www.who.int/docs/default-source/coronaviruse/situation-reports/20200620-covid-19-sitrep-152.pdf?sfvrsn=83af-f8ee_4. Accessed March 5, 2021.

[25] NHC of the PR of China. 2021 Novel Coronavirus Pneumonia Diagnosis and Treatment Plan (Provisional 7th Edition) http://www.nhc.gov.cn/yzygj/s7653p/202003/46c9294a7dfe4cef80dc7f5912eb1989.shtml. Accessed March 5, 2021.

CHAPTER

19

Evidence-based biomedical information systems: The road ahead

Rune Pedersen[a] *and Luis Marco-Ruiz*[b]

[a]Norwegian Centre for E-health Research, The Arctic University of Norway, Tromsø, Norway

[b]Norwegian Centre for E-health Research, Tromsø, Norway

The need for evidence-based systems

Health information technologies (eHealth) are viewed as the means to unleash the potential of health data for improving patients' lives and better use of healthcare system resources [1]. Both the World Health Organisation (WHO) and the EU recognise technology as an important facilitator for modern, patient-centered, and effective healthcare [2]. The implementation of comprehensive information systems in healthcare has proved to be a complex path with risks that often compromise the effectiveness of the investments performed [3–5]. One main reason is that usually health reforms, and the eHealth interventions they encompass, are not properly planned, monitored, and evaluated [6]. For example, implementing a new EHR or switching to another one is one of the most disruptive events a hospital can experience, affecting practically every employee and workflow at a hospital [7,8]. EHR implementation is usually beneficial for improving data quality, increasing efficiency, supporting research and reducing errors [9]. This is identified in the strategy of 'one citizen—one journal' [10] as a key objective, in Norway, and in the Meaningful Use initiative in the United States [11].

The need for healthcare services will increase in the future as a result of the population growth and increased life expectancy [12]. At the same time, forecasts from Statistics Norway show that the population composition in the municipalities will change over the next ten years, and the average age of the population will increase dramatically [13]. This means that the government needs to look at how the health and care services can meet these challenges. In this regard, a crucial part of the solution for the future of delivering health services implies the use of technology [14]. However, finding the optimal strategy for the selection of technologies and organising the adoption of them is a complex task that, when not optimally managed, has led to unsatisfactory results of billionaire investments in ICT infrastructures [3,4]. To reduce the risks on the investments in health informatics infrastructures, decision

https://doi.org/10.1016/B978-0-12-823413-6.00010-0

makers need to build their own evidence for making solid decisions on how to best approach the adoption of ICT interventions in their specific context.

In the following, we explain the Nordic context and how evidence can be built to guide future health ICT infrastructures. The remainder of the chapter is structured as follows. First, we use the Nordic context to justify the need for more solid methodologies for building evidence when approaching the adoption of new technologies and planning large health informatics interventions. Second, we elaborate on the need for relying on open platforms for allowing biomedical informatics evidence building. Third, we draft what such a framework may look like and discuss open topics that need to be covered by researchers when creating such a framework.

Norway as a representative of the challenges in European nations

The Nordic healthcare domain is complex, highly specialised, transferable between countries, and subject to massive technology investments [15]. The implementation and adoption of eHealth influence the relationships between work organisation and technology, which in turn transform each other [16]. All the Nordic countries have established major eHealth initiatives to improve and streamline services.

Currently, in a second digitisation wave, countries aim to replace today's text-based EHRs with systems that record structured data [15]. The health authorities see this as a necessary step to achieve a more transparent and educational health system, to be able to share data across sectors, and to reuse data for research [17]. This is in line with overall healthcare strategies in Norway related to realising the three goals in the government's whitepaper 'one citizen—one journal' [18]:

(1) healthcare professionals must have easy and secure access to patient and user information;
(2) citizens must have access to simple and secure digital services; and
(3) data must be available for quality improvement, health monitoring, management, and research.

'One citizen—one journal' will be realised through three strategic and parallel measures:

(1) further development of existing solutions in the specialist health service in the north, west, and south-east health regions;
(2) establishment of the *Helse Plattformen* (in English, Health Platform) in the Central health region;
(3) and establishment of a national solution for comprehensive collaboration and a joint journal for municipal health and care services (outside Central Norway) EIEJ/AKSON.

The status in Norway is a good representative of both Nordic and European countries. It is a somewhat heterogeneous context across regions and health levels which is divided in two main levels:

(a) Secondary healthcare systems
At the moment, there are two main approaches towards hospital EHRs depending on the region. On the one hand, three out of four regions have opted for implementing

DIPS Arena (DIPS AS, Bodø, Norway), an openEHR-based EHR. On the other hand, the central region has chosen to implement a centralised system called *Helse Plattformen* as a general platform for integrating data from hospitals, GPs, nursing homes, and home-care services [3].

For specialised healthcare, *Helse Plattformen* provides rich functionality for making data available, and it relies on a technology widely adopted in other countries [3,19]. Its original estimated cost is 3 billion NOK. While the directorate has done a great effort to design a strategy that can fulfill the three requirements mentioned, there have been discussions related to the approach for reaching them. *Helse Plattformen* is based on a central platform that relies on a proprietary information model for persisting healthcare data (Epic Care, Epic Systems, Verona, Wisconsin, United States). This has led to discussions on the challenges that may involve the third objective related to data reuse for quality improvement explained in section "Open platforms for evidence-based adoption of health information systems" [3,5,20–22].

Another approach followed by the Northern Norway health region is the implementation of DIPS Arena, which will be introduced in 3 out of 4 health regions in Norway. DIPS Arena started its development as an EHR system in 2012 and has been piloted in some services at small scale. In the Northern Norway health region, the project FRESK will start implementing DIPS Arena at large scale in 2021. Additionally, the electronic chart solution Metavision, will be implemented in 2021 and integrated with DIPS Arena. The main benefit of introducing these systems is, according to the decision makers, that the electronic health record (EHR) will facilitate reusing data from the EHR between different clinical systems and facilitate Computerised Clinical Decision Support (CCDS). DIPS Arena is a process-based open platform EHR built on the openEHR specification [23]. It uses archetypes as clinical information models (CIMs) to structure clinical data and forms.

Currently, these hospitals rely on a previous system known as DIPS Classic where clinical information is written mostly as free text, thus hampering data reuse for CCDS and research. This demands structuring the clinical content of DIPS Classic. To that end, the definition of archetypes as information models that shape the data stored in the EHR is needed. In addition to exchange and reuse of data, structured data is the basis for CCDS, which is another main objective for implementing DIPS Arena. During its development and pilots, the approach followed by DIPS has shown some challenges further discussed in section "Open platforms for evidence-based adoption of health information systems" that are related to time lags between platform and archetypes developments and the alignment with standard terminologies [19,24–28].

(b) Primary care

Primary care in Norway is highly dependent on the municipalities, and it is based on a reimbursement model where GPs are paid depending on the patients who are assigned to them. Municipalities have the freedom to decide which EHR system they adopt. This has led to heterogeneity in the EHRs that are currently in use. Aware of the challenge, the Directorate of eHealth has powered an initiative called AKSON Journal that intends to unify the information systems used by GPs, emergency units, home-care services, nursing homes, and health stations for youth [29]. AKSON has an estimated cost of 22 billion NOK [29–31], and it aims for providing a joint medical record for all municipalities.

Although the needs for AKSON are acknowledged in the Norwegian healthcare context [31], the strategy designed to develop and deploy it has received criticism from different medical informatics experts and researchers in Norway. The main critics relate to the lack of enough stakeholder involvement and not facilitating the integration of many vendors. AKSON Journal was transferred to Kommunenes Sentralforbund (KS), the Norwegian Association for Municipalities early in 2020. Based on this the new program will change direction. These issues are further explained in the section "Open platforms for evidence-based adoption of health infromation systems" [5,19,21,22,32–34]:

Towards next generation EHR: Current challenges and requirements

First generation EHRs were developed as a document storage, management, and retrieval system for the provision of clinical care, but not for front-end to back-end seamless data extraction, subsequent analyses, or virtually real-time reporting. The rationale for acquiring a new EHR system often concerns the needs to satisfy more advanced and meaningful use requirements and the need for more usable systems that improve clinicians' performance. Modern EHR systems achieve this by, for example, helping clinicians to reduce the risk of errors using Computerised Clinical Decision Support (CCDS) [35]. However, there may exist a significant difference between the visions of policymakers and primary health providers [35]. Mandatory implementation of a comprehensive EHR does not guarantee that healthcare service providers will perceive the benefits of the system [3]. In the early days, following the original concept of document store [36], many EHRs followed a natural evolution towards the implementation of proprietary models more focussed on providing support for care delivery than on enhancing data quality for reuse in research and health quality measures [19,37].

Consequently, many of them still do not commit to a specific standard information structure. This lack of standardisation translates to high costs for migrating to a competitor vendor and a lack of freedom to evolve the persistence information model for facilitating data reuse and interoperability [5,20]. This situation is commonly known as *vendor lock-in*. Several studies have documented some of these challenges such as the payment for each transaction that standardises EHR extracts, the lack of control over the data schema evolution, and the barriers for extending the EHR core functionality using a standard API [3,5,19].

Partly because of a need for EHR adoption and partly for boosting standardisation, the United States has seen significant investments to substitute home-grown EHRs with vended EHRs during the Meaningful Use initiative [5,11,19]. This replacement of home-grown EHRs in favor of commercial systems was partly motivated by the high costs involved in adapting and maintaining homegrown systems for the Meaningful Use (MU) requirements [19]. However, after several years, commercial EHRs have not been able to provide all the functionality needed by organisations and, currently, they still allocate resources to maintain previous home-grown systems for accessing historical data, for using them as downtime systems and for accommodating some functionality not covered by commercial systems [19].

The situation experienced in Norway with the ongoing adoption of DIPS Arena has been the same. The previous version of the EHR, not openEHR-based, known as DIPS Classic is maintained for guaranteeing access to historical data during the transition towards DIPS Arena, and both systems have been in parallel use at several locations. These experiences have shown that the migration process from one EHR to another one is a complex hazardous task [38]. In the following section, we elaborate on the challenges involved:

(a) Transformation of legacy notes into structured standard forms: First generation EHR systems contained a huge amount of free text notes [39], while new EHRs often opt for structured information for data quality reasons. Increased structured data quality is needed to enable its secondary use in research and CCDS. In order to truly allow for generating knowledge from data using data driven algorithms, these data need to be represented in a way that minimises ambiguity [25,40]. Otherwise, if only free text notes are used in EHRs, it may be very difficult and expensive to improve the quality of data, so it can be used for quality assurance and research. Although NLP (Natural Language Processing) has certainly advanced a lot by applying computational statistics [41,42], its best advantages are not in completely transforming long free text notes into valid machine interpretable information schemas [39]. The reason is that the use of NLP tooling requires human supervision to ensure the validity of its outcome [42] because clinical notes are often messy, biased, and dependent on the clinician's own preferences [43]. Conversely, their best results are achieved by assisting the entry of free text, by recommending predefined selections (e.g., of terminology codes) that assign precise semantics to clinical documents [42,44]. This sets a need for structuring the information at the point in time when it is first recorded for clinical use.

A clear structure of clinical documents is needed not only for defining their syntactic structure but also for annotating information schemas with terminologies so that each section of the schema is bound to a common standard definition that defines their semantics [45]. Clinical data represented by means of standard information models (e.g., archetypes) and terminologies dramatically increase data quality. Data quality is needed to assure the secondary use of data for research and quality control [39]. However, data quality requires structuring using clinical information models (e.g., archetypes) validated by clinicians and working with terminologies in order to ensure that the meaning of data is precise and machine understandable [25,26,28,46].

In this first instance, using a structured data capture process in the EHR may hamper the system's usability from the clinical user perspective. However, the possibilities of high-quality data are later seen in the form of real-time CCDS, research, and quality measures. Therefore, the chosen solutions need to be carefully designed as user centered design to find a good balance between flexibility in writing clinical notes and structuring the information to enable its efficient reuse [21,39]. This also means that users shall be educated in the rationale for having high data quality, so they are willing to accept some ways of recording their clinical notes.

(b) Lack of control on the stored data: Standardisation in the form of structured data annotated with clinical terminologies has faced issues in some contexts. An important issue raised is the barrier imposed by some vendors to extend the vended EHR with functionalities needed by the hospital using standards such as the HL7 FHIR API [19]. This hampers research and secondary uses for research and CCDS, ultimately affecting the patient's health. Innovation to offer functionality not covered by commercial systems is still implemented as part of EHRs regarding visualisations, display functions, and CCDS [19,38]. This evidences the need for advancing to the open platform model where an ecosystem of health information technologies [47] can be implemented allowing more freedom on the development of both home-grown solutions; compatibility of home grown systems with commercial products; and health trusts to keep control on their data.

(c) Delay in benefits perception by clinical users: Another complexity in the biomedical informatics domain is that success is often measured as clinical satisfaction, but it may take several years until the benefits of a new introduced technology are perceived [48]. This could be alleviated by more involvement of clinical stakeholders during the design and implementation of health IT interventions. Unfortunately, the interaction between stakeholders, including clinical ones, is not fluent, and this is causing a lack of innovation, long projects, heavy procurements, and saturation of the market, concentrating it in just a few dominant suppliers [5,47]. End clinical users are often not part of the digital systems they use, and they cannot influence large vendors [5,47]. In addition, small companies cannot enter the sector due to oversaturation [5]. Some authors advocate for enabling *platform ecosystems* that allow different vendors to enter the market and specialise providing loosely coupled highly specialised solutions while still interoperating [47]. Clinicians can influence these smaller complementary solutions while they often cannot influence large commercial systems.

(d) Health information interventions that are designed as a mere software project often underestimate the complexities of the biomedical domain. The inclusion of biomedical informatics researchers, technology champions, and clinical users helps to analyze reliable experiences about technology, development opportunities, and pitfalls [19]. In other words, biomedical informatics interventions need to work in an evidence-based manner. In this regard, several stakeholders in the Norwegian eHealth realm have demanded to change direction from a vendor-specific solution to an open platform approach. In this lies a desire to move away from the idea of a single main supplier to an ecosystem of products from various suppliers that compete and advance the functionality of health information systems (HIS). At the same time, this change requires a major shift in how the milestones for the development of biomedical informatics interventions take place.

In Norway, various researchers, vendors, and medical associations have advised the Directorate of eHealth to focus on developing the foundation for adopting standards that facilitate data representation in interoperable formats and open architectures. This would mean to adopt the role of a counselling organisation and mediator to make different stakeholders develop their solution within a common generic framework. In this way, the strategy would approach to the open platform concept where different stakeholders commit to some platform or, in this case, specifications of a platform for developing an ecosystem where some of them work on the main platform that makes long-term data storage, while others create highly specialised software that extends the platform functionality. In the following sections, we explain how the challenges discussed may be addressed relying on the open platform approach using openEHR.

OpenEHR as an open platform ecosystem

Open platforms are defined as software systems that are based on open standards and well-specified APIs, so the functionality of the platform can be extended without modifying the source code [49]. According to Eisenmann, an open platform needs to guarantee the following [49]:

(a) nonrestrictions are placed on participation in its development, commercialisation, and use; and

(b) any restrictions set on the implementation, compliance with standards, etc., are nondiscriminatory for any present or future platform participant.

Open platforms allow third parties to work and extend the functionality of the system without the original implementer. There are four levels of openness [49]: (a) proprietary; (b) licensing; (c) joint venture; and (d) shared. OpenEHR complies with the fourth level, the so-called *shared model* for open platforms that is characterised by the control of the specifications held by many firms and representatives (in this case, the openEHR International Board), and several firms providing the available platform implementations [50].

Regarding the roles mentioned by Eisenmann, openEHR is open with respect to the following:

(a) end users because any organisation can access the openEHR specifications and develop openEHR-compatible systems or use the platform for storing data (open source or not);
(b) any organisation can download the specifications and open-source software and use openEHR platforms for storing data;
(c) any organisation can add to the specs and open-source implementations subjected to the rules specified by openEHR and its SEC committee.

It is important to understand that an open platform does not necessarily need to be open source. In the case of the *shared model* for open platforms adopted by the openEHR community, the specifications are an open standard and implementers develop their product based on them. Some implementers are commercialising their product as compiled code packaged into a product [51,52] (Better Platform, Better d.o.o., Ljubljana, Slovenia), while others provide it as an open-source system [53]. Fig. 1 represents the organisation of the openEHR community. Members of the openEHR community (individuals and vendors) and the openEHR foundation elect the openEHR International Board which controls different programs (affiliate, specifications, modelling, software, etc.) through the not-for-profit company openEHR International (CIC). Some vendors implement the specification offering several implementations, and other suppliers use these platforms to develop their own openEHR compatible systems (bottom of Fig. 1).

The concept of open platform has many parallelisms with digital ecosystems [47]. Digital ecosystems, in their most open technical dimension, can be defined as a core platform with external components operated by various networked organisations and some governance bodies that constructively define the rules for all the platform participants [47,54]. Thus, this flavour of digital ecosystem adds to the concept of open platform by better contextualising the governance of it and its role on the society. Based on this, we believe that the ideal context for operating a health information network based on openEHR should be an *evidence-based open platform ecosystem*. In the following section, we explain this vision.

An evidence-based open platform ecosystem

Some advocates for centralised solutions believe that an open platform should rely on central data storage and standardise some parts of the EHR using clinical standards and terminologies, but that the architecture can remain proprietary. However, a true open platform that behaves as a digital ecosystem should allow for backward compatibility and the absorption of complements into the core platforms, not placing any restrictions on the participation or use [47,49].

The approach adopted by vendor proprietary models (centralised or not) standardising on demand some portions of the EHR as extracts does not fit this definition and does not provide

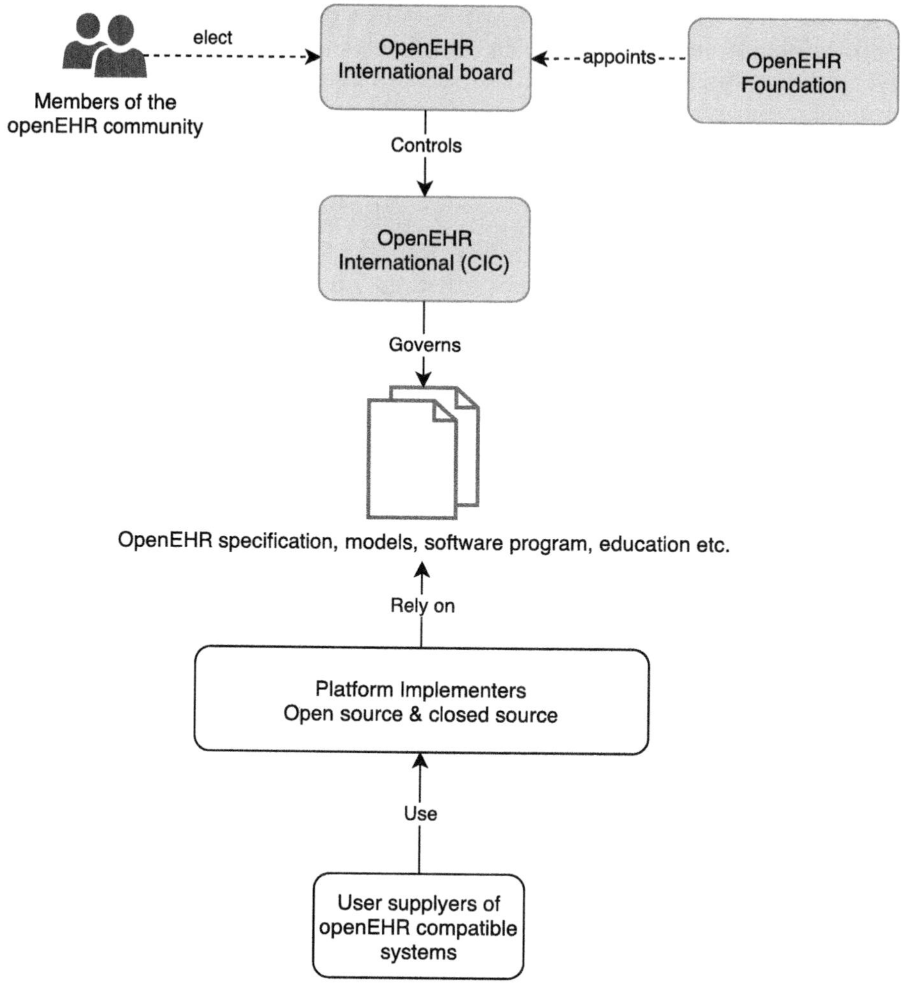

FIG. 1 Organisation of the openEHR community under the open platform paradigm.

the benefits of evidence-based open platform ecosystems. Colicchio documents some of these challenges such as the payment for standardisation as per transaction basis, lack of control over the data schema evolution for extending the EHR functionality with new applications, and overpayments due to the concentration of the market in just a few large EHR vendors [5,20]. In addition, lack of control over data and its schema impacts the transition from one vendor to another due to the high dependence of healthcare services on EHRs [22,33]. These issues have been a matter of concern in the development of EHRs in Norway [22,33].

Clinical data is no longer just a derivative resulting from providing medical care, but it is a highly valuable strategic asset [55]. For this reason, data access mechanisms should not incur further costs beyond the implementation of these mechanisms and their maintenance. This means that data and information schemas must be publicly owned and not subject to any vendor-imposed constraints. This can be done in two ways: first, by adopting EHRs that commit to open platform architectures and, second, for legacy systems not relying on these architectures, by allowing access to them with very generic (and minimally structured) models that are compliant with the specification of the open platform ecosystem API (e.g., openEHR in our case).

Relying on an open platform ecosystem allows for different technologies to be tested and evaluated in different scenarios of the healthcare domain. This allows health organisations to develop understanding on which technology is better suited for their purposes and discuss what should be used for minimising the risks at a high level. In addition, this approach gives small vendors the possibility to develop specialised products that enhance the functionality of EHR open platforms, thus favouring free competition and fuelling the national health IT sector [20].

Provided that national suppliers of EHR solutions should be part of the procurement, the research communities also see that an ecosystem with multiple suppliers of EHRs to the municipalities creates greater transparency and free competition and facilitates higher speeds for system replacement [56].

The last two decades have led to a good level of maturity in both technologies and standards [37]. In our view, the current challenges do not come from choosing one implementation technology, standard, or terminology. Current challenges come from the need of appropriately leveraging them and optimally using them for improving the clinical workflow, data reuse, and computerised clinical decision support (CCDS). Different approaches can be followed for this, but the evidence-based open platform digital ecosystem is a sustainable approach for developing large health IT interventions. In the following, we will focus on openEHR as open platform digital ecosystem architecture. We will also present our view on how it can facilitate evidence-based large health IT interventions. We do not state that openEHR is the best or only option for an open platform approach that enables evidence-based decision, but it is the standard used in Norway for structuring hospitals EHRs and thus the one closest to us.

Open platforms for evidence-based adoption of health information systems

The key in a health information system that is compliant with the open platform feature is the involvement of the users, platform providers, and complementary producers in the continuous evolution and improvement of health information systems [49]. In the case of openEHR, the specifications do not only define a full specification for an EHR as an open platform, but they also rely on a model-driven approach for adapting the platform to the existing clinical areas. This translates to the following:

(a) the need of the clinical information models to drive the development of HIS and
(b) the possibility to have numerous vendors that are compatible among themselves.

This compatibility allows for running different interventions in parallel developed by different implementers. This parallel way of procuring interventions allows for evaluating and discriminating their success, thus building evidence to approach larger interventions. eHealth is extremely dependent on the context; therefore, this evidence must be built for each context allowing stakeholders to learn and develop solid experiences for guiding future interventions. We name this approach *bottom-up evidence building*. This contrasts with the *top-down* approach, which relies on larger interventions led by a reduced group of decision makers who decide on an integral strategy for subordinated end users. The latter has, in many cases, led to challenges and failures based on presumptions and misunderstandings of requirements due to the complexity of the biomedical domain [57].

Bottom-up evidence building intends to minimise risks, but it requires an investment in the long term for observing the ecosystem and building the evidence necessary for optimally designing local, regional, and national health information infrastructures. To build this evidence, it is necessary to distribute resources for smaller projects that resolve more local/regional problems with a high involvement of GP offices, hospitals, health trusts, municipalities, and researchers. The latter need to ensure that evidence produced is documented and that it is free of biases. To ensure compatibility, the general architecture, clinical information models, and terminologies need to be agreed by representatives from the majority of stakeholders.

This approach has both positive and negative effects. On the one hand, it creates many small initiatives, some of them replicating each other in functionality and, consequently, replicating their costs. On the other hand, there is a positive effect in making the actual users feel the platform and health IT intervention as theirs. This empowers users to propose changes and use their experiences to propose new features for the evolution of the system. In this way, users, vendors, and researchers can gradually build their own evidence in their context.

The rationale of this approach is to work in the medium-long-term letting stakeholders to experiment and share experiences. These lessons learned should lay the foundations for the design and implementation of larger national infrastructures, such as the adoption of an open platform that benefits from the abilities built and the lessons learned in the development of their smaller interventions.

We do not state that the open platform ecosystem approach is adequate for any context. In some cases, a new hospital needs to rapidly adopt a brand new EHR and a packaged product may be chosen for time and resources constraints [57]. However, in the long term, if the country or region considers health data as a strategic asset, the open platform model can bring the advantages previously discussed, minimising the risks involved in large top-down interventions [30,58].

A grounded example for adopting evidence-based open platform ecosystems is the discussion often witnessed by governments, technologists, vendors, and health informatics associations. The complexity in the biomedical informatics domain involves data persistence, project management, software architectures, terminologies, IT governance, and so on. The experienced reader will surely recall discussions and decisions based on 'technology hypes'. This means that, often, the adoption of clinical information standards and terminologies is not only led by objective requirements arising from evidence gathered across health institutions [19,21,43].

Conversely, decisions on health information architectures are, in many cases, based on the popularity of a specific technology and what similar countries do. It is not possible for decision makers to become experts or even find objective advisors on these topics due to the high level of specialisation that they require. In addition, what is the optimal choice in one context may not be optimal in another context. Therefore, the only way for decision makers to rely on the best possible evidence is to build evidence and experience gradually by themselves. This gradual self-learning will help to determine what the best choices for their projects are. Besides experiences from users and implementers, funding organisms need ways to measure and justify further investments. To that end, technology assessment frameworks adapted to eHealth interventions and eHealth indicators can be used [59,60].

Case studies: The co-constructed national approach in Norway

In the following section, we further discuss two case studies in Norway. The first is the ongoing discussion about the AKSON platform and the critics it has received. The second is the work performed in Norway to build common CIMs as openEHR archetypes, the organisation, the time invested, the benefits and the challenges.

AKSON and the controversy on top-down approaches

AKSON is a project that will provide a new joint EHR solution for the municipalities in primary care. AKSON will communicate with other digital solutions throughout the Norwegian healthcare system to provide a holistic view of patients. According to the Directorate of eHealth, AKSON will, through better access to information, reduce the risk of mistakes being made in the health service. It is estimated that this alone will provide NOK 3.7 billion in socio-economic gains. A number of actors and professionals in the sector have criticised the plans for implementing AKSON, while others, including the Norwegian Nursing Association, are positive on its implementation and believe it will be beneficial.

One hundred eighty five out of 291 municipalities have signalled that they will support further work on AKSON by signing a nonbinding declaration. A separate steering group has been established for AKSON with representatives from various health companies, municipalities, and health workers organisations. According to the Directorate of eHealth, 44 municipalities have been closely involved in the pre-procurement phase. More than 200 health workers from these municipalities have participated in workshops where the plans for the program were discussed. According to the plan, the first municipality will deploy AKSON in 2025. The goal is for AKSON to be operational in most municipalities by the end of 2030, and for 60% of GPs by 2033. However, as far as GPs are concerned, this is highly uncertain.

It is further estimated that more than 150,000 health workers in the municipalities will use the system, which is planned to be organised as a separate publicly owned joint stock company owned by the municipalities. It has not yet been decided who will develop and deliver the technology. The investment costs for AKSON are estimated at NOK 11.2 billion. The total costs including introduction and operation until 2024 are estimated at NOK 22 billion. It is still uncertain who will cover its cost, but all indications are that the municipalities themselves will share the cost.

The Directorate of eHealth aimed to develop a national platform with open APIs and common standards at the same time as a joint procurement for EHR is planned for the municipal health service AKSON. The project involves most of the Norwegian municipalities, and it is expected to be complex due to its size and coordination needs. At the same time, Directorate of eHealth has chosen the concept for procurement of municipal systems based on one main supplier for the EHR with a selection of sub-contractors or third-party vendors.

The choice of one main EHR supplier is based on the principle that interoperability between the services across the municipality will be easier by relating to as few information models as possible. Based on this, AKSON has received criticism, both from knowledge environments that are critical of the choices made based on a lack of evidence that this works and from an overall strength of national EHR suppliers. There is varying feedback on whether users consider that they have been heard and that they have contributed to the development of AKSON. Some participants have stated that they were not heard or that their suggestions were not included [22,34]. As of today, the Norwegian Medical Association does not support AKSON, and it is uncertain whether GPs will be a part of the procurement of AKSON.

The debate around large government projects such as AKSON and the Health Platform has been a matter of debate recently in Norway [22,29,30,33,61]. Much of the criticism is because the Directorate of eHealth is following a too strict top-down approach [30]. In summary, the critics regard the lack for a plan for independent evaluation [34], the poor involvement of research communities [21,22,34] and user representatives [22], the lack of control over data, the potential vendor lock-in which may hamper free competition [30,33], and the lack of an open platform ecosystem approach [30]. At the same time, provided that national suppliers of EHR solutions want to be part of the procurement, the research communities also see that an ecosystem with multiple suppliers of EHRs to the municipalities creates greater transparency and free competition and facilitate higher speeds for system replacement [33].

In the discussion, the term platform is frequently used to describe the procurement. The Directorate of eHealth describes the chosen solution as an open platform where one main EHR provider, including their sub-contractors, integrates with a common foundation with open APIs and with national solutions such as the patient's drug supply list and core medical record. Even though some of the digital ecosystem features are present, some requirements from stakeholders, such as more user involvement, evaluation, and clear vendor independence, are missing. To address these requirements, AKSON should commit to the Open Platform principles [30,47]. Based on the criticism, and the pressure described, the process of AKSON has taken a new direction in 2021. The program to procure new EHR systems for the municipalities has now been transferred to KS, the central association for municipalities. So far, the new program KS Journal has started a process of drawing an architecture that is more in line with an open platform ecosystem.

In this context, openEHR can bring a specification for the architecture of AKSON that would truly commit to the open platform principles and enable seamless interoperability with hospital information systems. Although the experience in archetype development and adoption has shown that the amount of time and resources was much higher than expected [24], the models for primary care are easier to define and require fewer amounts of details than those for specialised care. Thus, relying on openEHR for the architecture of the open platform ecosystem is a reasonable approach. The development of openEHR should not alienate other standards since many vendors are currently implementing HL7 FHIR. Therefore, providing support for HL7 FHIR is necessary. FHIR is a standard developed for interoperability, but it does not define a complete architecture for the EHR.

With this in mind, some architectures have opted for using openEHR as a platform specification for storing clinical data and implemented transformers available through Web services that transform between openEHR and HL7 FHIR, when a client of the platform wants to interoperate using HL7 FHIR [62]. AKSON would benefit from considering the benefits of these architectures where the strengths of both standards are leveraged. In such an approach, openEHR should be used for the persistence model of the open platform, while interoperability among different clients and the platform is allowed in both openEHR and HL7 FHIR.

OpenEHR adoption in secondary healthcare: Opportunities and challenges

In the period between 2008 and 2012, the National ICT Health Trust (National ICT 2011) initiated several projects with a focus on semantic interoperability, which resulted in a recommendation to build an openEHR-based infrastructure for the specialised healthcare sector [63]. The Norwegian EHR vendor DIPS ASA gained interest in the openEHR approach and was assigned with the contract for the new EHR in 2012. The plan was that the new EHR should be deployed in 11 hospitals across Norway.

In a 25-year period, this vendor had gathered high-level expertise in developing ICT systems for the EHR domain. Three out of four health regions in Norway use the EHR system from DIPS ASA, including about 80,000 healthcare workers and 4.3 million patients across the country (DIPS 2015). Due to the complexity of maintaining a large EHR portfolio, DIPS ASA started to experiment with model-driven development in 2006 [24]. This process ended in 2011, when a tender for EHRs in several regions made DIPS ASA to change direction and ended with the decision to use the openEHR architecture for their next-generation EHR [24].

Based on the adherence to the openEHR framework, the vendor can concentrate on developing the technical architecture of the new EHR, while clinicians focus on modelling the clinical content using openEHR archetypes. The development process started in 2012 and was performed in close collaboration with the Northern Norway Regional Health Authority. The region was the first to sign a contract for the new openEHR-based EHR system (2012). In the years to come, two more regional health authorities bought DIPS EHR as their future system, thereby acquiring an openEHR-based system. As a result, the vendor has run development projects in three health authorities.

The cross-regional impact of the openEHR-based system adoption resulted in the need for national coordination of archetypes development. In 2013, the National ICT Health Trust decided to establish the National Editorial Board for Archetypes (in Norwegian, NRUA) with the overall goal of coordinating the work of archetype development. Currently, NRUA consists of five full and part time employees, supplemented with a varying number of representatives from each of the four regions (all together around 15 persons). For this matter, a primary task for NRUA was to recruit clinicians and other EHR users for developing archetypes. The focus was on the consensus processes, reviewing and approving archetypes while increasing the understanding of the openEHR model.

The work towards improving national archetypes progressed slowly the first year due to the effort of establishing the NRUA organisation. In the following years, the process accelerated after establishing a closer collaboration with the four health regions. NRUA also established a close relation with openEHR International, which coordinates several archetype repositories worldwide through the International Clinical Knowledge Manager (CKM). Through this collaboration, NRUA has translated existing archetypes from the international CKM, and the International CKM has adopted archetypes developed in Norway.

After many years of gaining competence, the Norwegian CKM is unique compared to CKMs in other countries due to the high number of clinicians included in archetypes development and validation. Of the almost 400 reviewers recruited, almost one-third are clinicians. In addition, Norway is the only country conducting this work at a national level, having contributors participating from all over the country, and with a standardised selection of clinicians that contribute to the consensus process [64]. The number of archetypes approved in Norway by September 2020 was 113, and more than 370 were in progress. The overall objective is to establish a repository of 1000–2000 archetypes to provide a sufficient number of clinical standards for a large-scale EHR system. The total number of contributors is 648. Contributors are broadly divided between healthcare personnel and medical informaticians. For the consensus processes, there are a total of 399 participants and 40 translators. A total of 40 templates have been developed through 62 public and private incubators and projects framed within the Norwegian CKM.

There are several representatives from the vendor DIPS ASA participating in the consensus work. This provides DIPS with an understanding on the design of archetypes. In this way, the vendor understands what are the requirements and priorities of clinicians. This also leads to a deeper understanding of the challenges related to this work and the complexity of the archetypes, particularly the interdependence between the archetypes and the EHR system. A close cooperation between the authority and the vendor is crucial.

The work to publish validated archetypes at a national level has been very challenging for several reasons. First, in the initial years of NRUA, the EHR system was not ready to adopt and test the archetypes for clinical use. So far, not all the archetypes developed by NRUA have been put into clinical practice, so we still lack evidence on the benefits of the openEHR CIM (Clinical Information Modeling) approach. Moreover, developing the archetypes is just the initial task for the EHR to comply with all clinical requirements before being operational. For example, archetypes need to be embedded in each clinical process workflow of the EHR, user interfaces need to be developed and tested, data input mechanisms need to be configured, etc. [65,66]. Therefore, the potential of openEHR has not yet been fully exploited, and it will take time to observe the benefits of it. A new project started in 2021 to focus on the adoption and development of archetypes for the EHR in specialised clinical domains. Only when that project has structured a significant part of the EHR and the EHR has adopted them, will we be able to build evidence on the openEHR intervention as a whole.

Still there are some benefits that the development of archetypes has already brought. The most appealing benefit of the work done by NRUA is not the development of archetypes as technical artifacts but the definition of the semantics they convey (independently from the standard of specification). The increasing number of clinicians involved has made it possible to develop highly consistent CIMs that act as a gold standard for other CIMs to follow (openEHR-based or not). This means that the Norwegian CKM, beyond producing openEHR archetypes, has elicited consistent CIMs that are useful for facilitating the development of CIMs in other formalisms. The reason is that the resource intensive work related to describing which data items should be present for each CIM and their exact intended meaning is already done. This knowledge is something that is stable among clinical areas irrespectively of the standard. Thus, their alignment to other formalisms (e.g., HL7 FHIR) is much less resource demanding because it does not involve the number of clinical reviewers needed to reach consensus.

Another challenge present in the adoption of openEHR is the use of ontology-based terminologies in combination with archetypes for binding standard definitions to each archetype element and content [25]. The regional implementation program FRESK in the

North Norwegian Health Authority has experienced that the work to mark the semantic content of archetypes with SNOMED-CT codes is a time-consuming manual process. This is partly based on the richness of the semantic content in the archetypes, the granularity, and a lack of good collaboration between CIM modelers and terminologists for eliciting guidelines about the optimal use of both [27,28,67]. Based on this, it has been decided that new national archetypes must be bound to SNOMED-CT codes before being nationally approved. This will increase the cost of eliciting archetypes since NRUA will need to include highly knowledgeable terminologists that define value sets and optimal bindings to archetypes. Furthermore, there are no clear guidelines for leveraging the use of openEHR and SNOMED-CT [26,28].

Principles for developing an evidence-based open platform ecosystem

In the previous sections, we have explained the challenges faced by complex health IT interventions and the reasons for a change of paradigm from a top-down large project approach to a bottom-up evidence-based approach. To finalise this chapter, we propose a list of principles that evidence-based interventions based on open platform ecosystems should commit to:

- Consider health data as a strategic public asset.
- Minimise risks by splitting the project into smaller interventions. Pursue a step-by-step bottom-up approach.
- Funding should be distributed allocating smaller budgets to many interventions. For those showing success and added value, it should be sustained in the long term. Methods for measuring success should be defined before the intervention.
- A general framework should be established that allows maximum collaboration and minimum intrusion. However, at the same time, objective evaluation by regional and national committees is needed. Committees should be formed by representatives of several levels [68].
- Do not be afraid to replicate interventions. This duplication will allow comparing interventions and work based on real evidence. It will pay off in future investments
- Involve users, researchers, and health trusts [38]. Empower them and make them be in control of the digital ecosystem evolution.
- Aim for long-term evolution and be time flexible based on the health value provided. Prioritise user satisfaction rather than medium-term rigid deliverables.
- Allow mixed approaches where some municipalities or health trusts may choose to implement themselves, some may outsource everything and some others may have a mixed approach.
- Large proprietary commercial platforms are not necessarily bad. They may be the optimal choice in some contexts. However, decision makers need to ensure that data ownership and control stay in the public sector and that these data can be accessed in high-quality formats at no extra costs.
- Think about data reuse and leverage with usability.
- Maintain home grown systems if they provide value, but offer their information through a standard API.

Conclusions

As is done in the biomedical domain, biomedical informatics interventions must be challenged by evidence following the scientific method. Naïve optimism has already cost billions in health informatics projects that were based on technology hypes and silver bullets. It is time for biomedical informatics to walk the road towards enabling evidence-based interventions. However, there is not a single way to achieve this and each intervention is highly dependent on its context. The most sensible approach is by developing one's own experience involving clinicians and technicians from health trusts and healthcare organisations. Funding should be allocated for small interventions that allow them to pilot different alternatives in parallel and share their experiences in national forums. Interventions should be independent but commit to open platform ecosystem architectures for ensuring their compatibility. Decision makers should plan for the long term and understand that evidence production takes time. In this way, the healthcare system as a whole can accumulate experience and walk towards better and objective decision-making in biomedical informatics interventions.

References

[1] K. Zoi, D. Kalra, P. Wilson, H. Martins, DigitalHealthEurope Recommendations on the European Health Data Space. Supporting Responsible Health Data Sharing and Use Through Governance, Policy and Practice [Internet], 2021, p. 18, Available from: https://digitalhealtheurope.eu/wp-content/uploads/DHE_recommendations_on_EHDS_July_2021.pdf.

[2] C. Peterson, C. Hamilton, P. Hasvold, From Innovation to Implementation—EHealth in the WHO European Region, first ed., vol. 1, WHO Regional Office for Europe, Copenhagen, Denmark, 2016. 1 vols. Regional Office Europe 1 https://www.euro.who.int/__data/assets/pdf_file/0012/302331/From-Innovation-to-Implementation-eHealth-Report-EU.pdf.

[3] M. Hertzum, G. Ellingsen, The implementation of an electronic health record: comparing preparations for epic in Norway with experiences from the UK and Denmark, Int. J. Med. Inform. 129 (September) (2019) 312–317, https://doi.org/10.1016/j.ijmedinf.2019.06.026.

[4] NHS Connecting for Health, Wikipedia, 2021. https://en.wikipedia.org/w/index.php?title=NHS_Connecting_for_Health&oldid=1001303315.

[5] T.K. Colicchio, J.J. Cimino, G. Del Fiol, Unintended consequences of nationwide electronic health record adoption: challenges and opportunities in the post-meaningful use era, J. Med. Internet Res. 21 (6) (2019), https://doi.org/10.2196/13313, e13313.

[6] A. Couffinhal, J. Cylus, R. Elovainio, J. Figueras, P. Jeurissen, M. McKee, et al., International Expert Panel Pre-review of Health and Social Care Reform in Finland [Internet], Helsinki, Finland, 2016, Available from: https://julkaisut.valtioneuvosto.fi/bitstream/handle/10024/79070/RAP_2016_66.pdf?sequence=1&isAllowed=y.

[7] K.B. Johnson, J.M. Ehrenfeld, An EPIC switch: preparing for an electronic health record transition at Vanderbilt University Medical Center, J. Med. Syst. 42 (1) (2017) 6.

[8] M. Krousel-Wood, A.B. McCoy, C. Ahia, E.W. Holt, D.N. Trapani, Q. Luo, et al., Implementing electronic health records (EHRs): health care provider perceptions before and after transition from a local basic EHR to a commercial comprehensive HER, J. Am. Med. Inform. Assoc. 25 (6) (2018) 618–626.

[9] W. Priestman, S. Sridharan, H. Vigne, R. Collins, L. Seamer, N.J. Sebire, What to expect from electronic patient record system implementation; lessons learned from published evidence, J. Innov. Health Inform. 25 (2) (2018) 92–104.

[10] Meld. St. 9 - Én innbygger - én journal [Internet], Det Kongelige Helse - og Omsorgsdepartement; 2012–2013, p. 60, Available from: https://www.regjeringen.no/no/dokumenter/meld-st-9-20122013/id708609/.

[11] D. Blumenthal, M. Tavenner, The 'meaningful use' regulation for electronic health records, N. Engl. J. Med. 363 (6) (2010) 501–504, https://doi.org/10.1056/NEJMp1006114.

[12] United Nations, Department of Economic and Social Affairs, Population Division, World Population Ageing [Internet], 2017, p. 124, Report No.: ST/ESA/SER.A/408, Available from: https://www.un.org/en/development/desa/population/publications/pdf/ageing/WPA2017_Report.pdf.

[13] Statistics Norway, Key Figures for the Population, 2021, Accessed October 2021, Available from: https://www.ssb.no/en/befolkning/nokkeltall/population.

[14] M. Rigby, S. Koch, D. Keeling, P. Hill, A. Alonso, E. Maeckelberghe, Developing a New Understanding of Enabling Health and Wellbeing in Europe, European Science Foundation, 2013. http://www.esf.org/fileadmin/Public_documents/Publications/Health_Wellbeing_Europe.pdf.

[15] C. Nøhr, S. Koch, V. Vimarlund, H. Gilstad, A. Faxvaag, G.A. Hardardottir, et al., Monitoring and benchmarking eHealth in the Nordic countries, Stud. Health Technol. Inform. 247 (2018) 86–90.

[16] B. Flyvbjerg, The Oxford Handbook of Megaproject Management, Oxford University Press, 2017. 625 pp.

[17] S.M. Meystre, C. Lovis, T. Bürkle, G. Tognola, A. Budrionis, C.U. Lehmann, Clinical data reuse or secondary use: current status and potential future progress, Yearb. Med. Inform. 26 (1) (2017) 38–52, https://doi.org/10.15265/IY-2017-007.

[18] Der Kongelige Helse-og Omsorgsdepartament, "Stortingsmelding nr. 9 (2012-2013) «Én innbygger – én journal»." Stortingsmelding. Stortingsmelding nr. 9 (2012-2013) «Én innbygger—én journal», 2012, 2012. https://www.regjeringen.no/no/dokumenter/meld-st-9-20122013/id708609/.

[19] T.K. Colicchio, J.J. Cimino, Twilighted homegrown systems: the experience of six traditional electronic health record developers in the post-meaningful use era, Appl. Clin. Inform. 11 (2) (2020) 356–365, https://doi.org/10.1055/s-0040-1710310.

[20] F. Aminpour, F. Sadoughi, M. Ahamdi, Utilization of open source electronic health record around the world: a systematic review, J. Res. Med. Sci. 19 (1) (2014) 57–64. https://www.ncbi.nlm.nih.gov/pmc/articles/PMC3963324/.

[21] R. Pedersen, K. Malm-Nicolaisen, Store IT-prosjekter må ha med forskning fra A til Å—Debatt og kroni…, archive.vn, December 21, 2020, Dagens Medisin edition, 2020. https://www.dagensmedisin.no/artikler/2020/02/11/store-it-prosjekter-ma-ha-med-forskning-fra-a-til-a/. http://archive.vn/Wn9TM.

[22] A.G. Storvik, Forskere fra flere fagmiljøer sier de ikke blir hørt i Akson-prosjekt…., Dagens Medisin, 2020. September 11, 2020, Dagens Medisin edition Https://www.dagensmedisin.no/artikler/2020/09/11/forskere-fra-flere-fagmiljoer-sier-de-ikke-blir-hort-i-akson-prosjektet/. http://archive.vn/kQluO.

[23] T. Beale, S.Heard, D.Kalra, D. Lloyd. n.d. "EHR Information Model." OpenEHR. www.openehr.org/programs/specification/releases/1.0.2.

[24] B. Christensen, G. Ellingsen, Evaluating model-driven development for large-scale EHRs through the OpenEHR approach, Int. J. Med. Inform. 89 (May) (2016) 43–54, https://doi.org/10.1016/j.ijmedinf.2016.02.004.

[25] L. Marco-Ruiz, K. Nicolaisen, R. Pedersen, A. Makhlysheva, P.A. Bakkevoll, Ontology-Based Terminologies for Healthcare—Impact Assessment and Transitional Consequences for Implementation, first ed., vol. 1, Norwegian Centre for E-health Research, Tromsø, 2017. 08-2017 https://ehealthresearch.no/files/documents/Prosjektrapporter/NSE-rapport_2017-08_Ontology-based-terminologies-for-healthcare.pdf.

[26] L. Marco-Ruiz, R. Pedersen, Challenges in archetypes terminology binding using SNOMED-CT compositional grammar: the Norwegian patient summary case, Stud. Health Technol. Inform. 245 (2017) 1332.

[27] L. Marco-Ruiz, R. Pedersen, Challenges in archetypes terminology binding using SNOMED-CT compositional grammar: the Norwegian patient summary case, in: Proceedings of the 16th World Congress on Medical and Health Informatics (MedInfo2017), IOS Press, Hangzhou, 2017. in press:in press.

[28] L. Marco-Ruiz, R. Pedersen, The patient summary case: challenges in archetypes terminology binding using SNOMED-CT compositional grammar, in: ETELEMED 2019 : The Eleventh International Conference on EHealth, Telemedicine, and Social Medicine, 2019, pp. 49–55. Athens (Greece).

[29] Direktoratet for e-helse, Akson—felles kommunal journal og helhetlig samhandling—ehelse, 2021, Government. ehelse.no. January 4, 2021 http://archive.vn/YOj6c.

[30] B. Bygstad, Hva er problemet med Akson? Det er i utakt med moderne digital innova…., archive.vn, 2020. August 5, 2020, Dagens Medisin edition Https://www.dagensmedisin.no/artikler/2020/08/05/hva-er-problemet-med-akson- -det-er-i-utakt-med-moderne-digital-innovasjon/. http://archive.vn/ouugj.

[31] J.O. Skisland, Visjonen om «Én innbygger—én journal» kan realiseres—Debatt og kr…, 2020, November 18, 2020, Dagens Medisin. http://archive.vn/8wp6j.

[32] M. Aanestad, Akson, elefanter—og Vipps—Debatt og kronikk—Dagens Medisin, Dagens Medisin, 2020. December http://archive.vn/WiRuu.

[33] G. Tømmerbakke Siri, A.G. Storvik, Forskningssenter advarer mot Akson:—Langt på vei avskiltet - Nyhete..., archive.vn, December 21, 2020, Dagens Medisin edition, 2020. Https://www.dagensmedisin.no/artikler/2020/08/17/forskningssenter-advarer-mot-akson- -langt-pa-vei-avskiltet/. http://archive.vn/EO9YI.

[34] A.G. Storvik, Forskere om Akson:—Som å lage koronavaksine uten plan for utvikling..., Dagens Medisin, 2020. November 4, 2020 Https://www.dagensmedisin.no/artikler/2020/11/04/forskere-om-akson- -som-a-lage-koronavaksine-uten-plan-for-utvikling-og-testing/. http://archive.vn/NuWBI.

[35] B. Rothman, J.C. Leonard, M.M. Vigoda, Future of electronic health records: implications for decision support, Mt Sinai J. Med. J. Transl. Pers. Med. 79 (6) (2012) 757–768.

[36] R. Nelson, N. Staggers, Health Informatics—E-Book: An Interprofessional Approach, Elsevier Health Sciences, 2016.

[37] R.S. Evans, Electronic health records: then, now, and in the future, Yearb. Med. Inform. (2016) S48–S61, https://doi.org/10.15265/IYS-2016-s006. Suppl 1 (May).

[38] G.-H. Severinsen, L. Silsand, G. Ellingsen, R. Pedersen, From free-text to structure in electronic patient records, Stud. Health Technol. Inform. 265 (August) (2019) 86–91, https://doi.org/10.3233/SHTI190143.

[39] C. Safran, Update on data reuse in health care, Yearb. Med. Inform. 26 (1) (2017) 24–27, https://doi.org/10.15265/IY-2017-013.

[40] A. Rector, D. Sottara, Chapter 20—Formal representations and semantic web technologies, in: R.A. Greenes (Ed.), Clinical Decision Support (Second Edition), Academic Press, Oxford, 2014, pp. 551–598. http://www.sciencedirect.com/science/article/pii/B9780123984760000208.

[41] "Advances in Natural Language Processing | Science." n.d. Accessed September 29, 2016. http://science.sciencemag.org/content/349/6245/261.

[42] P.M. Nadkarni, L. Ohno-Machado, W.W. Chapman, Natural language processing: an introduction, J. Am. Med. Inform. Assoc. 18 (5) (2011) 544–551, https://doi.org/10.1136/amiajnl-2011-000464.

[43] J.A. Greene, A.S. Lea, Digital futures past—the long arc of big data in medicine, N. Engl. J. Med. 381 (5) (2019) 480–485, https://doi.org/10.1056/NEJMms1817674.

[44] D.A. Hashimoto, G. Rosman, D. Rus, O.R. Meireles, Artificial intelligence in surgery: promises and perils, Ann. Surg. 268 (1) (2018) 70–76, https://doi.org/10.1097/SLA.0000000000002693.

[45] V.Stroetmann, D.Kalra, P.Lewalle, A.Rector, J.M.Rodrigues, K.A.Stroetmann, S.Gyorgy, U.Bedirhan, M.Virtanen, P. Zanstra, European Commission, Semantic Interoperability for Better Health and Safer Healthcare, Deployment and Research Roadmap for Europe, 2009, European Commission, Directorate-General for Communications Networks, Content and Technology, 2009. http://bookshop.europa.eu/en/semantic-interoperability-for-better-health-and-safer-healthcare-pbKK8009453/;pgid=Iq1Ekni0.1lSR0OOK4MycO9B0000QJkBOYqB;sid=ijW-cXiXYwwicXnKizoIC-Uf9KPbgDocb9mU=?CatalogCategoryID=OG4KABst1uEAAAEjnZAY4e5L.

[46] L. Marco-Ruiz, P. Carlos, J.A. Maldonado, L. Panziera, R. Chen, J. Gustav Bellika, Publication, discovery and interoperability of clinical decision support systems: a linked data approach, J. Biomed. Inform. 62 (August) (2016) 243–264, https://doi.org/10.1016/j.jbi.2016.07.011.

[47] B. Bygstad, A. Dulsrud, Digital Ecosystems as a Unit of Scientific Analysis. A Sociological Investigation, Hawaii International Conference on System Sciences (HICSS), 2020.

[48] Z.S. Morris, S. Wooding, J. Grant, The answer is 17 years, what is the question: understanding time lags in translational research, J. R. Soc. Med. 104 (12) (2011) 510–520, https://doi.org/10.1258/jrsm.2011.110180.

[49] T.R. Eisenmann, G. Parker, M. Van Alstyne, Opening Platforms: How, When and Why? Platforms, Markets and Innovation, Edward Elgar Publishing, 2009. https://www.elgaronline.com/view/9781848440708.00013.xml.

[50] OpenEHR—Organisational Structure, n.d. Accessed March 3, 2021. https://www.openehr.org/governance/organisational_structure.

[51] DIPS Arena, DIPS ASA, Bodø, Norway, Accessed September 27, 2021.

[52] Ocean Informatics Pty Ltd. n.d. Multiprac ShEHR. Ocean Informatics Pty Ltd. Accessed November 1, 2013. http://www.multiprac.com/.

[53] EHRbase—Ready for Action, Accessed September 27, 2021, https://ehrbase.org/.

[54] E. Ostrom, Governing the Commons: The Evolution of Institutions for Collective Action, Cambridge University Press, 1990.

[55] P.H. Conway, J.M. VanLare, Improving access to health care data: the open government strategy, JAMA 304 (9) (2010) 1007–1008, https://doi.org/10.1001/jama.2010.1249.

[56] K. Boudreau, Open platform strategies and innovation: granting access vs. devolving control, Manag. Sci. 56 (10) (2010) 1849–1872, https://doi.org/10.1287/mnsc.1100.1215.

[57] K. Malm-Nicolaisen, L.M. Ruiz, E.R. Evenstad, R. Pedersen, Efforts on Using Standards for Defining the Structuring of Electronic Health Record Data: A Scoping Review, Linköping University, 2019, p. 8.

[58] K. Malm-Nicolaisen, R. Pedersen, A.J. Fagerlund, Open or closed: a project proposal for investigating two different EHR platform approaches, Stud. Health Technol. Inform. 265 (August) (2019) 207–212, https://doi.org/10.3233/SHTI190165.

[59] Nordic Council of Ministers, Nordic Ehealth Indicators: Organisation of Research, First Results and the Plan for The … Place of publication not identified, Nordic Council of Ministers, 2013.

[60] G.-H. Severinsen, L. Silsand, A. Ekeland, PP136 how to apply health technology assessment to large scale E-health processes, Int. J. Technol. Assess. Health Care 35 (S1) (2019) 63–64, https://doi.org/10.1017/S0266462319002526.

[61] Akson, Elefanter—Og Vipps—Debatt Og Kronikk—Dagens Medisin, 2020. Archive.Vn. December 21, 2020 http://archive.vn/WiRuu.

[62] Ehrbase/Fhir-Bridge, 2021. Java. ehrbase, 2020. https://github.com/ehrbase/fhir-bridge.

[63] National-ICT, National [Norway] ICT's action 41: Use of archetype methodology for the definition, availabiliy and use of clinical information models in health-based Information Systems, 2012, p. 37.

[64] Ulriksen, Gro-Hilde, How to Involve the Users in the Large Scale Work with OpenEHR Archetypes in Norway? | Request PDF, ResearchGate, 2017, Accessed January 7, 2019, https://www.researchgate.net/publication/315862619_How_to_Involve_the_Users_in_the_Large_Scale_Work_with_OpenEHR_Archetypes_in_Norway.

[65] L. Marco-Ruiz, E. Bønes, E. de la Asunción, E. Gabarrón, J.C. Aviles-Solis, E. Lee, V. Traver, K. Sato, J.G. Bellika, Combining multivariate statistics and the think-aloud protocol to assess human-computer interaction barriers in symptom checkers, J. Biomed. Inform. (2017), https://doi.org/10.1016/j.jbi.2017.09.002.

[66] G.-H. Ulriksen, R. Pedersen, G. Ellingsen, Infrastructuring in healthcare through the OpenEHR architecture, Comput. Supported Coop. Work 26 (1) (2017) 33–69, https://doi.org/10.1007/s10606-017-9269-x.

[67] L.J.A. Marco-Ruiz, R.K. Maldonado, J.G. Bellika, Multidisciplinary modelling of symptoms and signs with archetypes and SNOMEDCT for clinical decision support, in: Studies in Health Technology and Informatics, IOS press, Madrid, 2015.

[68] G.-H. Ulriksen, R. Pedersen, G. Ellingsen, The politics of establishing ICT governance for large-scale healthcare information infrastructures, Int. J. Soc. Organ. Dyn. IT 6 (1) (2017) 48–61, https://doi.org/10.4018/IJSODIT.2017010104.

CHAPTER

20

Norway, Sweden, and Finland as forerunners in open ecosystems and openEHR

Hanna Pohjonen

Rosaldo Oy, Helsinki, Finland

Historical background and the way to organise healthcare services

Norway, Sweden, and Finland may be considered as forerunners in adopting openEHR and moving towards open digital health ecosystems. The rationale behind this is around the view that these countries have been early adopters of electronic health and social care records. Some of these early adoptions include the first paperless hospital of the world in Norway [1] (in 2006), as well as various eHealth services that have been developed for both professionals and citizens in the Nordic countries, such as virtual doctoral visits that have been reimbursed in Finland for many years.

Primary care has been mainly organised by municipalities either in healthcare centres (Sweden and Finland) or through family doctors/GPs (Norway). These services are financed mainly by municipal taxes, and subsequently, these municipalities have had the freedom to select their own IT systems which has led to a fragmented system base.

Specialty care/secondary care has been organised in hospitals owned by larger entities called counties, regions, hospital districts, or joint municipal authorities. Tertiary care has been organised by university hospitals that have had a unique role in healthcare education and in coordinating the activities in their catchment area. Hospitals and university hospitals have received their financing from regional taxation (Sweden), from governmental taxes (Norway), or as a combination of governmental and municipal taxation (Finland). Moreover, the counties, regions, hospital districts, and joint municipal authorities have had the freedom to choose their own IT systems, resulting in a fragmented system base in secondary and tertiary care as well. Furthermore, the best-of-breed approach in selecting dedicated systems such as an intensive care or maternity application has led to a variety of systems and the need for an extensive amount of integrations even within single hospitals.

Roadmap to Successful Digital Health Ecosystems
https://doi.org/10.1016/B978-0-12-823413-6.00011-2

The role of occupational health services differs between Nordic countries. There is a legal requirement in all Nordic countries for the employer to organise and mainly finance occupational health and safety services. In Sweden and Norway, the focus is primarily on preventing work-related causes of illness and on securing a safe work environment, whereas in Finland, occupational health services may also cover a large part of primary care or even some parts of secondary care services. Occupational health and safety services are primarily delivered by private healthcare service providers. Private healthcare providers do naturally have the freedom to choose their own IT systems.

Quote: *'Nordic countries have healthcare systems and healthcare financing that resemble each other. Healthcare is to a large extent publicly funded and organized by public authorities. Private healthcare is mainly seen as supplementary. In Finland however occupational health and safety services are often extended to cover a major part of primary and secondary care services. Decentralized decision making has resulted in a fragmented system base and extensive integration work in order to share data'*. ***Hanna Pohjonen***

Nordic drivers for data sharing

Cross-organisational data sharing, or data exchange, has historically been the objective in all Nordic countries. There have been several drivers for sharing data between the systems and organisations including:

- The best-of-breed approach that has resulted in a need to transfer data between the systems within the same organisation. Even in monolithic EHR installations, there have been other systems in use for dedicated areas such as intensive care, operating theatre, maternity, laboratory, and imaging.
- Recognition that cross-organisational patient pathways need to benefit from shared data between the systems in primary and secondary/tertiary care. The patient is in the central role, and data needs to follow the patient regardless of the organisation it was produced in. Decentralized decision-making has, however, resulted in a fragmented system base in the Nordics. More recently, the need for sharing social care data has been examined, and subsequently, this has led to social care organisations in Finland gradually starting to get connected to the national document sharing infrastructure. The integration of health and social care services is being discussed in the Nordics, and Finland has set this as one of the goals for its health and social care reform. Furthermore, as health care and social care become more integrated, there is an increased pressure to further develop and foster mechanisms to share data more effectively.
- There has been a need to use telemedicine tools due to long distances (the length of Norway, Sweden, and Finland is > 1000 km) and limited resource availability (radiology, pathology, etc.). This has understandably increased the need to share data in a trusted and secure way.
- Nordic citizens are typically keen on participating in their own welfare and care and having a possibility to access their own data. Citizens have the right to view their own data and also determine the level of sharing between professionals. The laws regarding consent and denials for data exchange differ between the Nordic countries, but in all Nordic countries, the citizen is in a key position to determine the use of his/her own data.

Evolution of data sharing: From documents to structured openEHR data

The Nordic countries have been early adopters of healthcare IT and are pioneers in broadband telecommunication networks. Sweden set up the Sjunet network in 1997 to facilitate data sharing between counties and municipalities in a trusted and secure way. Finland established so-called link-databases under a pilot law to enable data sharing across specialty and primary care in 1998. The 'link-databases' were regional, and each time patient data were produced locally, a link was generated to the 'link-database'. With patient consent, it was possible for a health professional to retrieve data from the local system for viewing purposes. The 'link-database' can be considered as an early version for an IHE XDS (Cross-Enterprise Document Sharing) registry. The Nordic countries also consolidated IT systems and built large regional solutions for electronic health records (EHRs) and imaging to facilitate data sharing, which would not have been possible in a multivendor environment. Sweden and Denmark were amongst the first to implement electronic prescriptions in Europe [2].

The need to share data finally resulted in national document-based data sharing infrastructures in all Nordic countries. These included the National Archive of Health Information (Kanta) [3] in Finland and the National Patient Overview (NPÖ) [4] in Sweden. All healthcare organisations were connected to national sharing infrastructures to enable the exchange of patient data with the patient's informed decision/consent. Document-based sharing is being used by health professionals as everyday work; in Finland, social care data are also gradually being exchanged. Additionally, whilst adopting document-based data sharing, the legal environments were also renewed to enable sharing to occur. It should be noted that the legal readiness for sharing is one of the prerequisites in adopting open ecosystems as well.

Benefits from document-based sharing are, however, somewhat limited. The need for complementary structured sharing then started to emerge following this realisation. Instead of *inter*-operability, healthcare stakeholders began to seek *intra*-operability, along with a semantically coherent data model for the various applications, to gain more benefits and to decrease the need for integrations. IHE XDS and HL7 v3 in connection with HL7 CDA r2 (Clinical Document Architecture) are examples of health data exchange schemas adopted for document sharing in the Nordics.

Document-based data sharing and structured openEHR-based data sharing are clearly complementary. The main difference is how the data can be utilised, that is, data processing methods that determine data usage possibilities. openEHR adoption enables the use of content-based queries (using AQL, archetype query language) and supports queries both on the individual and at the population levels. This is essential for analytics, automation, and decision support. Documents are for viewing purposes mainly, as it is impossible to easily combine data from various document-based data sources. For instance, XDS queries are based on administrative metadata only. Document-based sharing enables viewing of documents connected to different organisations, which is an organisation-centric approach. This approach is not suitable for the development of new innovations for utilising patient-centric, semantically coherent data from various sources. However, document-based sharing allows for any (read-only) content to be shared and there is no need for harmonisation, unlike in structured data sharing.

Table 1 illustrates a summary of Nordics' patient data exchange [5] methods. This shows that Denmark relies on document-based sharing, whereas in Sweden, Norway, and

TABLE 1 A summary of patient data exchange in the Nordics [5].

		Sweden	Norway	Denmark	Finland
Level of data exchange	Ongoing national openEHR modelling	Starting[b]	✓		✓[a]
	National CKM		✓		
	Regional/national openEHR projects	✓	✓		✓
	Local openEHR vendors	✓	✓		✓
	National data exchange (documents)	✓	✓	✓	✓
	Regional data exchange (documents)	Through the national data exchange	Through the national data exchange	✓	Through the national data exchange
	XDS-based data exchange	In imaging	✓	✓	Mainly in imaging

[a] *Related to UNA core.*
[b] *Vendor-driven until now.*

Finland, both document data sharing and structured data sharing exist. Each of these three countries is benefiting from a number of openEHR-based projects and local vendors, as well as archetype modelling initiatives. Norway is, however, the only country with its own established Clinical Knowledge Manager (CKM) supported by a national archetype governance programme.

As an example, Finland has had the Kanta architecture, with a general concepts and information model, since 2010 [3,6,7]. Kanta connects all public and private healthcare providers and pharmacies and is gradually connecting social care providers. A central data repository allows storing and sharing of electronic document-based health and social care data. Kanta has published the Clinical Document Architecture-Application Program Interface (CDA API) source code in their GitHub service to enable the formation and reading of CDA documents. The national patient data repository went live in 2010. All prescriptions have been issued and dispensed electronically since 2017. As a patient, you can browse your own medical records as well as prescriptions. The Finnish personal health records (PHRs) were made accessible to all citizens in 2018 [8]. The PHR allows citizens to store their own data with the help of various apps.

National openEHR introduction in Finland

Professionals started to look for structured data sharing, however—to complement the current way to share. There was a desire to combine data from various local systems—including both health and social care data to build an overview for different use cases: a community consisting originally of 19 hospital districts, 5 cities, 292 municipalities, and other stakeholders initiated a project called 'UNA Core' in 2015. Its intention is to establish a modular multivendor ecosystem and create an environment able to implement a customer-oriented, effective health and social care service [9]. UNA's mission is to support and reform health and social care practices bottom-up, by sharing data to service coordination [10].

UNA Core is an openEHR-based data integration and management solution connecting health and social care legacy systems, Kanta, and various other local data sources. UNA Core harmonises the legacy data with the openEHR platform and makes the data available in computer-processable formats for various applications. The UNA Dashboard application is the first user-visible application in place providing a combined health and social care-related overview of the person. These variable data exchange methods in use in the Nordics are summarised in Table 1.

Two perspectives to openEHR

(1) openEHR-based data exchange
openEHR projects typically start as data exchange projects and gradually develop into a modular EHR. Many projects around the world start by building a common openEHR data repository, then separating and harmonising the data gradually from local systems. Applications are built as functional modules to utilise these data. Typically, the first functionalities are various dashboards providing patient overviews to different types of users or different views to support any number of other use cases. New functional modules utilising the openEHR data repository are created one after another, including a laboratory summary, medication management, infection control, or chronic disease management. This leads to a multivendor environment and open ecosystems.

(2) Modular openEHR-based electronic health and social care records
The openEHR data model is revolutionary in the healthcare IT market as it makes applications semantically interoperable and enables the provision of an openEHR-based modular health and social care record. The future Nordic electronic health and social care record will not be a vendor-specific monolith, where the user interface, business logic, data repository, data model, and workflow are bundled in one proprietary software system. In future, it will be easier to split electronic health and social care functionalities into separate modules provided by multiple vendors.

Modularity is enabled, since openEHR-based electronic health and social care records allow the user interface, business logic, data repository, data model, and workflow to be separated from each other. Each functional module stores the data to a common openEHR data repository using standard APIs. The data are immediately available for another module, taking into account data privacy and security. This is how organisations can achieve flexibility with their procurement processes as it is possible to acquire the functionality that is necessary at that moment and from the vendor with the best-of-breed module. Instead of one broad monolithic suite, electronic health and social care records can be built from multiple functional modules. Some modules can also be built in an agile way by health and social care service providers themselves. The monolithic approach and open modularity are compared in more detail in Table 2 [11].

TABLE 2 Comparison of the monolithic approach and open modularity [11].

Monolith	Open modularity
• One vendor approach • User interface, business logic, data repository, data model and workflow in one integrated software • Specific system functionality cannot be changed to an equivalent one from another vendor • Cumbersome version updates • Proprietary data model • Data stored in a proprietary internal database • Heavy and complicated data migration when the system is changed • Internal care pathway • Analytics based on internal patient/customer data	• Can consist of functional modules from several vendors • User interface, business logic, data repository, data model and workflow have been separated • Modules are replaceable • Modules are independent • Enables frequent version updates by module • Agile development of innovations • End-users actively involved in module development • Open semantically harmonized data model • Data stored in a common clinical data repository • Migration not needed when the system is changed • Enables modelling and management of cross-organizational care pathways • Population level analytics

Nordic implementations

Both perspectives to openEHR have been used in the Nordics, even if the latter related to modular EHRs has been the dominant way. Finland started with the data exchange approach with UNA Core sharing openEHR-based health and social care data nationwide. UNA Core collects and harmonises patient and customer data from several IT sources, including Kanta, to be available for any openEHR-based module/application. The first application is UNA Dashboard providing combined patient and customer overviews. There are end-user pilots running in elderly care, acute care, specialty care, and management of the ward-level dashboard using the UNA Dashboard application. The future vision is to create an ecosystem of modules such as decision support and various artificial intelligence applications utilising patient and customer data via UNA Core. UNA Core is illustrated in Fig. 1.

Finland's approach was not a typical method, as openEHR data are not stored on the platform but kept only for a limited period, i.e. when the patient or social care customer is active. Activation may be triggered by any scheduled appointment, a hospital stay, or emergency. On the other hand, Finland is also following the other openEHR approach where legacy systems are being modernised step by step, functionality after functionality.

Sweden and Norway have proceeded by way of renewing their legacy EHRs gradually to openEHR-based applications. The procurement of the first openEHR platform is only now under way in Sweden, separately and not connected to any EHR renewal. In Norway, all openEHR implementations are related to modular EHRs. In fact, the dominant EHR market leaders in these countries chose openEHR as their own data model and started to modernise their solutions as their own internal decision. This is not necessarily the approach taken in other countries where openEHR projects have typically not been vendor-driven to a large extent. In Norway, the dominant EHR vendor evaluated possible new data models at the same time as the end-users. Both of them, yet independently, came to the same conclusion that openEHR was the right way to go.

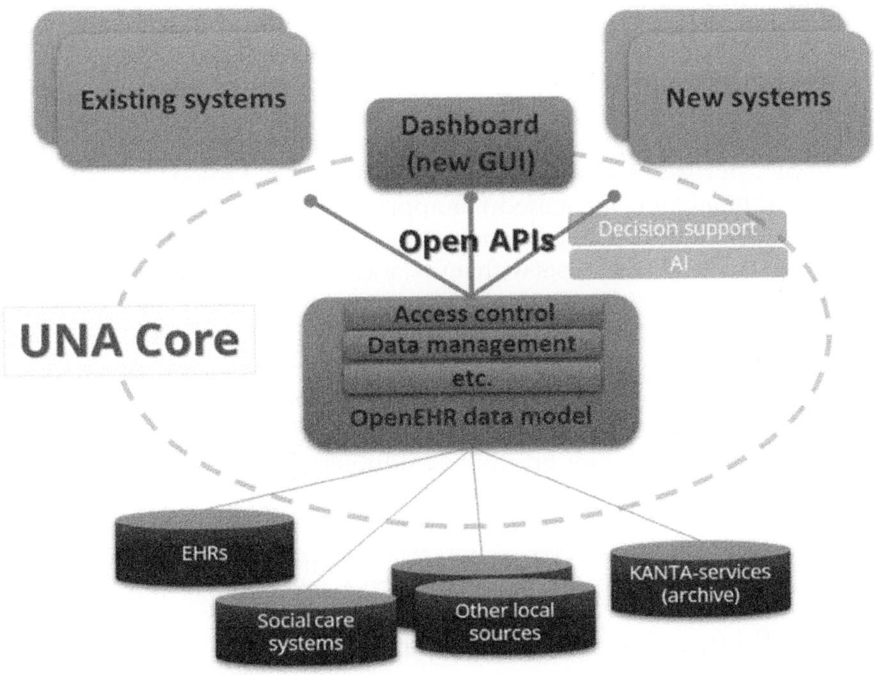

FIG. 1 UNA Core connecting different data sources, harmonising the data to openEHR, and providing it for different use cases. *Source: Una Oy [10].*

There have been important pioneering developments in Norwegian specialty care to model and execute certain patient pathways using the openEHR Task planning specification. In Sweden, on the other hand, the dominant EHR vendor created the openEHR guideline definition language (GDL) to facilitate different decision support scenarios. In Finland, there is a growing interest in both these developments and in connecting decision support to cross-organisational pathways.

Benefits of openEHR

What made Nordic companies select openEHR as their internal data model?

One of the reasons for a Nordic EHR software company to select openEHR as their internal data model was the inability to react to ever-changing needs with their monolithic approach and proprietary data models. Civil society is facing continuous change, and IT solutions must be flexible to adapt. There will be changes in organisation and service delivery responsibilities, the level of health and social care domain integration, utilisation of the possibility for cross-border health care in the European Union, and changes in laws and interpretations of the law. There are also changing expectations and capabilities of persons to participate in their own care, prevention, and self-care. There are continuously emerging new treatment options, personalised medicines, guidelines, and care protocols.

We now need a comprehensive view of patient data, analytics based on large amounts of data, cross-system and cross-organisational patient pathways, that adapt to changes in a flexible way. We also need an easy way to add innovative niche functionalities to benefit from changes in technology and know-how. Access to a considerable amount of data is needed, instead of the narrow data set provided by message-based integrations. This is hard to solve using a monolithic approach—a more disruptive approach is needed.

Application development using a modular approach is also more straightforward, and international expansion is simpler due to open platform thinking. Applications are developed faster and with less development costs with an established data model and shared services on the platform, thus enabling the developers to focus on functionalities. The applications are portable across open platforms, simplifying international expansion for small and big vendors. It is also possible to add niche modules from other vendors to your own ecosystem offering, or to develop modules in an agile way together with end-users increasing the probability of success.

For a Nordic vendor, openEHR is also a differentiator against the US monolithic giants offering an all in one solution but forcing adopters to compromise regarding specific functionalities.

Quote: *'The Nordic openEHR market is characterized by the dominant EHR vendors changing their portfolios from monoliths to openEHR-based applications—unlike many other markets'.* ***Hanna Pohjonen***

Why is openEHR attractive to Nordic healthcare delivery entities?

One of the biggest benefits of openEHR is the possibility to select best-of-breed functionalities for various purposes and dedicated areas creating a multivendor environment. The end-customers are frustrated with closed monoliths, and there is a desire for change. Healthcare delivery entities see possibilities in utilising open data for various innovations including secondary use of data. The dominant Nordic EHR players, former monoliths themselves, decided to modernise their solutions, and they received end-customers' support for that. Some end-users have even started to collaborate with the vendors in common research programmes.

Cross-organisational care pathways, as well as decision support, have been of special interest in the Nordic countries. openEHR offers tools for modelling standardised pathways, or parts of them, to personalise or reuse these for other pathways' definitions. The capability to integrate health and social care pathways is also attractive. openEHR-based decision support is seen as having the potential to bring extensive benefits for various scenarios.

IT professionals working in healthcare delivery entities appreciate simpler interoperability and less integration challenges between various systems and, in the longer run, the lack of the need for data migration. There is no vendor lock-in, and health and social care organisations can decide how to utilise data. Application modules can be replaced and updated independently as needed. Adding innovations is easy and even encouraged.

Managers consider openEHR to be cost-effective because of the step-by-step approach for renewing health and social care IT. Up-to-date management views, including population-level health status, can be provided as well as analytics, enabling for a better utilisation of resources with a focus on preventive actions. openEHR also allows for the optimisation of care pathways across regions leading to increased effectiveness.

Country maturity assessment

In this section, an evaluation framework for openEHR is discussed in the Nordic context. The purpose of applying this framework is to assess how ready or mature a country is to adopt openEHR and open ecosystems.

Market maturity, or the preparedness of a market to take openEHR-based solutions into use, is very important to understand when initiating openEHR activities. Even if you started openEHR activities in a country whilst the overall level of openEHR market maturity within the country is low, it would be difficult to be successful in the long run. A careful evaluation of the maturity level is essential prior to an openEHR-based solution provider considering entry into a new country, or prior to a hospital district considering the implementation of an innovative EHR renewal strategy based on an open ecosystem. When assessing country maturity for open ecosystems' and openEHR adoption, we need to consider various collective perspectives, including but not limited to the following:

- How digital is the country and its health and social care services in general?
- How common is data sharing in general?
- If shared, what is the geographical dimension of sharing? Organisational, regional, national, or even cross-border?
- What kind of content is shared?
- Who is the data shared with?
- What is the role of the citizen in the sharing process?
- Is the country legally ready to enable sharing?
- Are there regional or national development road maps for open ecosystems?
- What is the level of harmonisation in health and social care processes?
- Is the vendor community ready for open ecosystems? What is the maturity of the solutions in use? What is the culture inside the solution providers' enterprises? Are openEHR consultancy services available? Do software developers exist with openEHR capabilities?
- What is the current scope of openEHR activities? Is there a CKM and archetype/template governance programme in place? Is openEHR included in the RFP specifications? How big is the vendor community?

Fig. 2 shows the results of assessing market maturity for openEHR in the Nordic countries [5]. I have chosen certain axes and given each country numerous maturity measures along these axes. For example: the axis describing the usage of IHE XDS (Cross-Enterprise Document Sharing [12]) as a data sharing protocol is scored as follows: [5]

1. XDS projects in imaging
2. XDS projects beyond imaging, separate XDS consumers for viewing
3. National metadata definitions
4. XCA between XDS registries
5. Document access control based on XDS support profiles
6. XDS-compatible clinical applications like EHR
7. XDW in use for sharing workflow

A similar structure has been made for each of the axes.

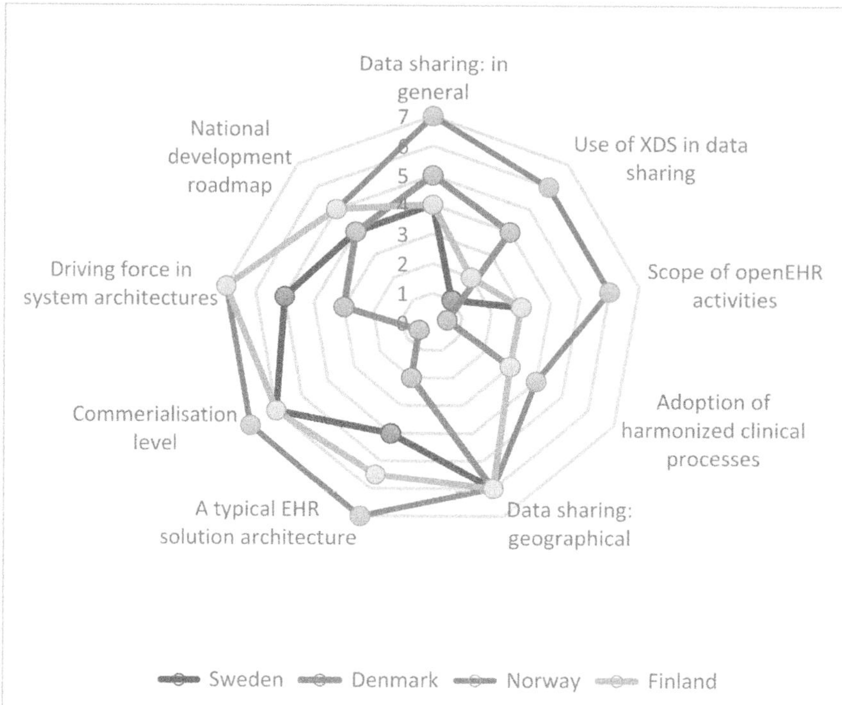

FIG. 2 Comparison of maturity levels for openEHR in the Nordic countries [5].

The figure clearly shows that data exchange is common in all Nordic countries and the sharing culture is strong. This is favourable for openEHR and indicates that these countries are potentially fast-growing markets for openEHR. Most successful openEHR implementations have been completed in regions with prior experience in document-based data exchange. Norway is one step ahead, but Sweden and Finland are developing fast. Denmark is not active in openEHR or open ecosystems, but considering its long history of data sharing and the developments in the neighbouring countries, the situation may evolve in the coming years.

Quote: *'Data exchange is common in Nordic countries making it a fast-growing market for openEHR'*. **Hanna Pohjonen**

Stepwise approach is characteristic for the openEHR road map in Nordic countries

Legacy system modernisation

Legacy EHR vendors are well-positioned to move towards openEHR-based ecosystems in the Nordics. They made their own decision to change the internal data model to openEHR, and they also started to drive the localisation of openEHR archetypes. In Sweden and Finland, the localisation work was originally vendor-driven, whilst in Norway, a collaboration of four specialty care regions and vendors exists within a national governance modelling programme.

TABLE 3 Modular openEHR-based EHR in Nordic countries [5].

Country	Geographical coverage	Examples of modules
Modular EHR		
Sweden/several counties	At county level, step-by-step transfer from monoliths to modules	Several modules covering primary care and specialty care functionalities, generic clinical decision support module
Norway/specialty care	3/4 Regions transferring step by step from monoliths to modules (covers 5/6 university hospitals, altogether around 70 hospitals)	Specialty care modules
Norway/primary care	Cloud-based service for private GPs, 4.5 million patients. In addition, local on-premise GP applications.	Core functionalities of primary care. Physiotherapy module.
Finland	On healthcare organisation level: transfer step by step from monoliths to modules	Core functionalities of health care (structured journal), various overviews for different specialty areas like child growth curve

The dominant EHR players are renewing their existing installed base and gradually turning legacy functionalities into openEHR-based functional modules. Table 3 illustrates the progress in different countries [5]. In Norway, only one specialty care region selected a US monolith, and the remaining regions are involved in this renewal. All regions are involved in the archetype definition work. Almost 50% of primary care in Norway is using openEHR-based applications. Most of the counties in Sweden are also developing from legacy to openEHR together with their existing EHR vendor. There are two dominant EHR vendors in Finland. One of them is modernising their legacy system and renewing the installed base step by step via upgrades. Big procurements pursuing open ecosystems are also expected.

In addition to the dominant players, there are also innovative agile openEHR challengers in all of these countries. Some also operate across Nordic country borders. One of these agile vendors has set up a cloud-based service for private GPs in Norway. The goal is to create Nordic openEHR applications or modules which can be utilised in open ecosystems across the Nordics.

New open ecosystem-based procurements

Finland is facing big health and social care IT renewals in the coming years. UNA, the Finnish nationwide IT development collaboration for public health and social care organisations, plans, manages, and resources collaborative projects on behalf of hospital districts and municipalities. The UNA vision is to enable the implementation of a modular ecosystem needed to move towards modern, open, and agile health and social care IT applications. This requires it to promote an ecosystem approach, allowing competition and enabling new innovations through true modularity and open APIs. This collaboration began with the UNA Core project described previously.

The upcoming health and social care IT renewals in Finland will also adopt a step-by-step approach to meet the specific needs of hospital districts or service providers. The usual practice is to separate data from applications and harmonise the data to a semantically coherent data format. Migration of legacy data can also be on-demand. The legacy EHR can then be changed step by step acquiring new EHR modules replacing specific functionalities. The legacy system stays in parallel until it is ready to be removed. It is not necessary to replace the whole system in order to change one functionality.

The order of modules to be acquired and the pace of renewal happen according to the specific local needs. This allows for a cost-effective way to renew, with a focus on meeting urgent needs. It also allows for innovations to be added, even when the existing legacy system is still in place. It is important to notice that some systems will always be non-openEHR-based and integrate with standards like HL7 FHIR. In Finland, the national CKM is still deficient, even with ongoing archetype work.

In summary, the typical road map to openEHR in the Nordics is to gradually change the existing legacy systems to openEHR-based solutions. This is strongly related to the software development of legacy EHR vendors. In addition, there will also be new procurements, primarily in Finland, to purchase parts of health and social care ecosystem. Fig. 3 summarises the typical steps for a regional or national stakeholder on its way to open ecosystems [5].

Success factors for a digital ecosystem and openEHR-based projects—A Nordic experience

The Nordic experiences associated with each country's specific openEHR adoption are presented in Boxes 1–3, as a summarised overview to highlight context before identifying the collective success factors.

The experiences in the Nordic market provide several factors considered to have influenced the success of an open ecosystem or an openEHR-based project. Unless there is a sufficient level of progress being made in each of these areas, there is a probability that any one of these projects will face significant delays. Factors influencing success are as follows:

- The data sharing culture needs to be present in the market. Experience gained on document-based sharing is beneficial, preferably with broad scope content and in large geographies.
- The legal environment should be favourable for sharing, and the necessary data privacy and security services should be in place.

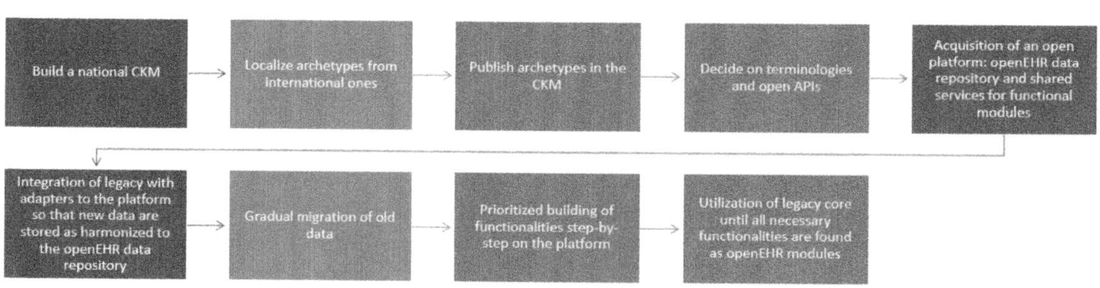

FIG. 3 A typical road map to open ecosystems [5].

Box 1 Summary of Norway's openEHR adoption.

Factsheet Norway

- Leading the way to open ecosystems in the Nordics.
- National recommendations for openEHR.
- Archetype modelling nationally led collaboration between all four specialty care regions.
- National CKM providing the archetypes to the vendors.
- More recently, social care-related modelling.
- Large vendor commitment.
- Three out of four specialty care regions changing their legacy systems to openEHR (vendor-driven legacy modernisation).
- Almost 50% of the GP market already using openEHR-based applications.
- Several pilots in modelling and management of patient pathways in specialty care (openEHR Task planning).

Box 2 Summary of Sweden's openEHR adoption.

Factsheet Sweden

- The openEHR community started with the lead of Östergötland County.
- The Swedish Association of Medical Informatics is in a key role in archetype localisations.
- No national CKM yet.
- Most of the counties changing their legacy systems to openEHR (vendor-driven legacy modernisation).
- openEHR-based clinical decision support in use within several counties.
- The first end-customer led openEHR platform project ongoing in Östergötland County.
- openEHR-related academic research.

Box 3 Summary of Finland's openEHR adoption.

Factsheet Finland

- UNA Core project for national health and social care data exchange.
- New legislation coming to facilitate sharing of health and social care data.
- UNA Core collects and harmonises patient and customer data across several IT sources for use with any application.
- The first application is UNA Dashboard providing combined patient and customer overviews.
- The future vision is to create an ecosystem of applications utilising patient and customer data via UNA Core.
- End-user pilots running in elderly care, acute care, specialty care, and management of the ward-level dashboard.
- The national CKM is still deficient, even with ongoing archetype work.
- Many healthcare organisations are upgrading their systems to openEHR (vendor-driven legacy modernisation).
- Large health and social care IT procurements are coming in the next few years.
- A proposed health and social care reform is under development, where health and social care sector integration is one of the key objectives, setting lofty expectations for sharing architectures.

- Clinical processes should be somewhat harmonised, or it should be understood that harmonisation is beneficial.
- There should be a broad enough understanding amongst the end-users and decision-makers about the benefits of an open ecosystem.
- There should be readiness within the vendor community for collaborating and for acting in open ecosystems. It is good to recognise that the vendors will need to have capabilities for transforming their legacy systems to openEHR-based offerings and would need to allocate resources accordingly. This cost will require financial investment that needs to be justified through customer projects.

In Nordic countries, there are always multiple stakeholders in every project. Ministries are involved, through their guidance and control role. Hospital districts or hospitals are involved as providers of secondary or tertiary care services. Municipalities or cities are involved through their responsibility to organise and finance the services or as providers of primary care or social care services. Health and social care professionals are involved as users of applications and data. In order to build a true ecosystem, it is important to increase awareness amongst all stakeholders. A monolithic solution may seem lucrative unless there is sufficient understanding of the benefits of modularity and openEHR.

Finland has been successful in using an openEHR eLearning environment for educating a broad number of stakeholders about openEHR benefits [5]. The eLearning environment describes openEHR and openEHR market in an understandable way, and it is built according to the pain points reported by end-customers. Besides learning, it is also a social forum to communicate and to discuss. In Finland, it has been used by health and social care providers, software companies, consultants, and in-house IT companies to achieve buy-in for openEHR amongst all stakeholders.

The biggest challenges in the Nordic openEHR projects

The biggest challenge associated with the software development projects for the modernisation of legacy systems has been the slow pace of functional module development. This is, however, understandable due to the vendor-required transition where parts of the offering are still legacy and new functionalities are added to the environment as openEHR-based modules.

In data exchange projects, the primary challenge revolves around harmonising legacy system data. In Finland, this has been easier since the legacy system vendors have understood the importance and benefits of openEHR. Harmonising data is a cumbersome task especially when some parts of the system base, like social care systems in Finland, are at the end of their product life expectancy.

If the rules and processes for an ecosystem are ignored whilst planning open ecosystem procurements, this can risk a procurement process and may result in a nonworking ecosystem. An ecosystem-related procurement process differs from a traditional procurement process, as innovative capabilities are required. A successful procurement is based on a well-defined architecture and rules. Open ecosystems also need to be orchestrated, but not in a traditional way. The main processes and rules of the open ecosystem need to be defined, including decision-making processes (who decides what, including approval of new applications to the ecosystem), change processes, and the business model of the ecosystem.

It is also important to understand that there are tasks required to be performed at the national level that require owners and resources for the national development activities. In Norway, the modelling work is not led by the governmental bodies but by specialty care regions themselves and this may create a risk for suboptimisation. In Finland, the discussion about the CKM owner is still ongoing. As multiple vendors are already localising international archetypes, there remains a risk to standardisation due to the localisation differences across vendors. This is best overcome by the development of national standard data value sets using jointly agreed data elements that best indicate value for multiple uses throughout the data supply chain. In Sweden, modelling work was led by the dominant EHR vendor until recently, when this changed due to a key role being played by the Swedish Association of Medical Informatics.

References

[1] J.-T. Lium, H. Laerum, T. Schulz, A. Faxvaag, From the front line, report from a near paperless hospital: mixed reception among health care professionals (in eng), J. Am. Med. Inform. Assoc. 13 (6) (2006) 668–675.

[2] McKesson, e-prescription—Making It Happen, 2018, 18 January, Available: https://www.ocp.pt/frontend/contents/e-prescription_Policy_Paper.pdf.

[3] V. Jormanainen, Large-scale implementation and adoption of the Finnish national Kanta services in 2010–2017: a prospective, longitudinal, indicator-based study, Finn. J. eHealth eWelfare 10 (4) (2018) 381–395, https://doi.org/10.23996/fjhw.74511.

[4] N. Sellberg, J. Eltes, The Swedish patient portal and its relation to the national reference architecture and the overall eHealth infrastructure, in: M. Aanestad, M. Grisot, O. Hanseth, P. Vassilakopoulou (Eds.), Information Infrastructures Within European Health Care. Health Informatics, Springer, Cham, 2017, https://doi.org/10.1007/978-3-319-51020-0_14.

[5] H. Pohjonen, Rosaldo openEHR eLearning Environment, 2021, Available: www.rosaldo.fi.

[6] T. Suna, Finnish national archive of health information (KanTa): general concepts and information model, FUJITSU Sci.Tech. J. 47 (1) (2011) 49–57.

[7] Kanta, Architecture, 2020, 19 January, Available: https://www.kanta.fi/en/system-developers/architecture.

[8] P. Gratzel-von-Gratz, Digital Finland: Towards an Ecosystem of Health and Wellbeing Apps, 2019, 19 January, Available: https://www.mobihealthnews.com/news/emea/digital-finland-towards-ecosystem-health-and-well-being-apps.

[9] E. Kujansuu, UNA Program: Together to the Future, 2016, 19 January, Available: https://www.kuntaliitto.fi/sites/default/files/media/file/Kujansuu_UNA120417_1.pdf.

[10] J. Rannanheimo, UNA Core-OpenEHR Based Data Exchange in Finland, 2020, Available: https://www.openehr.org/news_events/event_reports/11.

[11] H. Pohjonen, What Is a Modular openEHR Based Electronic Health Record? How Does It Differ From a Traditional Model? 2020, Available: https://www.openehr.org/news_events/event_reports/10.

[12] Integrating the Healthcare Enterprise (IHE) International. www.ihe.net.

CHAPTER

21

The Brazilian digital health system: Building the digital transformation to engage country citizens

Heimar de Fatima Marin[a], *Juliana Pereira de Souza Zinader*[b], *Frank James da Silva Pires*[c], *and Jacson Venancio de Barros*[d,e]

[a]Federal University of São Paulo (UNIFESP), ICT in Health Research at CETI, Brazilian Internet Steering Committee, NIC.Br, Certificate Program in Health Informatics at Sírio Libanês Hospital (HSL), Sao Paulo, Brazil [b]Innovation in Digital Systems of Datasus, Institute of Informatics of the Federal University of Goiás (UFGO), São Carlos School of Engineering, Ribeirão Preto Medical School and Chemistry, Institute of São Paulo (USP), Sao Carlos, Brazil [c]Information System and Operation, Department of Health Informatics of the Unified Health System, Ministry of Health (Datasus/SE/MS), Health Informatics by Sírio Libanês Hospital Program, Sao Paulo, Brazil [d]Department of Health Informatics of the Unified Health System, Executive Secretariat of the Ministry of Health (Datasus/SE/MS), University of São Paulo (USP), Sao Paulo, Brazil [e]Department of Medical Informatics of the USP Medical School, Sao Paulo, Brazil

Introduction

Health systems across countries are facing several challenges never experienced before by governments, patients, providers, and citizens. Factors such as an ageing population, and a technological revolution that offers different options for disease treatments, were overcome by the biggest challenge that has been used to cope with the pandemic of COVID-19.

It is advocated that information and communication technology (ICT) has the potential to support healthcare delivery. Technology is the fundamental resource that can be used to provide the population with the best and adequate information and knowledge, to maintain health status and enhance the quality of life. With the COVID-19 pandemic, ICT is playing a crucial role to engage citizens and offer qualified information. Thus, country leaders across

Roadmap to Successful Digital Health Ecosystems
https://doi.org/10.1016/B978-0-12-823413-6.00003-3

the globe are aware that they must deploy the digital process and select among a range of ICT, the best strategies available to address needs of each region, considering the diversity of situations and degree of spread and severity of the virus. Technology resources are the important tools to support health promotion. To achieve better results, national strategies must be in place and be considered as the major strategy for the nation, involving all levels of stakeholders and citizens.

At this chapter, we will share information on how the digital transformation has been conceived in Brazil to engage citizens in a country with enormous diversity of culture, economy, industry, and technology development.

The Brazilian healthcare system

Brazil, the most populated country in South America, has 211,755,692 inhabitants, according to the Brazilian Institute of Geography and Statistics [1]. It occupies 8,510,345,538 Km^2 in territorial division, containing macroregions, federation units, mesoregions, microregions, immediate geographic regions, intermediate, geographic regions, and municipalities, with a demographic density of 22,43 inhabit/m^2. The most recent survey showed that the Brazilian territorial structure kept its 5570 municipalities. In addition to these municipalities belonging to the 26 states, there is a federal district (Brasília) and the isle of Fernando de Noronha that belongs to the State of Pernambuco. In 2020, there were a total of 10,631 municipal districts [1]. These numbers are substantial when the healthcare model is considered to achieve goals and open access to all citizens—from the capital of each state to the most distant and small municipality where, most of time, there is just a healthcare basic unit with no hospital.

It is worth emphasising that the Brazilian healthcare system considers health as a universal right assured to each citizen by Constitution. The Unified Healthcare System (Sistema Único de Saúde—SUS) started in the years of 1979 and 1980 as an initiative from several health providers and clinicians that were engaged in the health organisations across the country. The objective was to create a public system able to solve challenges in the care delivery to the population in such a diverse and geographic dimension [2].

In the 1980s, Brazil faced an economic and political crisis and, along with this economic crisis and democratisation, the public health movement proposed a reform agenda. On the proposal, the guidelines included in the Federal Constitution of 1988, recognised health as a citizens' right. Consequently, the universal and public Unified Health System (SUS) was created [3].

By that time, the member of the meeting named as 'the Eighth National Health Conference' held in March 1986 discussed the reformulation of the national health system and its financing possibilities. The proposal was based on the United Kingdom National Health Service (NHS) in place at that time. Therefore, it became official on 17 May 1988 by the Brazilian Federal Constitution which states that it is the duty of the state to guarantee health to the entire Brazilian population [4].

By principle, the fundamental goals were promoting health, preventing sickness, treating the sick and injured, and attacking serious disease. To achieve these goals, the system must be supported by three levels of government: federal, state, and municipal. Consequently, the number of cities, municipalities, and organisations represents a necessary control to maintain the health system and deploy the model.

In 2020, Brazil was the only country with over 200 million inhabitants to provide the universal coverage on public health [5]. After more than 30 years of its implementation, the principle remains [6] as:

Universalisation: health is a right of citizenship for all people, and it is up to the state to ensure this right, and access to services must be guaranteed to all people, regardless of sex, race, occupation, or other social or personal characteristics.
Equity: the objective of this principle is to reduce inequalities. Although everyone has the right to services, people are not the same and consequently have different needs. Hence, equity means treating the unequal unequally, investing more where the need is greatest.
Comprehensiveness: this principle considers all people healthcare needs. For this, it is important to integrate actions, including health promotion, disease prevention, treatment, and rehabilitation. Together, the principle of integrality assumes the articulation of health with other public policies, to ensure integration between the different areas that have an impact on the health and quality of life of individuals.

With the development of science, economy, and technology, Brazil as many other countries in Latin America faced the necessary reform agenda as consequence from health policies on financing, decentralisation (federal to municipalities), the production of supplies, and labour management. The growing open market after the decade of 1990 brought a significant contribution from the private sector in the country in several activities including health sector.

Despite the expanded public services, the Brazilian healthcare system relies also on the private hospital beds, laboratories, and other private services. Health insurance and healthcare plan companies expanded in the country. Therefore, in 2000, the National Health Agency was created to regulate contracts and organise the private health plan market.

In summary, the Brazilian health policies present advances and contradictions as many other segments in the country [7,8]. As highlighted by Machado [9]:

> 'Brazil has a universal public system which covers more than 200 million people but remains limited by financial difficulties and inequalities. The country also has a dynamic private sector, which includes private providers to the SUS and a significant health plan and insurance segment, which serves nearly 50 million people, that is 25% of the Brazilian population, who also use the SUS'.

Information and communication technology in Brazilian healthcare facilities

Since 2008, the Organisation for Economic Co-operation and Development (OECD) has led an effort to provide countries with reliable statistics to compare information and communication technology (ICT) development and policies in the health sector [10], to assist governments in understanding the barriers and incentives to use and to understand the results on economic and social benefits from their application. A model questionnaire was completed and published in 2013 [11].

Once available, Brazil adopted the OECD model survey. The initiative was carried out by the Regional Center for Studies on the Development of the Information Society (CETIC.br). The research planning was initiated since January 2012 and in 2013, Brazil became one of the first countries to pilot the draft model survey questionnaire. The survey has been conducted using a probabilistic sample of public and private healthcare facilities. It also included healthcare professionals: physicians and nurses. Moreover, the study has institutional and methodological support from an expert group composed by representatives from the government, academia, organisations from civil society, and international agencies. The results have enabled a mapping of the ICT infrastructure; the use of ICT systems and applications; and activities, motivations, and barriers preventing the use of ICT by healthcare professionals [12].

Thus, since 2012, yearly, the ICT in health survey investigates the access and use of information and communication technologies (ICTs) in Brazilian healthcare facilities. The sample comprised executive and ICT directors (CEO and CIO), physicians and nurses included at the National Registry of Healthcare Facilities (CNES), kept by the Department of Informatics of the Ministry of Health (DATASUS, Ministry of Health).

According to the definition adopted by the National Registry of Healthcare Facilities (CNES), healthcare facilities can be broadly defined as all locations designated for the provision of collective or individual healthcare actions and services, regardless of their size or level of complexity. The ICT in health survey took into account the information adopted by the CNES to identify the healthcare professionals analysed in the study. These professionals worked in healthcare facilities providing care to patients from the Unified Health System, from any private institute or both. The identification of physicians and nurses was based on the Brazilian Occupational Classification (CBO) held by the Ministry of Labor and Employment (MTE).

To maintain the indicators and be able to keep a longitudinal data collection, it was decided to exclude isolated clinics, mobile units (e.g. boats on the Amazon River to reach small indigenous villages), and establishments with nonphysician and nonnurses—just trained health agent. Consequently, the sample was selected by probability sampling technique considering the country region and the type of establishment (outpatient, diagnosis and therapy services, inpatient up to 50 beds, and inpatient with more than 50 beds), including both private and public facilities [12].

The data are collected through two structured questionnaires: the first is answered by administrative professionals from the facilities (CIO or CEO) and the second questionnaire is applied to the healthcare professionals (physicians and nurses). It is important to highlight that to assure the complete understanding of all questions on the survey by the subjects, some terms were avoided such as electronic patient record and electronic medical record. The idea was to guarantee the quality of responses even after a cognitive evaluation of the complete questionnaire. Thus, it was decided to use the functionalities terms included on the computer systems.

As stated, to keep comparable data over the years, the overall goal of the survey was to understand the stage of ICT adoption in Brazilian healthcare facilities and their appropriation by healthcare professionals. Within this context, the study had the following specific objectives:

1. **ICT penetration in healthcare facilities**
   - Identify the ICT infrastructure available in Brazilian healthcare facilities.
   - Investigate the use of ICT-based systems and applications to support care services and management of facilities.
2. **ICT appropriation by healthcare professionals**
   - Investigate the ICT skills of professionals and the activities carried by them with the use of ICT.
   - Understand the motivations and barriers related to the adoption of ICT and its use by healthcare professionals.

As an annual survey, the questions examine data and information related to the last 12 months of implementation and adoption of ICT-based systems and services. The last edition published on 2020 results from data collected in 2019.

For healthcare facility analysis units, the results are presented for domains defined according to the following variables and levels:

- **Administrative authority:** Corresponds to the classification of institutions as public or private.
- **Type of facility:** This classification is associated with four different types of facilities, based on the type of care delivery and size, in terms of beds—outpatient, inpatient (up to 50 beds), inpatient (more than 50 beds), and diagnosis and therapy services.
- **Region:** Corresponds to the division of Brazil into macroregions (North, Central-West, Northeast, Southeast, and South), according to IBGE criteria, already mentioned before.
- **Location:** Refers to whether a facility is in a capital or in noncapital cities of each federative unit.
- **Primary healthcare unit (PHU) identification**: Refers to the PHU and not PHU classifications.

To the analysis for healthcare professionals (physicians and nurses), the following characteristics were added to the domains above:

- **Age group:** Refers to the age group of the professional, divided into three ranges, depending on the sample group:
  - for nurses was considered up to 30 years old, from 31 to 40 years old, and 41 years old or older.
  - for physicians was up to 35 years old, from 36 to 50 years old, and 51 years old or older.

In 2019, the interviews were conducted in 2427 healthcare facilities, reaching 63% of the planned sample of 3827 facilities. Related to the professionals, 1732 physicians and 2458 nurses were interviewed [13].

Since 2013, the results show a growth trend in the availability of computers and the Internet in Brazilian healthcare facilities. At inpatient institutions and those that provide diagnosis and therapy support services, ICT was universally available in 2019. Among outpatient facilities, 95% had computers and 91% had Internet access. This result indicates a deficit: about 4400 facilities did not have computers and 8200 did not have Internet access. Disparities in access to ICT infrastructure are striking between administrative jurisdictions and the country's geographic regions. It is estimated that 50,000 public facilities (92%) had computers and 46,300 (85%) had Internet access, while in private facilities access was universal (100%) [13].

The results for 2019 showed progress on the availability of computerised systems to record health data in 82% of facilities, compared to 2018, when this percentage was 73% [14]. Public facilities, inpatient facilities with up to 50 beds, and those in the North and Northeast region presented the lowest proportions of electronic systems. That means: an estimated 18,000 facilities still did not have electronic systems. While nine out of ten private facilities (88%) had electronic systems to record information about their patients, this proportion was 74% among those in the public sphere. In other terms, it shows that the resource was not available to about 12,000 public healthcare facilities in 2019. The presence of electronic systems was also uneven among facilities located in noncapital cities (79%) compared to those in capital cities (92%) [13].

The questionnaire comprises investigation related to the information security and resources privacy and security of the patient data recorded. The results showed that biometric-protected access to electronic systems presented the greatest positive variation compared to the previous edition (from 8% to 16%). Inpatient facilities (more than 50 beds) and facilities that provided diagnosis and therapy services presented the best results in terms of the adoption of security tools. In 2019, two new tools were included: data loss protection/prevention (DLP), present in 22% of healthcare facilities, and two-factor authentication, installed in 7% of facilities [13,14].

Although accurate information is fundamental to provide healthcare delivery, the COVID-19 pandemic brought an urgent sense of timely access to updated information as a relevant factor in a dynamic global health decision-making scenario. The use of ICT has been essential to reconcile agility, accuracy, and reliability in the sharing of information between the public and private networks of the Brazilian health system and between federated units. Thus, it was an urgent call to all governments at the three levels (ministry, states, and municipalities) to expand the role of ICT in health for the population.

For example, telemedicine was allowed if the presence of a physician was guaranteed on both sides of the connection. In other words, it was forbidden to practice telemedicine between patient and physicians or nurses. However, with the pandemic reaching all cities and creating a state of emergency, the Brazilian federal government enacted Law No. 13989 of 15 April 2020 [15], authorising the use of telemedicine during the pandemic and allowing physicians to care for their patients through virtual media. This measure also allows healthcare professionals who are at higher risk for the virus to provide care virtually online.

Brazil, as other countries, confronts the challenge of having many professionals transferred to the front-line care of patients with COVID-19. At the same time, others were on leave because they have been contaminated. Consequently, another extra difficulty was added to the scenario: the impossibility of expanding the health workforce within a short time frame to meet the growth of this new demand. Thus, the use of technologies and telemedicine has made it possible, and patients could receive care orientation and diagnosis by telemedicine [16].

However, as already presented, since the access to computer and Internet is not equally spread and stable in the country, the use of telemedicine still faces barriers to its expansion due to gaps and inequalities in ICT access by healthcare facilities and citizens.

Among programs and schemes to act in response to a different challenge, in 2019, the Ministry of Health launched the Ordinance No. 1434 that institutes the Connect SUS Program, composed of the National Health Data Network (RNDS) and the informatisation of healthcare, developed by the Department of Informatics SUS (DATASUS). Connect SUS, since November 2019, started with the Pilot Project in Alagoas, a state in the Northeast region. The major objective was enabling citizens to know their trajectory in the SUS, such as which vaccinations they took, the consultations performed, exams, hospitalisations, treatments, and medications. In addition, Connect SUS provides information related to the nearest health facilities organised from the RNDS already implemented in the country. As a consequence, the connectivity of primary healthcare units (PHUs) had to be updated or established, since PHU is considered as the first level of assistance or the entrance door to the healthcare system for the population [17].

It is worthy to emphasise that in the face of epidemiological outbreaks and pandemics, primary healthcare represented an even more distinguished role due to the nature of its main functions. The local PHU has the information of local areas supplying links between patients and health teams, monitoring families and following up on suspected or mild cases.

Another feature to be considered by the healthcare system is the professional training and capacity building of human resources to use computerised systems supporting clinical practice. Therefore, the ICT in health research also investigates the use and appropriation of ICT by healthcare facilities and clinicians. It is also assumed that a proper training of these professionals is an important factor for the advancement of digital health, as well as their engagement in the adoption of new devices and systems in patient care routines. Thus, the Connect SUS program also demanded training courses in ICT and data analysis to providers who work at municipal health departments.

The availability of computers for use by physicians and nurses has remained relatively stable in recent years. In 2019, 88% of physicians had computer access at health facilities and 94% had Internet access. In addition, among physicians who worked in public facilities, 23% said they did not have computers available, a high percentage compared to those who worked in private facilities (3%). Among nurses, 90% had computers and 92% had Internet access. The difference in computer availability between nurses from private and public institutions was slightly lower compared to that between physicians, considering that 86% of nurses in public facilities and 98% of those in private facilities reported having computers available. The use of computers in patient care influences how medical prescriptions are written. In 2019, 14% of doctors said they wrote their prescriptions only by hand, a proportion that remained the same in relation to 2018. At the inpatient facilities with more than 50 beds, 75% of physicians wrote prescriptions in electronic format [13].

The appropriation of ICT in the work of physicians and nurses can also be observed in how they perceive the impact of these technologies on their activities. In 2019 [13], most nurses (60%) perceived neither an increase nor a decrease in the workload with the use of computers and the Internet. Those who perceived an impact were divided equally among those who believed the workload had increased (22%) and those who said it had decreased (17%). However, it is important to emphasise that the perception that computer use increased workloads was more common among nurses who worked in public facilities (25%) and in PHUs (37%) when compared to those in private facilities (16%).

In 2018, nurses used electronic system functionalities less often compared to the physicians who use these for prescription and surgery reporting. Nurses used more functionalities related to administrative activities, such as generating requests for materials and supplies, and for booking appointments, tests, or surgeries. Functionalities used during patient care were also among those most used by nurses, such as listing medications being taken by a specific patient and listing lab test results for a specific patient [14].

Similar to many other countries in the world, Brazil also presents a high-level use of mobile phones. According to the Regional Center for Studies on the Development of the Information Society (Cetic.br) at the Brazilian Internet Steering Committee (Comitê Gestor da Internet no Brasil—CGI.br), 74% of users aged 16 years old access the network exclusively by cell phone. Among Internet users aged 16 or over, who attend school or university, the cell phone also appears as the tool most frequently used (37%) to attend classes and remote educational activities. The use of the device as the main resource for participation in activities is greater among users of classes DE (54%), compared to the percentage of those in classes C (43%) and AB (22%) [18]. According to the Brazilian Institute of Geography and Statistics,[a] Brazil has five social classes based on the income and salaries. Thus, the Class A receives over 20 minimal salaries (MS). The current MS are R$ 1030.00 (about USD 215). Class B receives from 10 to 20 MS; Class C from 4 to 10 MS, Class D from 2 to 4 MS, and Class E up to 2 MS.

In healthcare, the use of mobile and apps is increasing undeniably. Expansion in the connections may be associated with an increase in apps made available by the Informatics Department of the Brazilian Unified Health System (DATASUS). Some of these apps include MedSUS, which presents a list of medications indicated by SUS; modules for managers, such as e-SUS Gestor and the e-Gestor AB for primary care; and an app for users—Meu digiSUS—which provides personal and clinical information present on the National Registry of SUS users, on the Citizen's Health Portal and on the National Registry of Healthcare Facilities (CNES). (https://datasus.saude.gov.br/wp-content/uploads/2019/08/Catalogo-de-Produtos-DATASUS.pdf).

Greater availability of patient data in electronic format is highly important for decision-making and continuity of care in the public and private sectors, especially with the exchange of information. In the context of the COVID-19 pandemic, the Ministry of Health is monitoring the development of the National Health Data Network (RNDS), a platform that allows electronic health record systems to connect and exchange information about patients' health to enable the potential control of the pandemic in the country, since all public and private laboratories must notify the Ministry of Health of the results of COVID-19 tests within 24h through the RNDS.

In 2011, Brazil began to expect OpenEHR to be used. This strategy was based on the adoption of an electronic medical record by healthcare facilities and its consolidation into a large database with OpenEHR as the data wrapper. This strategy proved to be ineffective, as Brazil was still having problems digitising its administrative and assistance processes. In addition, the Ministry of Health did not have the expertise to support an electronic medical record in its environment. This scenario also faced the regulatory bodies that questioned the sustainability of the project, and consequently, it was necessary to present a flexible strategy, of low operational cost, that could show added value in a short period of time.

For this reason, technical professionals on DATASUS sought the state of the art about information security and interoperability standards. Thus, with knowledgeable consulting, the decision was to use blockchain and *Fast Healthcare Interoperability Resources* (FHIR), respectively. Still in this context, it was not expected that all establishments would be able to exchange information. Consequently, simple informational models were created, which would allow data to be loaded from information already consolidated in our environment.

From those decisions, simplified models emerged that reflect a care contact, focused on primary care services, discharge summary, medications, immunisations, and laboratory tests, among others. For each type of entry, an informational model was provided, informed by a unique identifier of the individual and the representation of the applied context. With these models, it was possible to transport much of the existing data, sent by the establishments

[a] IBGE. Brazilian Institute of Geography and Statistics [Available from: https://www.ibge.gov.br/busca.html?searchword=classes+sociais.

themselves and serialise them within the conceptual architecture of RNDS. From then on, all information is shared between patients and service providers.

It is also worth mentioning that with the use of FHIR, it was feasible to end up with point-to-point integrations. Currently, it is possible to use the entire technological framework to create standardised messages for the exchange of information. The proposal is that DATASUS starts to build message models, so that all systems already exist and being used across the country can interoperate with each other. Finally, it should be noted that the adoption of FHIR, within the concept of RNDS, is to harmonise all information that may represent a care contact and be shared between all levels of care delivery to the population.

The sum of all messages that represent the different types of assistance contacts generates a timeline of the patient, and this can only be observed at any level of care. Currently, DATASUS congregates a collection of microdata (represented by this set of messages) that form an individual's clinical history, which can collaborate with the health professional in the performance and decision-making process, having not only a picture of the patient's health situation, but also a video on what led the patient to reach their existing health status.

The National Health Data Network (RNDS)

The National Health Data Network (RNDS) was created in accordance with the guidelines of the Digital Health Strategy, based on the National Health Information and Informatics Policy—Ordinance MS No. 589/2015 [19] and the document e-Health Strategy for Brazil through the Resolution No. 19/CIT/2017 [20]. It was institutionalised by Ordinance GM/MS No. 1434, of 28 May 2020 [21], which constitutes an information platform of structured high availability in a cloud, whose objective is to promote the exchange of information between the different points of the healthcare network, allowing the transition and continuity of care in the public and private sectors.

It can be considered that the development of RNDS was underpinned by an important initiative: the project *Informatiza APS*. This specific project approved by the Resolution 2.983 was fundamental to create the financial program to increase the units dedicated to providing computerised and connected primary healthcare units (PHUs) to the population across the country. The program allowed to transfer economic resources and to obtain better quality on the information transmitted to the Ministry of Health.

The 2019 Program *Informatiza APS* major premises included the electronic medical record system to be used throughout the primary care network to health, preferably the electronic health record of the citizen available by the Ministry of Health, or another system compatible with the data model adopted by the Ministry of Health. Thus, it was the responsibility of states and municipalities to assure the integration with the database of primary healthcare information system, based on the most recent model of the layout standards of interface data (LEDI) of communication between the systems and according to the technical specification of the national system repository. The electronic medical record system must meet the requirements defined by the Ministry of Health, including for interoperability purposes to receive financial support, and be integrated by Program *Informatiza APS* and transfer data and information with quality to allow analysis and evaluation of the potential impact on the health population. Otherwise, the resources will be evaluated with the potential suppression of financial transfer [22].

Throughout its evolution, RNDS aims to become the health information highway, as the platform of availability, safe and flexible, which will comply with the ethical use of health data, thus allowing the emergence of new services, innovation, research, and development that results in benefits for the population and for Brazil.

As the evolution of the first document on the e-Health Strategy [20], DATASUS launched the Digital Health Action, Monitoring and Evaluation Plan for Brazil [23] and the Digital Health Strategy for Brazil 2020–2028 [24], which states:

> 'By 2028, RNDS will be established and recognized as the digital platform for innovation, information and health services for the whole of Brazil, for the benefit of users, citizens, patients, communities, managers, professionals and health organizations'.

The Digital Health Strategy for Brazil (ESD 28) relies on three interrelated pillars:

1. Seeks to strengthen, consolidate, expand, and extend the actions proposed on the Digital Health Action, Monitoring and Evaluation Plan for Brazil [22], mainly those related to the National Healthcare Data Network (RNDS), considering the urgency to open the door to interoperability among all health information systems deployed in the country [25].
2. Intends to construct and reinforce organisational, legal, regulatory, and governance framework to promote integration between stakeholders.
3. Aims to implement a conceptual, normative, educational, and technological environment to promote collaboration from different partners (public and private health organisations) to optimise resources, addressing the decision-making process made by the Triparty Intergovernmental Commission (CIT).

The Triparty Intergovernmental Commission (CIT) comprises the level of the organisations (ministry, state, and municipalities) responsible for the public policy to manage the Unified Health System. It is a permanent forum to define strategies, programs, projects, and the use of financial resources [26].

The stakeholders of the ESD28 are the CIT, the Ministry of Health—Informatics Department of the Brazilian Unified Health System (SUS) (DATASUS), are the National Health Surveillance Agency, the National Private Health Insurance and Plans Regulatory Agency, State and Municipal Departments of Health, Industry and Technology Sector, professionals' associations and councils, academy, technical and scientific societies, citizens and patient advocacy groups, among others.

As in any strategic plan, priorities must be in place to achieve results. The ESD28 [24] decided to focus on seven priorities:

1. Governance and leadership to ensure the development and involvement of all potential participants.
2. Digitalisation of the three levels of care delivery, adopting electronic health records and hospital information systems across facilities.
3. Support to improve healthcare delivery through the best practices, applying telehealth and smartphone applications to engage citizens, providers, and clinicians.

4. Consider the user as the centre of the healthcare process, engaging patient and citizen to promote health status and optimal habits of living, promoting digital literacy and supporting systems for a health data care register in an integrated and interoperating construction.
5. Training healthcare providers offering educational programs in health informatics reinforcing the research area and the professional recognition.
6. Enhance the RNDS to promote collaborative work to deploy systems according to the technological standards, models, policies, and regulations that are recognised by the national healthcare system.
7. Guarantee that the ecosystem allows innovation to expand health interconnectivity serving as an open laboratory under the defined policies, regulations, and standards established, creating a virtuous cycle where the progress will find opportunity to grow and be updated.

In addition, digital health initiatives must be aligned with the General Data Protection Law (LGPD) as the set of good practices that offers security for users of digital health services. The General Data Protection Law in Brazil was released only by No. 13,709 in 2018 [27]. The law was based on related policies already applied in many countries, including the European Union's General Data Protection Regulation (GDPR) and the legislation in the United States Government Department of Health and Human Services Office of the National Coordinator for Health Information Technology [28].

The LGPD major contribution is related to the ownership of data. Organisations, hospitals, and any healthcare facility used to be the 'guardian' of the patient data since that time, by law, the patients are the owner and have free access to their own data. Thus, the institutions are no longer solely responsible for keeping those data. With the LGPD, the patient/citizen must be informed and provide consent to data processing operations [29].

To comply with the Brazilian LGPD, the RNDS was developed to promote accessibility and interoperability of information in a secure and controlled manner. The technological architecture provided by the RNDS represents a layer of federated interoperability in which several applications, such as the various electronic patient records (also denominated as *PEP: Prontuário Eletrônico do Paciente*) deployed in the country, patient and consumer portals, and applications for mobile and smartphones available to citizens, health professionals, and managers, are able to exchange information through a service bus [30].

Along with requirements of the RNDS, it must be emphasised that health data are being collected, processed, and stored according to confidentiality and security standards proportional to its sensitivity. Hence, a premise of the RNDS is that all access to data must be tracked. Consequently, the RNDS can unequivocally identify which data were accessed and which professional and health establishment accessed the data and information related to a specific patient or event.

Additionally, considering that all data available by the RNDS are documents that were previously sent to the RNDS, either by systems hosted on the data centre maintained by DATASUS or by other information systems in place, it is worth mentioning that it is also possible to track where this document originated and who sent it to the RNDS. To meet this premise, blockchain technology was chosen for controlling the information diligence. Having verified the authorisation for access and the pertinence of access to data, the integration with the RNDS will take place through services (web services), where the message travels in an encrypted form. On the other hand, if a data packet is intercepted, the message will be unreadable for the interceptor [30].

To attend those premises, the RNDS architecture works in three high availability zones; if one zone is unavailable, there will be two others as redundancy. Still in this aspect, in cases of increased requests, the RNDS architecture will readjust the infrastructure resources automatically with a self-scheduling plan to adapt to the new demand.

As mentioned, the RNDS was structured for decentralised operation between nodes but interconnected in a federated way. Hence, the architecture chosen for representing the citizen's timeline was blockchain. Such technology was structured in two main contexts: ledger and private data collection. The structure ledger is the blockchain itself. It contains the indexes for pointing out clinical documents. The ledger is reproduced among all organisations in the blockchain chain. The RNDS considers each organisation as one of the 26 states and the federal district of the Brazilian Federation, with the flexibility to increase the number of organisations according to the volume of information generated [30].

In the private data collection, the storage of clinical documents adopted the Fast Healthcare Interoperability Resources (FHIR—HL7 organisation) standard entirely and respects nonrepudiation. The smart contract is the development structure responsible for interacting with the components of the blockchain as well as ensuring the consent for the release of the citizen's clinical documents and information. There are still components for interacting with integrators such as the FHIR Gateway, the entry to the network, or the electronic health record service that is responsible for validating the business rules of clinical documents.

The main sources of government data are the CNS (National Health Register), CNES (National Registry of Health Facilities), SIA (Outpatient Information System), SIH (Hospital Information System), SIPNI (National Program Information System of Immunization Program), SISAB (Health Information System for Primary Care), TISS (Supplementary Health Information Exchange for Private Providers), BNAFAR (National Database for Pharmaceutical Care Actions and Services), and RTS (Health Terminology Repository). The main applications are Connect SUS, e-SUS APS (e-SUS Primary Care), and PEPs (electronic patient records).

The raw information from the databases is extracted by an ETL process (extract, transform, load), sent to an S3 repository (Amazon simple storage service) and processed by Apache NiFi to send the data to the service layer. The data from S3 also go through an extractor in the standardisation layer that prepares it to be sent to the data analysis level (analytics sandbox) in which it passes through a machine learning engine (ML engine) and by an artificial intelligence engine (AI engine) for later storage in an Analytics database (Analytics DB) and for sending to the CMD (minimum data set) base. This entire process is performed by the application programming interface (API) of analytics.

The service layer, in turn, receives data that first pass through a FHIR interface (FHIR gateway) and distribute it to the following services: e-SUSAB service (service e-SUS APS), CMD service (minimum data set service), notification service (e-SUS notification service), EHR service (electronic health record service). After processing, the data are sent to a staging DB (intermediate storage database), which also receives health data from the CMD database via EHR listener.

The EHR listener also triggers the sending of data to the EHR service by e-SUS notification data publication/subscription events, e-SUS APS, and SIPNI. These data will be waiting in the queue until it is absorbed by the integrator to be sent to the blockchain. The integrator also sends the data to an Intake Layer (Kafka) for storage in S3.

Interoperability with public and private laboratories was established as the first major achievement of the RNDS. From the month of March 2020, when the new coronavirus pandemic in Brazil started sharply, until the month of November 2020, the RNDS received approximately more than 4 million COVID-19 test results.

Currently, it is already possible to carry out safe and controlled sharing of individual results of laboratory tests related to COVID-19 performed in different laboratories. To make the connection process to the RNDS viable, DATASUS made the technical documentation and accreditation system for the laboratories, available on the Services Portal, based on the best practices and interoperability standards. In addition, in a joint action with the Brazilian Association of Emergency Medicine (ABRAMED), DATASUS held a series of meetings and live gatherings with staff from the main laboratories in the country to provide clarifications, provide additional information on the connection with the RNDS, as well as open channels to support the integration process. The connection of public and private laboratories to the RNDS is a relevant milestone in the fight against COVID-19 and, in addition, it will bring long-term benefits as a fundamental reference for Brazilian digital health.

Final considerations

Brazil still has several challenges to enhance the connectivity and information exchange across facilities, regions, services, and citizens—the reason source of its creation and deployment. Interoperability is a continuous process to be achieved considering the differences and dimensions of the country.

However, it must be emphasised that the RNDS is the foundation for building an environment of total interoperability in the context of health, data exchange, and governance, considering that collected data must be used to design or redesign citizen care delivery to improve health population. RNDS is a journey. The foundation is created and now it must start to exchange information for the continuity of care. It is well known there is much to do, such as the standardisation and regulation of the next informational models, which includes technical governance.

References

[1] Brazilian Institute of Geography and Statistics—IBGE. n.d. Available on: https://www.ibge.gov.br/en/cities-and-states.html?view=municipio. 2021. Accessed on March 30, 2021.

[2] G. Vecina Neto, A evolução da assistência à saúde no mundo e no Brasil até o SUS, in: G. Vecina Neto, A.M. Malik (Eds.), Gestão em Saúde, Guanabara Koogan, Rio de Janeiro, 2014, pp. 3–14, ISBN: 978-85-277-1708-3.

[3] C.V. Machado, J.S. Paim, Reforma Sanitária Brasileira: contribuição e crítica, Edufba, Editora Fiocruz, Salvador, Rio de Janeiro, 2008.

[4] Brasil. Constituição, Constituição da República Federativa do Brasil, 1988, Available at http://www.planalto.gov.br/ccivil_03/constituicao/constituicao.htm. Accessed on March 12th, 2021.

[5] Folha de S.Paulo (Ed.), Brasil é único com 'SUS' entre países com mais de 200 milhões de habitantes, 2019. 10 de outubro de. Accessed on March 17th, 2021.

[6] Brazilian Ministry of Health. Virtual Library in Healthcare. n.d. Available at: http://bvsms.saude.gov.br/ultimas-noticias/3295-lei-n-8080-30-anos-de-criacao-do-sistema-unico-de-saude-sus, Brasilia, 2021. Accessed on March 17, 2021.

[7] G. Ligia, M.-R. Adriana, P.A. de Carvalho Amand, D.R.M. Cantanhêde, M.G. Branco, S.I. Soares, et al., Universal health system and universal health coverage: assumptions and strategies, Cien. Saude Colet. 23 (6) (2018) 1763–1776. [Internet]. June [cited 2021 Mar 17]. Available from: http://www.scielo.br/scielo.php?script=sci_arttext&pid=S1413-81232018000601763&lng=en. https://doi.org/10.1590/1413-81232018236.05562018.

[8] M.C. Vieira, Health policies in Argentina, Brazil and Mexico: different paths, many challenges, Cien. Saude Colet. 23 (7) (2018) 2197–2212. [Internet]. July [cited 2021 Mar 17]. Available from: http://www.scielo.br/scielo.php?script=sci_arttext&pid=S1413-81232018000702197&lng=en. https://doi.org/10.1590/1413-81232018237.08362018.

[9] C.V. Machado, L.D. Lima, T.W.F. Baptista, Políticas de saúde no Brasil em tempos contraditórios: caminhos e tropeços na construção de um sistema universal, Cad. Saude Publica 33 (Supl. 2) (2017), e00129616.

[10] J. Adler-Milstein, E. Ronchi, G.R. Cohen, et al., Benchmarking health IT among OECD countries: better data for better policy, J. Am. Med. Inform. Assoc. 21 (2014) 111–116.

[11] OECD, Strengthening Health Information Infrastructure for Health Care Quality Governance: Good Practices, New Opportunities and Data Privacy Protection Challenges, OECD Publishing, Paris, 2013.

[12] Brazilian Internet Steering Committee—CGI.BR, Survey on the Use of Information and Communication Technology in Brazilian Healthcare Facilities —ICT Health 2013, CGI, São Paulo, 2014.

[13] Brazilian Internet Steering Committee—CGI, Survey on the Use of Information and Communication Technologies in Brazilian Healthcare Facilities: ICT in Health 2019, CGI.br, São Paulo, 2020. Available in https://cetic.br/media/docs/publicacoes/2/20201123084414/tic_saude_2019_livro_eletronico.pdf. Accessed on March 19th, 2021.

[14] Brazilian Internet Steering Committee—CGI.BR, Survey on the Use of Information and Communication Technology in Brazilian Healthcare Facilities—ICT Health 2018, CGI, São Paulo, 2019.

[15] Brazilian Ministry of Health, Dispõe sobre o uso da telemedicina durante a crise causada pelo coronavírus (SARS-CoV-2), 2020, Law n. 13.989, de 15 de abril de 2020 Brasília, DF. Recuperado em 15 agosto, 2020 http://www.planalto.gov.br/ccivil_03/_ato2019-2022/2020/Lei/L13989.htm.

[16] J. Portnoy, M. Waller, T. Elliott, Telemedicine in the Era of COVID-19, J. Allergy Clin. Immunol. Pract. 8 (5) (2020) 1489–1491, https://doi.org/10.1016/j.jaip.2020.03.008. Epub 2020 Mar 24 32220575. PMC7104202.

[17] Brazilian Ministry of Health. n.d. The Conecte SUS Program. Available at: https://www.gov.br/saude/pt-br/assuntos/noticias/conecte-sus-avanca-em-todo-pais-com-a-implantacao-da-rede-nacional-de-dados-em-saude, Brasilia, 2020. Accessed on April 5, 2021.

[18] Brazilian Internet Steering Committee—CGI. n.d. Panel ICT-Covid-19. Available at: https://cetic.br/pt/noticia/celular-e-o-dispositivo-mais-utilizado-por-usuarios-de-internet-das-classes-de-para-ensino-remoto-e-teletrabalho-revela-painel-tic-covid-19/#:~:text=O%20uso%20do%20dispositivo%20como,%25)%20e%20AB%20(22%25). São Paulo, 2020. Accessed on April 6, 2021.

[19] Brazilian Ministry of Health, Institutes the National Health Information and Informatics Policy (PNIIS), 2015, Ordinance MS No. 589, of May 20, 2015 Retrieved on June 17, 2020, from https://bvsms.saude.gov.br/bvs/saudelegis/gm/2015/prt0589_20_05_2015.html.

[20] Brazilian Ministry of Health. Estratégia de e-Saúde para o Brasil 2017, n.d. Available on: http://portalarquivos.saude.gov.br/images/pdf/2017/julho/12/Estrategia-e-saude-para-o-Brasil.pdf, Brasília, 2017. Accessed on April 1, 2021.

[21] Brazilian Ministry of Health, Institutes the ConecteSUS Program and Amends Ordinance Consolidation No. 1/GM/MS of 28 September 2017, to Institute the National Health Data Network and Addresses the Adoption of Standards for Interoperability in Health Care, 2020, Ordinance MS No. 1434, of May 28, 2020 Retrieved on June 17, 2020, from http://www.in.gov.br/web/dou/-/portaria-n-1.434-de-28-de-maio-de-2020-*-259414569.

[22] Brazilian Ministry of Health, Brasília, 2017. Institui o Programa de Apoio à Informatização e Qualificação dos Dados da Atenção Primária à Saúde—Informatiza APS, por meio da alteração das Portarias de Consolidação n° 5/GM/MS e n° 6/GM/MS, de 28 de setembro de 2017 Ordinance no. 2.98, 11 November 2019.

[23] Brazilian Ministry of Health, National digital health strategy action, monitoring and evaluation plan, 2019–2023. Brasília, in: Approved on the 34th Ordinary Meeting of the Digital Health Strategy Steering Committee, 2020. on December 13th, 2019.

[24] Brazilian Ministry of Health, Executive Secretariat, SUS Computer Department, Brazilian National Digital Health Strategy 2020–2028 [Electronic Resource], Ministry of Health, Executive Secretariat, SUS Computer Department.—Brasília: Ministry of Health, 2020, p. 128.

[25] Brazilian Ministry of Health, Implements the Conecte SUS Program and amends Consolidation Ordinance No. 1/GM/MS, dated September 28, 2017, to Establish the National Healthcare Data Network and Provide for the Adoption of Health Interoperability Standards, Federal Official Gazette, Brasília, DF, May 2020. Ordinance No. 1.434, dated May 28, 2020. Available at: https://saudedigital.saude.gov.br/wpcontent/uploads/2020/06/portaria-n%C2%BA-1.434-202.pdf. Accessed on: 08 April. 2021.

[26] Brazilian Ministry of Health. Sistema Único de Saúde -Comissões, Brasília, 2009. Available on: http://bvsms.saude.gov.br/bvs/sus/comissoes.php. Accessed: April 8, 2021.

[27] Brazilian Ministry of Health, Lei Geral de Proteção de Dados Pessoais (LGPD), Presidência da República, Brasília, DF, 2018. Ordinance n° 13.709, 14 August 2018 Available in http://www.planalto.gov.br/ccivil_03/_ato2015-2018/2018/lei/L13709compilado.htm. Accessed: April, 8th, 2021.

[28] B. Kaplan, PHI protection under HIPAA: an overall analysis, in: A.B. Dallari, M. GFC (Eds.), LGPD na Saúde, Thomson Reuters Revista dos Tribunais, São Paulo, 2021, pp. 61–86.

[29] T.L. Oliveira, G.N. Neves, J.N. Almeida, J.P. Souza-Zinader, J.V. Barros, LGPD e Saúde Pública: a rede nacional de dados em saúde e a proteção de dados pessoais, in: A.B. Dallari, M. GFC (Eds.), LGPD na Saúde, Thomson Reuters Revista dos Tribunais, São Paulo, 2021, pp. 159–170.

[30] J.V. Barros, S. HNS, R.M. Baptista, P. FJS, J.P. Sousa-Zinader, National health data network: interoperability experience in the healthcare network, in: Brazilian Internet Steering Committee—CGI.br (Ed.), Survey on the use of information and communication technologies in Brazilian healthcare facilities: ICT in Health 2019, CGI. br.: TIC Saúde, São Paulo, 2020, pp. 251–263. 2019, 2020.

CHAPTER

22

Benefits of using SNOMED CT in the UK National Health Service (NHS)

Monica Jones[a,b]

[a]Health Data Research UK (HDR UK), Animo Consultancy Ltd, Leeds, United Kingdom
[b]Global eHealth Collaborative (GeHCo), Melbourne, VIC, Australia

Why SNOMED CT?

So firstly, we must start with answering the question of why to implement a clinical terminology such as SNOMED CT? There are several reasons but those most apparent are its global scope in terms of contribution and utility. It covers a wide range of disciplines and specialties as well as being ever growing and improving. As clinical terminology meaning representation means it has multiple purposes and there is a full breadth of functionality supported. In relation to the use case described throughout this chapter, it was made mandatory as an NHS Information Standard (in England) [1] with implementation required in primary care from 2018 and secondary care from 2020. This also helped to accelerate the uptake as well as justifying financial investment.

The problem ... the meaning

Healthcare is complex and its language even more so and a simple solution is not appropriate. Representation of clinical information is complex so clinical terminologies such as SNOMED CT are complex for this reason. Words and phrases change meaning according to the context in which they are used too and the place where they are used including which clinical disciplines and specialities are using them and the time when they are used. There are some disadvantages to using SNOMED CT such as legacy term, some lack of transparency and limited access, and complexity such as full post-coordination is not yet a practical proposition and undefined boundaries. These disadvantages are outweighed by the benefits of SNOMED CT which has a future-proof structure, can be improved and made fit for purpose, inherently multilingual which means it has a broad coverage [2].

Roadmap to Successful Digital Health Ecosystems
https://doi.org/10.1016/B978-0-12-823413-6.00023-9

SNOMED CT is also a multiaxial classification, but it does not have the characteristics of a statistical classification (which is what we usually mean in healthcare when we refer to a classification), in that it does not have 'other' and 'not otherwise specified' buckets such as ICD-10. It is also a reference terminology with reference features which are supported by retrieval mechanisms inherent in the logical based structure of the terminology. Finally, it is an interface terminology supporting human interface through synonyms, reference sets, navigation hierarchies, and search support resources.

Use of SNOMED CT in an NHS hospital

There are a number of NHS trusts that have implemented SNOMED CT in line with the directives of the Personalised Health and Care 2020 framework for action [3] and the goals stated in the local digital roadmap to support the health and care community in achieving the ambition of working 'Paper Free at the Point of Care'. In addition to the strategic use of SNOMED CT, a key driving goal of the implementation is to improve the efficiency of recording patient episode information in ICD-10 for disease and OPCS-4 × for procedures (classification terminologies) to support NHS reporting. The Healthcare Resource Groups (HRGs) together more detailed patient information is then used for statistical and epidemiological analysis, reimbursement for care provided and planning of health and care services. SNOMED CT contains more detailed terms than the classifications and is used by clinicians in an electronic patient record (EPR) system to record, at the point of care, detailed information about a patient including but not limited to conditions, findings, and procedures.

The example used in this chapter is an NHS trust that decided to move to an electronic patient record system, replacing their patient administration system (PAS) in the process. They worked closely with their chosen supplier to adapt the existing supplier product to utilise SNOMED CT throughout for both procedure and diagnosis recording. It was the first instantiation of this system in the UK, and this chapter will provide an overview of these achievements. The NHS trust is a modern, progressive trust, with a focus on supporting the local district. The main hospital site is situated just two miles south of the town centre. They also operate many community services out of other sites across Rotherham. The Urgent and Emergency Care Centre (UECC) deals with around 105,000 patients per year, and there are approximately 72,000 inpatient and 500,000 outpatient attendances each year. There are 560 inpatient hospital beds.

Using SNOMED CT for procedure and diagnosis recording

The aim throughout implementation is for clinical ownership of the content in the patient record and for that content to be of quality (completeness, level of detail, and accuracy). The rationale for requiring SNOMED CT, while accepting the trust would be an early adopter, was to implement the national objectives of collecting data once, at source, and facilitating other process from that data. There are several current benefits from utilising SNOMED CT including capturing point of care procedure and diagnosis codes, driving the classification coding of episodes of care and supporting the move towards a paperless system. In addition, the ability to support reporting requirements utilising SNOMED CT such as national emergency care data set reporting has allowed NHS strategic alignment. There are, however, additional specific areas where the functionality of SNOMED CT can be leveraged within a trust to achieve additional benefits in the future. These include clinical decision support, monitoring adherence to standards and guidelines, auditing of care delivery, patient cohort identification and case finding, analytics to support areas such as resource planning, transfer of care—discharge and admittance, and of course further systems interoperability.

Once the decision to use an EPR with SNOMED CT was made, it was decided that users would be able to select from predefined lists that were approved by the clinicians in each specialty. This was achieved by creating SNOMED CT subsets which are then used by the system to provide a restricted list to users. The reason for using subsets for data entry was to improve the speed of data entry, to get standardisation within the hospital on recording and so that data quality could be assured from the outset. This trust currently has around 60 subsets, one which covers all diagnoses and the remainder relating to the different specialties' procedures—each being available in the different clinical settings and selected by the system when a user logs into a particular clinic.

Two different approaches were taken when developing the subsets: one for the procedure subsets and another for the diagnosis subset and are described below. The procedure subsets were all developed through working collaboratively with the relevant consultant or clinical director in that specialty. These subsets are loaded into the system by the administrative staff who managed that module while others were created with the assistance of the UK Terminology Centre using criteria defined by the trust and the terms identified through an application using the criteria.

The procedure subsets are specialty specific and location based. They are used for scheduling and resource allocation. When a patient is listed for a procedure, the clinician decides where they would like to carry out the procedure, for example theatre, day-case unit, or special clinic, which in turn selects the correct subset they then select the appropriate clinical term from that subset. These two pieces of information are then used to book resources in the appropriate location.

The diagnosis subset was created by applying a set of rules that were defined by the trust. This produced a subset of around 65,000 concepts and contains all the descriptions for diagnoses. The preferred term is displayed as the default display description and the system allows only one description to be used as a default description for each concept; however, the search for a particular diagnosis can be performed on all the SNOMED CT diagnosis descriptions. The procedure subsets have been in operational use from day one across the trust. It was reported that there were a few problems, mainly in relation to missing content; however, these issues were quickly resolved as they were able to change the content of the subsets immediately. Once the clinicians mastered the ability to search for the terms, they required they were generally satisfied with the approach. They are also able to create individual subsets referred to as 'favourites' as well as searching the whole of the concept hierarchy if required.

There was a need to schedule physical procedures, so it was decided to implement procedures first. The system needed to schedule theatre time and resources. The procedures were separated into three main groups: in-patient procedures, day-case procedures, and clinic procedures (carried out in outpatients). With the introduction of the system, the trust is now able to better monitor theatre usage and resources, for example equipment and staff. As every procedure undertaken is being captured in several clinical areas, both clinical staff and management have a better view of exactly what is being performed. Some procedures which may be classified as 'new and innovative' can now be tracked and staff can understand the levels at which these are being done. Efficiencies have been achieved for cases that are straightforward in terms of identifying the classification codes. In these cases, the SNOMED CT entered term can be translated through software to the content for either the commissioning data sets (CDS) or for clinical coding. While the latter are checked by a clinical coder, the speed of this is increased with the automatic generation of the classification codes. The potential now exists to better support clinical audit, both through doing some analysis of the data in the system, but also identifying cases.

Implementation issues

As the subsets were location-based initially, if a procedure was carried out at different locations, which is often the case, concepts appear in multiple subsets. This not only adds unnecessary maintenance work, but because the subsets were created in isolation, different concepts may have been chosen to represent the same thing. The diagnosis subset was created in a different way due to the lessons learned from the procedure's subsets development. In the case of the procedure's subsets, it was led by the clinicians; however, the diagnosis subset was led by the informaticians and approved by the clinicians. Diagnoses also need to be available to more than just that specialty. As the effort was placed into getting the new system live along with the subsets, limited thought has been given to how the subsets will be maintained long term, bearing in mind that if the subsets are to be kept in line with SNOMED CT, then they may need to be reviewed every time a new edition is issued (currently every 6 months).

Possible methodologies for this were briefly discussed, but with all the efforts still focussed on the implementation, it was accepted that this is for the next phase of work. The trust has documented how the different types of subsets were created and the criteria used, but it is not yet known if this is in sufficient detail when it comes to maintaining the subsets, especially if different individuals do the work. The trust has access to system reports that allow them to analyse data that has been entered by individuals to identify data quality issues and anomalies. The trust can also review each clinician's favourites list and, when they see very generic terms on the favourites list, staff can liaise with the clinician to find out the reason behind it and try to replace them with specific terms, if possible. The trust also has direct access to the subsets installed in the system and can change them on the fly which has proved extremely useful if a clinician notices a term is missing. There have been some issues with the cross-maps used to translate from the SNOMED CT term to the classification code. These are being resolved as they are identified in collaboration with the National Classification Service. The trust does need to be able to make a temporary change to a local system until the new national maps are issued to ensure their current data are mapped as required. This facility is not currently available and so can require some postprocessing to data.

Training

Depending on the roles, members of staff are trained with various modules and total training time varies. Doctors are given seven hours of classroom-based training on the system of which a limited period is dedicated to SNOMED CT. Anecdotal reports from clinicians are that the SNOMED CT part is relatively easy, and all agree that once their favourites list is populated, it is generally not an issue. When they experience problems finding some concepts they are looking for, in many cases these concepts are not in the subset. These problems are easily solved by adding the missing concepts to the subsets. In addition, clinicians who identify content that is not in SNOMED CT can request additional content through the UK National Release Centre (UKTC) to have the new content added. Training takes place before the user goes live on the system because of the amount of notice required to release clinical staff from normal duties with end users initially being trained in a 'TEST' environment which during roll-out could be different to the 'LIVE' environment. Subsequently, a 'TRAINING' environment was created and all three versions of the EPR are now synchronised to have the same configuration as the live system. This makes training and development very effective. Where the development of the subsets is concerned, they were developed as they went along. Advice, guidance, and assistance were given ad hoc by the UKTC, who also helped develop some of the subsets.

Scope extension

The work to date has created a strong foundation from which to expand capabilities. An independent review was carried out in 2018 which gave several recommendations to enhance the implementation further. There were other areas of terminology use which were outside of the scope of the report that would benefit from a review of use of the NHS Dictionary of Medicines and Devices (NHS dm+d) [4], current use of laboratory codes for orders and results. It also recommended expanded use of SNOMED CT outside of procedures and diagnoses and terminology use outside of the core EPR system supports longitudinal rather than episodic patient records.

Improved data entry

Using SNOMED CT data entry lists (subsets) has improved the speed of data entry, achieved standardisation of recording within the hospital, and improved data quality and subsequent analysis. Some clinical specialties have championed these benefits, and obstetrics and gynaecology now have several front-end forms that save time for data entry, as well as reports that enable them to undertake clinical research and audits. Previously, these were undertaken by clinicians trawling through notes to identify appropriate patients. These are used to better understand and design the service offered, as well as improve patient care. Improved data quality has resulted in a saving of approximately £28 K per annum to the trust. Figs. 1 and 2 show how SNOMED CT is implemented in an EPR.

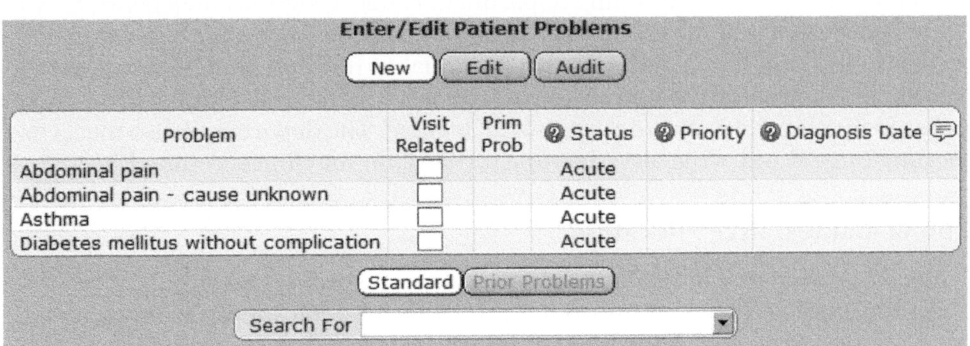

FIG. 1 EPR clinical problem screen.

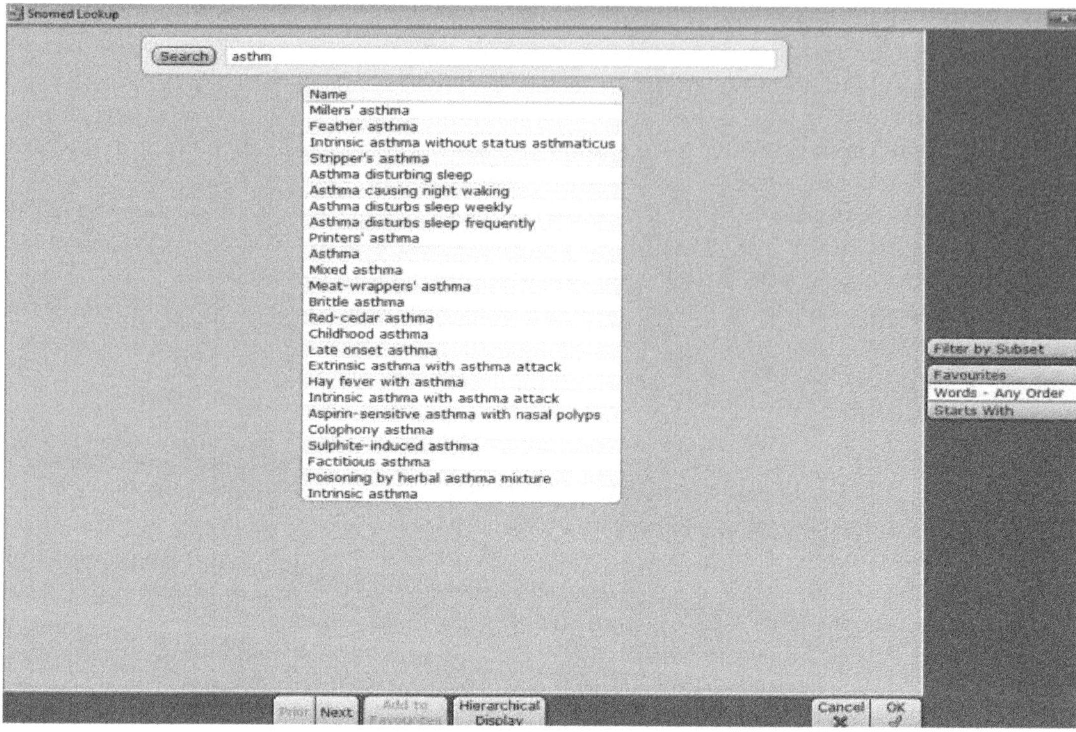

FIG. 2 EPR SNOMED CT lookup.

Clinical coding benefits

Outpatients use SNOMED CT for diagnosis and procedures, which are entered by clinical staff. This trust uses a clinical encoder as a 'bolt-on' to its EPR to undertake automatic processing of outpatient data. Mapping of SNOMED CT procedures to classification OPCS 4.8 codes is now 95% automated, and these are then put through a grouper that allows creation of the respective Healthcare Resource Groups (HRGs) [5]. Clinical coders still do a final review for any obvious irregularities. These process improvements have led to a savings of £30 K per annum for the trust. The clinical coding department has seen significant benefits; they no longer rely on agency staff and have less staff attrition and their roles have evolved. The clinical coding role is now closely aligned with the data quality role and clinical assurance [6]. Staff morale is higher, they enjoy their more holistic role, and the overall depth of coding in patient records has increased and is above the national average. The department also meets its financial reporting targets and achieves a five-day turnaround for clinical coding [7].

Scheduling and resource allocation

The procedure subsets are also utilised for scheduling and resource allocation. Choosing the correct SNOMED CT procedure from a predefined list allows clinicians to book the resources in the appropriate location and has resulted in a saving of approximately £28 K per annum. The trust is now able to better monitor theatre usage and resources, for example equipment and staff. As procedures undertaken are being captured in several clinical areas, both clinical staff and management have a better view of exactly what is being performed.

Mapping

As well as publishing SNOMED CT and the classification terminologies, there are mapping files from SNOMED CT to ICD-10 and SNOMED CT to OPCS-4 × for the purpose of supporting the derivation of classification terms directly from the SNOMED CT concepts which have been recorded by the clinician in the EPR. These mapping files consist of a set of complex rules that often provides a single classification term for a given SNOMED CT term, but sometimes, according to the rules of the classifications systems, additional information is needed to identify the correct classification term which is not available in a single SNOMED CT term and therefore a choice of classification terms is sometimes presented requiring additional input to refine. It should be noted that not all SNOMED CT terms will have a map to the classification terminologies as they may be 'out of scope' of the intent and purpose of the classification terminologies. Where no map is available either the SNOMED CT concept is not reportable or manual investigation by a specialist clinical coder would be required to review the patient episode to identify the correct classification term. The strategy is to capture information in the EPR utilising SNOMED CT and to map to the classification terminologies using the clinical classification encoder to fulfil the patient episode reporting requirements. Understanding this strategy is key to understanding the implementation approach.

Analysis

This trust now has many years of data at the level of detail provided through using SNOMED CT and visual access to case mix data which enables much more accurate trend analysis and thus improved planning and commissioning. The potential now exists to better support clinical audit, both through doing analysis of the data in the system and identifying patient cohorts.

Clinical dashboards through a knowledge portal can be used to visualise activity across trust from waiting times to patient conditions. For example, the chief complaint can be seen very quickly. SNOMED CT has enabled a longitudinal record to be visible as well as a more holistic view of the patient, for example if the patient is feeling isolated. This provides data in real time on the trusts progress against its service delivery plan. Initially, the trust did not have the ability to accurately know the services it had previously provided and thus what was agreed in commissioning was estimated; this resulted in the trust delivering services for which it did not receive accurate funding. Now the level of services provided are within a 2% tolerance to the plan agreed with commissioners, so all income is received. Fig. 3 shows how SNOMED CT can be displayed in a clinical dashboard.

Benefits for the patient

Improved visibility of the clinical outpatient activity leads to much better patient care. The trust cuts outpatient waiting lists by 65% in some areas, as well as recorded a drop in the number of missed appointments (DNAs) per month. Discharge summary is massively improved with a better level of information. This is beneficial to both general practice in primary care and to patients.

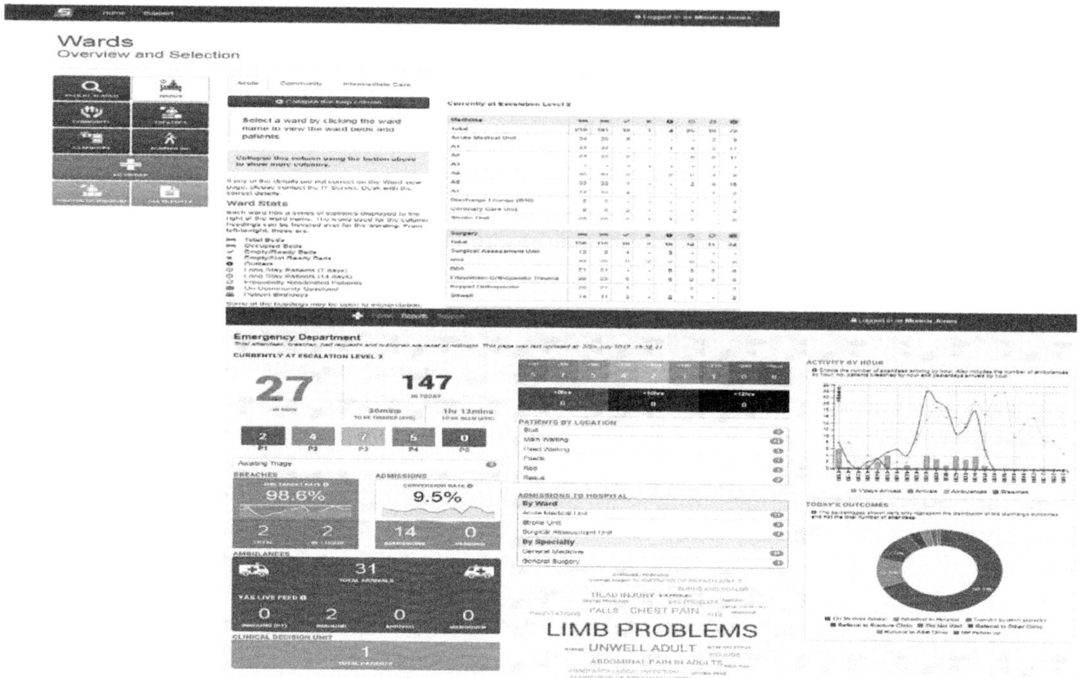

FIG. 3 Clinical dashboards.

Challenges and lessons learned

Having SNOMED CT specialism within a trust is important and enables appropriate training and support. A mixture of face-to-face training, follow-up checks, and work shadowing in the early phases of go-live was undertaken. Clinicians requested a smaller choice of SNOMED CT clinical terms in data entry lookup lists. A clinical lead from each speciality was needed to help determine the correct content that should be used. The SNOMED CT preferred term is displayed as the default display description as the system allows only one description to be used for each concept; however, the searching can be performed on all the SNOMED CT descriptions. The time to upgrade does need to be factored into plans which involves checking for changes in subsets and favourite folders, and this is considered essential and no different to any other data set change.

To get the full benefit from an electronic patient record, the content must be recorded using a clinical terminology. It supports true digitisation. It is designed by clinicians for clinicians, so the safety case is supported by contextual information, that is it is easier to understand unintended consequences. Interoperability is better now that primary care is using SNOMED CT in the UK. Implementing mandatory national data sets for audit and performance management is easier [8].

Implementing a comprehensive EPR is a big change for clinicians, and implementing SNOMED CT at the same time is an extra challenge. However, it does mean the use of clinical terminologies is native to the EPR. In this case, initially clinicians complained there was too much choice using SNOMED CT, so the trust concentrated on setting up the users' favourites and subsets to lower the number of terms in the look up. It is important to a clinical lead from each speciality to help determine user favourites and subsets. In this case, it was set up as part of postimplementation phase.

Conclusion

This exploration and overview of implementation of SNOMED CT in the UK NHS has highlighted the many benefits of adopting a clinical terminology as a national information standard. The experiences described for a particular NHS trust provide real examples of good practice and pragmatic implementation solutions that will allow others to have the confident and competence to tackle this important step as part of the roadmap to successful health ecosystems.

References

[1] SCCI0034: SNOMED CT is the information standard for clinical terminology in the NHS Mandate, May 2020.

[2] J.J. Cimino, Desiderata for controlled medical vocabularies in the twenty-first century, Methods Inf. Med. 37 (1998) 394–403.

[3] Personalised Health and Care 2020: Using Data and Technology to Transform Outcomes for Patients and Citizens: A Framework for Action, 2014.

[4] NHS Dictionary of Medicines and Devices (NHS dm+d) 2021.

[5] NHS Digital Key Messages about the role of SNOMED CT, the classifications and the coders. 2019.

[6] Classification maps and different use cases for SNOMED CT. 2020.

[7] NCCQ—the National Clinical Coding Qualification (UK) 201-19 Syllabus, Framework, and Bibliography. 2019.

[8] NHS Futures Collaboration Platform—Good practice case studies. Case Study 1: 2013. Case Study 2: 2018.

CHAPTER

23

Primary health reform for the digital era

Noah Crampton[a,b], *Anson Dong*[a,c], *Payal Agarwal*[a,d], *Catherine Ji*[a,b], *and Onil Bhattacharyya*[a,d]

[a]Department of Family and Community Medicine, University of Toronto, Toronto, ON, Canada [b]Family Medicine Department, Toronto Western Hospital, Toronto, ON, Canada [c]Family Medicine Department, Sinai Health System, Toronto, ON, Canada [d]Institute for Systems Solutions and Virtual Care, Women's College Hospital, Toronto, ON, Canada

Principles of primary care

In 1978, the World Health Organisation (WHO) adopted the Declaration of Alma-Ata which established primary healthcare as an essential service. The key features were refined in the Astana declaration forty years later [1]. These features include meeting people's health needs through comprehensive preventative, curative, rehabilitative, and palliative care throughout the life course, aimed at individuals and populations as a central element of integrated health services. Primary care is a narrower concept, which includes services delivered to individuals and panels of patients, but not necessarily for populations within a defined geographic area. This term is widely used within high-income countries. Primary care is defined as 'first-contact, continuous, comprehensive, and coordinated care provided to populations undifferentiated by gender, disease, or organ system' [2]. The distinct features of primary care services are as follows:

- accessibility as the first contact for new health needs,
- continuity of care across the life course,
- comprehensive care for most health needs, and
- coordination of care that may be provided by others.

Strong primary care reduces total healthcare costs by improving access to preventive services, more timely diagnosis and treatment and more equitable care, focused on the needs of the whole person [3–6]. This chapter will focus on the narrower concept of primary care in its discussion of health reform.

Roadmap to Successful Digital Health Ecosystems
https://doi.org/10.1016/B978-0-12-823413-6.00009-4

An opportunity for primary care transformation

In 2015, the World Health Organisation (WHO) founded the Primary Health Care Performance Initiative (PHCPI), in partnership with the Bill & Melinda Gates Foundation and the World Bank Group among others. As part of this effort, they have developed a conceptual framework which links the system factors and inputs that go into the delivery of primary care to patients and the outcomes we aim to improve (see Fig. 1) [6].

It highlights features of broader health systems that affect the structure and function of primary care and how it influences service coverage and population health outcomes as well as overall system performance. This section touches on the organisation of primary care providers, overall health spending (including provider payment and financial coverage for services), and human resources. Next, we discuss information systems, which are an important input, but they act as a catalyst through which the healthcare workforce and related facilities deliver effective services.

Primary care in Canada

In international comparisons of the quality of primary care based on measures such as access, coordination, and comprehensiveness, Canada ranks in the middle of industrialised countries [3]. However, there is substantial regional variation because healthcare provision is managed by provincial governments while the federal government transfers funds to the provinces to cover part of the costs. Until the end of the 19th century, Canadians paid physicians privately for healthcare and physicians largely functioned as free market businesses paid out of pocket by individuals at the point of care [7].

The first Medicare insurance program in North America was established in the province of Saskatchewan in 1947 with strong resistance from physicians, who organised a provincial strike. They continued operating as independent contractors who were able to charge the provincial government on a fee-for-service basis under the new public payer system [8,9]. After multiple phases of reform, a national approach to health insurance was developed, and the Canada Health Act was passed in 1984 mandating comprehensiveness, universality, accessibility, and portability as criteria for federal funding of provincial healthcare delivery [8]. Canadian primary care continues to be structured with a public payer and providers acting as private contractors. Australia has adopted a similar system.

In Canada, healthcare is financed through a public insurance program funded by the federal government and regulated and administered through the provincial or territorial governments. Canadian healthcare functions as a one-payer system with heterogeneous managers of provincial healthcare provisions, much like separate divisions in a corporation. Based on the Canada Health Act, there is universal access to essential services provided by physicians and hospital-based care with no cost. Canada spends 10.79% of its annual GDP on healthcare, which is more than the United Kingdom (UK) at 10.25% but less than countries like the United States and Germany, at 16.96% and 11.65%, respectively [10]. Public health insurance in Canada generally does not cover outpatient medications, dental care, and paramedical services such as physiotherapy and psychotherapy, apart from some limited public options

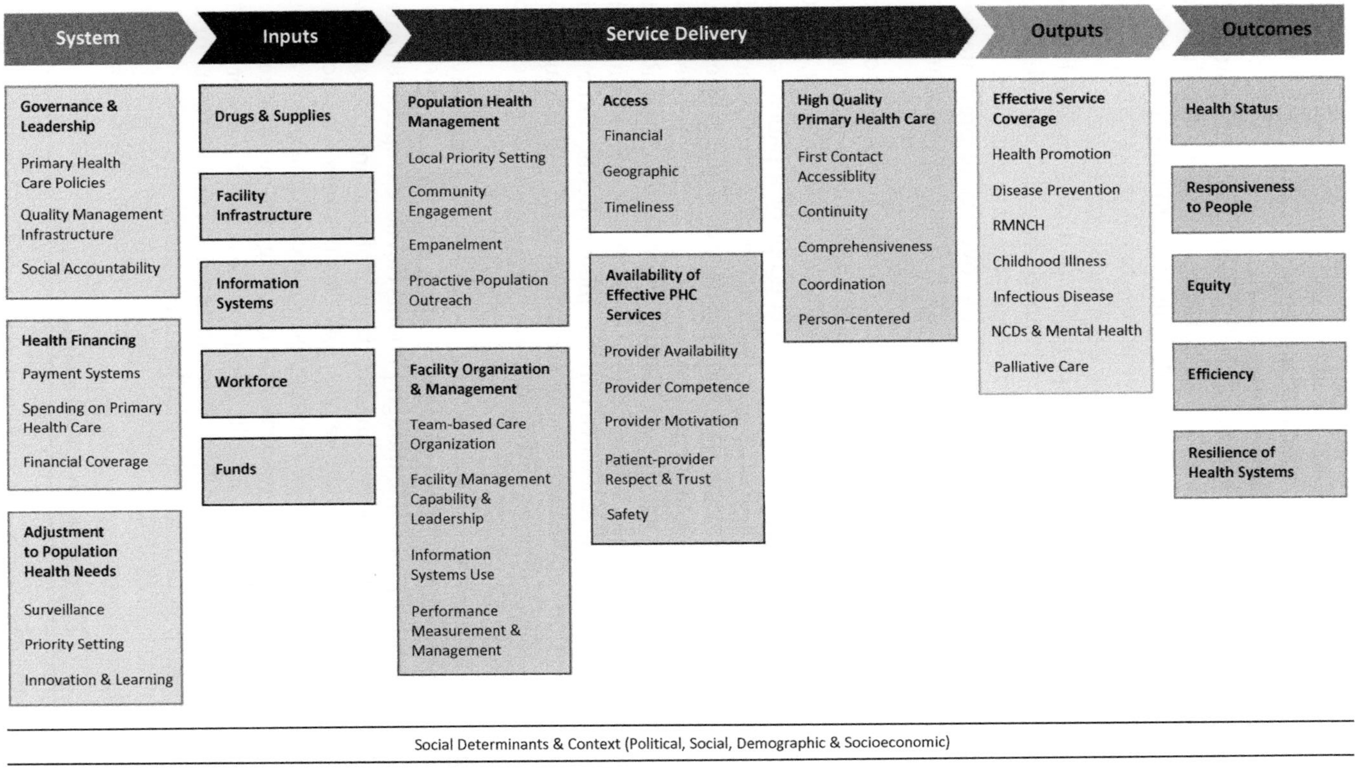

RMNCH: Reproductive, Maternal, neonatal and child health, NCDs:

FIG. 1 Primary health care performance initiative conceptual framework.

for specific at-risk populations. The extended coverage for these additional services is usually available through private insurance purchased by individuals or employers.

Canada has one of the lowest physicians to citizens ratios out of high-income countries. The 2017 Commonwealth Fund Survey reported Canada having only 2.7 working physicians per 1000 people, lagging behind countries such as Australia, Denmark, New Zealand, and Italy and only slightly ahead of the United States (US) at 2.6 [11]. And though Canada has more family physicians than specialists, access to primary care is worse than comparable nations, in part because it has fewer doctors overall. This leads to supply issues for patients, with 15% of Canadians reporting they do not have access to a regular family physician [12]. In rural or remote regions in Canada, the supply is even more constrained.

While there is agreement in Canada on the importance of access, comprehensiveness, coordination, and continuity across the life course in primary care [13], progress has been uneven. Universal coverage ensures financial access to primary care but there was little focus on providing patients access to weekend and after-hours coverage or preventative medicine [14]. Over the past decade, there have been major efforts at reforming primary care to better serve patient needs.

Historically, the mix of fee-for-service payment for providers and free choice for patients resulted in a focus on brief office visits with limited accountability, and more recent efforts have improved the comprehensiveness and coordination of care. In 2009, the College of Family Physicians of Canada (CFPC) introduced the concept of a patient-centred medical home to guide the provision of high-quality comprehensive care for Canadians. For most patients, a family physician is relied upon to provide primary care and is often the first point of care. In more remote areas, primary care may be provided by an independent nurse practitioner or a registered nurse under remote supervision by a family physician. Recent health reforms have promoted team-based care, where family physicians work with interdisciplinary health professionals to provide comprehensive care for their patients. Some provinces have introduced blended capitation models where physicians are paid for managing a roster of patients and are offered some outcomes-based bonuses, creating an incentive for population health management [15,16].

Primary care in Canada aims to provide 'cradle-to-grave' comprehensive care, but the scope is region dependent, with rural physicians providing a wider range of services [17]. It is the first contact point for new problems, though patients can access specialist care when urgent by presenting to the emergency room. Canadian family physicians also offer patients enhanced services including providing palliative care, low-risk obstetrics, addictions medicine, emergency medicine, and more. Family physicians also help coordinate care across specialty and community-based services through referrals and ongoing communication.

Primary care has the highest volume of visits and provides care at a lower cost than other specialties, which makes it the best sector to explore information systems as an input to health care and their progress and potential to transform health systems.

Digital health and primary care

Digital tools have enormous potential to enhance the performance of primary care because they make it easier to collect, store, access, analyse, and communicate information.

They can improve access and continuity by allowing patients and providers to exchange information across different locations at different time points. They can improve comprehensiveness by allowing physicians to look up information to address a wider range of clinical issues or reach out to specialists for specific advice. They can improve coordination by allowing people in different institutions to access and revise a shared record of all relevant information about a person. There are tools for each of these functions, but they are not widely available, not well integrated with each other, and models of primary care have not adapted to incorporate them.

We define digital health as 'any interaction between patients and/or their circle of care occurring virtually, using any form of communication or information technologies, with the aim of facilitating or maximising the quality and effectiveness of patient care' [18]. We have reviewed a range of health IT frameworks and adapted them to describe core components of digital health in primary care (as shown in Fig. 2 and described in Table 1).

These core components are reflective of the structures needed to support current primary care activities. A more holistic view, consistent with primary healthcare as defined above, would include an IT infrastructure to enable a health ecosystem that could be organised around primary care. This would facilitate the sharing of information and coordination of public health, and health services in schools, care homes, acute care facilities, as well as maternal health, dental care, housing, employment, financial support, and judicial systems.

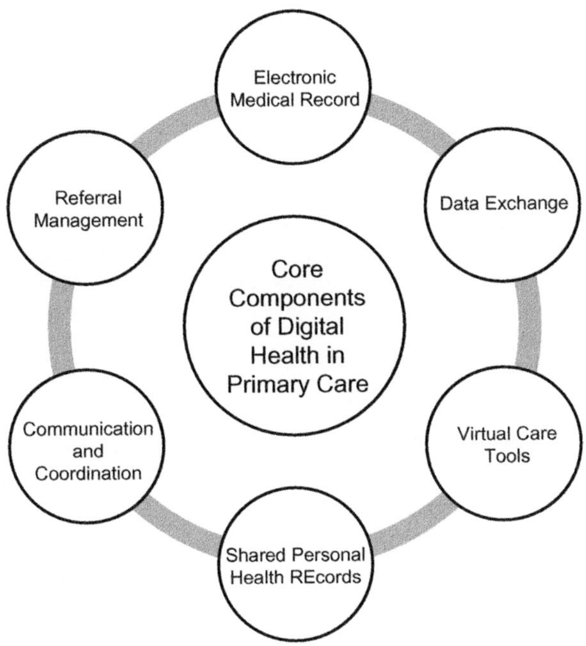

FIG. 2 Core components of digital health in primary care.

TABLE 1 Definitions for the 6 core components of digital health infrastructure in primary care.

The 6 core components of digital health infrastructure in primary care	
Electronic Medical Record	The electronic medical record (EMR) is typically internal to an organisation. EMRs function to support the basic workflows required for an interaction between a patient and provider including medicolegal documentation, writing prescriptions and, reference letters. In the past, most EMRs did not include ways to engage patients or the ability for patients to access their health data; however, some have now developed these capabilities.
Data Exchange	Digital health systems require mechanisms to exchange data between applications and organisations. This requires protocols to guide how the data are exchanged. Protocols to guide permissions to exchange the data are also required in accordance with legislative regulations. Ideally, data exchange includes structured data elements, to support ongoing care as well as analysis of the data.
Virtual Care Tools	Virtual care tools enable healthcare provision over space (i.e. not in person) and/or over time (i.e. asynchronously). This can include voice, video, or text-based interactions. These tools may or may not be integrated into other parts of the system like EMRs. Patients and providers typically require a device at their end, along with an Internet connection to engage with these tools.
Shared Personal Health Records	Personal health records (PHRs) provide a comprehensive snapshot of a patient's health and related data, such as diagnostic tests and imaging. Ideally, PHRs enable data exchange with multiple organisations and applications. Typically, patients have full control of this record, with the ability to share with other members of their care team. This also includes patient health portals, which are typically linked to the EMR of a single institution.
Communication and Coordination	Applications that enable members of a care team both within organisations and across organisations to communicate with each other and actively coordinate the care of patients. These platforms may include patients and caregivers, though many currently do not. Platforms can support asynchronous messaging within an organisation or the development of care coordination plans that can be accessed and revised by various members of a care team. ePrescribing is often the first form of structured communication between providers (in this case, physicians and pharmacists).
Referral Management	A platform to manage the movement of patients through the system. This includes centralised referral pathways as well as electronic consults (e-consults). Patients should have access to monitor the status of their referrals. Intelligent analytics can triage patients and ensure they are matched to the best resource for their needs.

Digital health in Canada and regulation

Canada is not a leader in the implementation of digital primary care, but its recent development is instructive as it shows the potential for rapid improvement. In 2019 Commonwealth Fund International Health Policy Survey of Primary Care Physicians in eleven high-income countries, Canada ranked near the bottom in health information technology use by primary care physicians. For example, only 22% of Canadian primary care physicians offered patients the option to communicate via email or secure portals compared to 91% in Sweden [19].

Furthermore, less than a third of Canadian primary care doctors had electronic access to patient clinical summaries, laboratory test results, and complete medication lists.

However, there is potential for improvement through the following digital health technologies: electronic medical records and electronic health records (EHRs), laboratory and diagnostic testing systems, patient portals, and virtual care.

The first electronic medical records and electronic health records were developed along advances in computing technology in the 1960s and 1970s; however, wider physician adoption did not occur until the 1990s [20]. In Canada, this was accelerated by the creation of Canada Health Infoway in 2001 which transferred funds to provinces and primary care clinics to subsidise adoption of EMRs in primary care and in hospitals. As of 2019, the proportion of Canadian primary care physicians using EMRs is 86%, second last behind Switzerland in the survey discussed earlier while physicians in countries such as New Zealand, Norway, and the UK reported 100% usage.

The UK started testing and offering incentives to primary care physicians in the 1970s. Companies began offering free computers to physicians in 1987, and by 1990, the general practitioner contract began stipulating the use of computers. Today, the UK continues to encourage the use of EMRs through their remuneration systems [21]. Their GP2GP service uses a standardised record system that allows patients' general practitioners to transfer EMR records to each other in cases of transfer of care [22]. Much like financial information, patients are able to access as well as transfer their health information transforming primary care. In addition, the NHS has plans to go fully digital by 2024 with the scaling of services such as electronic prescribing [23,24].

Virtual care tools: These were usually used by urban physicians in Canada to provide care in remote rural communities with few doctors. Otherwise, most patients seeking virtual care would have had to rely on private telemedicine companies and pay out of pocket for episodic care. In the US diverse insurance network offerings, virtual care was slowly expanding, increasing from 35% of hospitals in 2010 to 76% in 2017 [25]. While the federally sponsored Medicare programs did not offer comprehensive coverage for virtual care, telemedicine was more prevalent than in Canada. In Germany, which has multiple payers, they have instituted the Digital Healthcare Act which allowed patients to be covered for low-risk medical devices which could be used for remote monitoring or diagnosis [26]. With the advent of the pandemic, virtual care has taken off globally in countries including the USA and China [27].

Shared personal health records: In primary care, this includes laboratory tests, diagnostic imaging, and drug dispensing information. With the improvement of Internet infrastructure and access to the web, countries are moving towards the web by providing patient portals in which to access health information and communicate with care providers. While Canada is in its infancy in providing these services, countries such as Denmark and Estonia already have strong patient access to health information [28]. The lack of portability and patient access in Canada creates information silos whereby patient information is stuck within different health regions or hospital systems.

In comparison, Denmark has established an eHealth portal for patients as far back as 2003 where patients and care providers could share information such as hospital EHR data, shared medication records, personal medical histories, and previous hospital and primary care physician contacts. Similarly, Estonia has implemented a patient portal in the context of developing electronic access to all sectors of government in the 1990s [28,29]. The country utilised its existing competencies in developing electronic portals for other government services by successfully rolling out the national program. Patients opting-in were provided eHealth ID cards and even those that chose not to participate were still included in the national information database in cases of emergencies when their choice could be waived.

Communication, coordination, and referral management: The latter two components are often bundled together with shared personal health records or EMRs and in some ways are dependent upon the delivery of a virtual care service. As patients become more familiar with digital applications, some community family physicians in Canada have integrated messaging features in their EMRs. This pace of innovation is encouraging. Of course, there are many digital platforms one could use to coordinate care in Ontario, but like EMRs, they are generally fragmented and heterogeneous.

Communications also extend to other members of the patient's care team, such as the newly available PrescribeIT tool that allows physicians to electronically manage patient prescriptions and directly links to EMRs, formularies, and pharmacists [30]. Up until the first decade of the 21st century, referrals for patient care were largely done via fax and paper forms and this resulted often in inefficiencies and long wait times for patients to access specialist care [31]. In January 2010, Ontario piloted then expanded an electronic referral system which provided a secure web-based platform in which primary care providers could consult specialists virtually, providing a more seamless experience for patients. These efforts to improve communications, care coordination, and referral management have also been successfully trialled in other jurisdictions such as the US, with similar successes, yet there has not been substantial economic analysis on their benefits [32].

In the field of digital health, there is no central regulating body in Canada for software used in clinical care, apart from medical devices which are regulated by Health Canada [33]. Regulation has now been achieved with the combined input of professional groups, regional and provincial health authorities, as well as independent certifying bodies. If the technologies can comply with provincial privacy laws [34] and satisfy the standards of medical colleges and professional bodies, then digital health companies can usually enter the market without further issues. The low barrier to entry does not mean that the landscape is particularly competitive.

Digital health companies are often dominated by parent companies that essentially operate as monopolies or oligopolies such as those with a strong presence in telecommunications and groceries [35]. While it may seem that regulation is not yet optimised, Canada is not alone in its struggle to regulate the evolving field of digital health. While the UK has developed its own recommended health apps library to guide clinicians and patients, the US Food and Drug Administration has developed formal guidelines on how to approach the regulation of mobile apps. True regulation with standard setting and enforcement is still in its infancy [36]. Table 2 provides a summary of the current state of primary care and digital health across 4 provinces, with a focus on virtual care.

Comparing digital primary care across Canadian provinces

In the absence of any national mandates, we find substantial variation in maturity of the approaches to digital primary care across provinces in Canada (see Table 1). Even

TABLE 2 Primary care and digital health in four Canadian provinces.

	British Columbia	Alberta	Ontario	Nova Scotia
Primary Care Structure and Remuneration	• Mostly FFS, very few on capitation • < 2% work in multidisciplinary team-based practices	• Largely FFS billing, blended capitation at teaching sites • 41 primary care networks (3800 physicians and 1000 allied health providers)	• Half in blended-capitation models • Around a quarter in multidisciplinary team-based practices	• Half FFS and half capitation • One-third are in multidisciplinary team-based practices
Policy and Strategy	• Provincial Digital Health Initiative has a board with cross-sector governance • Digital infrastructure development strategy office	• Under revision • No specific digital health regulatory body or strategy	• Digital Health division in 2019 • Multiple provincial implementation partners funded by government	• Regional Health Authorities oversee healthcare • These bodies have no official oversight over digital health
HIMS	• Shared EMR for virtual care teams servicing rural areas and telephone helplines • Provincial portal into data repositories for provider use • Prototype of secure gateway for citizens launched	• EMR selected by community family physician: shared EMR for hospitals • Province-wide laboratory information system • Patient portal for basic laboratory results	• EMR selected by family physicians; no shared EMR for hospitals • Province-wide laboratory information not accessible by all doctors • No provincial patient portal (some hospitals have their own systems)	• No shared EMR but have adopted 2 systems with Zoom post-COVID • No province-wide laboratory information system • Has patient portal, My Health Analysis, for patient test results, but physician optional
Covered Virtual Care Modalities	• Many billing codes for virtual care since 2012 • Since COVID, added fees for telephone visits • Covers video, asynchronous text, phone, remote monitoring • Remote monitoring programs offered through regional health authorities	• Billing codes for telephone and video (premium) since COVID not adjusted for complexity • No formal virtual care platform prior to COVID-19 (telemedicine only at approved sites) • No formal remote monitoring programs	• Billing codes for telephone and video with time modifiers since COVID • Billing for provincial platform (OTN) at approved sites prior to COVID included virtual premiums • Select premiums added for virtual in August 2020 • Remote monitoring for specific diseases	• Only hospital-based telehealth services pre-COVID • Virtual care codes for phone and video since COVID • Previously, video and telephone used for remote consultations by specialists • Secure messaging portal discontinued by vendor • No formal remote monitoring programs

in a single-payer system, the primary care varies in structure and delivery across the country. For example, in provinces such as Nova Scotia and Alberta, physician remuneration is primarily fee-for-service (FFS), whereby family physicians are paid per clinical interaction. In Ontario, there are a range of payment models from FFS to blended capitation, where family physicians are compensated by the number of patients rostered to them with minimal fees per interaction. Physician remuneration affects uptake of virtual care tools, where physicians paid by FFS are more likely to use modalities that are reimbursed, while physicians paid on a capitated basis may consider how efficient use of different tools can help grow their roster of patients, irrespective of whether the service is directly reimbursed.

Likewise, the structures to guide digital health strategy and implementation vary by province, with British Columbia (BC) and Ontario choosing very different approaches. Ontario Ministry of Health partnered with the Ontario Medical Association to promote uptake of EMRs and later certification of EMRs but did not consider how the various information systems interacted. The Ministry then established a digital health division in 2019, which has made suggestions around desired features of the systems but did not set up incentives or enforcement mechanisms. BC, on the other hand, is organising a provincial effort to scale up health information architectures. It involves leaders from multiple sectors including physicians, hospital administrators, government, and patients. Furthermore, the province is looking beyond the traditional request for proposal system and attempting to develop its own open standards-based multivendor ecosystem. These two provinces illustrate two potentially important components of digital health policy in having a dedicated leadership body, a vision, and a means to implement.

Virtual care has seen the most growth in the last year. Prior to the pandemic, British Columbia publicly funded video visits for patients through a range of platforms, and Ontario did so through the Ontario Telemedicine Network, which was not integrated with EMRs. Most other areas provided video visits through specific technologies at approved telehealth sites, mostly for people in rural areas. Canadians were otherwise required to pay out of pocket (or through employer-based insurance) for virtual care. Physician remuneration for services varies substantially, with Ontario and BC compensating at a similar rate for virtual care versus in-person care, while in Alberta, they are offered a reduced rate.

Finally, the four provinces have varying degrees of regulation and integration of health information and management systems. The only province with a province-wide electronic health record system is Alberta. The province has recently further integrated its systems with the ongoing rollout of its EPIC-based ConnectCare EMR in addition to its MyChart patient portal [37,38]. This progress may seem advanced in comparison with the other three provinces, but often access to these systems is limited to hospital-based physicians.

In Alberta, most family physicians work in private clinics in the community and thus most do not have access to either system. Instead, family physicians in the community rely on their private clinic EMRs. In their efforts, policymakers in all provinces have different approaches to digital health: ranging from the hands-off approach in Alberta currently to the strategic board assembled in BC aimed at developing digital infrastructure on a provincial level. BC also goes a step further in the adoption of open standards and building technology to their needs rather than the traditional procurement system focusing on buying commercial solutions from vendors. In our next section, we do a deeper dive on the province of Ontario to understand the evolution of digital health within a region.

The digital foundation of Ontario's primary care system

Ontario has invested in the digitisation of healthcare to improve effectiveness and efficiency in the last two decades [39]. In 2003, the government established OntarioMD, which certified electronic medical records from private vendors for community physicians, and supported implementation. The proportion of Ontario's primary care physicians using EMRs went from 3% to 86% now, because of government funding and OntarioMD's leadership in EMR certification [40]. It also established eHealth Ontario, which oversaw implementation of a digital identity management system for patients and providers in Ontario, linked organisational health information systems with provincially developed digital assets, and introduced tools to share patient data within the circle of care across multiple institutions. Thanks to its efforts, the Ontario Lab Information System (OLIS) data can be downloaded directly into a patient's electronic chart and the ConnectingOntario service allows physicians to view up to date clinical information on their patients from a range of participating institutions. Physicians using EMRs certified by OntarioMD are enrolled in a service called Hospital Report Manager in which reports from hospital visits such as discharge notes are automatically electronically embedded into a patient's primary care record. Taken together, these newly widely used digital health tools show the government's ability to support the EMR and data exchange components of primary care digital health.

However, many initial digital health policy decisions which seemed reasonable at the time now constrain efforts to leverage health information to enhance primary care performance. One example is providing subsidies to primary care providers to support a private EMR vendor marketplace rather than deploying a single government selected EMR with its own proprietary platform. A clear benefit to this approach has been giving the flexibility to autonomous physicians to select the platform that best suits their needs and practice. At the time, the value of digital information exchange between healthcare organisations was less clear. The government did not mandate that the various EMR vendors be able to exchange digital information and map data between them. This hands-off approach to industrial policy has made it very challenging to implement current care integration strategies because data exchange between different silos of the healthcare sector is so difficult.

The recent COVID-19 pandemic has introduced new ideas that will shape Canadian digital health for years to come. In the process, we see successes and failures as well as key lessons that highlight a path to improving primary care by modernising our digital health capabilities. Two key case studies of digital primary care implementation, examples from Ontario, Canada's largest province, illustrate the interdependency of the components of digital health infrastructure in primary care. Ontario's primary care sector and the historical dynamics have generated the current digital health landscape. Both case studies support two core functions of primary care: access through use of virtual care tools in response to COVID-19 and coordination of care across the continuum through data exchange.

Case #1: Adoption of virtual care in Ontario primary care as a result of the COVID-19 pandemic

Prior to the onset of the COVID-19 pandemic, use of virtual care in Ontario was being explored in several instances in primary care. This included:

- the phone call and email follow-ups to patients when a physician deemed it more appropriate to do so than require the patient to come in for an in-person appointment;
- billing codes for use of government-funded video chat technologies with proper built-in security features such as the Ontario Telemedicine Network (OTN);
- private vendors operating 'walk-in medicine' style telemedicine platforms outside the sphere of government-funded healthcare; and
- the pilots of secure text messaging technologies with patients.

Only a fraction of physicians were using video visits. Physician billings from 2015 showed that only 0.15% of clinical encounters were remunerated virtual care visits [41]. Recently, Digital Health Canada, the association representing health informaticians across Canada, published a virtual care maturity framework which describes three phases of maturity: basic, emerging, and advanced use of virtual care [42]. These are shown in Fig. 3, and unquestionably, Ontario was at a basic level of maturity prior to the pandemic.

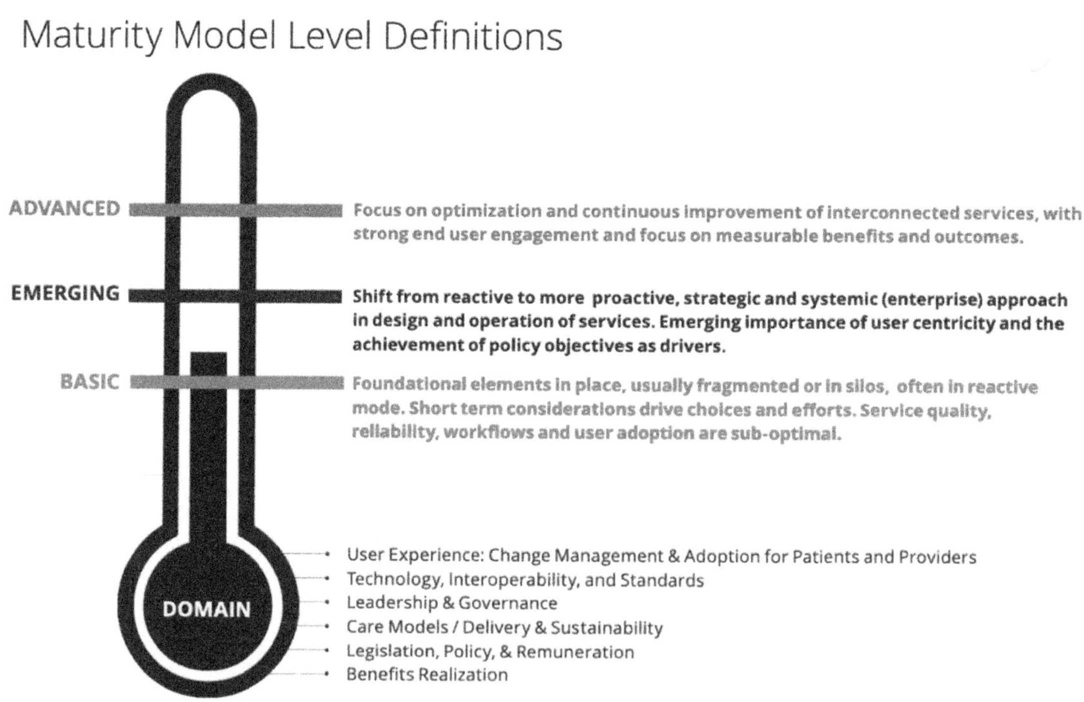

FIG. 3 Maturity model level definitions. *Credit: Digital Health Canada. Virtual Care in Canada: Maturity Model framework. [Internet]. Available from: https://files.constantcontact.com/d93e3261201/160831a0-a439-41c4-8275-42dd9300ad5a.pdf.*

Until February 2020, the Ontario government was assessing the value of the various virtual care options available to primary care clinicians. The government tolerated the provision of healthcare by private telemedicine companies, even if it was unclear whether these complied with Canadian Medicare laws. The Royal College of Physicians and Surgeons of Canada issued a guideline report on how to ensure safe adoption of virtual care in Canada. [43] Then, in the middle of March 2020, the province declared a state of emergency due to the COVID-19 pandemic, and the value of virtual care became very clear. Without the appropriate personal protective equipment available and an inability to adhere to infection protection and control measures, most primary care offices in Ontario decided to close to in-person visits, to ensure safety of patients and staff. This occurred before any substitute compensation measures to conduct care virtually were offered by the government.

Within the span of a few weeks, virtual care policy and operations had shifted dramatically. The spread of the virus meant in-person clinics had to be suspended and it was unclear how physicians, particularly those in fee-for-service models, would provide care and bill the government for their services. The Ministry realised that supporting virtual care was essential, as was done in Ontario during the SARS epidemic in 2002 [44]. The government approved billing codes for physicians to conduct clinical appointments with patients virtually, either on the phone or through video chat software, such as OTN. While it took several weeks to update the antiquated billing system to accommodate these new virtual care billing codes, adoption of virtual care skyrocketed during the one-month time span of April 2020 when the codes went live.

In Ontario, the proportion of virtual visits went from 0.7% before 2020 to 71% in the second quarter of 2020, and in the same time periods, the proportion of doctors billing for virtual care went from 4% to 86% [45]. Furthermore, the government realised that its own video platform could not meet the demand, so it allowed private vendor video chat services that used to charge patients or employers directly to bill the government for their care. As such, physician and governmental organisations amended their guidelines on virtual care almost overnight. They provided a list of video chat software that met privacy and security regulations, if clinicians asked for routine patient consent to virtual care. This list initially included video conferencing tools that were not built for healthcare, such as FaceTime, Skype, and Zoom [46,47]. These systemic changes quickly led to a proliferation of new virtual care options, with new video conferencing platforms that were directly embedded in physicians' EMRs.

Through the first and second waves of the pandemic in Ontario, clinics conducted care through a combination of both virtual and in-person appointments, with periodic increases in the proportion of virtual care appointments. For instance, at the Toronto Western Hospital Family Health Team in Toronto, patients were offered telephone or video appointments versus in-person appointments with physicians first, according to the recommended rubric seen in Fig. 4. If the clinic's receptionists were unsure about a specific patient, they conferred with the physicians working that day, who decided if it was appropriate and safe to bring in a patient for an in-person assessment. In their decision-making process, physicians took the following questions into account:

- Does the patient feel the benefit of an in-person visit exceed the risk of leaving their home?
- Would you be putting the patient at risk by asking them to come to the office for something that could be handled over the phone or virtually?

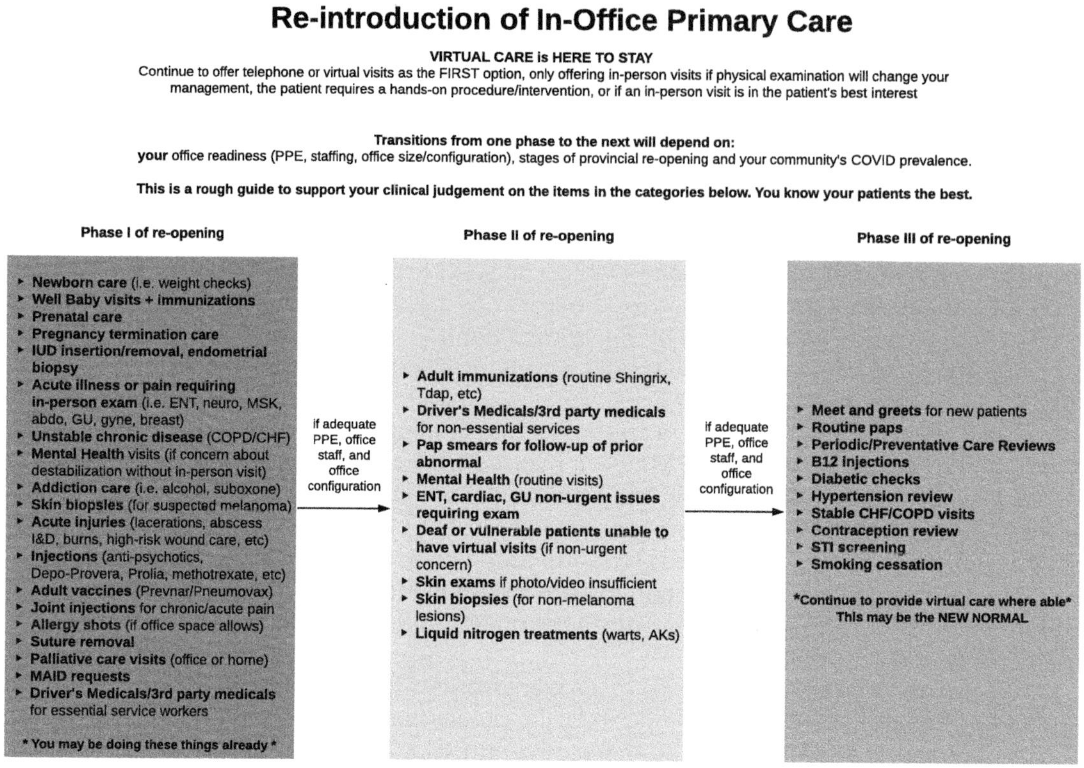

FIG. 4 Virtual care rubric.

- Is the medical benefit to the individual patient worth the risk to you and your office staff by having them come for an in-person assessment?
- Could further delay in provision of in-person care result in a worse outcome for the patient?
- Will offering care remotely lessen the burden on acute care hospital facilities?

The rapid uptake of digital tools for virtual care in Ontario was mirrored in most high-income countries and highlights several common factors. First, key stakeholders become much more amenable to compromise in a time of crisis, and this flexibility enables creative solutions to be trialled and scaled in a safe enough way because the benefits outweigh the risks. In this case, we see that funding virtual care led to a reduction of in-person medical appointments, which likely reduced the spread of COVID-19 within a clinic's physical space. This was widely implemented despite the unclear costs and the potential liability regarding the security of the video chat tools used. Second, crises can rapidly accelerate ongoing changes and achieve a new equilibrium. For instance, Ontario was gradually trying to expand virtual care from a very low baseline prior to March 2020. But with the introduction of the virtual billing codes, and with patients becoming accustomed to convenience of a virtual visit, it is unclear if the government could ever revert back to the status quo prior to the pandemic.

Outside of a pandemic, it is unclear if the quality of primary care is maintained or improved with patients accessing care more often virtually, and what the cost implications might be. Governments in Ontario and elsewhere are reviewing the available information to determine if it is worthwhile to continue to fund virtual care in the same way.

Case #1 key messages and lessons learned

- Prior to the COVID-19 pandemic, virtual care was a minor feature of primary care in Ontario.
- Digital tools can be rapidly adopted in a flexible health policy environment during a crisis, like a global pandemic.
- After widespread adoption of virtual care, it is essential to constantly reevaluate their potential for continued success in a postcrisis period. Embedding regularised evaluative mechanisms as virtual care matures can ensure continued alignment with a health system's strategic priorities.

Case #2: Integrating care through data exchange

In the increasingly digital structure of the healthcare system, high-quality patient data will allow us to take advantage of advances in computation and novel algorithms to improve care. Proven benefits of integrated care, facilitated by data integration, include improved outcomes and reduced cost of care [48]. However, getting access to data from different organisations' servers in a timely manner and linking them is challenging in systems where institutional practices and legislation have resulted in heterogeneous patient data that are siloed. In this case, we explore the core component of data exchange and its effect on the other components of communication and coordination and shared personal health records. Specifically, we review the barriers to digital integration and innovations that may support information sharing and coordination between primary care and other entities within the health system.

Given the sensitivity of patient health information, laws and regulations on healthcare data, its use and exchange have been enacted by national and subnational governments across the world. In Ontario, the Health Information Protection Act and the Personal Health Information Protection Act or 'PHIPA' are the legislations passed in 2004 that outlined how a patient's personal health information is to be collected, used, and disclosed, both within and between a patient's circle of care [49]. It defines who are health information custodians, agents, and owners, and how patient-level data are handled by these different defined groups. This is similar to the Health Insurance Portability and Accountability Act or 'HIPAA' in the United States.

Generally, a clinician and the organisation they work for are responsible for keeping the data integral and secure, and they are data custodians or stewards. Patients are owners of the data held by these organisations, and they are entitled to access their data, although this can occasionally be a complicated process.

Regrettably, how patient data flows within a patient's circle of care (such as primary care physicians and their specialist colleagues) remains stuck in the analogue past. Most consultation notes by specialists are sent by fax machines to the primary care physician; these paper-based notes and all the key patient data described therein are scanned as PDFs, and none of that data can then be entered into their EMR and aggregated or analysed by that physician. This leads to costly omissions of key pertinent information and duplications of efforts. And yet, with the unique patient identifiers of the common format health card number in Ontario, known as the OHIP number, province-wide data mapping and integration is technically feasible, even if it is costly to achieve.

Data flows between different clinicians and organisations have only recently been addressed by governmental initiatives in Ontario. Initiatives that took nearly a decade to come to fruition include certain tools integrated into EMRs which allow physicians to pull down data from the provincial laboratory test tracking system (e.g. OLIS) as well as key data pertaining to hospital admission and discharges in the provinces (e.g. HRM). Furthermore, a separate, web-based portal called 'ConnectingOntario', which is not integrated into community practitioner EMRs, allows physicians to access other types of patient data, such as dispensed drugs, imaging reports, and consultation notes from hospital-based clinicians. This required significant investment from the government to bridge and map the disparate health information systems used by different organisations. This is because the database structures, schemas, models, and interfaces have all been developed by private vendors with very limited standardisation, since regulation to mandate data standardisation was minimal in Ontario and Canada more broadly.

The next level of digital integration maturity includes pushing and merging data between healthcare organisations in a seamless and timely way that respects data stewardship and data ownership regulations. Ad hoc negotiated data sharing agreements between different organisations with complex consenting procedures built-in are not practical, even if software tools to facilitate these have been developed and marketed. In Ontario and many other jurisdictions, there are no clear payers for these tools, and thus health organisations, which are run as nonprofit entities with limited budgets, have no incentive to invest in solving these data integration issues on their own.

The provincial government has recently recognised that previous legislation limits the sharing of patient data that is in the patient's interest. They passed an amendment to PHIPA in 2020 which allows patients and patient-facing consumer electronic service providers—or vendors who facilitate transmission of patient health information—to access and digitally transmit PHI according to PHIPA rules. And importantly at the centre of this amended legislation is a patient's ability to provide preemptive consent to share data among the circle of care, no matter where their health data reside.

New primary care models designed by physician organisations and cross-sector integration models designed by the Ministry of Health aim to align incentives and digital infrastructure across data silos. For instance, the College of Family Physicians of Canada's 'patient-centred medical home' and 'patient-centred medical neighbourhood' describe how patient data should ideally flow as a patient traverses the healthcare system [50,51]. Similarly, the government is promoting 'Ontario Health Teams' as a cross-organisational structure in which organisations within a specific geography come to an agreement on how they will collectively care for a patient population living in their shared jurisdiction, including how digital infrastructure can be retooled to have high-quality patient data available at each health system touchpoint for a specific patient. [52]

But even with these new models in place, the obstacles to appropriate, timely data sharing—both technical and contractual—are enormous. In terms of technical barriers, standard platform, technology-neutral data repositories, and formalisms with high levels of expressivity will be required to allow seamless data mapping and exchange between each organisation's customised, procured health information system (HIS) and their data silos. Alternatively, a less promising approach is to deploy a hodgepodge of application programming interfaces (APIs) and software development kits (SDKs)—technical devices employing health standards such as HL7 and FHIR, between disparate health data repositories and electronic and personal health record systems [53].

Either approach will require a major investment to establish and maintain as new health information systems are developed and existing ones are upgraded. Thankfully, this issue has been recognised in other jurisdictions. The Office of the National Coordinator for Health Information Technology (ONC) in the United States has recently approved new health system regulations to implement the 21st Century Cures Act of 2016 provisions on data interoperability, as a revision to the HITECH Act [54]. These new regulations will mandate digital health vendors to comply with facilitating data exchange between them; in fact, most American digital health vendors have already seen the writing on the wall and have recently joined burgeoning technical interoperability networks such as CommonWell Health Alliance and CareQuality to align their data standards, schemas, and APIs for data exchange [55,56]. The hope is that the Ontario Ministry will eventually introduce similar regulations or at least be incentivised to do so by the Government of Canada.

In the interim, healthcare organisations are already considering the fundamental framework for data exchange, instead of simply tolerating the status quo of individual interinstitutional data sharing agreements. One of the most important policy decisions is whether to proceed with a centralised or decentralised approach. In a centralised approach, an overarching organisation supersedes the accountability structure of an individual organisation, which may undermine institutional autonomy and could compromise an institution's reputation in case of a failure of the centralised structure. The government-developed tools of OLIS and ConnectingOntario previously described are robust versions of this approach, and clinician training and acceptance of the terms and conditions aim to mitigate the risks.

However, the goal of this next phase of integration may not lend itself as neatly to a centralised approach as previously. The noble goal of sharing an organisation's granular, real-time data with other organisations in the interest of optimising patient care across the continuum of care constrains an institution's autonomy, as they are agreeing to the sharing of that institution's most sensitive data holdings. Depending on the culture of a particular organisation, these institutional concerns could be a significant challenge. In Ontario, mandating rapid exchange of granular patient data among untrusted organisations is likely to encounter significant opposition by local healthcare groups and stakeholders. Furthermore, the centralised approach to data exchange may not be viewed favourably by Indigenous peoples, who are developing their own models of data governance.

In contrast, a decentralised tool would employ novel technical approaches to ensure that data exchange occurs in a trustless fashion between all the independent organisations.

Being trustless means that the participants involved do not need to know or trust each other or a third party for the system to function. The data sharing system is designed and developed to function as desired regardless, which means that one health organisation does not need to ensure a new centralised body's tool is developed and maintained to the standards of security of each organisation it integrates with.

The issue is that these tools remain relatively experimental and there are few real-world cases to justify moving towards decentralised data sharing and exchange at this time. However, Ontario is designing and developing 'permissioned blockchain technology' to explore how this might be done. Blockchains are defined as immutable, cryptographically secured shared data ledgers between nontrusting parties. The technology embeds decentralisation, which allows for federated data sharing and built-in health system stakeholder consent mechanisms, such as for the patient and the provider, without a sole point of failure. Once validated and mature, these will not forfeit institutional autonomy and will address concern about poorly designed and maintained centralised tools. Note that permissioned, private blockchains are access-controlled, unlike public blockchains such as Bitcoin, which are where the data are completely transparent and viewable by anyone.

An example of a permissioned blockchain-based data sharing technology in development in Ontario healthcare is the eConsent Gateway proof of concept developed by the University Health Network (UHN) in Toronto. This tool was developed to ensure unquestionably valid patient consent permitting data transfers between different organisations and data holders. The blockchain stores the patient consent, therefore turning on the individual APIs which allow transfer and mapping of the data according to the patient's immutably documented consent. This mechanism is depicted in Fig. 5.

At UHN, a proof of concept of data collection for research activities was conducted and functioned as intended. One can envision that further roll-out of this technology, for instance by embedding in an Ontario Health Team's digital health policy, can lead to integration of primary care data silos with data silos from other subsectors of the health system such as hospitals, pharmacies, and long-term care homes. In short order, a near real-time updated, time-stamped edited, community-wide, longitudinal patient record could follow a patient as they traverse the health system, with the patient in control of which information is shared

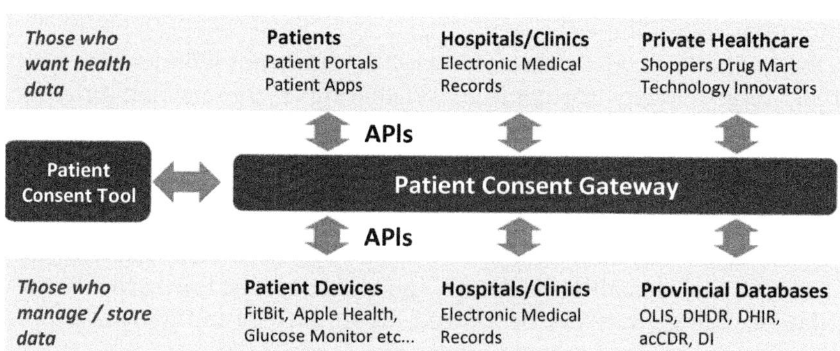

FIG. 5 Permissioned Blockchain-based data sharing mechanism.

with which party in their circle of care. One of the data sharing options for which a patient could consent to might also include providing their anonymised, granular data to a population database. Overseen by a sanctioned data trust, the aggregated data would be computationally analysed, and previously unknown health system findings and trends might become actionable by policymakers.

The current digital tools provide a new opportunity to meaningfully address the integration challenge for primary healthcare within the wider health system and community. Limited funding and poor alignment of interests among key system stakeholders make trialling and scaling of these integration technologies a significant challenge. Despite their promise, there is a need for modelling and evaluation of risks/benefits of these tools and for visionary leadership to advocate for these technologies to be developed in a safe and secure way. With the centralised and decentralised examples already provided, it is possible for Canadian incremental pragmatism to embrace a bold digital redesign of the healthcare system and its capacity for data exchange and interoperability. Regardless of approach taken, healthcare systems around the world will need to focus on maintaining the values of patient-centred care as they solve for the challenges of digital integration in healthcare.

Case #2 key messages and lessons learned

- Data exchange and care coordination are central to reaping the benefits of an integrated care system with primary care at the centre.
- There are enormous regulatory, structural, and technical challenges in many jurisdictions that need to be overcome in order for the integration benefits to be realised.
- There are trade-offs with various approaches to address these challenges. But the notable recent, decentralised data sharing technology of blockchain may offer one solution that if implemented among a network of partnered organisations can readily deal with several of the most complex barriers to data exchange and interoperability.

Digital health infrastructure as a force shaping the primary care transformation

While primary care systems aim to improve health outcomes, experience of care and its management costs while maintaining equity during the COVID-19 pandemic has accelerated the uptake of digital tools with great potential to improve access and coordination of care. Although we have focused on our experiences with rapid digitisation in our local primary care context, there are several recent and upcoming national and international guidelines around strategies to improve the digital health infrastructure available to primary care and the broader healthcare system [57–60]. These efforts will need to be supported by a willingness to invest resources into the development of technologies with the right set of functions and implementation, as well as crafting incentives for patients and providers to adopt new technology to enhance primary care.

In our experience with rapid virtualisation of primary care under the pressures of COVID-19, we see the impact of digital health infrastructure becoming a core component of primary care delivery. While the impact on routine quality care is unclear, virtual care maintained access to care during a time of physical distancing and achieved a core goal of pandemic response [61]. Now we must work towards discovering the full scope and possibilities of virtual care as well as its limitations.

Finally, we examine the prospect of using a digital infrastructure to coordinate and integrate patient care through shared patient records. When reviewing Ontario's prior attempts, we identified the need for strong governance and leadership, on top of financing and investing into the development of useful technologies that draw on cutting-edge innovations. Not only does this require policymakers attuned to the nuances of digital health strategy and implementation, but this also requires the interest of stakeholders such as patients, individual physicians, as well as facilities such as clinics or hospital networks, to buy in and adopt these services to ensure interoperability and optimal use.

We have presented the Canadian experience in virtualisation of care during COVID-19 and recent efforts to improve care coordination through our infrastructure. Yet the use of a digital health infrastructure to innovate healthcare has been a longstanding global initiative as outlined in WHO's Digital Health Playbook and Global Strategy on Digital Health for 2020–2025. Their latest strategic objectives emphasise the importance of global collaborations, national digital health strategies, stronger digital health governance, and the use of digital health infrastructure to enhance care centred around people.

There is much to do to enhance collaboration as well as create a national movement for better governance to encourage policies that will result in patient-centred digital health infrastructures to strengthen health systems. The urgency of change associated with the COVID-19 pandemic could be used to develop a national digital health policy. Thus, the issue of maturity in digital health infrastructure in primary care is not just one of modernisation of technology, but rather the mix of incentives, implementation supports, workforce, workflows, and accountability mechanisms needed to improve health outcomes for the entire population.

Guiding Canada into the future

This chapter provides a deep dive into the potential of digital health in Canada. The Canadian organisation of healthcare delivery at the provincial level results in differences which highlight the range of options within a given context. Contrasting the choices made between provinces and between countries can help build a broader evidence base to create digital ecosystems that improve the health of populations in other countries as well.

Not only will better infrastructure allow for better care, but it also will allow for broader implementations of learning health systems [62] from which we can truly utilise real-time analytics to further push the boundaries of care for our patients. Digital health can be transforming for our healthcare systems and patients deserve no less.

References

[1] International Conference on Primary Health Care, Declaration of Alma Ata, World Health Organization, Geneva, Switzerland, 1978.
[2] L.K. Muldoon, W.E. Hogg, M. Levitt, Primary care (PC) and primary health care (PHC), Can. J. Public Health 97 (5) (2006) 409–411.
[3] B. Starfield, L. Shi, J. Macinko, Contribution of primary care to health systems and health, Milbank Q. 83 (3) (2005) 457–502. [Internet]. Available from: https://www.ncbi.nlm.nih.gov/pmc/articles/PMC2690145/.
[4] L. Shi, The impact of primary care: a focused review, Scientifica 2012 (2012) 1–22.
[5] M.J. Hollander, H. Kadlec, Financial implications of the continuity of primary care, Perm. J. 19 (1) (2015) 4–10.
[6] Primary Health Care Performance Initiative, The PHCPI Conceptual Framework, 2021, [Internet]. Available from https://improvingphc.org/phcpi-conceptual-framework.
[7] B. Hutchison, J. Abelson, J. Lavis, Primary care in Canada: so much innovation, so little change, Health Aff. 20 (3) (2001) 116–131.
[8] D. Martin, A.P. Miller, A. Quesnel-Vallée, N.R. Caron, B. Vissandjée, G.P. Marchildon, Canada's universal health-care system: achieving its potential, Lancet 391 (10131) (2018) 1718–1735.
[9] Macrotrends, Canada Healthcare Spending 2000–2021, 2021, [Internet]. Available from https://www.macrotrends.net/countries/CAN/canada/healthcare-spending.
[10] The Commonwealth Fund, Number of Practicing Physicians per 1000 Population, Selected Health & System Statistics, 2020.
[11] Statistics Canada, Primary health care providers, 2019 [internet], in: Health Fact Sheets, 2019. Available from: https://www150.statcan.gc.ca/n1/pub/82-625-x/2020001/article/00004-eng.htm.
[12] A. Katz, N. Herpai, G. Smith, K. Aubrey-Bassler, M. Breton, A. Boivin, et al., Alignment of Canadian primary care with the patient medical home model: a QUALICO-PC study, Ann. Fam. Med. 15 (3) (2017) 230–236.
[13] G.P. Marchildon, B. Hutchison, Primary care in Ontario, Canada: new proposals after 15 years of reform, Health Policy 120 (7) (2016) 732–738. [Internet]. Available from: https://doi.org/10.1016/j.healthpol.2016.04.010.
[14] T. Kiran, A. Kopp, R. Moineddin, R.H. Glazier, Longitudinal evaluation of physician payment reform and team-based care for chronic disease management and prevention, CMAJ 187 (17) (2015) 494–502.
[15] R.H. Glazier, B. Hutchison, A. Kopp, Comparison of Family Health Teams to Other Ontario Primary Care Models, 2004/05 to 2011/12, Institute for Clinical Evaluative Sciences, Toronto (2015).
[16] P. Hutten-Czapski, R. Pitblado, S. Slade, Short report: scope of family practice in rural and urban settings, Can. Fam. Physician 50 (NOV) (2004) 1548–1550.
[17] J. Shaw, T. Jamieson, P. Agarwal, B. Griffin, I. Wong, R.S. Bhatia, Virtual care policy recommendations for patient-centred primary care: findings of a consensus policy dialogue using a nominal group technique, J. Telemed. Telecare 24 (9) (2018) 608–615.
[18] The Commonwealth Fund, Commonwealth Fund Survey, 2019, 2019, [Internet]. Available from https://www.cihi.ca/en/commonwealth-fund-survey-2019.
[19] R.S. Evans, Electronic health records: then, now, and in the future, Yearb. Med. Inform. (Supp 1) (2016) S48–S61.
[20] B. McMillan, R. Eastham, B. Brown, R. Fitton, D. Dickinson, Primary care patient Records in the United Kingdom: past, present, and future research priorities, J. Med. Internet Res. 20 (12) (2018), https://doi.org/10.2196/11293. [Internet]. Available from.
[21] NHS Digital, GP2GP [Internet], 2020, Available from: https://digital.nhs.uk/services/gp2gp.
[22] E. Wilkinson, Electronic prescribing, Pharm. J. (2019). [Internet]. Available from: https://pharmaceutical-journal.com/article/feature/a-blessing-and-a-curse-the-struggle-to-introduce-e-prescribing.
[23] NHS Digital, Electronic Prescription Service [Internet], 2021, Available from: https://digital.nhs.uk/services/electronic-prescription-service.
[24] American Hospital Association, Fact Sheet: Telehealth [Internet], 2021, Available from https://www.aha.org/factsheet/telehealth.
[25] S. Gerke, A.D. Stern, T. Minssen, Germany's digital health reforms in the COVID-19 era: lessons and opportunities for other countries, NPJ Digit. Med. 3 (1 PG) (2020). [Internet]. Available from: https://www.nature.com/articles/s41746-020-0306-7.
[26] P. Webster, Virtual health care in the era of COVID-19, Lancet 395 (2020) 1180–1181. North(10231 PG-1180–1181). [Internet]. Available from: https://pubmed.ncbi.nlm.nih.gov/32278374/.

[27] C. Nøhr, L. Parv, P. Kink, E. Cummings, H. Almond, J.R. Nørgaard, et al., Nationwide citizen access to their health data: analysing and comparing experiences in Denmark, Estonia and Australia, BMC Health Serv. Res. 17 (1) (2017) 1–11.

[28] R. Mill, Developing an integrated e-health system in Estonia, Est e-Health Found (Box 1) (2016). [Internet]. Available from: https://www.integratedcare4people.org/media/files/CaseProfileEstonia.pdf.

[29] I.T. Prescribe, The PrescribeIT Advantage [Internet], 2021, Available from: https://www.prescribeit.ca/71-about-us/192-the-prescribeit-advantage.

[30] C. Liddy, M.S. Rowan, A. Afkham, J. Maranger, E. Keely, Building access to specialist care through e-consultation, Open Med. 7 (1) (2013) e1–e8. [Internet]. Available from: https://pubmed.ncbi.nlm.nih.gov/23687533/.

[31] V.G. Vimalananda, G. Gupte, S.M. Seraj, J. Orlander, D. Berlowitz, B.G. Fincke, et al., Electronic consultations (e-consults) to improve access to specialty care: a systematic review and narrative synthesis, J. Telemed. Telecare 21 (6) (2015) 323–330.

[32] Government of Canada, Medical Devices Regulations (SOR/98–282), 2021, [Internet]. Available from https://laws-lois.justice.gc.ca/eng/regulations/sor-98-282/.

[33] Government of Ontario, Personal Health Information Protection Act, 2004, 2004, S.O. 2004, c. 3, Sched. A [Internet]. Available from https://www.ontario.ca/laws/statute/04p03.

[34] OntarioMD, Vendor Market Share: Market Share by Funded Physician [Internet], 2020, Available from https://wwwdev.nonprod.ontariomd.ca/emr-certification/certified-emrs/vendor-market-share.

[35] M. Jogova, J. Shaw, T. Jamieson, The regulatory challenge of mobile health: lessons for Canada, Health Policy 14 (3) (2019) 19–28. Available from: https://www.ncbi.nlm.nih.gov/pmc/articles/PMC7008689/.

[36] M. Avdagovska, T. Stafinski, M. Ballermann, D. Menon, K. Olson, P. Paul, Tracing the decisions that shaped the development of MyChart, an electronic patient portal in Alberta, Canada: historical research study, J. Med. Internet Res. 22 (5) (2020) e17505.

[37] D.C. Baumgart, Digital advantage in the COVID-19 response: perspective from Canada's largest integrated digitalized healthcare system, NPJ Digit Med 3 (1) (2020) 114, https://doi.org/10.1038/s41746-020-00326-y. [Internet]. Available from:.

[38] Government of Ontario, Ministry's Submission to Mr. Clark on Digital Health, 2016, [Internet]. Available from https://files.ontario.ca/1._ministry_of_health_and_long-term_care.pdf.

[39] Canadian Medical Association, CMA Physician Workforce 2017 National Results, 2017, [Internet]. Available from: https://surveys.cma.ca/en/viewer?file=%2fdocuments%2fSurveyPDF%2fCMA_Survey_Workforce2017_Q22_ElectronicRecords-e.pdf#phrase=false.

[40] Canadian Medical Association, CMA Health Summit—Virtual Care in Canada: A Discussion Paper, 2019, [Internet]. Available from https://www.cma.ca/sites/default/files/pdf/News/Virtual_Care_discussionpaper_v2EN.pdf.

[41] Digital Health Canada. n.d. Virtual Care in Canada: Maturity Model framework. [Internet]. Available from: https://files.constantcontact.com/d93e3261201/160831a0-a439-41c4-8275-42dd9300ad5a.pdf.

[42] Royal college of Physicians and Surgeons of Canada, Virtual care: Recommendations for scaling up virtual medical services – Report of the virtual care task force, 2020, [Internet]. Available from https://www.cma.ca/virtual-care-recommendations-scaling-virtual-medical-services.

[43] Ministry of Health and Long Term Care of Ontario, Bulletin. Fees for Telephone Services During the S.A.R.S Emergency, 2003, [Internet]. Available from http://www.health.gov.on.ca/en/pro/programs/ohip/bulletins/4399/bul4399.aspx.

[44] R. Glazier, M. Green, F.C. Wu, E. Frymire, A. Kopp, T. Kiran, Shifts in office and virtual primary care during the early COVID-19 pandemic in Ontario, Canada, CMAJ 193 (6) (2021) E200–E210, https://doi.org/10.1503/cmaj.202303.

[45] OntarioMD. n.d. Virtual Care: Covid-19 guide. https://www.ontariofamilyphysicians.ca/tools-resources/timely-trending/novel-coronavirus-2019-ncov/oma-virtual-care-covid19.pdf.

[46] OntarioMD, Virtual Care is Here to Stay, 2021, [Internet]. Available from https://ontariomd.vc/.

[47] S. Rocks, D. Berntson, A. Gil-Salmerón, et al., Cost and effects of integrated care: a systematic literature review and meta-analysis, Eur. J. Health Econ. 21 (2020) 1211–1221. Available from https://doi.org/10.1007/s10198-020-01217-5.

[48] Government of Ontario, Personal Health Information Protection Act, 2004, 2004, [Internet]. Available from https://www.ontario.ca/laws/statute/04p03.

[49] College of Family Physicians of Canada, Patient's Medical Home, 2021, [Internet]. Available from https://patientsmedicalhome.ca/.

[50] College of Family Physicians of Canada, Patient's Medical Neighbourhood, 2020, [Internet]. Available from https://patientsmedicalhome.ca/files/uploads/PMN_BAG_ENG.pdf.

[51] Government of Ontario. n.d. Ontario Passes Legislation That Puts Patients at the Centre of an Integrated Health Care System, April 18 (n.d.) 2019, https://news.ontario.ca/en/release/51917/ontario-passes-legislation-that-puts-patients-at-the-centre-of-an-integrated-health-care-system.

[52] C. Peng, P. Goswami, G. Bai, A literature review of current technologies on health data integration for patient-centered health management, Health Informatics J. 26 (3) (2019) 1926–1951. Available from https://doi.org/10.1177/1460458219892387.

[53] The Office of the National Coordinator for Health Information Technology, 21st Century Interoperability, Information Blocking, and the ONC Health IT Certification Program, [Internet]. Available from https://www.healthit.gov/curesrule/.

[54] Commonwell Health Alliance Available from https://www.commonwellalliance.org/. Accessed on October 26, 2021.

[55] CareQuality, Available from https://carequality.org/.

[56] Canadian Medical Association, Virtual care: Recommendations for Scaling Up Virtual Medical Services, 2020, [Internet]. Available from https://www.cma.ca/new-report-roadmap-improve-virtual-care-canada.

[57] World Health Organization, WHO Handbook for Digitalizing Primary Health Care, 2020, (in-press) [Internet]. Available from https://www.who.int/reproductivehealth/publications/handbook-digitalizing-primary-health-care/en/.

[58] World Health Organization, WHO Guideline: Recommendations on Digital Interventions for Health System Strengthening, 2019, [Internet]. Available from https://www.who.int/reproductivehealth/publications/digital-interventions-health-system-strengthening/en/.

[59] World Health Organization, Digital Health Platform Handbook: Building a Digital Information Infrastructure (Infostructure) for Health, 2020, [Internet]. Available from https://apps.who.int/iris/handle/10665/337449.

[60] R.S. Bhatia, C. Chu, A. Pang, M. Tadrous, V. Stamenova, P. Cram, Virtual care use before and during the COVID-19 pandemic: a repeated cross-sectional study, CMAJ Open 9 (1) (2021) E107–E114.

[61] R.M. Califf, J.P. Briggs, T.S. Coster, K.B. DeSalvo, R.L. Fleurence, P. Kaufmann, et al., Transforming evidence generation to support health and health care decisions, N. Engl. J. Med. 375 (24) (2016) 2395–2400.

CHAPTER

24

Caribbean/PAHO—Jamaican case study

Monica Jones[a,b]

[a]Health Data Research UK (HDR UK), Animo Consultancy Ltd, Leeds, United Kingdom
[b]Global eHealth Collaborative (GeHCo), Melbourne, VIC, Australia

Background

The Pan-American Health Organisation/World Health Organisation established a framework of 'Information Systems for Health' (IS4H) which included an integrated system of policies, legislation, governance, human, financial and information technology resources, health indicators, data sources, data management processes, information products, and the effective dissemination and use of information. This integrated system is reflected in the four IS4H strategic goals that drive and enable effective healthcare, namely:

(1) Management and governance;
(2) Data management and information technologies;
(3) Knowledge management and sharing; and.
(4) Innovation and performance.

This strategic framework was adopted by participants at the *High-Level Meeting on Caribbean Information Systems for Health* in Kingston, Jamaica, in November 2016 as the framework for Caribbean regional and countries IS4H strategies, plans, and investments. There are a number of *Technical Advisory Groups (TAGs)* that also provide national technical leadership and recommendations on matters related to these strategic goals. Health information systems, along with information sharing, have been proven to improve clinical practice by reducing staff errors, improving automated harm detection, monitoring infections more effectively, and enhancing the continuity of care during physician handoffs. A significant number of opportunities exist for the Hospital Information Management System (HIMS) to contribute to savings for both the health system and the Jamaican economy.

Focusing on the ways in which people are cared for across the whole healthcare system, from their GP to the hospitals, and when they leave hospital and may need extra care in the community before going home is described as 'whole pathway' and takes a comprehensive

Roadmap to Successful Digital Health Ecosystems
https://doi.org/10.1016/B978-0-12-823413-6.00026-4

view of every service that a patient needs; sadly these are often not linked up, and patients may fall between the gaps and signs are missed. The aim is to have a system that spots the warning signs earlier and is joined up. In order to make use of the data that is available in health and social care for patients, public, and staff and beyond, it needs to be recorded it in a way that makes sense. It is the duty of everyone involved in direct care to keep a true and accurate record for every patient. In the 21st century, this should be digital in a proper 'Electronic Patient Record' and use clear clinical terminologies. Once that happens things get better. Good-quality structured data for direct care happens, good-quality data for secondary uses such a population health management happens and good-quality data for innovation and research also happens.

Since the COVID-19 pandemic outbreak, regional, national, and international population health management needs have become accelerated. We have all been working hard to address this urgency identifying new data requirements across all aspects of healthcare. Using real-time, real-world evidence, we have a better chance of tackling the enormous challenges that we all face. It is important to join up the clinical systems across the country to make the patient 'journey' better. The global pandemic made us see the absolute need to link a person's class I, case management, and laboratory data for care. The use of unique identifiers/registries as a main service within the platform is paramount. Information systems where everyone involved in a patient's care can see the patient record to avoid duplicate tests so they can get diagnosis and treatment quicker are required. There are also opportunities to have de-identified records for analysis to help understand wider population needs, both now and in the future.

What are the advantages of taking an open standards approach vs. a traditional vendor lead approach? It's clear that we are reaching a pivotal point, where health and social data needs to be better connected, combined, and shared. As the focus of health systems moves towards supporting wellness—anytime, anywhere—an open platform environment is required to connect and share data, at scale, within and among enterprises and systems. The optimal platform will separate content and technology and will be vendor neutral, distributed, and modular incorporating third party as well as legacy systems.

This demands a new way of thinking about data. Rather than keeping it locked away in siloed systems, a decentralised and networked infrastructure could unify disparate information from multiple sources and make sense of it. This means capturing and linking all relevant data regardless of where it is stored. OpenEHR is a cost-effective and innovative way of achieving this and avoiding traditional 'vendor lock-in'. This health and care information platform described separates the architecture into different layers that organise transactions and interactions: the data layer, the application layer, and the logic layer (Fig. 1).

What are the benefits of using OpenEHR?

- ✓ An OpenEHR platform solution may be put in a single hospital as any traditional EHR is, but can also be across a city, region, or whole country such as Jamaica … providing continuity of care. *Its joined up*
- ✓ It puts the patient at the centre of everything. *It's safe*
- ✓ A big problem with EHRs is 'vendor lock-in'. It becomes too expensive to change even though the system is not up to the job. OpenEHR avoids this. *It gives flexibility*
- ✓ The separation of data models from the technical layers allows the platform to be built and deployed independently, removing one of the major sources of cost at a stroke. *It's cheaper*

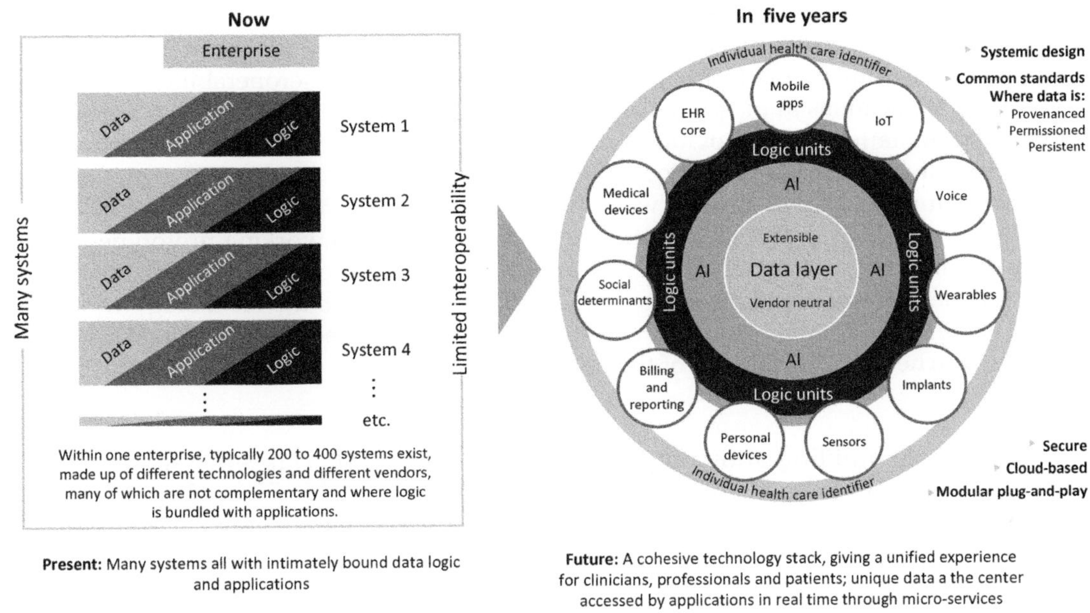

FIG. 1 The future architectural layers [1].

✓ It also allows the people who know their own data and workflows, to be in the driving seat when specifying Hospital Information. *It's built by clinicians for clinicians*
✓ Technical advances of OpenEHR lead naturally to a plug-and-play platform economy, in which any developer can produce a component, as long as it conforms to open standards. *It's easy to use*
✓ This puts the customer back in charge of their own system environment, while enabling incremental procurement of new components. *It gives you back control*

Internationally, there are numerous examples of steps being taken towards more open and interoperable systems. Estonia, for example, has almost fully digitalised its health system and is currently integrating health and social data for clinical practice support and research. Several of the Nordic countries, including Norway and Sweden, are establishing permission-driven consumer and clinician accessible health records that follow the patient, irrespective of care location. In the UK, the Salford Royal NHS Foundation Trust has deployed an open EHR clinical data repository alongside a core patient record system to meet changing clinical requirements and to incorporate patient-recorded outcome measures. In Germany, the AOK, a major health insurance provider, is developing a digital health network built around data interoperability to make health information available nationwide for its 25 million members. What these examples have in common is a steady progress towards system change built around meeting end-user needs, vendor neutrality, data persistence, and demand for flexibility and data fluidity. None of these examples are big-bang changes; rather, they're staged building blocks for the future.

Jamaica have taken a 'stepwise' plan of action to get to the future. The OpenEHR project aimed to plan, design, build capacity, and implement the required components of the information architecture of the national digital eHealth Architecture for the health sector, with emphasis on data design and specifications using OpenEHR clinical modelling, related resources, and tools with an open platform approach. OpenEHR and its resources will enable the vision of the Ministry of Health and Wellness for (semantic) interoperability across various eHealth applications in the public and ultimately the private health sector. It enables the use of an open platform and provides a large repository of predefined data elements ready for use including for queries, reporting, analytics, and research.

Over the duration of this project, we demonstrated the advantages for the MOHW in a stepwise, affordable way. The specification and procurement of a working prototype are underway. Having done a situational analysis, we defined priorities for applications, highlighting critical elements of the platform to be put in place and prove the technology works for Jamaica putting the country ahead of the rest of the Caribbean, leading by example as well as improving the health and wellness of the nation.

Capacity building and competence in the openEHR reference model

Initially, an educational framework and associated resources for use by the Jamaican Ministry of Health and Welfare for the continuation of workforce capacity building was developed. Capacity-building strategies focus on the health workforce who occupy the roles given in Table 1.

The openEHR standard defines both a methodology for defining and curating clinical content; and the specification of an open API and query language that allows openEHR content to be stored and queried in a compliant Clinical Data Repository (CDR). OpenEHR also provides support for terminologies allowing the inclusion of any appropriate terminology, e.g. SNOMED CT, LOINC, HL7 FHIR. These can be integrated with openEHR models via

TABLE 1 Educational framework and associated resources.

Workforce category	Examples and roles
Policy makers	Those making decisions about the implementation, purchase, or design of electronic health record systems or the use of terminologies and information models in healthcare systems. This includes managers, CIOs, and CXIOs.
Technical implementers	These are individuals selling, designing, building, or maintaining electronic health record systems or implementing terminologies within those systems. This also includes those who are designing or implementing information exchange processes or standards. This includes IT vendors, software developers, implementers, and maintainers of systems
Clinicians	Those involved in direct patient care such as doctors, nurses, pharmacists, and allied health professionals who are experts in the meaning to be represented in healthcare systems by information models and terminologies.
Health informaticians	Those involved with the development of clinical content from a health record or clinical perspective (i.e. not the IT technical perspective). These individuals are responsible for designing the health data, including defining data fields to be used in records and terminology value sets such as subsets of SNOMED CT. This includes health information managers as well as clinical informaticians.

terminology bindings. Terminologies can be bound to both openEHR archetypes and templates through Clinical Decision Support through its Guideline Definition Language (GDL) [2]; and Workflow and Task Planning through its Task Planning Visual Modelling Language (TP-VML) [3]. openEHR workforce capacity building and its adoption requires continuing access to a Clinical Knowledge Manager (CKM), tooling to develop archetypes and templates for an open platform.

Tooling for openEHR adoption

A Clinical Knowledge Management (CKM) system provides a repository and tools that help develop the knowledge domain and further openEHR community's collaborative efforts. Beyond being a repository, a CKM combines features, attributes, and extensive experience in clinician engagement, online asynchronous collaboration, crowd sourcing, quality control, and the very complex implementation environment, which requires strict artefact governance, ongoing maintenance, and distribution. The International CKM is freely available but its IP is privately owned by Ocean Health Systems [4]. There are other examples of CKMs, which are serving the needs of particular geographies. These include the Apperta CKM [5] serving the UK and Ireland, the Norwegian Nasjonal IKT CKM [6] serving Norway, and the Slovenian Ministry of Health CKM [7] serving Slovenia.

An open platform architecture is where the applications are separated from the data upon which they operate. The data is stored in a shared repository, known as a Clinical Data Repository (CDR), on the platform, which also provides shared services useful to many applications. Applications communicate with the platform via an Application Program Interface (API), simplifying the task of building applications. The best example of an open platform is the Internet itself. A proprietary platform that includes a limited number of open APIs is not an open platform as it does not satisfy the definition.

There was a need for Jamaica to organise a technical capability with openEHR tools, including GitHub's EHRBase, oriented to application development over the openEHR platform APIs, and as a parallel activity to doing things with the active GnuHealth environment. The adoption of an open platform facilitates ready data retrieval using the Archetype Query Language (AQL) [8] for use in any application. AQL is a semantic query language unlike query languages, such as SQL or XQuery. This means that AQL queries can be run over multiple openEHR Clinical Data Repositories irrespective of their technology and internal physical data structures. Therefore to understand how to query an openEHR system via AQL, you only need to know what archetypes and templates have been registered with that system. Tooling is available to simplify the creation and visualisation of AQL queries.

Skill development process, use of software, and openEHR tools

Establishing requirements is a first step for any system development. Traditionally, this task was undertaken to suit a specific application. This practice has resulted in many duplicated efforts, numerous data silos, and where connectivity between systems was a requirement, this generated additional development costs associated with the need to map data. The adoption of the openEHR approach reduces these costs significantly whilst enabling the development of requirements to suit a well-connected national digital ecosystem. This approach provides a consistent infrastructure that supports national/regional data collections as well as software vendors to develop their products at a greatly reduced expense. The artefacts developed have multiple uses to suit many use cases, thus reducing any need to duplicate efforts. Each artefact generates patterns which can also be made use of for the generation of new artefacts. These features ensure that the establishment of data collection and data use system requirements is achieved faster at a greatly reduced cost.

Firstly, there is a need to provide a big picture overview of the openEHR adoption roadmap and its specific features covering topics such as the relationship to the national vision of a well-connected sustainable health system able to optimise population health, followed by an overview of where openEHR fits within the national roadmap for digital transformation activities. How openEHR differs from other traditional information systems was explained and an introduction to the CKM was provided. The use of openEHR and the attainment of potential benefits is subject to an appropriate use and governance of health data, data and metadata standards, patient demographics, and unique identifiers. This requires the establishment of a suitable national support infrastructure, including hands-on activities.

Design and develop technical documentation for core archetypes

Healthcare is an industry which is information dependent, and to use that information, it is essential that all data captured and shared be able to be attributed to a specific source (an individual healthcare provider or a healthcare organisation) to provide authentication and be associated directly to an identifiable individual. For this reason, identity management is a key piece of underpinning infrastructure for any electronic health record system. As part of data about people and organisations, there is also a need to consistently capture and represent demographic details which do not relate to an individual episode of care but are relevant to each episode of care. These details do not change frequently (usually) and are relevant across all healthcare domains. These demographics include data such as addresses, birth and death details, and occupations.

For healthcare providers, information about qualifications and professional registration is essential, and for organisations, the location, specialties offered, and department contacts are core data. This part of the national programme aims to establish a consistent data for the identification and demographic details of individuals, healthcare provider individuals, and healthcare provider organisations. These data underpin national/internationally consistent identify management requirements to support the Caribbean. Identity data elements were reviewed from existing forms used in the healthcare environment in Jamaica, and initial mind maps were created to define the scope of data required explaining the rationale for each data element. Consideration was given to the code system(s) which are currently used or could potentially be used. The data included the following data for all entities—individuals, provider individuals, and provider organisations include an Identifier, eCommunication, and an address. Data for all individuals (including provider individuals) include the Name (used by individuals of all types), the Person (used by individuals of all types), Relationship, and Demographics. Data for individual providers include Registration and Qualifications.

Other archetype development activities

The focus adopted to guide these activities was the information continuum associated with individual patient journeys through the health service delivery system. This requires the identification of data collection needs at every point of care by examining current practice and evaluating the data use life cycle using a step-by-step approach as listed and explained below. Existing data collection practices were identified from forms now in use; these provided a useful foundation for the following capacity-building activities. Data to be collected or made available for decision making at every point of care at various levels of detail need to be considered for each patient journey. Every process, operation, activity, or procedure at any point of care consists of any number of operations or steps linked to data collection or data use. All transports (any movement by a person, document, equipment, sample, information transfer) along any type of patient journey, needs to be considered to identify connectivity requirements, decision-making points, and endpoints for data use such as storage. There are a series of questions to ask about each data collection/use activity in terms of purpose, place, sequence, person, and means, to identify data collection and data use needs. The information gathered can then be used to eliminate, combine, re-arrange, or simplify an activity.

These questions are:

- ➢ What is the purpose and what is achieved? (value proposition)
- ➢ Why is the activity necessary, what else could or should be done?
- ➢ Where is it being done, why in this location, where else might it be done, where should it be done?
- ➢ When is it done, why at that specific time, when could or should it be done?
- ➢ Who is doing it, why that specific person, who else could or should do it?
- ➢ How is it being done? Why is it being done in that specific way, how else could or should it be done?

Systematically posing these questions about every data collection process and data use is essentially the first step towards problem-solving and identifying requirements. These questions need to be posed by anyone developing a template for a specific use case. It's common for people to say, 'we've always done it this way'. Using this recommended approach enables one to question, be innovative and make changes to current practices to improve efficiency and effectiveness of new systems, thus reducing costs and achieving improved outcomes. Once a template has been developed, all missing archetypes or desirable archetype extensions or modification should be noted in the accompanying mind map for follow-up. This process delays the use of the template within an application and needs to be completed prior to its use by the software developer. Such potential delays need to be considered a part of the associated application development process. The openEHR specifications state that:

> ***Each archetype*** *is a set of constraints on the reference model, defining a subset of instances that are considered to conform to the subject of the archetype, e.g. 'laboratory result'. An archetype can thus be thought of as being similar to a LEGO® instruction sheet (e.g. for a tractor) that defines the configuration of LEGO® bricks making up a tractor. Archetypes are flexible; one archetype includes many variations, in the same way that a LEGO® instruction might include a number of options for the same basic object. In terms of scope, archetypes are general-purpose, re-usable, and*

composable. For data capture and validation purposes, they are usually used at runtime by templates. Archetypes also form the basis of semantic querying. Queries are expressed in a language, which is a synthesis of SQL (SELECT/FROM/WHERE) and W3C XPaths, extracted from the archetypes.

***An openEHR Template** is a specification that defines a tree of one or more archetypes, each constraining instances of various reference model types, such as Composition, Section, Entry subtypes, and so on. Thus, while there are likely to be archetypes for such things as 'biochemistry results' (an Observation archetype) and 'SOAP headings' (a Section archetype), templates are used to put archetypes together to form whole Compositions in the EHR, e.g. for 'discharge summary', 'antenatal exam', and so on. Templates usually correspond closely to screen forms, printed reports, and in general, complete application-level lumps of information to be captured or sent; they may therefore be used to define message content. They are generally developed and used locally, while archetypes are usually widely used. A template is used at runtime to create default data structures and to validate data input, ensuring that all data in the EHR conform to the constraints defined in the archetypes referenced by the template.*

Archetypes and templates stored in the CKM are the result of developments to suit specific research projects and applications. Once accepted by a custodian organisation such as the openEHR foundation, then rules of governance and quality assurance for all artefacts included in the international CKM are managed via a version control mechanism. New archetypes need to conform to the style guide and be reviewed by an archetype editor before it can begin its first international review round.

The development of templates is about emphasising and reducing variations via pattern recognition for re-use between different use cases. This enables the adoption of standard procedures using tools based on extensively researched and fundamental theoretical underpinnings that remain constant whilst maximising available options for data use and re-use. This is a key feature of the value of adopting the openEHR approach as it results in significant time reductions for the identification of user requirements, data sets, application design, and development time.

Further development

The Ministry of Health and Wellness, Government of Jamaica (MOHW/GOJ), planned and executed an investment loan from Inter-American Development Bank (IDB; JA-L1049). The programme objective was to improve the health of Jamaica's population by strengthening comprehensive policies for the prevention of noncommunicable disease (NCD) risk factors and for the implementation of a chronic care model with an improved access to strengthened and integrated primary and hospital services networks, that provide more efficient and higher-quality care. The policy-based operation of this Hybrid Programme is the first operation of a series which to be made up of two contractually independent and technically linked loans. The policies in the Programmatic Series will consolidate regulatory measures to address the preventable causes of NCDs and to reorient health systems to address prevention and control of NCDs through a people-centred primary health chronic care model. The investment component, in turn, will finance activities to consolidate integrated health networks and improve the management, quality and efficiency of health services. While the PBP will benefit the Jamaican population at-large, the Investment Loan will have approximately 800,000 potential direct beneficiaries who reside in the catchment areas of the health services networks that will receive investments.

Investment programme components

The Investment Programme implemented by the Ministry of Health and Wellness (MOHW) had two major components and an allocation to support programme administration and evaluation, organisation, and consolidation of integrated health services networks. This component financed the purchase of medical equipment and the improvement of infrastructure for primary healthcare services in the catchment areas of three priority hospitals to increase their capacity in health promotion and disease prevention, especially regarding chronic, noncommunicable diseases. The investments focussed on strengthening the diagnostic and screening capability as well as the clinical and resolutive capacity of health clinics. This component further financed the upgrading and/or expansion of three hospitals selected on criteria relating to strategic role in the national hospital network, supply-demand gap analyses, and physical needs assessment. The hospitals will benefit from infrastructure upgrading and/or expansion as well as modernisation.

Strengthening Primary Care was also a key subcomponent is to increase the physical capacity for service provision at the primary care level in three priority geographical areas. Approximately ten health centres have been identified to receive investments in medical equipment and infrastructure refurbishment and expansion. In addition, there was provision to increase the Capacity and Efficiency of Hospital Services. This subcomponent addressed urgent needs to enhance patient safety and services in three hospitals whose catchment areas contain the health centres identified.

Finally, Improvement of Management, Quality and Efficiency of Health Services was to provide technical assistance to design and implement the Chronic Care Model (CCM) in the participating health services networks; to review and develop care pathways and protocols; and to prepare change management, continuous quality improvement, and social media marketing for behaviour change strategies. It will also finance the implementation of the fourth Jamaica Health and Lifestyle Survey. This component will further support:

- **(i)** the creation of a strong foundation for a digital health ecosystem, including the adoption of standards for interoperability, system architecture, updated governance structure, and other key elements;
- **(ii)** the design and implementation of a sustainable Electronic Health Record (EHR) platform focusing on digitalisation of key processes within the improved CCM;
- **(iii)** the strengthening of telehealth/telemedicine/telementoring capacity to include chronic care management, and the establishment of norms and processes for its institutionalisation.
- **(iv)** the strengthening of telehealth/telemedicine capacity through the expansion of the ECHO model, the inclusion of chronic care in the platform, and the establishment of norms and processes for its institutionalisation.

Capacity building

Information systems for health are critical to primary health care and the CCM, as well as complex care. The World Health Organisation (WHO) recognises the need for information systems for health as part of the Package of Essential Noncommunicable (PEN) Disease Interventions for Primary Health Care in low-resource settings. Additionally, research shows that digital tools can make important contributions to the provision of chronic care when patients and providers are connected to share information, compare this information to knowledge found in evidence-based standards, and monitor results through regular feedback and interaction. Adoption of health information technology has produced mortality rate reductions for complex patients whose diagnoses require cross-specialty care coordination and extensive clinical information management in hospital settings and improvements in resource allocation efficiency. Health information systems, along with information sharing, have the potential to improve clinical practice by reducing staff errors, improving automated harm detection, monitoring infections more effectively, and enhancing the continuity of care during physician handoffs. A significant number of opportunities exist for the Hospital Information Management System (HIMS) to contribute to savings for both the health system and the Jamaican economy. Conservative estimates suggest if the United States healthcare system were interoperable, at least US$77billion would be saved annually [9].

Information architecture

The primary function of this role was to plan, design, build capacity and implement the required components of the information architecture of the national eHealth (Digital Health) Architecture for the health sector, with emphasis on data design and specifications using OpenEHR clinical modelling, related resources, and tools with an open platform approach. OpenEHR and its resources will enable the vision of the Ministry of Health and Wellness for (semantic) interoperability across various eHealth applications in the public and ultimately the private health sector [10]. OpenEHR enables the use of an open platform and provides a large repository of predefined data elements ready for use, including for queries, reporting, and analytics.

Scope

This case study describes the work and the key responsibilities which included:

- Providing OpenEHR capacity building for data specifications, clinical modelling through archetype selection and development, template design, and establishing associated value sets to the point of readiness for the design and implementation of an electronic health records platform. This included demographics and the journey of the NCDs patient across primary and secondary care services.
- Establishing access to the tooling and resources required for the development of archetypes and templates and a Clinical Knowledge Management (CKM) system.
- Technical contribution to the national architectural design and workforce capacity-building support for software developers, enabling them to make effective use of modelling outcomes and facilitate early integration of EHR with Laboratory and Pharmacy.

- Support the MOHW to determine what data needs to be pulled from existing systems and HL7 message standard version capability as well as HL7 FHIR standard. This required capacity building regarding mapping of data to a shared concept representation.
- Providing advice and support on governance processes, including Data/Terminology, Data dictionary and data standards for medical, laboratory, and medication terminology management in a manner that facilitates early integration of EHR with Laboratory and Pharmacy systems.
- Advice on the adoption of an Open EHR Clinical Data Repository (CDR) or similar technology and support the development of specifications and procurement of hardware, software, and services for the technology neutral database design to be adopted nationally to optimise the support for data sharing and data analytics.
- Establishing identity management protocols and the establishment of patient, provider, and health institution unique identifiers registries.
- Establishing a demo environment of an Open EHR and demonstrate the core and extended functionalities.

New roles were created. The Digital Health Standards Programme Coordinator organised stakeholders within and beyond the MOHW to adopt and build capacity on Digital Health Standards (including openEHR specifications, HL7 FHIR, and IHE Profiles) and collaborate to establish priority clinical service areas for application of the standards. Organise all resources and tools required for template building (for example archetypes and term sets), and transform existing data forms to build templates which are stored in the appropriate format and ready for consumption by an application. Manage the processes for template building and support capacity building and the requisite review processes. The Digital Health Standards Officer will provide training and capacity building in Digital Health Standards, the use of related resources and tools, terminologies and data mapping; support the work of the Digital Health Standards Programme Coordinator to build capacity on Digital Health Standards (including openEHR specifications, HL7 FHIR and IHE Profiles) according to priority clinical and public health areas identified for application of the standards; support the Coordinator in organising and managing all resources and tools required for template building (for example archetypes and term sets) and participate in transforming existing data forms to build templates which are stored in the appropriate format and ready for consumption by an application.

The Coordinator and Officer received technical guidance primarily from the Senior Supplier and also liaised with the Digital Health Engineer and Full Stack Software Engineers contracted via a separate but related initiative.

Discussion

The primary function of this role was to plan, design, build capacity and implement the required components of the information architecture of the national eHealth (Digital Health) Architecture for the health sector, with an emphasis on data design and specifications using OpenEHR clinical modelling, related resources, and tools with an open platform approach. OpenEHR and its resources will enable the vision of the MOHW for [semantic] interoperability across various eHealth applications in the public and ultimately the private health sector. OpenEHR enables the use of an open platform and provides a large repository of predefined data elements ready for use, including for queries, reporting, analytics, and research.

The workplan and joint road map were produced, and a Project Initiation Document (PID) was approved by a Project Board, which involved the extended team and other relevant consultants and stakeholders where highlighted reports were produced. A risk and issue log was maintained throughout. Education and skills building was a major part of this project activity covered identity management, data specification (using identity management examples), Provider and Organisation identity management, Provider and Organisation—Functionality and Migration, Identity management—functionality and migration, Flavours of Null, Choosing a Code system, Data Mapping, Data Quality, Jobs and Skills for a national Digital Health Program, Tasks for identity management—individuals and Tasks for identity management—providers and organisations. Draft design of technical infrastructure requirements had close links to the architecture project in so close liaison with development of the business and information architecture are required. Three roles were identified to support, namely a health informatics/OpenEHR specialist consultant; a digital health integration engineer and finally an identity management engineer consultant.

A Open EHR Clinical Data Repository demo environment was established with a local CKM and CDR for archetype/template development and use in Open EHR specification compliant applications. A 'sandpit' environment was set up with pathways for NCDs being use cases, building on exemplars used in the UK as well as Jamaican pathways for PAHO Long-Term Condition Management 'Passport' and archetypes. There was also support to develop outline architecture & project plan for OpenEHR CDR / Open Platform design & implementation (Phase 1 and Phase 2 including budget definition 3–5 years). Functionalities were demonstrated with high-level documentation of the methodology. This will lead to a cohesive proposal for CKM for Jamaica and potentially for all the English-speaking Caribbean. Further work was also accelerated on establishing the workforce, i.e. editors, mappers, data spec and quality officers for ongoing provision. Job Descriptions were written for data governance organisation skills and responsibilities, including CKM Editor, Terminology and Mapping, Data Quality Officer, and Data Standards Officer roles.

The National Terminology Infrastructure activity focused on documentation on a national terminology infrastructure review, to update and manage consolidated terminology infrastructure in an acceptable format [11]. To help in choosing clinical coding system documentation for ICD-10 and ICD-11 was provided, including skill and infrastructure requirement analysis. Evidence to support this was required in a data extract form existing systems to analyse the feasibility of approach and resources required to address gaps in skills and infrastructure. Documentation of the development of governance principles included a process for the identification of required areas to be mandated and governed as support to the establishment of a national/regional governance organisation from inception to final approval.

Throughout the project, there was activity developing and managing own archetype/template development and their use in Open EHR specification compliant applications ongoing. A report on training programme with confirmation that selected local workforce demonstrates knowledge transfer and is able to manage its own archetype/template development, and their use in Open EHR specification compliant applications. This then led to alignment of training requirements to Phase 1—Proof of concept; shared OpenEHR system as an initial form of the intended national Electronic Master Patient Index (EMPI), shared health record and clinical data repository that is intended to demonstrate the development cost characteristics of using OpenEHR, clinical terminology, i.e. SNOMED CT, classifications, i.e. ICD-10, Interoperability standards, i.e. HL7 FHIR and related technologies, by building a representative set of integrations, applications, data sets, and reporting functionality sufficient to inform cost estimation for long-term development. This will expose the features and challenges of installing the platform within the Jamaican healthcare IT environment; Provide certain core functionality of strategic utility to MOHW, including Patient, Professional, and Facility registries, specifics of the Chronic Care Model for NonCommunicable Diseases management, basic COVID-19 reporting and analytics functionality. The Phase 1 system will not attempt to satisfy the volume, scaling or performance requirements of the ultimate operating infrastructure, but will exhibit acceptable performance consistent with the limited user and transaction loads intended in phase 1. There was a need to develop a seamless transition to incremental implementation of OpenEHR and open standards approach. Knowledge transfer is essential to ensure the management of its own archetype/template development, and their use in Open EHR specification compliant applications.

Conclusion

This exploration and overview of implementation of OpenEHR in the Caribbean has highlighted the many benefits of adopting open standards as a national information standard. The experiences described for Jamaica as a country provide real examples of good practice and pragmatic implementation solutions that will allow others to have the confident and competence to tackle this important step as part of the roadmap to successful health ecosystems.

References

[1] E&Y Health Sciences and Wellness 2019. How will you design information architecture to unlock the power of data?, 2019.

[2] openEHR provides the Guidelines Definition Language (GDL) as an extension to the openEHR standard. GDL allows the representation of clinical guidelines in an open and computable format, 2020.

[3] openEHR provides the openEHR Task Planning Visual Modelling Language (TP-VML) to enable workflow and task planning to be managed. This is the most recent extension to the openEHR specification, 2021.

[4] The International CKM available at is freely available but its IP is privately owned by Ocean Health Systems, 2021.

[5] These include the Apperta CKM serving the UK and Ireland, 2021.

[6] The Norwegian Nasjonal IKT CKM serving Norway, 2021.

[7] Slovenian Ministry of Health CKM serving Slovenia, 2021.

[8] Archetype Query Language (AQL) is a declarative query language developed specifically for expressing queries used for searching and retrieving the data found in archetype-based repositories, 2021.

[9] M.J.A. Salomi, P.B. Claro, Adopting healthcare information exchange among organizations, regions, and hospital systems toward quality, sustainability, and effectiveness, Technol. Invest. 11 (03) (2020) 58–97, https://doi.org/10.4236/ti.2020.113005.

[10] Information Systems For Health (IS4H) Information Systems For Health (IS4H)—PAHO/WHO | Pan American Health Organization, 2016.

[11] ISO 13606-2:2019 Health informatics—Electronic health record communication—Part 2: Archetype interchange specification ISO—ISO 13606-2:2019—Health informatics—Electronic health record communication—Part 2: Archetype interchange specification, 2019.

CHAPTER

25

Addressing workforce knowledge gaps in digital health

Evelyn Hovenga[a,b,d] *and Helen Hovenga*[c]

[a]Digital Health, Australian Catholic University, Melbourne, VIC, Australia [b]eHealth Education Pty Ltd, Melbourne, VIC, Australia [c]Telix Pharmaceuticals Limited, Melbourne, VIC, Australia [d]Global eHealth Collaborative (GeHCo), Melbourne, VIC, Australia

A whole of ecosystem approach

'The future is already here—it's just unevenly distributed', William Gibson, a famous science-fiction writer [1], once said. Health data is created and documented every day, but the forms of recording, usage, and levels of understanding of its value vary greatly. The form in which data is recorded varies from memory to pen and paper to excel spreadsheets, and also to and between health systems. The degree to which the data can be used for decision making also varies.

What we do know is that those enlightened to the power of data and how this relates to the digital ecosystem can efficiently achieve the best health outcomes. Healthcare providers and health IT suppliers need their workforce enabled to make any digital health ecosystem successful. The biggest workforce knowledge gap we have identified is a disconnect between those familiar with the health data science, terminology, and knowledge domains and their relationship with digital information and communication processing technologies, their scientific principles, and methods.

Yet health data represent the primary asset that needs to be well managed. In this digital era, 'data' are now viewed as the new 'oil' in terms of importance [2]. It is essential that our future generation of health professionals are suitably prepared for working in a knowledge industry that is undergoing major disruption and transformations, due to various technological upheavals, similar to those experienced following the invention of, for example, steam engines, cars, trains, planes, the Internet, or the international adoption of containers for the management of world trade. The entire culture, strategy, healthcare delivery, management, funding, and educational initiatives to best suit this digital era need to include workforce capacity building, to ensure every health-related facility has staff with the necessary knowledge and skills to implement and achieve desired strategic outcomes.

Roadmap to Successful Digital Health Ecosystems
https://doi.org/10.1016/B978-0-12-823413-6.00007-0

Globally, we need governments, health data authorities, regulators, healthcare executives, leaders and businesses, health service providers, health professional educators, and individuals to embrace their digital health ecosystem. They need to be enlightened and engaged in their approach to benefit from the many new digital health advances. This requires a deep understanding of the way they view and interpret data that are provided, created, generated, digitally processed, used for multiple purposes, and reported. Such a realisation early will ensure that the current random creation of any digital health ecosystem, developed by those with an uneven distribution of awareness across each workforce, will instead be created with intention, purpose, alignment, standards, consistency, and clear governance of these valuable health data assets. This capability is needed within and across every entity, and will be reflected in their culture, their strategy, and their people.

Building the digital capability of entire organisations within any industry requires intention, focus, clarity of purpose, and a whole of system approach. `Capability' building refers to not only building capacity and curiosity in people (which is most critical), but essentially this needs to be embraced within a whole organisation or nation's DNA. The approach needs to ensure digital capability is reinforced in every structure, every role, every system, and process and in every experience.

It is only with this whole of system approach, as shown in Fig. 1, that it will become deeply embedded. The aim is for digital capability to be sustainable and constantly evolving. Such programmes within each entity are precursors to the achievement of an intentional digital health ecosystem, evenly distributing the future which is already here.

Fig. 1 outlines the need to define the strategic directions as the first step. It is then critical to define and understand the goals and measures to be used to achieve the output of performance (so you know when you get there). The inputs are an entire network of interdependent actions. This starts with clarity of the organisational structure and roles, identifying the types of people for these roles, which need to be decomposed to identify the skills and development activities to be put in place, to enable the delivery of the chosen strategy. The processes, tools, and systems then provide repeatable, aligned, and consistent methods that reinforce and uphold the capability across the organisation. Finally, the enablers reference the culture of everyday behaviours, focus, and actions that ensure that the desired capability is a habit that is embedded and sustainable across the organisation.

Previous chapters have highlighted many digital health-related issues and challenges in need of transformation. Numerous stakeholders within the healthcare ecosystem are in need of transforming their business operations. We have identified many products that are no longer fit for purpose. A successful transformation requires us to acquire many new approaches to knowledge

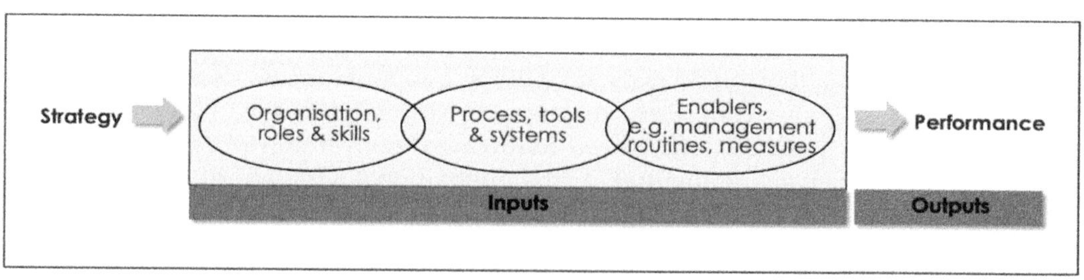

FIG. 1 Whole of system approach for delivering strategy through capability building.

and skill acquisition, especially in integrated areas of health data governance, management and use, technology design, procurement, health IT governance, application development and use.

Healthcare data needs to be viewed as a valuable asset as it:

- Has intrinsic value,
- Costs a significant amount of money to collect, store, and make available for use,
- Forms the basis for core clinical decisions and core day-to-day business operations,
- Drives health policy and funding,
- Informs research and public health initiatives.

Data needs to be actively managed at the executive level as a strategic resource of any healthcare organisation [3].

The workforce needs to be well prepared to successfully implement a digital transformation. This requires an agreed vision and strong leadership enabling innovation and autonomy at all levels. Health workforce capacity building needs to adopt multiple educational and training methods able to support practical and timely local transformations. This chapter begins with the importance of strategic drivers for any entity with any digital health ecosystem focusing on a human resource management (HR) perspective. This is followed by a discussion on how to decompose roles into knowledge and skill requirements, which form the foundation for educational and professional development programme design.

Defining the vision, then the strategy

What does a digital health workforce need to be able to achieve? Here is one example of one healthcare facility's strategic aspiration for Digital Health delivery:

Drive service improvement and delivery of outcomes through improved data governance, collection and reporting.

This performance outcome focuses on the rigorous collection, reporting, and analysis of patient-reported outcomes and experience measures gathered throughout the patient journey. This will be enabled through strategic organisation-wide data governance for efficient, high-quality business and clinical decisions, reporting and data analytics capability. Health data management and governance needs to be led by a Health Informatician with the right skill set, supported by data stewards to build capability and empower clinicians to drive data requirements and usage, and help the workforce understand the impacts and outcomes for correct data input, processing, and output. It is necessary to always seek to clarify what data and/or knowledge is needed by when, for who, and for what purpose to ensure accurate and complete information and curated knowledge are available for use by service providers when and where needed.

Strategically, this requires:

- An examination of the possibility to automate tasks that optimise the use of workforce skill for high-value activities that result in employee engagement, efficiency, and cost-saving.
- A digital transformation enabled through an engaged workforce, a positive and supportive culture, an effective organisational structure, and capability building.

- Continued support for building our workforce enabling them to adopt digital thinking, implement digital capability, and embrace technological transformation and innovation.
- A structure that allocates accountability for digital health within job roles and empowers employees to engage with and deliver the digital strategy.

Defining 'performance'

What outcomes do we need people in Health to deliver when it comes to Digital Health? Here are some examples of the types of measurable indicators of success:

- Improved accessibility and quality of care for (specific groups of or all) patients (measured by increased usage along with improved outcomes and patient feedback)
- Easy, convenient, and secure health information exchange internally (within one organisation) and externally with ecosystem partners (measured by analytics, automated reporting capability)
- Increased engagement and empowerment of patients and improved patient experience (measured by feedback)
- Improved workflows that drive operational efficiencies (measured by cost, time, resourcing)
- Sustained improvement in data quality for information management, clinical governance, knowledge enrichment, decision making, improved patient outcomes, and research use (measured by accuracy, rework rates, quality of data).
- Innovative ideas and technology that enables safer, faster, cheaper healthcare service delivery along with better outcomes overall (measured by cost, time, resourcing, take-up rates of new technology).

Organisation roles and skills

Once Vision, Strategy, and desired Performance of the workforce responsible for health service delivery are clear, there is a need to define the ideal technology infrastructure needed to support these activities in a functional, workflow-compatible, highly automated, and cost-effective manner. This requires opportunities for those responsible for data/information collection, and the many different data 'use cases' to collaborate with software engineers and application developers to first and foremost design the local ecosystem architecture and develop a digital health transformation programme based on current system imperatives and priorities. Organisational structures need to be enabled to deliver the Strategy as the next step in the process.

A well-written and well-thought through strategy should automatically point to the types of broad activity that will be needed, the potential interdependencies, and where the accountability should lie. Allocating roles starts at the most senior level of accountability, whose behaviour and leadership styles directly influence organisational culture. The level of seniority for any role is decided upon the level of impact expected from the activity, i.e. Will the activity impact the whole organisation or discipline, or one function or area only? If it impacts the whole organisation, an additional senior role may be required, or a change in current senior role accountabilities may suffice.

In the above strategy example, there are four references that can be used to define roles:

1. Accountability for the entire digital ecosystem functionality within the health service (organisation, network, or jurisdiction);
2. Accountability for data governance across the organisation's digital health ecosystem needs to be defined;
3. Data stewards need to be referred to;
4. Health and health-IT professional accountabilities.

Data stewards need to fully understand data science principles, data standards, and terminologies and be familiar with the relevant governance policies, procedures, and rules. Their focus needs to be on points of data collection and data exchange to ensure meaning (semantics) is retained at all times. They become champions of any entity's data governance programme and can provide an enterprise leadership role by mentoring others. Some positions may include multiple roles, depending on the size of the organisation. These are roles that will need to immediately be considered when designing the organisational structure.

The allocation of numbers (headcount) depends on the amount of work expected to be generated by the activity and the diversity of the required tasks. Some tasks may only apply for specific periods of time, while others remain constant based on, for example, the number of recipients estimated to be provided for by each type of service. It should also be noted that ideal organisational structures outline positions only and are defined based on activity, not on current people, personalities, existing roles, and/or existing skill levels. The new structure may also require the reallocation of resources and possibly additional expenditure. If an organisation is not willing or able to take this action and ensure a reasonable resource allocation, then the strategy and organisational priorities may need to be revisited.

Once key roles are defined, then the implications for all other roles across the organisation will need to be considered such as the heightened level of responsibility for any employee who gathers data, asks questions of patients, uses systems, undertakes diagnostic, therapeutic, or caring activities, documents patient health status assessment observations, writes reports, undertakes research, maintains systems, develops applications, provides a support or administrative service, or partners with other health services. Everyone working in the health industry or accessing any service provided is likely to be impacted by any digital health transformation.

One of today's benefits and challenges is the passion, interest, and intuitive approach to technology and systems of the upcoming workforce who are digital natives or immigrants (born in this digital era or who have adopted the new technologies). The challenge is with those who are impatient with technology that is not intuitive, human-centred, or logical; yet this defines many of today's systems. Such an expectation may lead them to drive change and improvements to systems in the future, which must fit with agreed strategic directions. Others may have a greater understanding of the foundations within the systems we use as they know what data was collected using pen and paper and how useful the information was or wasn't depending on the person collecting it, the assumptions made, the shorthand used, or the training they had.

Effective use of data relies on knowing the need for standards, the impact of quality data, what the data is made up of, what assumptions are included, where the data came from as well as how these data relate to the systems we use. As William Gibson wrote 'When you want to know how things really work, study them when they're coming apart' [1].

An additional challenge is the varied understanding and exposure to the foundations within the systems in use. Just as we have seen the smart phone democratise the use of computers, health records also need to be democratised and patient-centred, to enable them to view and maintain their own health information and manage their own health. Digital savviness should not be interpreted as being equal to the possession of the necessary digital health knowledge and skills required for the role. Workforce capacity must relate to the desired level of performance.

Identifying new knowledge and skill requirements

The digital health knowledge domain has great depth and breadth. It is the result of numerous new technological advances and their adoption, as well as the integration of numerous well-established knowledge domains of relevance to the health industry. The digital health or health informatics knowledge domain is not well defined or structured. There is no agreed framework or metadata structure or reference information model for this domain, which, due to ongoing medical and other technological advances, has experienced numerous changes in boundaries of practice over time.

Many attempts have been made to date, to define new workforce roles and their associated knowledge and competency requirements [4–15]. Hovenga and Grain have contributed to and published many of these activities [16–25] since the mid-1980s. The International Medical Informatics Association (IMIA) established a working group on education when it began during the early 1970s, it has been reviewed and updated, the latest version was published in 2010. These guidelines have continued to make important contributions, resulting in numerous publications since that time [26], they again being reviewed [24,27].

Universities, vocational and health professional organisations, and other education providers have developed various specific health informatics qualifications and short courses based on desired graduate outcomes and/or specified competency statements considered to best suit their target market [8,28–30]. Others have taken action to integrate digital health into various curricula, although this group continues to represent only a small percentage of educators.

It is our experience that most educators responsible for the development of new graduate health professionals or speciality fellowships and for IT graduates, are themselves unfamiliar with many of the foundational digital health specific, knowledge, and skill requirements. As a consequence, we continue to produce new health professionals educated in accordance with mostly disciplinary traditions. Most IT graduates need to learn about health industry-specific requirements 'on the job', through mentoring and trial and error. Another contentious issue is how to tell who is an expert, due to the absence of an agreed standard foundational knowledge and skill framework against which to evaluate the specific expertise of a 'Health Informatician' or the many potential specialisations with either a clinical or technical focus pertaining to this domain.

The Australian Digital Health Agency (ADHA) recognised this challenge and concluded that the education and training provided to the current and future health workforce must be re-shaped in order to meet the existing and emerging digital requirements [31]. It is of concern that its definition of the health workforce only includes those working at any point of care. It fails to recognise the need for digital transformation amongst the many higher-level authorities, including the Australian Institute for Health, Government funding authorities, policy developers, health service executives, insurers, regulators, clinical registry workers, clinical trial workers, vendors, and more, all of whom represent stakeholders within any nation's health ecosystem. Foundational jurisdictional infrastructures are in the best position to provide the necessary high-level strategic leadership. Maintaining the current status quo limits success.

The digital health knowledge domain

The IMIA body of knowledge [32], referred to in a previous Chapter, covers the competencies identified as referenced previously. The body of knowledge is used to select learning topics to be included in any proposed educational programme. Each of those topics can then be further decomposed based on knowledge depth or breadth and skills required along with levels of autonomy, authority, or responsibility relevant to specific roles. Each of those composite parameters then need to be identified in terms of cognitive knowledge level as described by, for example, Bloom's taxonomy [33].

This is a hierarchical order, ranging from the ability to remember, understand, apply, analyse, evaluate, and finally create. Its use enables educators to describe potential graduate roles, complete with desired learning outcomes, and relative performance assessment strategies. Foundational knowledge and skills at the remember and understand Bloom levels as a minimum, need to be integrated in every health professional and health IT undergraduate programme, to build on during higher-level digital health studies. In addition, an undergraduate programme dedicated to digital health based on these foundational knowledge and skills, with optional specialisations, can serve as a useful career starting point.

Professional development and micro-credentialling programmes need to have a stronger focus on specific roles and functions as these relate to individual clinical or health IT disciplines or to types of health services, supportive functions, or an entity's organisational structure and primary purpose. These types of programmes are well suited to 'just-in-time' learning strategies for the existing workforce. This is particularly useful during the transition phase as many of the staff do not have the time nor the interest to undertake additional formal university-based studies. These types of programmes need to address how best to implement practical solutions, building on current digital environments, but forming part of a bigger transformation agenda.

A digital health knowledge and skill framework

We first need to focus on foundational knowledge and skill requirements associated with using and managing data and digital technologies. This requires prerequisite core health, health industry, and computer science knowledge. Most health professionals do not have the skills to identify their own digital health-related knowledge and skill gaps or to identify specific talent requirements; they are simply not aware of such knowledge gaps. This influences demand for digital health educational programmes and is best addressed by in-house professional development strategies that enable its workforce to discover that the more they know, the more they realise what they don't know.

Adopting an educational approach to identifying knowledge and skill requirements enables the use of commonly occurring roles. Every organisation and every health discipline have their own set of knowledge and skill requirements based on the roles that need to be filled to meet performance requirements, as determined by the ideal organisational structure designed to deliver its strategic directions. Once these have been developed generically, they can be used as a common standard by organisations, as a useful starting point. Every role needs to be de-composed, in order to identify very specific competency or educational outcome requirements. Once known, then a local workforce capacity-building strategy can be developed based on available options.

From a big picture health industry perspective, we need to compartmentalise this by de-composing the digital health knowledge domain into various levels of disciplinary knowledge domains, pertaining to health, the health industry, computing and information science, and associated degrees of authority and responsibility. Ideally, this is represented as an ontologically structured digital health knowledge domain. Such a global framework, or metadata structure, will be most beneficial for those tasked with curricula design. An ontologically designed framework provides the necessary flexibility to define any specific role, identify level of responsibility, as well as the capacity to change and accommodate new knowledge advances over time.

A similar framework has been developed for the ICT industry. The Skills Framework for the Information Age (SFIA), now in its 8th edition, is internationally recognised, translated into multiple languages, and widely used by Human Resources Departments and educational facilities. Many of its generic descriptions for its categories apply to digital health, but these do not adequately cover the entire digital health knowledge domain [20], as described in the chapter on knowledge management. Nevertheless, this SFIA framework serves as a valuable example and proof of concept. SFIA links with a number of professionally defined bodies of knowledge and was developed to suit any type of industry, including health.

We need to be able to develop the workforce capacity to enable everyone to collectively address the many challenges identified in this text and solve numerous problems when any digital health transformation is being undertaken. An understanding of cause and effect of current issues provides the foundation for identifying skill and knowledge deficits that need to be addressed via suitably designed educational programmes. Such deficits need to be related to roles. Many current health industry roles will remain, but the knowledge and skill requirements associated with these roles will change, based on the type and extent of digital health integration required to achieve the desired role performance outcomes within a digital health environment. This is expected to result in various changes to boundaries of practice, impacting regulatory authorities. Failure to upskill the current workforce and/or to suitably prepare new graduates will continue to have major adverse cost and loss of benefit implications.

Building digital health workforce capacity

One major gap is the nonformal recognition of any digital health-related occupational categories. This absence limits not only the ability to identify potential globally recognised career structures but also limits our ability to obtain useful workforce statistics to support workforce planning methodologies. In addition, most health and health IT professionals, now undertaking any digital health-related roles, on a part- or full-time basis, continue to refer to their prior substantive disciplinary-focused positions. As a consequence, it is extremely difficult to assess the size of the digital health workforce or to estimate future educational demand.

The health workforce consists of multiple disciplines associated with unique knowledge domains. There are numerous clinical specialties and subspecialties as well as many health IT specialties, each with their own scope of practice. In addition, the members of any current health workforce represent many and varied work-related experiences. From an adult learning perspective, it is important to design educational activities in a manner that enables these adults to build on their past learning accomplishments and their own life experiences. That is, they make use of this frame of reference to gain new understandings. This means educators need to make use of the real-world examples these adults can relate to. The additional challenge is a common need to de-bunk outdated, incorrect, biased, or incomplete prior knowledge. We have shared many of our experiences and observations in this text, which may serve as useful examples to serve this purpose.

These variations in prerequisite knowledge and experiences on entry into any educational or professional development programme can be challenging for educators. One useful strategy is to engage participants by asking them to share and discuss their views and observations. Adult learners tend to be practical and focused on meeting their own learning objectives. To be fully engaged and successful, adult learners will choose their own preferred learning strategy, need to be self-motivated, and recognise relevance to their career prospects. Self-directed and life-long learning is to be encouraged.

One important knowledge area is the ability to manage privacy in accordance with legislative requirements and how these apply to digital systems within an ecosystem. In response, ISO's 2017 technical report [34] covers the privacy concept, the challenges associated with the protection of information, as components for content inclusion into a privacy education programme. Other similar developments in progress are knowledge and skill educational needs associated with the maintenance of security, and the use and governance of terminologies. These activities were undertaken in response to ISO's initial recognition of a need for healthcare organisations to adopt education programmes to address these skill gaps in healthcare organisations.

Process, tools, and systems—A human resource (HR) perspective

Once the ideal organisation structure, including roles and skills required, has been defined the most vital ingredient of all, the secret of successful implementation, sustainable performance and quality data, must be in place. What is this critical element? People.

People are fundamental to the health system, and their employment also represents the greatest cost. Having the right variety of people, with the right mindset, personal attributes, and purpose, as well as the right technical expertise is critical. The latter is the easiest part to develop. So what does it take?

At every stage in an employee's life cycle, from recruitment to retirement, there is a chance to build reinforcing systems that drive high quality, safe, and sustainable performance in people. These people systems will ensure you have the right people, with the right skills, the right purpose, and the right attitudes, in the right place at the right time. With the digital health strategy and performance measures in mind, review every step in the employee experience.

Identifying the role—Position descriptions and success profiles

With structures and roles identified, defining the role is vital. The best job descriptions include the role purpose, responsibilities, deliverables, measures of success, decision-making input, and rights. Key interfaces and key stakeholders are also outlined. It is ideal to utilise educational frameworks to write job descriptions and identify essential and desirable knowledge and skills. It is even more important that this is a live document that is discussed and mutually agreed annually, and that it is updated as the role evolves.

Beyond creating a job description, start with creating a success profile for the role. A success profile identifies not only the skills, experience, and knowledge needed for the role (what the role occupant does), but also the ideal competencies and behaviours that enable success (who you are in that role). An ideal success profile is created using your highest performing current incumbents in the roles you are profiling [35]. Competencies to select from are available in global competency frameworks such as the Korn Ferry Leadership Architect [36] or competencies [37].

Competencies needed will vary depending on the type of role, but some examples of the competencies of great people in health, who will advance the cause, are those who:

- care about the data and what it's used for,
- are curious, willing to unlearn and learn (which is harder when you are really smart),
- adaptable, patient-focused,
- resourceful, solution-oriented, and flexible,
- able to develop and empower others (team members and patients),
- are open to new ideas, innovative, and embrace change.

Team work and collaboration are also fundamental requirements for everyone working in digital health.

Recruitment to the role

Once the success profile is in place, recruitment is made a lot easier. In fact, 75% of companies that define Success Profiles say they are able to hire required talent, compared to 32% of organisations that don't build profiles. Organisations that identify Success profiles are 4.8 times more likely to rate themselves in the top 10% of the competitors for leadership talent [38].

Recruitment may require talent scouting, ensuring you are sourcing great people in digital health. It should always start by looking at internal talent. There will be varied levels of technical knowledge and skills as proven capabilities in some more junior roles within your current workforce. Given this fact, it may be more important to recruit (internally or externally) for character first, then for attributes that enable technical skills to be developed quickly and effectively. Attributes that enable quick uptake of technical skills include learning agility, curiosity, resilience, emotional intelligence, and inquiry. These are easier to recruit for than to develop, as opposed to technical skills which are the easiest to develop.

The method of recruitment is also important to ensure the right attributes are identified. For example, when looking for collaboration capacity, it is recommended that instead of asking for examples in a panel interview, to create a team exercise with other applicants (or team members) and see how well they collaborate with one another. If data input is a necessity in the role, provide a practical exercise to test their data entry skills. Also, to test digital literacy, observe their use of technology during the interview process and their curiosity. In combination with traditional recruitment methods, seeing the candidate's practical skills is a much more effective way to get the right person.

Induction and onboarding

Once the appointment has been made, the induction and onboarding process is the best opportunity to introduce information about the role, the processes, expectations, and ways of working while the employee is motivated and has the time to learn. The emphasis on data quality, how the data is used, the standards and reporting implications of data as well as any systems training, is ideally provided at this point in time. It is also an opportunity for the new employee to decide whether they are engaged or disengaged in the new role. In fact, employees who experience great onboarding are more likely to stay for 3 years and are 50% more productive [39]. The best onboarding experiences are those which create connection with information networks and personal relationships, so include a buddy in the process to be a partner in the in-house learning journey.

Performance plans—Objectives and key results (OKR'S)

In defining the digital strategy, every individual has a role to play to achieve the vision, the aspirational standards and to deliver on the key actions outlined in the plan. Setting the standards and expectations early helps to focus the employee on what is important in their role and how it contributes to the Strategy. Defining Performance plans, objectives, and key results together is best done when an employee first starts in their role with check-ins that occur at least every quarter to redefine and realign or recognise great delivery. This is a critical step that can often be under-estimated or under-utilised, but is a powerful way to reinforce and focus on maintaining digital health standards, delivery, and outcomes.

Reward and recognition

Pay is often considered to be a 'hygiene' factor in employee experience. What this means is, that while most people would prefer to have a pay increase, there is a base level that they are willing to work for. If the base is too low, they will leave; if the base level is Ok, it will not be the factor that impacts retention or motivation. Many formal people reward and recognition systems may be tied to industrial instruments, such as an industrial award or enterprise agreement with pay structures. These can be useful tools to encourage the right focus and behaviours for digital health success and career progression. Pay levels should acknowledge the levels of expertise, contribution, and responsibility in different roles. Beyond pay, a more controllable (and possibly more effective) way of recognising contributions made, may be in annual nominations and celebrations of the contribution individuals and teams make towards digital health. Doing so, publicises instances of great initiative and contributions, by demonstrating, through the use of examples, what 'great' looks like and what the team should aspire to in a very overt way.

Development plans

Using 'Success Profiles' provides a guide for the skills, knowledge, and experience that enable success. These can also be used to identify gaps in skills or competencies to be included in workforce development plans beyond recruitment, so that employees are provided with every opportunity to work towards their full potential. Just as the performance plan is put in place annually, development plans should be developed at least annually with quarterly reviews to check progress and adjust as needed. Whilst this process is already commonly implemented, the key difference for digital health is the sheer speed at which technology evolves. Development plans have traditionally focused on an employee doing a course that builds their technical skills. In digital health, a far more contemporary approach is needed. Solutions may include networking events, reading articles, researching, on-the-job learning, and being mentored as well as formal courses.

Talent identification, career progression, promotions, and succession

The digital space is one that inspires and attracts people who want to innovate, learn, grow, and progress. Deliberate programmes that identify talent, enable career progression, and plan for succession are not only useful to ensure there is a talent pipeline to fill vacancies as they arise internally, they also attract great talent externally. Talent identification requires a focus on those characteristics that do not change much over time and can be detected early, such as emotional intelligence, and those characteristics that flower over time as the person learns to deal with new situations [40].

The manner in which talent is identified is also critical. There will be roles that require deep technical expertise without people management responsibility. In these roles, high performers need to feel challenged by the work they do. Their delivery is what counts. Other roles require broader, more ambiguous roles better suited to those with high potential, which is very different to high performance although there are some rare individuals who combine high performance and high potential. Learning agility is proven to be a prime indicator of potential [40]. Organisations who educate leaders and implement assessments to objectively identify high potential are better equipped to promote the right people, with the right talent into roles.

Diversity, inclusion, and belonging

It is a well-known fact that diverse teams, managed well, are more innovative and higher performing than homogenous teams. With the speed of change and digital technologies, innovation is more critical than ever. Diversity alone, however, is not enough. Inclusion ensures that the variety of employees with different views, experiences, and backgrounds have a voice and are a part of decision making.

A combination of diversity, inclusion, and belonging is required. High belonging is linked to 56% increase in job performance, a 50% drop in turnover risk, and a 75% reduction in sick days [41]. Diversity, inclusion, and belonging is no longer seen as a `nice to have'—it is a business imperative. Best practice suggestions include:

- Make an oath to protect people in the workplace—see something, say something;
- Encourage inclusion and welcome diversity of thought at every opportunity—without judgement;
- Actively promote and support organisations that support diversity and equality;
- Build and protect a safe space where difficult conversations can happen.

Innovation requires a sense of safety to try new and different things, make mistakes, and learn.

Life stage changes that impact on the type of work and the way it is done

Finding talented and experienced employees in digital health is a challenge. Therefore, being open to the full depth and breadth of talent available is vital. In the past, organisations have recruited from a pool of applicants who are willing and able to work on site, in a standard week and with standard working hours. This is no longer viable, as talented employees expect far more flexibility. Desirable working conditions alter as life stages change along with working preferences.

In the UK, there is a law on Working Time Adjustment, which encourages employees to apply for work from a broad menu of when, where and for how long work is done. While this law is for current employees asking to shift to a more flexible model, all newly vacant roles should be tested to see what flexible options are possible. Doing so enables organisations to mine the depth and breadth of talent available regardless of an individual's life stage [42]. Hewlett [43] and others have recommended novel approaches to work that allows 'on and off ramps' to assist employees in all stages of the life course to manage home and work more effectively. Such gradual entrances and exits could be used by all employees, such as workers going back to school, those nearing retirement age, mothers and fathers, and employees pursuing personal goals unrelated to work. This is particularly attractive to female employees. By expanding the groups of employees who might use such policies, the range of talent available in digital health also expands.

Enablers

The above People systems provide the framework needed for all employees across the entire organisation to deliver the digital health strategy. The final ingredient, the enabler that provides focus, high performance, and then embeds the right behaviours sustainably is culture. Culture is the most paramount of enablers and is reinforced by everyday management routines, measures, and every day behaviours. The greatest influence on culture is leadership and employee engagement.

Leading a digital health workforce

Organisations face a radically shifting context for the workforce, the workplace, and the world of work [44]. Leaders need to learn how to maximise human potential within these changing work environments. Building an effective digital workforce will require a new approach to management and leadership within every health-related organisation. Research by Gallup has found that the quality of the manager or the team leader's performance explains 70% of the variation between high engagement and poor engagement in the workplace [45]. This digital age needs a completely different kind of leader, someone able to build teams, keep people connected and engaged, drive a culture of innovation, risk tolerance, and continuous improvement [46].

Yes, workforce leadership starts with frontline leaders, but it is also up to the people who lead and manage the managers who can then impact their employee's engagement. The level of engagement drives innovation, collaboration, performance, and sustainable outcomes. Without great leadership, there is no chance of unlocking the potential of employees to embrace and build their digital capability. Without the necessary insight regarding any leader's own digital health knowledge and skill gaps, there is little incentive to seek further education or undertake professional development in this discipline. It is important for everyone to learn more about the use of digital support to deliver healthcare and real or potential impact of digital health on overall performance. Leaders' digital capabilities is a critical need. Most healthcare organisations and other key high-level digital health ecosystem stakeholders need to build a new breed of younger, more agile, 'digital-ready' leaders [46].

The Boston Consulting Group listed four imperatives for leadership transformation [47]:

1. rethink the art of the possible, by setting imaginative aspirations and bold directions;
2. move from managing to enabling, by nudging for the right behaviours and leading by doing;
3. harness the full power of technology, by insisting on technical fluency and unleashing transparency; and
4. translate purpose into action, by personalising purpose and bringing humanity to work.

> *Leaders everywhere in the world have a tendency to name the wrong person manager and then train them on administrative things—not how to maximise human potential* [48].

Success in a digital world requires leaders to think, act, and react differently compared to leaders frequently described as incumbent 'dinosaur' leaders. Digital leaders need to be able to conceptualise virtual possibilities and handle ever-increasing cognitive complexity. Leaders need to be adaptive and collaborative and value contributions made by others with respect, based on try, fail, and try again principle. This requires confidence and resilience despite constant change, plus an ability to succeed despite ambiguous and challenging environments. Table 1 provides a summarised comparison between traditional and digital leaders behaviours.

Performance benefits from leadership and employee engagement

With the right leadership in place, employees are far more engaged. They have an emotional commitment and connection to the organisation and its goals to give their best efforts each day. The level of engagement is an indicator of the level of motivation an employee has to put in extra effort; it is a sign of how committed they are to staying.

TABLE 1 Traditional manager and digital leaders behavioural differences.

Dinosaur manager	Digital leader
Low tech—use email, paper, and copies in everyone	Mobile—always connected
Binary—black or white, can't work with ambiguity	Agile, able to rethink assumptions, open to new ideas
Secretive—information sharing is controlled	Engaged with leaders and grass roots—keeps in touch
Territorial—no engagement across the siloes	Collaborative, co-creates and unites
Linear—Standard Operating Procedure (SOP) driven rather than outcome driven	Sets clear expectations and adapts as needed
Passive—waits for others to get it wrong	Proactive—healthy dissatisfaction with status quo and acts
Resists change—does not adapt or change their thinking	Embraces change, curious, keen to experiment, constructively disruptive
Task focus vs culture and values focus	Purpose, values, and outcomes driven
Won't share people, knowledge, or resources	Networked, uses social media and other collaborative tools
Focuses on what over why—dictates rather than co-creates	Co-creates to agree on purpose, outcomes, and delivery
Does not have the talent to see talent	Sees strengths in everyone

Employee engagement is also about the way an organisation listens to, builds trust, and improves relationships with its employees. It's about sharing a common purpose through a culture of communication, involvement, and collaboration. But beyond the feel-good aspects, engagement makes great business sense. High engagement enables improvements in employee morale, attendance, retention, quality outcomes, and performance. It is also proven to improve patient loyalty, pride in the workplace, productivity, and profitability. It also reduces employee safety incidents, patient safety incidents, and theft.

The Gallup 2020 Q12 Meta-Analysis: 10th Edition [48], which outlines The Relationship Between Engagement at Work and Organisational Outcomes, found that the Median percent differences between top-quartile and bottom-quartile units were:

- 10% in customer loyalty/engagement
- 23% in profitability
- 18% in productivity (sales)
- 14% in productivity (production records and evaluations)
- 18% in turnover for high-turnover organisations (those with more than 40% annualised turnover)
- 43% in turnover for low-turnover organisations (those with 40% or lower annualised turnover)
- 64% in safety incidents (accidents)
- 81% in absenteeism
- 28% in shrinkage (theft)
- 58% in patient safety incidents (mortality and falls)

- 41% in quality (defects)
- 66% in well-being (net thriving employees)
- 13% in organisational citizenship (participation).

The case for a people focus is very clear. A digital health ecosystem can be created, evolve, and can sustainably deliver with an inspiring Vision and Strategy that people can align to. There need to be:

- clear and measurable performance goals,
- organisational structures and roles that clarify accountability,
- clear skill development, people systems that provide the framework for success, and
- an enabling culture that inspires through leadership and engagement.

Conclusion

Implementing a digital transformation requires a well-executed change management programme with a strong emphasis on people and workforce capacity-building strategies as described. We do need to be practical and continue to make use of past investments. As is the case with all technologies, there needs to be well times a replacement plan that forms part of the digital transformation vision and implementation strategy. As digital health ecosystems are made up of numerous other complex ecosystems, the strategy needs to be adaptive to best accommodate changes being made elsewhere that have a local impact, or that impact the potential for progress to be made by other stakeholders.

Top-down leadership and the establishment of foundational supportive infrastructures can have a strong positive ripple-down impact if done well. As we have learned from past failures, this needs to be done well so as not to limit progress for another 10 or 20 years. A comparable workforce capacity-building strategy must begin with the development of every high-level leader. In some instances, this may require the replacement of those who simply do not possess the necessary personal attributes.

All education providers need to examine how best to promote digital thinking and integrate digital health into current programmes as well as develop short high niche programmes well suited for 'just-in-time' learning. Workforce capacity-building strategies need to focus on meeting the needs of both the current and our future workforce. Knowing how to use technology is insufficient, the health workforce needs to have a fundamental understanding of the many technical complexities to enable them to communicate effectively with those who need to make it happen. Those charged with the technical development and implementation need to be able to attain all the necessary knowledge and skills to make this a success.

References

[1] W. Gibson, The Future Has Arrived – It's Just Not Evenly Distributed Yet, 2021, https://quoteinvestigator.com/2012/01/24/future-has-arrived/ (Accessed 24 November 2021).

[2] T.D. Aungst, R. Patel, Integrating digital health into the curriculum-considerations on the current landscape and future developments, J. Med. Educ. Curric. Dev. 7 (2020) 1–7, https://doi.org/10.1177/2382120519901275.

[3] K. Andronis, K. Moysey, Data governance for health care providers, Stud. Health Technol. Inform. 193 (2013) 299–315.

[4] D.R. Wholey, M. LaVenture, S. Rajamani, R. Kreiger, C. Hedberg, C. Kenyon, Developing workforce capacity in public health informatics: core competencies and curriculum design, Front. Public Health 6 (2018) 124.

[5] R. van der Vaart, C. Drossaert, Development of the digital health literacy instrument: measuring a broad spectrum of health 1.0 and health 2.0 skills, J. Med. Internet Res. 19 (2017), e27.

[6] G. Strudwick, L. Nagle, I. Kassam, M. Pahwa, L. Sequeira, Informatics competencies for nurse leaders: a scoping review, J. Nurs. Adm. 49 (2019) 323–330.

[7] A.H. Sapci, H.A. Sapci, Teaching hands-on informatics skills to future health Informaticians: a competency framework proposal and analysis of health care informatics curricula, JMIR Med. Inform. 8 (2020) e15748.

[8] F. Martin-Sanchez, D. Rowlands, L. Schaper, D. Hansen, The Australian health informatics competencies framework and its role in the Certified Health Informatician Australasia (CHIA) program, Stud. Health Technol. Inform. 245 (2017) 783–787.

[9] M. Kleib, L. Nagle, Development of the Canadian nurse informatics competency assessment scale and evaluation of Alberta's registered Nurses' self-perceived informatics competencies, Comput. Inform. Nurs. 36 (2018) 350–358.

[10] M.A. Kennedy, A. Moen, Nurse Leadership and Informatics Competencies: Shaping Transformation of Professional Practice, 2016, https://doi.org/10.3233/978-1-61499-738-2-197.

[11] L. Jidkov, M. Alexander, P. Bark, J.G. Williams, J. Kay, P. Taylor, H. Hemingway, A. Banerjee, Health informatics competencies in postgraduate medical education and training in the UK: a mixed methods study, BMJ Open 9 (2019), e025460.

[12] B.R. Holroyd, M.S. Beeson, T. Hughes, L. Kurland, J. Sherbino, M. Truesdale, W. Hersh, Clinical informatics competencies in the emergency medicine specialist training standards of five international jurisdictions, AEM Educ. Train. 2 (2018) 293–300.

[13] Global_Health_Workforce_Council (GHWC). Global Academic Curricula Competencies for Health Information Professionals: American Health Information Management Association (AHIMA); 2015. Available from: https://ifhima.org/wp-content/uploads/2018/01/AHIMA-GlobalCurricula_Final_6-30-15.pdf.

[14] L. Cusack, M. Smith, B. Cummins, L. Kennewell, L. Dennett, D. Pratt, Advanced skills for enrolled nurses: a developing classification, Aust. J. Adv. Nurs. 32 (2015) 40–46.

[15] R. Abdulla, K. Benjamin, S. Munir, N.S.O. Ahmed, A review of informatics competencies tools for nurses and nurse managers, Can. J. Nurs. Inform. 13 (2018) 1–24.

[16] U. Hübner, T. Shaw, J. Thye, N. Egbert, H.d.F. Marin, P. Chang, S. O'Connor, K. Day, M. Honey, R. Blake, E. Hovenga, D. Skiba, M.J. Ball, Technology informatics guiding education reform—TIGER, Methods Inf. Med. 57 (2018) e30–e42.

[17] E. Hovenga, H. Grain, Learning, training and teaching of health informatics and its evidence for informaticians and clinical practice, in: E. Ammenwerth, M. Rigby (Eds.), Evidence-Based Health Informatics, Studies in Health Technology and Informatics, IOS Press, Amsterdam, 2016, pp. 336–354.

[18] K. Saranto, E.J.S. Hovenga, Information literacy—what it is about? Int. J. Med. Inform. 73 (2004) 503–513.

[19] H. Peterson, U.E. Gerdin-Jelger, Preparing Nurses for Using Information Systems: Recommended Informatics Competencies, National League for Nursing, New York, 1988.

[20] H. Grain, E. Hovenga, Health informatics competencies—underpinning e-health, Stud. Health Technol. Inform. 168 (2011) 73–81.

[21] S. Garde, D. Harrison, E. Hovenga, Skill needs for nurses in their role as health informatics professionals: a survey in the context of global health informatics education, Int. J. Med. Inform. 74 (2005) 899–907.

[22] AHIEC, Health Informatics Education—Strategic Workplan, Australasian College of Health Informatics, 2009.

[23] AHIEC. Health Informatics Scope, Careers and Competencies V1.9: Australian Health Informatics Education Council; 2011. Available from: https://www.ahiec.org.au/docs/AHIEC_HI_Scope_Careers_and_Competencies_V1-9.pdf.

[24] J. Mantas, A. Hasman, IMIA educational recommendations and nursing informatics, Stud. Health Technol. Inform. 232 (2017) 20–30.

[25] E. Hovenga, H. Grain, Developing a common reference model for the health informatics discipline, Stud. Health Technol. Inform. 192 (2013) 122–126.

[26] J. Mantas, Biomedical and health informatics education—the IMIA years, Yearb. Med. Inform. (Suppl. 1) (2016) S92–S102.

[27] J. Mantas, E. Ammenwerth, G. Demiris, A. Hasman, R. Haux, W. Hersh, E. Hovenga, K.C. Lun, H. Marin, F. Martin-Sanchez, G. Wright, Recommendations of the international medical informatics association (IMIA) on education in biomedical and health informatics. First revision, Methods Inf. Med. 49 (2010) 105–120.

[28] Digital-Healh_Canada. Digital Health Competency Requirements for Professional Development, 2019. Available from: https://digitalhealthcanada.com/.
[29] A.L. Valenta, E.S. Berner, S.A. Boren, G.J. Deckard, C. Eldredge, D.B. Fridsma, C. Gadd, Y. Gong, T. Johnson, J. Jones, E.L. Manos, K.T. Phillips, N.K. Roderer, D. Rosendale, A.M. Turner, G. Tusch, J.J. Williamson, S.B. Johnson, AMIA board white paper: AMIA 2017 core competencies for applied health informatics education at the master's degree level, J. Am. Med. Inform. Assoc. 25 (2018) 1657–1668.
[30] H.D. Silverman, E.B. Steen, J.N. Carpenito, C.J. Ondrula, J.J. Williamson, D.B. Fridsma, Domains, tasks, and knowledge for clinical informatics subspecialty practice: results of a practice analysis, J. Am. Med. Inform. Assoc. 26 (2019) 586–593.
[31] ADHA. National Digital Health Workforce and Education Roadmap: Australian Government: Australian Digital Health Agency; 2020. Available from: https://www.digitalhealth.gov.au/about-the-agency/workforce-and-education, https://www.digitalhealth.gov.au/about-the-agency/workforce-and-education/Workforce%20and%20Education%20-%20Roadmap.pdf.
[32] IMIA, Body of Knowledge, IMIA Endorsed Documents: International Medical Informatics Association, 2009, Available from: https://imia-medinfo.org/wp/imia-endorsed-documents/ [cited 2019 22 February].
[33] Bloom's-Taxonomy, 2001, Available from: https://www.bloomstaxonomy.net/.
[34] ISO/TR-18638. Health Informatics—Guidance on Health Information Privacy Education in Healthcare Organizations 2017. Available from: https://www.iso.org/standard/63100.html.
[35] J.E. Orr, C. Sneltjes, G. Dai. The Art and Science of Competency Modeling: Best Practices in Developing and Implementating Success Profiles: The Korn/Ferry Institute; 2010. Available from: https://www.kornferry.com/content/dam/kornferry/docs/article-migration/Korn-Ferry-Institute_The-art-and-science-of-competency-modeling.pdf.
[36] Korn-Ferry. Leadership Architect TM Legacy Competency Mapping, 2014, Available from: https://www.kornferry.com/content/dam/kornferry/docs/article-migration/KFLA_LegacyCompetencyModelsMapping.pdf.
[37] Korn-Ferry's, 38 Competencies, 2014, Available from: https://neprisstore.blob.core.windows.net/sessiondocs/doc_10cd2119-d549-47ac-8784-215eb85cdc8f.pdf.
[38] DDI, Why Define Success?: Development Dimensions International Inc, 2021, Available from: https://www.ddiworld.com/solutions/success-profile.
[39] A.S. Hirsch. Don't Underestimate the Importance of Good Onboarding: SHRM; 2017. Available from: https://www.shrm.org/resourcesandtools/hr-topics/talent-acquisition/pages/dont-underestimate-the-importance-of-effective-onboarding.aspx.
[40] R.W. Eichinger, M.M. Lombardo, Learning agility as a prime indicator of potential, Hum. Resour. Plan. 27 (2004) 12–15.
[41] E.W. Carr, A. Reece, G.R. Kellerman, A. Robichaux. The Value of Belonging at Work: Harvard Business Review; 2019. Available from: https://hbr.org/2019/12/the-value-of-belonging-at-work.
[42] M.C. Still, The opt-out revolution in the United States: implications for modern organizations, Manag. Decis. Econ. 27 (2006) 159–171.
[43] S.A. Hewlett, C.B. Luce. Off-Ramps and On-Ramps: Harvard Business Review; 2005. Available from: https://www2.byui.edu/HMM10/retaining_employees/resources/9416.pdf.
[44] B. Walsh, E. Volini. Rewriting the Rules for the Digital Age: Deloitte University Press; 2017. Available from: https://www2.deloitte.com/us/en/insights/focus/human-capital-trends/2017.html.
[45] R. Beck, J. Harter. Managers Account for 70% of Variance in Employee Engagement: Gallup; 2015. Available from: https://news.gallup.com/businessjournal/182792/managers-account-variance-employee-engagement.aspx.
[46] S. Philpot, I. Roy. Leadership disrupted: Pushing the Boundaries- in Re-Writing the Rules for the Digital Age: Deloitte Insights; 2017. Available from: https://www2.deloitte.com/us/en/insights/focus/human-capital-trends/2017/developing-digital-leaders.html.
[47] D. Dosik, V. Bhalla, A. Bailey. A Lot will Change—So Must Leadership: Boston Consulting Group; 2020. Available from: https://www.bcg.com/en-au/publications/2020/importance-of-transformative-leadership-post-coronavirus.
[48] J. Clifton. The Relationship Between Engagement at Work and Organizational Outcomes Q12® Meta-Analysis: Gallup; 2020. Available from: https://www.gallup.com/workplace/321725/gallup-q12-meta-analysis-report.aspx?thank-you-report-form=1.

CHAPTER

26

Emerging digital health ecosystems

Evelyn Hovenga[a,b,c], *Helen Hovenga*[d], *and Heather Leslie*[e,f]

[a]Digital Health, Australian Catholic University, Melbourne, VIC, Australia [b]eHealth Education Pty Ltd, Melbourne, VIC, Australia [c]Global eHealth Collaborative (GeHCo), Melbourne, VIC, Australia [d]Telix Pharmaceuticals Limited, Melbourne, VIC, Australia [e]Atomica Informatics, Melbourne, VIC, Australia [f]Clinical Program Co-Lead openEHR International, London, United Kingdom

A functional EHR environment

The compilation of this text was guided by a shared digital health ecosystem vision formulated more than 30 years ago. This ecosystem vision is now gaining global traction and getting close to the 'tipping point'.

A primary visionary component of a digital health ecosystem for this author was the functional EHR environment as described by Thomas Beale in his 2001 manifesto [1]. Beale described this environment as consisting of eleven key points, each of which has been explored within this text. These points represent an excellent summary of a high-level, well-functioning digital health ecosystem supporting patient-centred lifelong care and the 'holy grail' of semantic interoperability. They are listed in Box 1.

Beale went on to describe the provider enterprise as a business responsible for all operational management functions that need to be linked with financial and human resource (HR) systems taking into account the variety of physical resource locations and the need to manage distributed queries. The interface visible at the enterprise level needs to be quite varied, to satisfy the needs of many stakeholders, including clinical users, population health planning, government reporting, insurance organisations, and health consumers themselves.

This farsighted vision has stood the test of time. It has underpinned the development of the openEHR architecture specifications. It has supported and continues to receive contributions from a growing number of people who share this vision globally and who are committed to demonstrating its value. Over the past 20 years, the predominant volunteer openEHR community continues to test, evaluate, and contribute to Beale's scientific principles and theoretical constructs by building and implementing the components required to create this highly complex adaptive environment.

How do we create a road map for a comprehensive ecosystem with numerous interdependencies? How can we identify and evaluate the benefits to be realised? Adler's potential

Roadmap to Successful Digital Health Ecosystems
https://doi.org/10.1016/B978-0-12-823413-6.00024-0

Box 1

Thomas Beale's vision as described in his 2001 manifesto.

1. An electronic health record (EHR) is a repository of patient-centred records containing contributions by carers in the course of a lifelong care process.
2. A minimal identification service for any health information environment containing sufficient demographic information on patients, healthcare professionals, and healthcare institutions, to disambiguate parties, including if extracts from the EHR are sent to other sites.
3. Controlled vocabularies are a key element in structured health data, enabling proper automatic processing such as decision support to be performed.
4. Formal models of clinical concepts are the next revolution in health information systems. [*This prompted the development of the openEHR Clinical Knowledge Repository*].
5. A query/update service as the primary interface to the EHR used by most applications and users. [*This prompted the development of the openEHR AQL.*]
6. Workflow systems that support individual patient health journeys, executing agreed models of care, and reminding the relevant parties of events to be performed.
7. Health activity and service management.
8. Multimedia and genetics data integrated with EHRs.
9. In the era of evidence-based medicine, guidelines and protocols are the decision-making and planning tools of the clinician. Their use needs to be recorded in the health record, so that other users can understand the reasons for decisions, and so that medico-legal investigations can be effective.
10. Decision support systems require disciplined patient data in order to function. They rely heavily on coded terms and are likely to rely equally heavily in the future on formal clinical models in order to navigate data intelligently.
11. Handheld computing devices are becoming more common in hospitals and allied health. The EHR needs to be available for viewing and potentially update via relatively low-bandwidth interfaces, which may also need to support asynchronous transfers.

success factors have also guided the approach taken for this text. These success factors apply to any level within a global digital health ecosystem.

To change the world, we must start with ourselves—we need to recognise our inherent biases and inclination to comply with current fads and trends, yet be open to the potential offered by revolutionary transformation, rather than settling for simpler and safer incremental innovation.

Implementing our suggested road map requires the following considerations:

- A multifaceted digital health ecosystem that is greater than the sum of its parts.
- A change in focus from health records prioritising billing and administration to a focus on metrics of real value, such as the number of patients impacted and jobs created in the sector resulting in a positive impact on health outcomes.

- Evolving public investment models that keep pace with advancements in all technologies, reflect the evolving technological landscape, have a digital focus, and support proven performance, real-world impact, commercialisation, and innovation.
- Policy initiatives focused on local and global access to a digital health workforce capable of addressing the significant challenges ahead.
- Grant, research, and developmental programs must be coordinated and be amenable to the process of innovation and collaboration in digital health.
- Artificial Intelligence and personalised medicine are at the very heart of the medical device and digital health developments. These activities need to be supported as they will play an increasing role across all aspects of digital health, especially their integration into existing and evolving clinical systems.

In recent years, we have witnessed the spectacular failure of many of the biggest 'top-down', government-driven, one size fits all approaches. Currently, we are experiencing a proliferation of uncoordinated and piecemeal 'bottom-up' digital health technical advances, yet these are still unable to reach desired levels of semantic interoperability.

This text has exposed the many silos, fragmentations, limitations, overlaps, and barriers that are collectively responsible for the current state of our digital health environment. A common feature of both the 'top-down' and 'bottom-up' approaches is the lack of an agreed, documented, and realistic vision, well-coordinated strategic objectives, and insufficient collaboration and knowledge sharing between stakeholders and relevant experts. In reality, we need both to succeed. Those operating at the top level need to establish and govern a critical, common infrastructure foundation that facilitates bottom-up innovation to meet local needs and integration with other developments along the entire health delivery supply chain.

This 'middle-out' approach was first described by Coeira in 2009 [2] as enabling the needs of health providers, the IT industry, and government to be brought closer together by creating a common set of technical goals and underpinning standards that can sit between them. He noted that this development of shared goals, standards development, and, support for standards implementation must be well-resourced. Such integration requires a 'whole of health' ecosystem compliant with an essential set of foundational, mandated global and national standards, each selected for a strategic purpose and supplied, governed, and maintained by top-level leadership. Top-level failures have a long-lasting adverse impact on many.

Like in nature itself, digital ecosystems can exist at many levels, reflecting the complexity and interdependencies of the health and care domain. Regardless of size or purpose, every digital ecosystem needs to implement a change management approach to ensure its services fit the current and future digital environment.

Defining the vision and then the strategy

When digital health ecosystem stakeholders initiate a strategy focusing only on current incremental change methods, we can predict that it will fail with reasonable confidence. More than 60 years of steady, incremental progress has not delivered semantic interoperability or thriving digital health ecosystems. It has never worked before, so no reason for it to work this time.

In contrast, Bob Johansen, a futurist and fellow of the Institute for the Future in Silicon Valley [3], explains that we need to think `future back—look long and work backwards'. Strategy requires a clear, future-focused vision and purpose.

Simon Sinek's book entitled `Start with Why' outlines an inspiring method to create an organisation's `Why', `How', and `What'. When a stakeholder is looking a few years into the future, they need to consider their purpose within the context of that future world. What is the aspirational vision that will have been achieved (the `Why')? [4] In support of the vision and purpose, values and beliefs can be defined to demonstrate what every stakeholder stands for and believes in (the `How'). It is then that the strategy, outcomes, and delivery are defined (the `What'). The right vision provokes insight and, as a result, foresight and inspiration. Foresight, even more than ideas, provokes action, as we see the pros and cons and, therefore, future possibilities. An example of such a vision is:

`Our patients survive and thrive' and 'our population is healthy' (Why).
We achieve this by... having a digital health ecosystem, (How).
created with intention, purpose, alignment, standards, consistency and clear governance (What).

This level of clarity defines the strategy as the DNA that is deliberately embedded as a foundational component of our aspirational digital ecosystem.

Once clear on the strategic vision, it is then vital to be realistic about the current state. What needs to be preserved? What needs to be adapted or enhanced? What is missing? This activity then identifies the gap between the ideal and the present, and the step-change actions necessary to achieve the desired outcomes. Objectives and actions can all be defined in the strategy.

The process of planning and implementing an appropriate digital health enterprise within the broader ecosystem includes several phases [5] presented below as a general overview of fundamental principles and to be used as a guide only. Every stakeholder needs to develop a similar list that best meets their entity's needs within their digital health ecosystem.

1. Assessing the current state and enabling environment.
   (a) Refer to some of the examples provided in this text, then assess your current state, and enable environment relative to different types of stakeholders.
   (b) Identify the stakeholders who form part of your ecosystem and determine interdependencies.
   (c) Identify data use cases within your immediate ecosystem from the perspective of each stakeholder, to determine the data supply chain needs.
   (d) Identify funding models and reporting requirements, existing technologies in use, legislation, regulations, healthcare policies, public health infrastructures, statistical data collections, registries, clinical trial participation, research participation, Internet access, electricity supplies, and smartphone use.
   (e) Document fundamental higher-level infrastructure requirements that need to be in place to support the provision of the best possible digital solutions.
   (f) Assess and secure investment options.
2. Establishing a shared understanding and strategic planning
   (a) Multidisciplinary and intersectoral collaboration to prepare an agreed vision.
   (b) Document impact targets and potential benefits for relevant stakeholders.

   (c) Identify key policy options and recommendations for crucial associated stakeholders to enable them to work collaboratively and address interdependencies.
   (d) Develop a workforce capacity-building strategy able to develop and support the vision for the future state.
3. Defining the future state
   (a) Develop strategic directions and timelines to achieve agreed goals.
   (b) Identify technological, legislative, regulatory, and governance requirements.
   (c) Undertake a risk assessment.
4. Planning enterprise architecture
   (a) Review and update procurement arrangements based on services to be provided and current technical infrastructures.
   (b) Identify the value proposition from adopting a knowledge-based digital health platform, such as the application-independent openEHR approach, including the application of agnostic clinical models and standardised data sets.
   (c) EHR specifications are generic, not application-based, and able to fit within a digital health ecosystem architecture.
5. Determining health content requirements
   (a) Identify clinical and informatics champions who can support knowledge-domain modelling and governance by speciality.
   (b) Examine the data supply chain for specific patient cohorts.
   (c) Identify types of data use cases to suit every level and secondary purpose within the healthcare ecosystem.
   (d) Develop decision-making algorithms to suit specific niche applications
   (e) Establish standard data value sets to suit all monitoring, funding, evaluation, public health, and best practice research needs.
6. Monitoring and evaluation (M and E), fostering data use
   (a) Identify automation and AI potential, including their foundational data and technology requirements.
   (b) Review and update standard data value sets to meet automation and AI needs.
7. Implementing, maintaining, and scaling.
   (a) Identify the many interdependencies and then select implementation priorities accordingly.
   (b) Monitor progress, detect new challenges, and undertake timely problem-solving to minimise the impact of any adverse events.

Implementation strategy

Such transformative activities need strong national leadership, alignment, and integration between the health and education sectors. This high-level road map implementation strategy to successful digital health ecosystems is shown in Fig. 1. This represents a practical model for designing and executing a global and national digital health transformation road map, illustrating the major phases, input, and output of resources and data sets to achieve digital health knowledge integration.

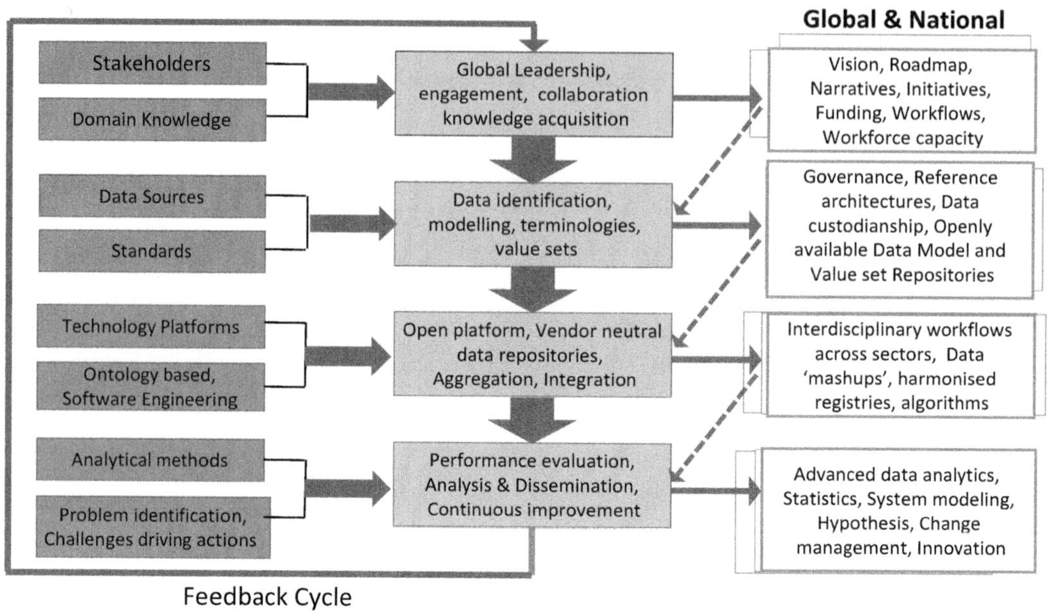

FIG. 1 Model for global and national digital health reformation road map. *Modified from P.R. Payne, Chapter 1: biomedical knowledge integration, PLoS Comput. Biol., 8 (2012) e1002826.*

Weaknesses in many healthcare systems and public health responses were exposed during the COVID-19 pandemic. As a consequence, the effects of this pandemic may reduce complacency, trigger awareness of the limitations on incremental change, and drive necessary reform. A panel of 13 experts participating in the global digital health summit held in Riyadh during August 2020 articulated the following seven key priorities [6]:

1. Adoption of health intelligence by the health and care sectors
2. Scaled up and sustainable interoperable digital technologies able to securely process trusted scalable data flows.
3. Support the adoption of artificial intelligence dependent upon rapid access to various data types and vast amounts of high-quality data to achieve acceptable accuracy and validity.
4. Effective communication about public health crises and risks based on timely dissemination of information, seamless digital integration using data visualisation tools.
5. Health data governance, quality, policy, regulation, and use.
6. Interoperable digital health systems designed and implemented to maximise quality and enable clinician and patient access.
7. Collaborative research and life-saving innovations' translations into widespread applications.

Top-down responsibilities:

(a) Leadership providing a supportive infrastructure for digital health ecosystem development.
   (i) Collaboration, innovation, driving investments, multidisciplinary, and multisector/departmental partnerships.

(ii) Transforming their own operational contributions to best meet new digital health revolutionary directions—smart automation, machine learning, and artificial intelligence requirements
(iii) Legislative/governance support
(iv) Policy initiatives:
1. Mandating key architectural ecosystem standards to be included in all local procurement processes.
2. Establishing key governance agencies/departments/processes
(v) Include social determinants of health (clean water, housing, food security, healthcare, and Internet access) within the health ecosystem build.

Bottom-up responsibilities:

(b) Leadership providing a supportive infrastructure by clinical discipline at all points of care.
(i) Collaboration, innovation, driving knowledge-sharing, multidisciplinary, and multisector/departmental partnerships.
(ii) Ensure all application developments reference and comply with the digital health ecosystem architectural requirements.
(iii) Identify specific use cases to improve patient-centred care, and identify and/or develop evidence-based clinical models that support best practice, research, and reporting automation for timely knowledge sharing and enabling decision support.
(iv) Identify the need for concept representation standard adoption and develop data value sets that support advanced analytics, including machine learning and artificial intelligence.
(v) Provide data stewardship
(vi) Include concept representation needs of social determinants of health (clean water, housing, food security, healthcare, and Internet access) within the health ecosystem build.

Current state and enabling environment—A summary of findings

This text began with the identification of currently held collective desirable health system and population health outcomes and explored a shared population health vision driving current health transformation. We noted that these desired health outcomes cannot be considered in isolation of the digital technologies now available to support the realisation of this 'health for all' vision. Our examination of a number of key stakeholder types and their historical origins that have guided their ongoing activities revealed that for most of these organisations to continue, they need to be fit for purpose in a digital world; this was followed by the presentation of a number of real-world examples of current digital health initiatives and/or shortcomings to provide our readers with a global view of the status quo.

Whilst writing this text, we have witnessed the need for faster response times as a result of the global COVID-19 pandemic. These latter events convinced many that we need to fully embrace technology to optimise the value of digital health across the whole of the health and care domain. We need to avoid re-inventing the wheel by working smarter and more collaboratively across multiple sectors and jurisdictions. Consolidating and coordinating efforts enable us to maximise beneficial returns on investments made, financially and personally, by sharing knowledge and lessons learned from our many collective experiences.

What are today's issues:

- Fragmented standards development organisations resulting in overlapping and competing standards,
- There are many and varied standards,
- Standards may not be compatible with each other,
- Many developers only access mandated standards,
- Many standards are not widely used, or implementations are noncompliant,
- Most industry standards remain proprietary,
- There are many categories of standards, based on purpose and desired functionality,
- Many standards are no longer suitable to meet this new digital era needs despite updates,
- Backwards compatibility may now compromise system usability,
- Software engineering methodologies have seen many advancements and will continue to do so in the future. Engineering solutions need to consider how to be independent of a technology paradigm that is liable to change
- Data and content representation methods have seen many changes,
- Most terminologies were developed to suit a number of historical statistical purposes,
- Many terminologies in use have limited clinical data and knowledge management suitability,
- Many lessons have been learned based on experience,
- Today's technical know-how requires us to change our thinking to maximise benefits now achievable to better suit the needs of patient-centred care, smart automation, machine learning, and artificial intelligence.

Future directions and purpose

Digital health represents a huge knowledge domain—it is broad, deep, and continuously evolving. No one person can know it all. There is a need to use multidisciplinary teams with members who collectively possess the required talent that best suits the tasks. We have described what a well-connected, optimally functioning digital health ecosystem looks like, the fundamental infrastructures required to support the provision and use of digital health solutions, and the foundational scientific and ethical principles that need to be followed. Everyone has a role to play. All health data need to represent the world's reality accurately and be regarded as a highly valued asset. This requires precise data collection, including the data context, throughout the data supply chain.

Every type of stakeholder should be responsible for the content and governance of their own knowledge domain. Enabling wider use requires our health knowledge domains to be represented digitally in computable formats. This requires global, national, organisational, and project leadership, engagement with multiple stakeholders, a commitment to collaboration, and a community of trust.

Leaders, at all levels for each stakeholder group, need to have a shared vision that is harmonised with their overall digital health ecosystem, develop strategic plans, take initiatives (and risks), and communicate effectively to manage their local transformation and fully embrace digital health. Their actions need to account for numerous interdependencies and will require top-down funding and a strong commitment to result in changes in information and workflows. Desired outcomes are dependent upon workforce (and population) capacity building to enable us all to be in the best possible position to maximise the adoption and use of digital technologies.

Strong national leadership can ensure that an effective digital health ecosystem supports timely, complete, and accurate information sharing, communication, and collaboration between researchers, healthcare providers, and their patients/clients, irrespective of location. Healthcare consumers will need support to make effective use of new technologies and their health information. Enabling them to partner in care and treatment option decision-making assumes an adequate level of health and digital health literacy.

Data need to be accurately represented at all times, requiring the adoption and governance of data standards, including shared clinical models and terminology value sets at all levels within our digital ecosystems. Every knowledge domain needs to be modelled to best enable our clinical information to be used and re-used in digital environments. The availability of computable information will enhance decision-making at every level, supporting many different use cases and benefitting everyone. Clinical models need to be based on open standards, freely available, governed, and linked to standardised value sets. Models need to relate to reference information models and integrate with EHR and ecosystem architectures.

Highly granular clinical data represent huge assets supporting knowledge discovery, evidence-based practice, clinical trials, and lifelong learning, and are used to capture the full benefits of decision support systems, robotics, artificial intelligence (AI), and precision medicine, and are essential to support patient-centred care. These values are best facilitated by smart data collection at points of care, supporting data use life cycles, data access, data linkages, sound knowledge management, statistical reporting, and the use of standard open digital platforms.

There is a need to adopt openly accessible knowledge-based technology platforms and to support vendor- and technology-neutral data repositories to maximise opportunities for data aggregation, linkage, reporting automation, and advanced data analytics. Knowledge-based platforms support interdisciplinary and cross-sector workflows, 'plug and play' of applications, generation of data registries accommodating larger sample sizes, and development and adoption of algorithms. Adopting a platform-based digital health ecosystem requires a technical workforce capable of using ontology-based software engineering skills.

Automated global public health information sharing, supply chain, and reporting enable more timely and comprehensive information to be accessed by decision-makers. This automation capacity depends on the consistent use of agreed data standards. Automation is also dependent upon streamlining, centralising, and, in some instances, globalising, routine and core organisational, operational processes. Expect this latter transformation to cause significant disruption for most organisations during the implementation phase, however, it should significantly improve national health system financial sustainability by automating many routine reporting processes.

If we can create and safely aggregate high-quality data, we are in a strong position to:

- undertake advanced analytics,
- identify best practice,

- monitor and evaluate clinical and organisational performance or undertake system modelling to continuously make improvements,
- undertake research,
- support clinical trials,
- test various hypotheses,
- address various ongoing challenges by taking action to drive change, and
- generate new knowledge and understanding of world realities.

All this is then reported back to the global leadership as part of the feedback cycle enabling the incorporation of new discoveries and support global pandemics to benefit the world's population as a whole.

Current global reliance on only a few competitive, proprietary, high-level infrastructure technologies, such as clouds, standard platforms, or large EMR systems, is a high-risk and costly strategy. The risk can be substantially mitigated by adopting publicly funded, open standards-based technologies and governed and managed for the public good as foundational national infrastructure. This new approach supports a strong focus on applications designed and built to operate on the shared national platforms, including any type of health information system, telecare, telehealth, wearables, health monitoring devices, health analytics, research, and population health planning. This drives local economies and provides market-driven competitiveness based on intelligent systems and usability rather than data silos or outdated business models.

Living and working within a supportive digital health ecosystem requires changes to roles and boundaries of practice for many healthcare professionals. Many new jobs will be created for people to drive or manage technical innovation, and workplace transformations, and implement new digitally enabled processes. Over time, a shift in roles is expected, a transition from processes operated by people to processes designed and audited by people. This will have profound implications for individual healthcare delivery organisations.

Key requirements

There needs to be a short, medium, and long-term vision. A digital health ecosystem consists of many interdependent components, each requiring strong leadership skills, a commitment to collaborative cross-sectoral practices, and niche expertise to successfully implement each component within the set time frames.

A national road map provides the blueprint for making this happen over a period of time in pragmatic, coordinated phases. Components include establishing foundational national infrastructure, data and information management governance, various technical tools, workforce capacity building, and a change management strategy. The duration of the road map will depend on many factors, especially the complexity of the digital health environment and its interdependencies. Each component go-live needs to build upon previous accomplishments and simultaneously accommodate many dependencies.

Standards development organisations need to collaborate more effectively regarding foundational standards suited to a 'core ecosystem architecture', with components intentionally designed to complement each other to attain semantic interoperability and optimise common technology solutions to patient-centred care. We need to globally adopt a published architecture whose component interfaces form the basis for independent application development and to move core services, data, and capabilities to a new paradigm based on a *knowledge-driven platform (KDP)*, within a coherent service infrastructure and modern application development technology environment. Architectural components need to complement each other.

Globally, we need to adopt standards for the following key architectural components:

- **Open Platform Architecture**—see Chapter 6, The knowledge-driven platform: Strategic technologies for a platform ecosystem approach.
- **Datatypes**—see Chapter 8, Health Data Standards Limitations
- **Terminology, concept representation, and definitions**—see Chapters 8 and 9, Health Data Standards Limitations and Quality data, Design, Implementation, and Governance
- **Unique identifiers**—see Chapter 12, Identity Management, Identifiers, and Their Governance
- **Governance**—see Chapters 9 and 12
- **Domain Information Models**—see Chapter 11, Modelling Clinical Knowledge
- **Workforce capacity**—see Chapter 25, Addressing Workforce Digital Health Knowledge Gaps.

We argue that the following shifts in healthcare systems can result in better global health:

From	To	Rationale
Fragmentation, local problem solving, unsustainable, costly health systems. 'Emergency break the glass' response systems.	'Always on' systems and collaborative partnerships, meeting the needs of all stakeholder use cases within any digital health ecosystem that can scale rapidly during epidemics.	Planned transformation is most effective when everyone adopts a logical ecosystem architecture based on global open knowledge-based platform adoption
Uneven disease and health surveillance	Strengthened ability to support patient-centred care anywhere, global, national, and local mechanisms for detecting infectious diseases.	Effective best-practice research and health burden detection capacity at all levels.
Waiting for system failure or infectious disease outbreaks	Integrated prevention and cost minimisation agenda	Targeted transformative interventions can reduce adverse events and pandemic risks.
Scramble for healthcare and health IT workforce capacity.	Targeted workforce capacity building programs and health systems ready to surge whilst maintaining essential services.	Health service delivery and epidemic management require the ability to divert healthcare capacity quickly without lessening core services.
Under investments in R&D for health IT, standards development, and infectious diseases, and costly clinical trials.	Transformative renaissance in health and health IT R&D, including open access to tested peer-reviewed standards.	Sharing of standards-based clinical models has demonstrated efficient global collaboration to underpin urgent health IT development such as for pandemics.

Table modified from M. Craven, A. Sabow, L. Van der Veken, M. Wilson. Not the last pandemic: Investing now to reimagine public-health systems: McKinsey & Company; 2021. Available from: https://www.mckinsey.com/industries/public-and-social-sector/our-insights/not-the-last-pandemic-investing-now-to-reimagine-public-health-systems.

A timeline to success

Adner described the timeline to success [7], based on considering:

- why things go wrong,
- the many identified co-innovation and chain adoption risks,
- ecosystem mapping that considers roles and relationships required to line up the many entities,
- decision facilitation to be made at every level within the health industry at the right time and place, based on the best available information and knowledge,
- the adoption and implementation of game-changing strategies in a manner that sequences success and considers the many interdependencies identified throughout his text.

Adner [7] also provides a wide lens toolbox considered crucial for multiplying our odds of success. His first step is about gaining a clear view of our collective *value blueprint*. This assists with articulating our shared vision of our value propositions driving investments as well as determining the way we can achieve it. Once we have agreed on the definition and vision for a global digital health ecosystem, we can begin to ask crucial questions about the role each one of us wants to play and when and how to make the change. Given that a global ecosystem consists of many complex systems that drill down to every possible health-related application, everyone needs to participate.

Our concluding recommendations and rationale

We would like to see a new coordinating standards body charged with the global responsibility to create a coherent set of health standards that will support a fully functional digital health ecosystem, as described above, and to commission tools that will support globally consistent implementation. This global responsibility includes the promotion and adoption of an open knowledge-based platform, clinical modelling/knowledge engineering to separate the data from the application, and a true model-driven architecture, including for process and logic that best suits a digital health ecosystem.

Such an entity needs to have a strong autonomous leadership group, facilitating cross-domain collaboration amongst experienced domain experts. Ideally, this organisation should operate as both a meritocracy and a do-ocracy—only involving people selected according to merit and expertise, and with a willingness to do the work. It specifically needs to avoid becoming stifled, or even paralysed, by decision-making processes made by elected officials ahead of experts with practical knowledge. If done well, this organisation could be a game changer for the digital health industry needs.

The platform concept underpins both the economic and technical strategic vision. It is essential to get the specifications and technical architecture right, especially within the context of the whole ecosystem approach—poor choices will result in failure, even if the individual approach is sound. A technically sound platform still has to be usable, or developers won't use it, and it has to be technically 'fit for use' or it won't be implementable. We have found that *de jure* standards are not always helpful and can sometimes be harmful if used in situations where they are not fit for purpose. Open-source solutions can be an advantage but should not be the overarching end goal—the open-source status of software is less important than the use of open standards, even if implemented in proprietary software. The chosen platform definitions must have industry credibility and be underpinned by scientific rigor and robust engineering principles. It does not matter too much where it comes from, but it does matter where it is going!

A successful digital health ecosystem needs to be built using a solid engineering methodology, compliant with scientific evidence, and underpinned by extensive implementation experience. More importantly, active participation and leadership from the broadest range of clinicians, especially clinician informaticians and local clinical champions, is essential to achieve a thriving digital health ecosystem that will improve global health outcomes and ensure health system sustainability.

References

[1] T. Beale, Health Information Standards Manifesto Revision 2.5, 2001, Available from http://workflow.healthbase.info/monographs/Tom_Beale_his_manifesto.pdf.

[2] E. Coiera, Building a National Health IT system from the middle out, J. Am. Med. Inform. Assoc. 16 (2009) 271–273.

[3] B. Johansen, Institute for the Future, Available from https://www.iftf.org/bobjohansen/.

[4] S. Sinek, Start with WHY to Inspire Action, 2016, Available from https://www.youtube.com/watch?v=HjriwYrGL28.

[5] WHO, Digital Implementation Investment Guide (DIIG): Integrating Digital Interventions into Health Programmes, World Health Organization, 2020. Available from https://www.who.int/health-topics/digital-health#tab=tab_1.

[6] B. Al-Knawy, M. Adil, G. Crooks, K. Rhee, D.W. Bates, H. Jokhdar, et al., The Riyadh Declaration: the role of digital health in fighting pandemics, The Lancet (2020). Available from: https://www.thelancet.com/journals/lancet/article/PIIS0140-6736(20)31978-4/fulltext.

[7] R. Adner, The Wide Lens, Penguin Group, New York, 2013.

Index

Note: Page numbers followed by *f* indicate figures, *t* indicate tables, and *b* indicate boxes.

S

T

CPI Antony Rowe
Eastbourne, UK
March 04, 2022